PHYSICAL GEOLOGY

PHYSICAL GEOLOGY

Nicholas K. Coch
and
Allan Ludman

Macmillan Publishing Company
New York

Collier Macmillan Canada, Inc.
Toronto

Maxwell Macmillan International Publishing Group
New York   Oxford   Singapore   Sydney

Cover photo by: Kirkendall/Spring

Senior Editor: Robert McConnin
Production Editor: Sharon Rudd
Art Coordinator: Raydelle M. Clement
Photo Researcher: Anne Vega
Text Designer: Cynthia Brunk
Cover Designer: Brian Deep
Illustrations by: Biographics, Inc.

This book was set in Galliard.

Macmillan Publishing Company
866 Third Avenue
New York, New York 10022

Collier Macmillan Canada, Inc.

Library of Congress Catalog Card Number: 90–62526
International Standard Book Number: 0–675–
   21034–8
Printing: 1 2 3 4 5 6 7 8 9   Year: 1 2 3 4

To Carol, Elaine, Jessica, and Kenneth

Preface

The study of geology has never been more exciting. Discoveries made in the last 30 years have greatly changed our understanding of Earth processes, landscapes, and resources. Technological advances have enabled us to study remote parts of the Earth as well as our planetary neighbors. Field and laboratory investigations have given us new insights into the origin of life and its evolution.

Geology, which began as a science devoted to finding mineral resources and fuels, has become increasingly more relevant to us as the population increases and more areas are settled. Today, geologists use their skills to find new sources of water, to combat soil erosion, to prevent landslides on highways, to predict and mitigate the effects of earthquakes and volcanic eruptions, and to minimize the effects of river and coastal erosion and flooding. Geologists work with other scientists and engineers to plan how we can obtain resources and build on land in a way that will preserve our environment for future generations.

Our purpose in this book is to show how Earth processes work and how this knowledge will be useful to you in everyday life. In our teaching, we have found that students appreciate geologic processes and materials more if they understand their application to everyday life. Thus, we have shown *in every chapter* how apparently academic topics—such as the viscosity of volcanic lava, or the type of rocks that underground water seeps through—affect humans.

We have used a traditional topic outline and have attempted to show in each chapter how geologists think (the scientific method). Plate tectonics, the unifying theory in geology, is introduced in the first chapter and developed throughout the book.

Chapter 1 describes the scope of modern geology, the unique problems that geologists face in studying the Earth, the scientific method and an example of its use, and a first view of plate tectonics and our planetary neighbors. Chapters 2–7 deal with the origin of minerals and rocks—the building blocks of the Earth—as well as the energy needed to power Earth processes. In Chapter 8 we introduce geologic time, discuss how geologists determine the ages of Earth materials and events, and give examples of how these techniques are answering some important questions: How old is the Earth? When did life evolve? What is the oldest rock?

In Chapters 9–16 we describe the processes that operate on the surface of the Earth and the landscapes that they produce. Chapters 17 and 18 describe how internal Earth processes deform rocks and raise them into mountains. Chapter 19 deals with earthquakes and seismology; it shows how geologists interpret the composition of the Earth and the geologic processes of the interior. In Chapter 20 we return to the plate tectonics theory, this time in detail, and show why this model for Earth's development is the one that best fits the available evidence.

In Chapter 21 we overview the Earth's resources. Many process-oriented aspects of resources are given earlier as examples in other chapters; for example, the role of weathering in developing aluminum deposits is treated in Chapter 9, and the deposition of copper by hydrothermal solutions is discussed in Chapter 4. In Chapter 21 we describe examples of other types of resources and address the problems of resource extraction and future supply.

Chapter 22 concludes the text by presenting what we have learned about our neighboring planets and what they, in turn, have told us about the early evolution of the Earth.

Our sequence of chapter topics (materials–surface processes–interior processes) has proven effective in our teaching because everyone is more intuitively familiar with surface processes. However, discussing internal processes before surface ones has worked as well. The weathering chapter has been written so that it can be used either after all the rocks have been introduced or before sedimentary rocks, as some may prefer. In a similar fashion, the chapter on geologic time can be discussed any time after rocks have been introduced. (We prefer doing so as early as possible, because geologic time is so important in understanding the evolution of landscapes.)

We have incorporated many learning aids in our chapters. **Key words** are boldfaced, defined at their first introduction, and collected in a glossary. Numerous examples from national parks are used to illustrate features. Some 600 illustrations—300 color photographs and 300 perspective renderings of geologic processes—illuminate the text. Boxed focus sections in each chapter highlight a fascinating aspect of the chapter subject material. Each chapter concludes with a detailed summary, annotated further readings, and review questions of varying difficulty. In the back of the book is a detailed index and a glossary which defines all key words.

We have shared equally in authoring this book. Our task was made easier by the help of many people. Several colleagues at Queens College provided information and constructive criticism on specific aspects of the text. These individuals included Eugene Alexandrov, Hannes Brueckner, Patrick Brock, Robert Finks, Peter Mattson, Charlotte Schreiber, and David Speidel. Other colleagues provided professional assistance: Carol Coch, U.S. Army Corps of Engineers, North Atlantic Division; Ervin Otros,

Gulf Coast Research Laboratory; Fred Wolff, Hofstra University; Henry Bokuniewicz, SUNY at Stony Brook; Robert Q. Oaks, Utah State University; Frank Fletcher, Susquehanna University; and Gerald Johnson of the College of William and Mary.

Help in reproducing portions of the text, tables, and illustrations was provided by Barry Dutchen, Muriel Divack, Sheila Berman, Bernadette Gatto, and Leonard Cinquemani.

Detailed evaluations of draft chapters were provided by Richard N. Abbott, Jr., Appalachian State University, NC; Gary C. Allen, University of New Orleans; Robert W. Baker, University of Wisconsin-River Falls; Robert Behling, West Virginia University; Michael Bikerman, University of Pittsburgh, PA; Andy R. Bobyarchick, University of North Carolina, Charlotte; Henry Bokuniewicz, SUNY Stony Brook; Harold W. Borns, Jr., University of Maine, Orono; Wallace A. Bothner, University of New Hampshire; Scott Brande, University of Alabama-Birmingham; Philip E. Brown, University of Wisconsin-Madison; Hannes K. Brueckner, Queens College, CUNY; H. Robert Burger, Smith College, MA; Nicholas Christie-Blick, Columbia University, NY; Habte Giorgis Churnet, University of Tennessee, Chattanooga; Peter Clark, University of Illinois-Chicago; Robert Decker, Dartmouth College, NH; George W. DeVore, Florida State University; Anne Erdmann, University of Minnesota; Robert H. Filson, Green River Community College, WA; Roy Grossman, Hunter College, NY; William R. Hackett, Idaho State University; Janet Hammond, Pasadena City College, CA; Glen R. Himmelberg, University of Missouri-Columbia; Gregory S. Holden, Colorado School of Mines; Ernst H. Kastning, Radford University, VA; Edward A. Keller, University of California, Santa Barbara; David T. King, Auburn University, AL; Cornelis Klein, University of New Mexico; Albert M. Kudo, University of New Mexico; Martin B. Lagoe, University of Texas; David P. Lawrence, East Carolina University, NC; Stephen P. Leatherman, University of Maryland; Thomas M. Lehman, Texas Tech University; Nancy Lindsley-Griffin, University of Nebraska-Lincoln; Steve Lipshie, Iowa State University; John D. Longshore, Humboldt State University, CA; David N. Lumsden, Memphis State University, TN; D. Brooks McKinney, Hobart & William Smith Colleges, NY; Henry O. A. Meyer,

Purdue University, IN; Ernest H. Muller, Syracuse University, NY; John Mylroie, Mississippi State University; Hallan C. Noltimier, Ohio State University; Robert Q. Oaks, Jr., Utah State University; Larry W. Prise, Portland State University, OR; Ivan Sanderson, Purdue University, IN; Christopher Suczek, Western Washington University; Alphonse Van Besier, University of Mississippi; Frits Van der Leeden, Gerahty & Miller Company, NY; David Verardo, Long Island University; Robert P. Wintsch, Indiana University; and Donald L. Woodrow, Hobart & William Smith Colleges, NY.

Our thanks to the people who worked this manuscript through its editorial and production stages. Final editing changes by Fred Schroyer were very helpful in improving our final manuscript. Anne Vega diligently obtained the photographs in this book. Brian Deep was responsible for cover design.

Cindy Brunk was responsible for text design. Sharon Rudd served as production editor and coordinated all aspects of our project during the production stage. Special thanks are due to Raydelle Clement, our art editor, for the painstaking care with which she supervised production of our line drawings and the constructive and patient way in which she responded to our many requests for modifications. The four-color illustrations throughout the book were rendered by Biographics, Inc.

Special thanks are due to our Developmental Editor, Janet Wagner, for editorial comments and organizational and presentational suggestions which significantly improved our final manuscript.

Finally, we reserve special thanks for our families and friends for their encouragement and patience during the preparation of this book.

Brief Contents

Contents

3

MINERALS 45

4

IGNEOUS PROCESSES AND
ROCKS 75

5

VOLCANOES 101

6

SEDIMENTATION AND
SEDIMENTARY ROCKS 133

7

METAMORPHISM AND
METAMORPHIC ROCKS 161

8

GEOLOGIC TIME 185

9

WEATHERING AND SOILS 215

10

MASS MOVEMENTS 239

11

STREAMS AND STREAM SCULPTURE 263

12

GROUNDWATER 303

13

GLACIERS, GLACIATION, AND CLIMATIC CHANGE   335

14

DRY REGIONS AND WIND ACTIVITY   369

15

COASTAL ZONES 397

16

OCEANS 429

17

DEFORMATION OF ROCKS 461

18

MOUNTAINS AND MOUNTAIN BUILDING 491

PHYSICAL
GEOLOGY

1
A Geological Perspective

Geologist-astronaut Harrison Schmitt examines rock outcrop at the *Apollo 17* lunar landing site, the Taurus-Littrow Valley. Courtesy of National Aeronautics and Space Administration.

A geologist walks the Moon's surface, collecting rock and soil samples during the *Apollo 17* mission. Another descends into a deep-sea trench in a two-person submarine. Another walks gingerly on the recently hardened lava of a Hawaiian volcano to collect a rock or gas sample. Still others seek energy resources—coal, petroleum, uranium, and geothermal; examine maps and data to find valuable metal deposits; and study ways to minimize landslide and earthquake damage. These are just a few of the varied activities that geologists carry out to study the Earth and its neighbors in space.

WHAT IS GEOLOGY?

Geology is the study of the Earth—its rocks, minerals, soils, water, and atmosphere. Geologists try to understand the processes by which these materials are formed and are constantly being changed. They attempt to explain the mechanisms by which mountains are created and then eroded by streams, wind, and glaciers. They investigate how our varied landscapes form, determine what the Earth looked like millions of years ago, and predict what it will look like millions of years from now.

Geologists provide the foundation studies that enable engineers to build highways, tunnels, and skyscrapers. Geologists can be found at work everywhere on Earth, studying rocks and landforms in the field, analyzing samples in laboratories, and observing geologic processes underwater in the oceans. One has even studied the rocks of our satellite, the Moon—astronaut/geologist Harrison Schmidt of the *Apollo 17* team.

Why Study Geology?

Geology has always been a practical science, and the information in this book will be very useful to you in the future. You may be a homeowner some day, and a knowledge of geology will help you make a proper choice of homesite. Consider these examples:

☐ Certain types of soils and slope conditions can result in land subsidence or even landslides, and geologic knowledge can help you avoid such problems.

☐ In small-town and rural settings, home water supply can be a major problem. With knowledge of the underlying geology, you will have a better idea of the quality and quantity of well water available.

☐ If you choose to live near a stream or along a lake or ocean beach, knowledge of stream and shoreline processes can help you avoid flood-prone areas.

☐ If you live in an area that has been mined or drilled, geologic knowledge can help you avoid hazards of land subsidence or well contamination.

Knowledge of geology also will give you a clearer understanding of larger-scale regional or global problems that occur as humans continue to modify natural geologic systems, such as the way streams flow or the erosion that moves beach sand along a shoreline. The harmful side effects that accompany many human modifications of geologic systems usually arise because the Earth's systems are interrelated—modifying one system can lead to changes in others. For example, paving over the land surface prevents water from naturally sinking into the ground and migrating slowly into streams. Instead, the rainwater runs quickly across the paved surfaces and enters nearby streams, causing the streams to rise over their banks and flood downstream areas. Removing vegetation and cutting into slopes causes the surface material to become unstable, resulting in landslides.

Geology is fun and exciting, even for those who never become scientists. Do you enjoy puzzles and mysteries? Earth's history is recorded in its accumulated rock layers, and reading it is like reading a fascinating, intricate detective novel. However, this "book of the Earth" is very different from any detective story you have read. First you must learn to read the language in which the "book" is written. Then you will discover that the first few chapters have many missing pages, and that the ink on the remaining pages is badly smudged. The middle of the book is more complete, but each time the story gets interesting a page or two has been ripped out.

The last few chapters are nearly complete, so that we can know how the story ends—or at least how it reached the present. The missing sections are frustrating, but geologists see them as challenges, and devise clever ways to retrieve the lost information. Many of the chapters in this book show the unique ways in which geologists go about solving these problems.

Geology also will increase your appreciation for the beauty and splendor of Earth's landscapes. You probably will visit some of our National Parks during your lifetime, so we will use examples from these parks to illustrate our geologic discussions. After reading this book, you will be able to answer many questions that puzzle park visitors who lack this knowledge:

- ☐ Why are the landscapes in Yosemite National Park so different from those in Grand Canyon National Park?
- ☐ How were huge underground caves hollowed out at Carlsbad Caverns National Park in arid New Mexico?
- ☐ Why are the layers of rock in Bryce Canyon National Park so varicolored and sculpted into such unusual shapes?
- ☐ Did the Colorado River carve the Grand Canyon, or was the canyon there first?
- ☐ Why are the volcanoes in the Cascade Range of the northwestern United States so different in shape and explosive nature from those in Hawaii Volcanoes National Park?

These are just a few of the things you will learn from this book.

In the rest of this chapter, we will introduce how geologists decipher the rock record. We also will provide a brief overview of Earth processes as a background for the more detailed discussions that come later in the book. Once you look at Earth as a geologist does, with an understanding of its dynamic behavior, your new perception will make learning more rewarding. We hope you will enjoy this journey into the past and that you will gain an increased understanding and appreciation for the geologic processes that affect us in the present and in the future. The adventure begins!

THE GEOLOGIST'S PERSPECTIVE

The study of geology requires a different way of thinking about certain topics. These topics include time, space (geography, past and present), scale, and special problems encountered in studying Earth processes and structures.

Time

The concept of *geologic time* is new to many nongeologists. Most people think about time in the usual human terms: generations, lifetimes, or centuries. A person aged 100 years is very old; something 3000 years old is ancient. In contrast, geologists deal routinely with much greater periods of time—usually millions or billions of years. An event that occurred 20 million years ago is viewed as relatively recent. On an Earth that is almost five billion years old, rocks that are "only" 200 million years old are thought of as young!

A geologic frame of reference is particularly important because many large-scale Earth processes act so slowly that long periods of time must pass before an appreciable change can be detected. To understand many geologic processes, you must adjust your time frame and think in terms of geologic time rather than in human-oriented historical time.

The study of Earth through geologic time raises an interesting problem: although the passage of time is *continuous* and *uninterrupted,* Earth's history is recorded in its rocks, which are *discontinuous, incomplete,* and *disrupted*. Episodes of nondeposition or episodes of erosion may have removed some of the rock record, resulting in the missing pages of our "book of the Earth" (Figure 1–1). Geologists therefore must use special techniques to determine what events occurred during time periods for which there is no local rock record.

Space

Another concept that may be new to you is that the present geography of Earth is only temporary. Surface features on the continents are slowly changing. In effect, the Earth has never before looked exactly as it does today, and it will not look exactly like this in the future.

For instance, changes in the Earth's geography, as reconstructed from the rock record, indicate that North America looked very different in the past (Figure 1–2). At one time there were seaways where

FIGURE 1–1
Earth's history interrupted. The record in the rocks sometimes is broken by an unconformity such as the billion-year hiatus between the Tapeats Sandstone and the Vishnu Schist in Arizona's Grand Canyon National Park. Photo by Nicholas K. Coch.

there is now land, and tropical volcanic islands where there are now small forested hills. Continents and oceans change shape and position. The Atlantic Ocean did not always exist. To fully understand how Earth processes operate, you must discard the image of unchangeable geography and realize that no feature of the Earth's surface, no matter how large, is permanent.

Scale

Geology differs significantly from other sciences. Many geologic phenomena are difficult to study because of their large size, as well as the great length of time involved in their activity. How could we recreate in the laboratory the formation of mountains, the erosion of a valley by a stream, or the shaping of a shoreline by ocean waves? The best we can do is to build *scale models* of such large-scale features and attempt to recreate the processes involved in forming them.

Figure 1–3 shows a scale model of upper New York harbor. Oceanographers and engineers have used this model to study how changes in present conditions could affect water flow and sediment erosion and deposition. For example, they have been able to determine the effects that dredging a deeper ship channel would have on the mixing of the Hudson River's freshwater with the ocean's saltwater in the harbor.

How good are these scale models? Do we really know the rates at which the forces acted? Are the materials we use in our models truly representative of the way Earth materials would behave in nature? Although we can successfully scale down the size of a river to fit onto a laboratory table, how can we possibly scale down *time*? We cannot wait hundreds or even millions of years for an experiment to be completed! These uncertainties make the results of our scale-model experiments only *approximations* of real phenomena. However, in many cases, these scale models provide valuable clues that guide geologists in gathering geologic data from the rock record.

Complexity of Geologic Processes

Geologic processes often are difficult to interpret because of the great number of factors involved. Unlike the controlled conditions in the laboratories of chemists or physicists, conditions within the Earth are extremely complex. For example, geochemical and mineralogical changes in a particular type of rock may be completely different, depending on whether water is present or absent, whether forces are applied slowly or quickly, whether organic material is present or not, or whether temperatures are high, low, or variable.

The complexity and variety of rock-forming processes is reflected by the wide variety of rocks found on Earth. In some cases, two or more processes may have acted simultaneously on a rock. A rock may have been subjected to several changes, but its

FIGURE 1-2
Geographic features of the North
American continent about 515
million years ago. After Dott and
Batten, *Evolution of the Earth,* 3d
ed., McGraw-Hill Book Co., 1981,
Figure 11.9, p. 237.

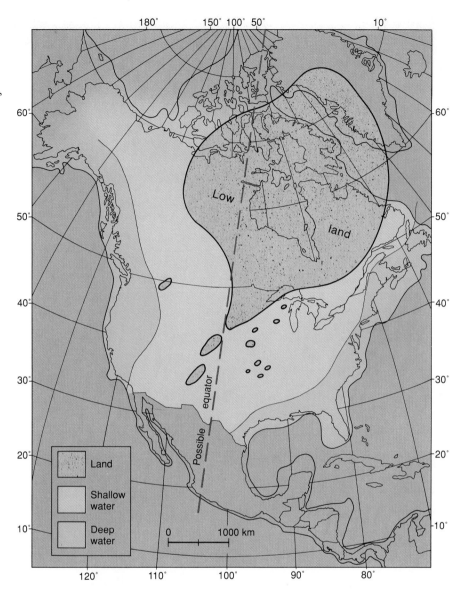

present appearance is most likely the result of the latest change (Figure 1–4).

How, then, can we detect earlier episodes in the history of rocks? Through patient examination of rocks in the field, we may be able to locate samples that reveal intermediate stages. Detailed laboratory study of all these samples can then determine— sometimes—the sequence of events that formed them. When the scale is increased from a single rock to a series of layered rocks, or to continental-scale processes, the magnitude of the problems that geol-

ogists face in their inquiries becomes even more apparent.

HOW DO GEOLOGISTS WORK?

Building scale models is one way of studying problems, but a basic conceptual approach was needed before geology could become a modern science. It took many years to develop an effective way of interpreting geologic data in an orderly, coherent man-

FIGURE 1–3
Scale model of upper New York harbor. Photo courtesy of U.S. Army Corps of Engineers.

(a)

(b)

FIGURE 1–4
Transformation of limestone (a) into marble (b) as a result of increased temperature and pressure. The final product has a very different appearance from the original. Macmillan Publishing/Geoscience Resources photo.

ner. The unique nature of many geologic problems led to development of new ways of studying the Earth.

Catastrophism

Prior to the early eighteenth century, most geological phenomena were thought to result from rapid, catastrophic events followed by long periods of relative inactivity. For example, fossil shells found in a rock high on a mountain were interpreted to be a record of rapid ocean flooding in which most of the world's mountains had been submerged in a matter of days or weeks. This approach to Earth history is known as **catastrophism**.

Uniformitarianism

In the latter part of the eighteenth century, a Scottish physician named James Hutton began to interpret geologic features in England in a way quite different from catastrophism. As a keen observer of natural phenomena, he saw features in rocks that strongly resembled those being developed in the sediments of present-day streams and coastlines. Mud cracks that he saw above the high-tide line along the shore, ripple marks that he observed in the sands of rivers and beaches, and layering that he noted in sand and mud—all had counterparts in ancient rocks. Some of these modern features and their rock equivalents are shown in Figure 1–5.

In 1785, Hutton proposed that much of what we see in rocks has been produced by processes comparable to those acting today. "The present is the key to the past" is often used to summarize this view that modern processes can be used to understand ancient Earth history. Further, most of the processes that Hutton observed operate very slowly and not in the catastrophic manner previously expected. Hutton thus suggested that Earth's history could be explained by relatively slow present-day processes acting over *a long period of time*. This concept is known

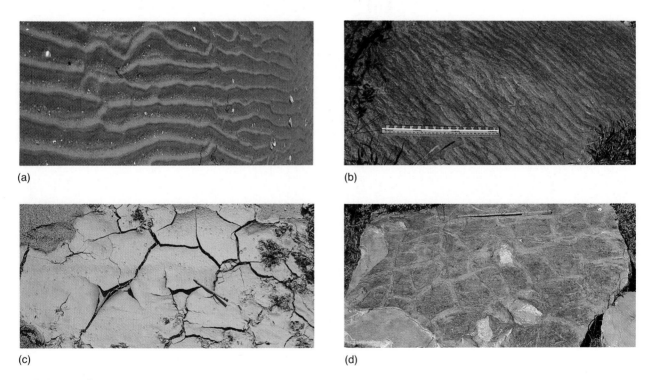

(a) (b) (c) (d)

FIGURE 1–5
Present-day features in sediments and their counterparts in sedimentary rocks. (a) Ripple marks in sand. (b) Ripple marks in sandstone. (c) Mud cracks in fine sediment. (d) Mud cracks in shale. Photos by Nicholas K. Coch.

DO GLACIERS MOVE? AN EARLY APPLICATION OF THE SCIENTIFIC METHOD

An early application of the scientific method was carried out in the Alpine area of France and Switzerland. In the Alps, snow covers the high mountain tops year-round, and small glaciers exist in the uppermost parts of some valleys. Before the eighteenth century, herders and farmers in this region saw that the angular boulders and cobbles strewn over the valley floors were very similar to the debris carried on and in the glaciers at the heads of the valleys. This observation led them (and some natural scientists) to suggest that the boulders had been deposited in the valleys by glaciers which at one time had advanced down the valleys, melted, and left the boulders as evidence of their former extent. Other scientists laughed at the idea that masses of solid ice hundreds of meters thick were capable of moving.

In the beginning of the eighteenth century, a simple experiment was designed and carried out by Swiss and French scientists to test the hypothesis that glaciers can move (Figure 1). Stone markers were placed on the bedrock walls of a valley that was filled by a glacier, and an accurately surveyed straight line of markers was set up on the ice surface between them. If the ice moved, the markers on its surface would change position relative to the two stationary markers on the valley walls.

The results? Markers on the ice moved down the valley, proving conclusively that glaciers do indeed move. Further, the experiment provided an unexpected dividend when it was noticed that the ice movement revealed by the new positions of the markers was *uneven*. Ice in the center of the glacier moved farther than the ice at its edges. As often is the case in science, the very observations that confirm or refute one hypothesis present new problems to be solved. No sooner did scientists show that glaciers move than they had to explain why the ice at the sides of a glacier moves less than the ice at the center. It was then suggested that friction between the ice and the rock of the valley walls was a cause of the differential movement, and this became a new hypothesis to be tested.

as the **principle of uniformitarianism**. The uniformitarian approach does not seem so revolutionary today, but its development laid the foundation for the geologic sciences, and Hutton is regarded as the father of modern geology.

The uniformitarian view of the Earth is quite different from catastrophism's view. From the perspective of uniformitarianism, fossils in rock exposed high on a mountainside would represent remains of life from an ancient ocean, not from a brief flood. These fossils were buried slowly as sediment accumulated on that ancient seafloor and were then slowly uplifted over millions of years to their present position.

Uniformitarianism does not imply, however, that true catastrophes do not occur. Catastrophic events such as volcanic eruptions, landslides, earthquakes, and floods certainly happen and are recorded in

FIGURE 1
An early experiment to see whether glaciers move.

Initial position
of markers

Final position

rocks. However, they are not considered to be as important to the development of large-scale features on the Earth's surface as are the processes that act slowly over long periods of time. Further, although uniformitarianism holds that processes acting today acted similarly in the past, it does not require that their *rates* or *importance* were identical to those of today.

For example, glaciers cover a much smaller part of the Earth's land surface than they did just 20,000 years ago when large ice sheets extended across Canada and well into the midwestern United States. Clearly, glacial processes were more important in the history of North America 20,000 years ago than they are today. The rate of glacial movement at that time may have been different from that of today's glaciers, but we have evidence that it was not catastrophic. The glaciers must have taken thousands of years to expand to their maximum size and thousands more to melt away.

The Scientific Method

Geologists and other scientists use an orderly, logical method of analysis called the **scientific method** to ensure that they determine the most reasonable explanation for problems under study. If the method is carried out properly, personal bias is reduced to a minimum. The basic steps in the scientific method are outlined as follows:

1. The method begins with **observation** of geologic material or processes and the recognition that something needs to be explained.
2. The second step is proposing a tentative explanation, more formally stated as proposing a **hypothesis.** Usually there is more than one possible explanation, and **multiple working hypotheses** are set up, any one of which may prove to be the correct answer.
3. The third step is setting up an **experiment** that will test the hypothesis or allow elimination of some of the multiple explanations. Scale models such as the one shown in Figure 1–3 are experiments designed to isolate certain variables and show how each contributes to the process being studied.
4. Step four includes modification or rejection of hypotheses in the light of data provided by experimentation; development of new experiments; and continuing revision until a satisfactory hypothesis is found that answers the initial questions.
5. When a hypothesis has successfully withstood the tests of numerous experiments over a period of years, it is then accepted as a **theory.**

An interesting example of an early application of the scientific method in geology is illustrated in Focus 1–1.

PLATE TECTONICS: EARTH AS A DYNAMIC PLANET

Geologists originally found it hard to understand the movement of glaciers and erosion of river valleys because of the large scale of these features and the enormous amount of time involved in their formation. Imagine how difficult it was to explain the origins of the largest features on Earth—the oceans, continents, and mountain ranges! It once was thought that ocean basins were merely parts of continents covered by water, but we now know that oceans and continents differ strikingly in composition, structure, and age.

Why are these features so different? Mountains are found both on the continents and in the ocean basins. Are there different kinds of mountains? Do they form by a single process, or are there as many mechanisms as there are different types of mountains? Why are there mountains in Colorado, Virginia, California, Alaska, and Hawaii, but not in Kansas?

The answers to these questions are sought by experts in the area of geology that studies the structure and development of the outer part of the Earth, a specialty called **tectonics.** Many hypotheses have been proposed to explain the features of the Earth (see Focus 1–2). However, most geologists now agree that the **plate tectonics theory** is the most satisfactory explanation for how oceans, mountains, and continents form. We will return to this concept often for help in understanding many kinds of geologic phenomena, but it is appropriate here to briefly outline its basic concepts.

Structure of the Outer Earth

According to the plate tectonics model, the outermost shell of the Earth, called the **lithosphere,** consists of a small number of rigid segments. The segments are called **plates** because they are relatively thin (80 to 150 km) when compared to their vast areas of millions of square kilometers (Figure 1–6). These lithospheric plates rest on a much less rigid (plastic) layer called the **asthenosphere** (Figure 1–7) and are capable of moving across this layer much as a raft floats on water.

Continents and oceans, as large as they may seem, are merely *passengers* on the plates. For example, the North American *plate* includes the entire North American continent and half of the Atlantic Ocean basin. As plates move, the Earth's geography can change drastically. Continents may change position,

BEFORE PLATE TECTONICS, GEOLOGISTS THOUGHT THAT. . .

The first primitive plate tectonics model was proposed only 30 years ago, but models of moving continents ("continental drift") date prior to the 1930s. The plate tectonics model quickly became a unifying concept that has revolutionized the way in which geologists look at the Earth. The widespread acceptance of this concept is due to the fact that it draws on information from nearly all areas of geologic research. The concept contains contributions by geochemists, geophysicists, and paleobiologists, and it answers a number of questions that geologists have pondered about the Earth for many years. For example, paleobiologists noted that identical land-animal fossils were found in the rocks of continents now separated by oceans. These animals could not have crossed an intervening ocean, so they must have lived together on a continent that subsequently broke apart.

Additionally, when coal with tropical plant fossils was first found in the mountains of Antarctica, it seemed quite puzzling. With the advent of the plate tectonics theory, we now know that the coal formed when Antarctica was part of a larger continent located near the Equator. That supercontinent subsequently broke apart, and Antarctica migrated to its present polar position. The plate tectonics model always is being refined as experiments designed to test its basic concepts lead to important new discoveries. Debate about the validity of the model and rapid advances in the understanding of our planet make this one of the most exciting times in the history of geology.

In the midst of this excitement, it is important to remember that plate tectonics is not a fact, but remains a theory. It is the most recent, comprehensive theory ever proposed for the origin of major Earth features, and so far it has survived all the tests that geoscientists have devised. Most geologists feel that this theory provides the best answers to nearly all former problems. However, during the past century, geologists felt the same way about other theories and hypotheses that attempted to explain how the Earth works, and these now are discredited. These models, described below, at one time or another received the acclaim now enjoyed by the plate tectonics theory.

The Static Earth Model

This hypothesis states that Earth as we know it is not greatly different from the Earth that originally formed—Figure 1(a). According to this model,

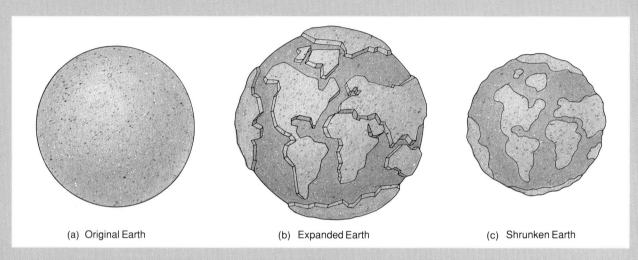

(a) Original Earth (b) Expanded Earth (c) Shrunken Earth

FIGURE 1
Comparison of (a) the "static Earth" with (b) the expanding Earth model and (c) the shrinking Earth model.

the mountains, oceans, and continents that first formed on the early Earth are the same ones that exist today, as are the rivers and lakes. Erosion may have lowered the mountains somewhat, and the material deposited by wind, streams, and glaciers may have filled in the lowlands a bit with sand, gravel, and mud, but in the scope of Earth history, these changes are insignificant, according to this model. The static Earth model dominated scientific discussion of mountain building for many years prior to the development of the principle of uniformitarianism as a way of looking at Earth processes.

The Expanding Earth Model

When geologists began to find fossils of ocean creatures on the highest peaks of the Alps, they recognized that Earth has indeed changed significantly throughout its history. Some argued that the fossils were the remains of Noah's flood, but this did not satisfy uniformitarianists, who stated that even extreme changes like the uplift of ocean basins must proceed at the same, generally slow pace as modern geologic processes.

After uniformitarianism became accepted, one group of scientists claimed that the Earth had been getting hotter since it first formed and consequently had been expanding. In the expanding Earth model, the surface of the planet is compared to a thin, brittle coating of dried clay on the surface of an expanding balloon—Figure 1(b). As the balloon expands, the coating of clay develops cracks. On the scale of the Earth, these cracks become the ocean basins and the intact remains of the coating become continents, such as Africa, and microcontinents, such as Madagascar. According to this model, as long as Earth continues to expand, new oceans can form at any time.

The Shrinking Earth Model

Other scientists argued that the Earth has been cooling, and thus shrinking, throughout its history—Figure 1(c). In this view, the globe is like a slowly deflating balloon. As the originally smooth surface shrinks, the creases and wrinkles that appear eventually form ocean basins and mountains.

The Pulsating Earth Model

Some geologists thought they had evidence for both expansion and contraction worldwide and developed a hypothesis that combined the expanding and shrinking Earth hypotheses. They stated that the Earth expands and contracts periodically like a beating heart and can change its major physiographic features during both stages. If expansion and contraction were related to worldwide heating and cooling, then Earth's thermal history must have been extremely complex.

The Plate Tectonics Theory's Success

Why have these hypotheses been discarded in favor of the plate tectonics model? For scientists, observations led to data that conflicted with some aspects of these models, and the conflicts were so severe that simple revision of the hypotheses was not possible. It is another example of the scientific method at work.

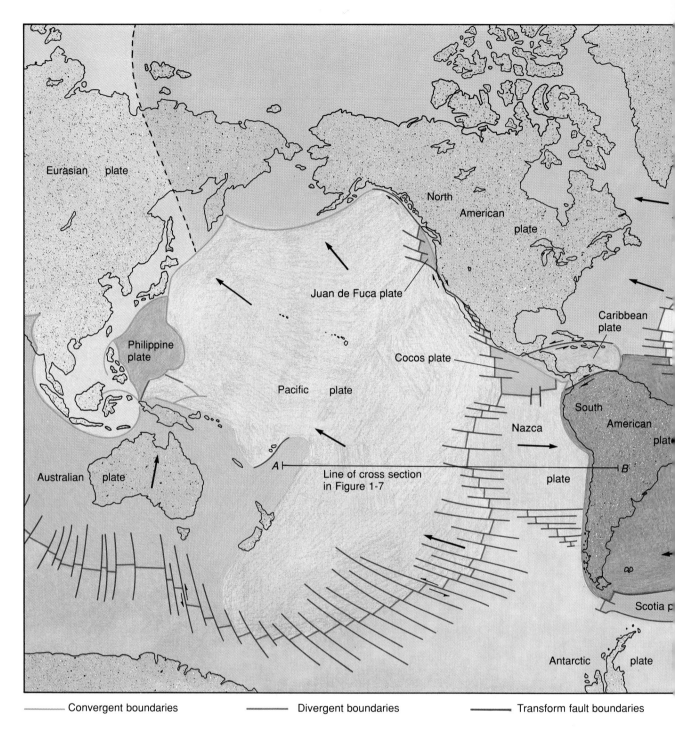

——— Convergent boundaries   ——— Divergent boundaries   ——— Transform fault boundaries

FIGURE 1–6
Earth's major lithospheric plates, showing directions of
movement, names, and types of margins. Note the line of cross
section *A–B*, detailed in Figure 1–7.

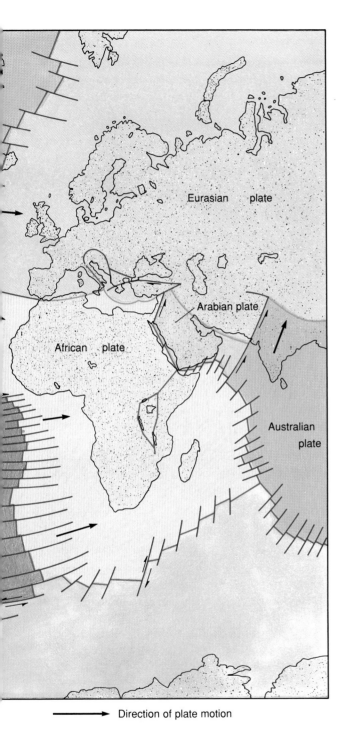

Direction of plate motion

rotate, split into smaller pieces, or collide and become attached to one another. New oceans may open and old ones may widen, change shape, or disappear. Evidence for some former plate movements are visible on ordinary maps; for example, the continents of South America and Africa can be fitted together much like the pieces of a jigsaw puzzle.

Many other examples of plate movements have been discovered by geologists. For example, the east coast of North America once was aligned east-west and located at the Equator (Figure 1–2), and an ocean once separated India from Asia.

Types of Plate Margins

The regions along the margins of plates are geologically active areas characterized by earthquakes and volcanoes. Lithospheric plates may interact with one another in three basic ways, creating three different types of plate margins: divergent margins, convergent margins, and transform fault margins (Figure 1–7).

Divergent margins are initiated when a single plate is split in two. As the plate fragments move apart, molten rock material rises from the asthenosphere and erupts onto the seafloor through cracks and volcanoes. As the molten material cools, it forms part of the floor of a new ocean. At the same time that the new rock is forming, continuing divergent plate motion moves it to either side of the margin. This "conveyor belt" movement of new oceanic rock away from divergent margins is called **seafloor spreading**. Underwater eruptions such as these produce elongated groups of undersea volcanoes, called **oceanic ridges,** which are found in all the ocean basins.

The Atlantic Ocean has grown in this way, and eruptions on Iceland and other parts of the Mid-Atlantic Ridge indicate that the ocean basin still is growing. We are moving farther from Europe every day as Atlantic seafloor spreading continues at a rate of about 2.5 cm per year. This seems incredibly slow, but if this rate is maintained for the next 100 million years, Europe and America will be 2539 km farther apart than at present! The Earth has all the time in the world.

FIGURE 1–7

Cross section along line A–B in Figure 1–6, showing the lithosphere, asthenosphere, and the types of processes active at convergent margins, divergent margins, and transform fault margins.

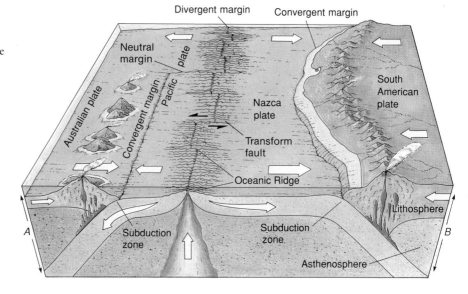

Things are quite different where two plates come together in what are termed **convergent margins**. At these margins, **subduction zones** form when a plate underlying an ocean (an oceanic plate) collides with another plate. In subduction zones, the ocean-floor material is then thrust downward into the asthenosphere, beneath the other plate, and the ocean gradually shrinks. This is thought to be happening today where the floor of the Pacific Ocean is being subducted beneath the west coast of South America. Some of the subducted material has melted and risen to produce the volcanoes that have built the Andes Mountains (the mountain chain shown on the South American plate in Figure 1–7).

The part of the subducted lithospheric plate that does not melt becomes part of the asthenosphere. When continents on opposite sides of a closing ocean collide, they are welded together to form a single supercontinent, and a mountain range forms at the "seam" or "suture" between the original continents. The Ural Mountains are thought to have formed in this way when Europe and Asia collided, and the Himalayas when India and Asia were welded together.

Transform fault margins form when two plates grind past one another without colliding head-on. Unlike divergent or convergent boundaries, there is little or no transfer of material between asthenosphere and lithosphere in transform fault zones, and

volcanic activity does not characteristically occur at this type of plate margin. Earthquakes, however, are numerous. The most famous example of this type of plate margin is California's San Andreas fault, which has resulted from the relative motion between the North American and Pacific plates (Figure 1–6). The Pacific plate is moving northward and the North American plate is moving westward, resulting in formation of an earthquake zone between them.

Geologic Cycles

The interaction between lithosphere and asthenosphere at convergent and divergent plate margins shows how the Earth continuously recycles materials, using them again and again. This recycling of geological materials is called a **geologic cycle**.

One type of geologic cycle is the **tectonic cycle,** which describes how plate tectonics movements recycle rocks. Our preceding description of the three types of plate margins can be summarized here by describing the operation of the tectonic cycle: when plates diverge, solid rock in the asthenosphere melts, becomes less dense, and rises. As the molten material cools, it solidifies into rock at divergent margins and is moved across the ocean floor by seafloor spreading. The rock may be subducted back into the asthenosphere when the ocean eventually closes. This cycle operates on a grand scale and may take hundreds of millions of years to be completed. In some cases, the tectonic cycle is not completed because plates may change direction of movement, or divergent margins may become inactive.

Several other important geologic processes also are cyclic. These cycles are smaller in scale, so they are completed in far less time than the tectonic cycle just described:

- The **water cycle** involves the journey of water from the oceans to the atmosphere and back to the oceans by means of rainfall, streams, and rivers.
- The **atmospheric cycle** is the periodic exchange of gases in the atmosphere with water, rock, animals, and plants.
- The **rock cycle** involves conversion of one type of rock into another.

The important thing to understand about these cycles is that an interruption at one stage of a cycle (for example, damming a river and preventing its water from returning to the ocean) has an effect on all subsequent stages of the cycle. We will examine these cycles in subsequent chapters.

THE SOLAR SYSTEM: EARTH IN ITS FAMILY OF PLANETS

Current excitement among geologists is due not only to the plate tectonics revolution but to technological advances that have made possible the study of other planets in ways only dreamed of 50 years ago. The best indication of how much the scope of geology has widened is the fact that geologists must explain not only how the mountains and oceans have formed on Earth, but how the solar system formed and evolved. Why are there different kinds of planets? Why are they arranged in such orderly distances outward from the Sun? If plate tectonics is a correct view of Earth's behavior, is it applicable to any of the other planets? Can it apply to Jupiter? What would a volcano be like on Saturn or Neptune?

The last chapter of this book focuses on advances in planetary geology, but some background is needed at this point to put the Earth in its proper perspective among the family of planets to which it belongs.

The Solar System: Population and Vital Statistics

Our solar system is composed of a star—the Sun—around which orbit planets, satellites, asteroids, and comets. The Earth, for example, follows a slightly elliptical orbit that keeps it about 150,000,000 km from the Sun.

The Sun. The Sun is a typical medium-sized star, a gaseous mass made up mostly of hydrogen and helium that acts as a huge nuclear reactor and emits enormous amounts of energy. How much energy? Enough to make the surface of the third planet away from it (the Earth) warm enough to sustain varied, thriving life; enough so that the Sun's surface temperature is estimated to be more than 6000°C. It is hard to comprehend fully the size of the Sun: although made mostly of gases, it accounts for more than 99% of the mass of the entire solar system, and its gravitational attraction keeps the planets and other bodies in their orbits. One measure of its size is its mass, more than one million times that of the Earth.

The Planets. The nine planets of our solar system are spheroidal objects, each of which travels around the Sun in a fixed, slightly elliptical orbit (Figure 1–8). All of these orbits lie close to the same plane, so that the shape of the solar system is that of a disc that bulges at the center because of the size of the Sun. Planetary orbits are not random but are spaced at regularly increasing distances from the Sun. Although each planet is unique in some ways, they all fall into two basic groups. The four inner planets (Mercury, Venus, Mars, and the Earth) are called **terrestrial planets** after *Terra,* the Latin name for Earth. They are relatively small but dense, have few if any satellites (moons), and are composed of rock material believed to be similar to that of the Earth.

The next four (Jupiter, Saturn, Uranus, and Neptune) are called **Jovian planets** after the Latin name for Jupiter. They are much larger than the terrestrial planets and are composed not of rock but mostly of frozen gases such as ammonia and methane. They also are much less dense and have several satellites each, as well as distinctive rings of particles not found around the terrestrial planets.

The ninth planet, Pluto, is small like the terrestrial planets, but is so far away that we still know little about its properties.

Satellites. Just as the planets orbit the Sun, smaller bodies called **satellites** orbit many of the planets. Our only satellite, the Moon, is the most familiar example. The Moon orbits Earth at a fixed distance, retained by gravity, just as the planets are retained by the Sun. Satellites differ widely in size and composition. The satellites of Mars—Phobos and Deimos—are irregular-shaped bodies only 43 and 24 km in longest dimension, whereas Ganymede and Callisto, the largest of Jupiter's satellites, are approximately the size of the planet Mercury (4800 km diameter).

Asteroids. Astronomical laws indicate that there should be another planet in the solar system, located between the orbits of Mars (the outermost terrestrial planet) and Jupiter (the innermost Jovian planet). Instead, this orbit is filled with a vast collection of rocklike debris ranging in size from dust to bodies larger than each of the two satellites of Mars. These particles are called **asteroids** and may represent the remains of an unstable terrestrial planet, or precursors of such a planet that never formed.

Origin of the Solar System

There have been many hypotheses for the formation of the Earth, but all agree that its origin was part of the overall process by which the entire solar system formed. The most widely accepted hypothesis is that of the solar system "condensing" from an enormous rotating cloud of gas and dust (Figure 1–9). The force of gravity acted within this cloud, attracting smaller masses to larger ones. Eventually, most of the

FIGURE 1–8
Solar system, showing the planets in their orbits around the Sun. At the scale of this diagram, the orbit of Pluto lies beyond the border of the figure.

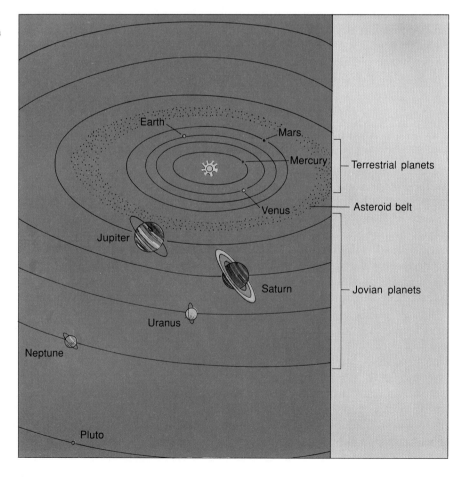

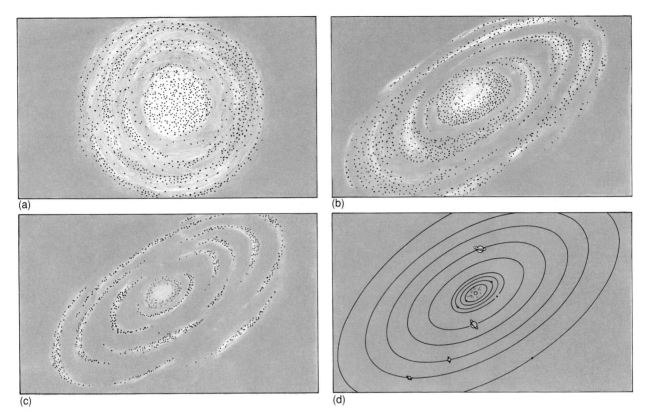

FIGURE 1–9

Origin of the solar system. (a) Initial state as a slowly rotating interstellar cloud of dust and gas. (b) Gravitational attraction causes concentration of mass at center of cloud, leading to more rapid rotation at the margins and the development of a discoid shape. (c) Enough heat is generated at the center of the cloud to initiate nuclear reactions—the Sun is born. (d) concentrations of dust at different parts of the disc produce the nine planets.

mass was concentrated as a huge cluster of discrete bodies at the center of this cloud, leading eventually to the nuclear reactions that converted them into the Sun. Smaller concentrations of matter became the planets and their satellites. We will examine this hypothesis in greater detail in the last chapter.

There's No Place Like Home

The processes that formed our solar system produced something very special on the third planet out from the Sun. There is literally nothing else like Earth in the solar system, and its uniqueness is vividly illustrated in Figure 1–10. Of all the planets, Earth alone has the oxygen-rich atmosphere and moderate sur-face temperatures necessary for life. Earth is the only planet with water on its surface. Only Earth has a single satellite that is so large in relation to its parent planet. (The other terrestrial planets either have no satellites—Mercury and Venus, or two—Mars; some of the Jovian planets have over a dozen each.)

Although we now can leave our home planet to visit other bodies in the solar system, Earth is still the only place where we can survive without protective clothing. We would be asphyxiated on any of the other planets; roasted on Mercury; boiled on Venus; frozen on Pluto; or be crushed by the gravity of the Jovian planets. We have looked for signs of life on Mars and Venus, but have detected no life in the solar system other than here on Earth.

FIGURE 1–10
Earthrise as seen from the lunar surface. Photo courtesy of National Aeronautics and Space Administration.

SUMMARY

1. Geologists view the Earth as a dynamic planet on which complex, interrelated processes (cycles) are constantly at work.

2. Earth's history fits into a vast framework of nearly 5 billion years called *geologic time,* compared to the 45 centuries of recorded human history.

3. The *principle of uniformitarianism* describes an Earth on which most large-scale changes take place slowly and over long periods of time by processes that are still active today.

4. Great changes in Earth's landscapes and the relative positions of continents and oceans have occurred throughout geologic time and are continuing at present.

5. Many of the processes involved are difficult to study because they are so complex. They operate at rates too

slow to permit observation within our lifetimes, and they occur on an extremely large scale, both on the surface and within the Earth.

6. Geologists use the *scientific method* to study the Earth, first observing Earth processes and materials, then proposing hypotheses to explain what they observe, and finally devising experiments to verify or refute these explanations.

7. Most geologists explain the formation of the ocean basins, mountains, and continents with the *plate tectonics theory.* In this model, the outermost part of the Earth is broken into rigid lithospheric plates that glide over an underlying, less-rigid layer called the asthenosphere. As the plates move, the oceans and continents that lie on them change position, shape, and size.

8. Plate margins reflect three major types of interactions between lithospheric plates. Oceanic ridges (spreading centers) are chains of volcanoes that form as lava rises from the asthenosphere at *divergent margins* where

plates are moving away from one another. *Convergent margins* occur where plates collide. When a plate bearing an ocean collides with another plate, a *subduction zone* forms in which oceanic lithospheric material is driven downward, back into the asthenosphere. When two continental plates collide, they weld together into a single continent, and mountains form along their original contact. *Transform fault margins* are places where two plates grind past one another with neither destruction nor creation of lithosphere.

9. Earth is one of nine planets in our solar system, which is itself an aggregate of planets, satellites, asteroids, and comets held in orbit by gravity around a central star that we call the Sun. The Sun is composed mostly of hydrogen and helium gas. It generates energy by nuclear reactions and contains most of the mass of the solar system.

10. The four innermost (terrestrial) planets are composed of rock material, are relatively small compared to the next four (Jovian) planets, and have few or no satellites.

11. The Jovian planets are made of frozen gases rather than rock, are much larger than the terrestrial planets, and have many satellites.

12. The solar system is thought to have formed from gravitational concentration of an enormous cloud of gas and dust, inception of nuclear reactions in the largest concentration of mass (the Sun), and consolidation of smaller bodies of mass into the planets and their satellites.

13. Planet Earth is unique in the solar system in having an oxygen-rich atmosphere, moderate surface temperatures, a partially water-covered surface, and a single satellite—the Moon—that is large in proportion to the parent planet.

REVIEW QUESTIONS

1. Why are some geologic processes more difficult to study than others?
2. How can a lack of understanding of the interrelationship of Earth processes lead to environmental problems? Can you cite examples from the area in which you live?
3. What are the basic differences between the uniformitarian and catastrophic concepts of Earth history? To what extent can they be reconciled?
4. What things must be kept in mind when applying uniformitarianism to the rock record?
5. What are Earth cycles? Give some examples.
6. Briefly outline the different ways in which lithospheric plates may interact with one another. Provide examples of each.
7. In the plate tectonics model, continents may become larger or smaller. Explain how this is possible.
8. How does plate tectonics influence the course of different types of Earth cycles? Give several examples.
9. The Andes Mountain system and the Mid-Atlantic Ridge are both chains of volcanoes, but, according to the plate tectonics model, they formed in very different ways. Explain the differences.
10. Why does the uniformitarian view of the Earth imply that the planet is older than the age implied by catastrophism?
11. Assume that the European and North American plates are diverging at a rate of 2.5 cm per year. If the spreading rate has always been constant, how old would your calculations make the Atlantic Ocean?
12. What are the major differences between the terrestrial and Jovian planets?
13. Why is it difficult to apply the principle of uniformitarianism to the origin of the solar system?

FURTHER READINGS

Faul, Henry, and Faul, Carol. 1983. *It began with a stone.* New York: John Wiley & Sons. (A history of geology as a science from prehistoric times to plate tectonics.)

Fenton, C. L., and Fenton, M. A. 1952. *Giants of geology.* Garden City, NY: Doubleday & Co. (An introduction to the personalities and contributions of the founders of modern geology.)

McPhee, J. 1981. *Basin and range.* New York: Farrar, Straus, and Giroux. (An entertainingly written account of the work of a modern geologist.)

Sullivan, W. 1974. *Continents in motion.* New York: McGraw-Hill Book Co. (An introduction to plate tectonics for the nonscientist.)

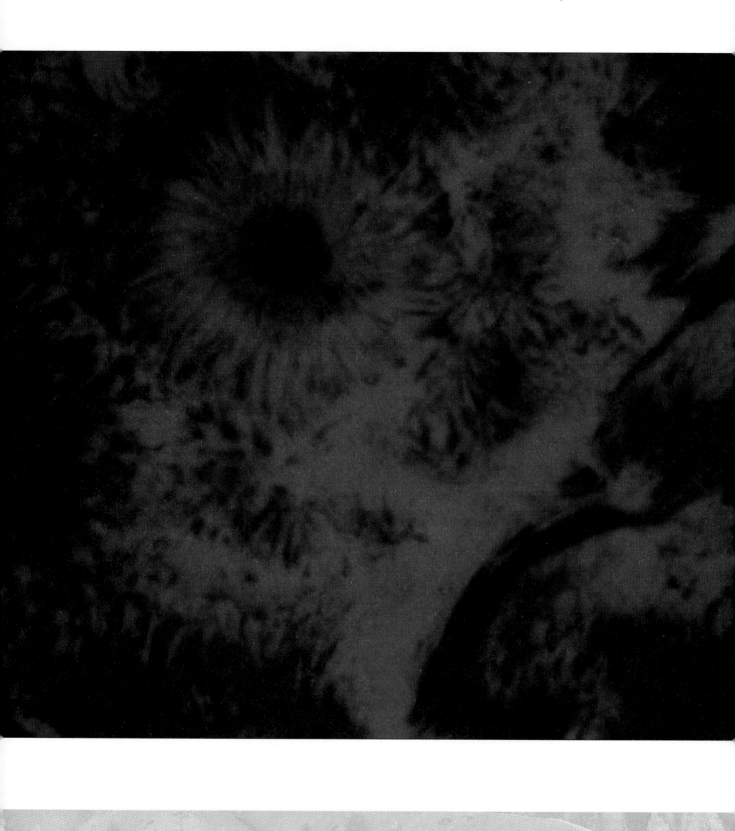

2
Matter and Energy

Sunspot activity on the Sun. Sunspots are exceptionally intense solar flares associated with magnetic disturbances in the Sun, and may extend thousands of kilometers into space. Photo by National Solar Observatory/Sacramento Peak.

In 1987, the Hawaiian volcano Kilauea erupted in what geologists consider a quiet way, sending streams of red-hot lava down its flanks toward the Pacific Ocean. By contrast, in 1980, Mount Saint Helens erupted explosively in the state of Washington, blasting tons of dust into the atmosphere but producing only minuscule amounts of sluggish lava. Why were these two eruptions so different? Are the lava and dust different in composition? What causes solid rock to melt inside the Earth and what then forces it to the surface?

In a very different geologic process, the Mississippi River delivers tons of sand, silt, and mud to the Gulf of Mexico every day. What makes the Mississippi flow thousands of kilometers from its source in Minnesota? What enables the river to carry its enormous sediment load? What is the sediment made of? Is it the same material as Kilauea's lava or the dust of Mount Saint Helens, or is it something completely different?

Millions of years ago, India separated from Antarctica, moved across the Pacific Ocean, and collided with Asia. What kinds of forces were powerful enough to move such enormous lithospheric plates, and what was their source?

These events are examples of the wide variety of processes that geologists study. To understand any geologic phenomenon, we must always ask questions similar to those raised above: *What are Earth materials made of? What makes the processes take place?* These two questions are fundamental to all of geology. Before looking at any particular process or rock, we will use this chapter to try to answer these basic questions.

WHAT IS THE EARTH MADE OF?

All Earth materials—gases, water, dust, soil, solid rocks, and molten lava—are composed of tiny, submicroscopic particles called *atoms*. To understand the Earth, we must first understand what an atom is and how atoms combine with one another to make the enormous variety of materials that our planet contains.

Atomic Structure

The word **atom** comes from a Greek word meaning "indivisible" because it was once thought that atoms were the smallest possible particles of matter. Today we know that atoms are themselves made of even smaller **subatomic particles,** and physicists continue to investigate these most basic building blocks of matter. Fortunately, an understanding of only three subatomic particles is necessary for us to explain the composition and nature of the Earth. These particles are electrons, protons, and neutrons:

> **Electrons** are particles that have very little mass but contain a small unit of electric charge designated as −1, or "negative 1."
>
> **Protons** have much more mass, approximately 1832 times that of an electron, and carry an electric charge that is equal but opposite to that of the electron: +1, or "positive 1."
>
> **Neutrons** have just a little more mass than a proton (1833 times the mass of an electron) but have no electric charge; hence they are "neutral."

Atoms are spherical. Their protons and neutrons cluster together in the center of the sphere to form the atom's **nucleus,** and the electrons orbit the nucleus at various distances (Figure 2–1). Most of the mass of an atom and all of its positive charge are thus found in the nucleus, whereas the negative charge is located in the electron cloud. The electrons move rapidly around the nucleus but are held in orbit by the positively charged protons, just as oppositely charged poles of magnets are held to one another. The number of protons in the nucleus of an atom is equal to the number of orbiting electrons, so that the overall electric charge of an atom is zero.

Elements

Atoms are not all the same. They may differ in the number of protons and neutrons they contain in their nuclei and in the number of electrons. The **atomic number** of an atom is the number of protons in its nucleus (always the same as the number of its orbiting electrons). All atoms with the same atomic number are said to be atoms of the same **element**. They have the same physical properties and behave

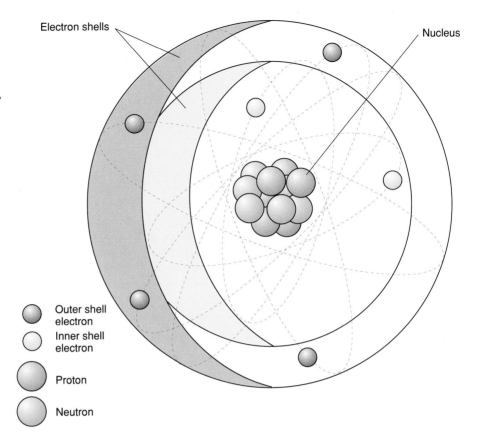

FIGURE 2–1
The structure of a carbon atom. Not all of the six protons and six neutrons can be shown because of perspective. The electrons orbit at different distances from the nucleus, two at the same distance close to the nucleus, and four farther away.

Electron shells

Nucleus

Outer shell electron

Inner shell electron

Proton

Neutron

the same way in natural processes. For example, all atoms of the element chlorine, which is green, gaseous, and extremely poisonous, contain 17 protons. All atoms of the solid, heavy, and unstable element uranium have 92 protons.

Of the 106 known elements, 88 have been discovered to occur naturally in the Earth and its atmosphere, and 18 more have been synthesized by physicists. However, only a small number of these play important roles in Earth processes; these elements are shown in color in the periodic table in Appendix A.

Isotopes

Atoms also may differ in mass as well as atomic number. The mass of an atom is measured in **atomic mass units (amu),** each of which is equivalent to the mass of a single proton. The contribution of electrons to the mass of an atom is so small that it is usually ignored, and the mass of a neutron is so similar to that of a proton that we will consider them to be equal. The **atomic mass** of an atom is thus equal to the number of protons plus the number of neutrons in its nucleus. For example, the carbon atom shown in Figure 2–1 has an atomic mass of 12 (6 protons + 6 neutrons).

All atoms of an element must have the same atomic number, but not necessarily the same atomic mass. Hydrogen atoms (atomic number 1) may have masses of 1, 2, or 3 amu depending on how many neutrons (0, 1, or 2) are present in the nucleus along with the single proton. Uranium atoms (atomic number 92) may have masses of 234, 235, or 238 amu. Atoms of an element that have different masses are called **isotopes** of that element. We will use a simple shorthand system for precise identification of atoms:

Atomic mass ⟶ $^{234}_{92}U$ ⟵ Chemical symbol
Atomic number ⟶ (uranium)

The different isotopes of hydrogen may thus be designated as ^1_1H, ^2_1H, and ^3_1H, and of uranium $^{234}_{92}U$, $^{235}_{92}U$, and $^{238}_{92}U$.

Structure of the Elements

Certain aspects of the structure of atoms are important to our understanding of Earth materials, and we will examine these by looking at some of the simpler elements (Figure 2–2). Hydrogen atoms are the simplest. Each hydrogen nucleus contains a single proton (with zero, one, or two neutrons), and this is orbited by a single electron. Helium (^4_2He) has two protons and two neutrons in its nucleus. Both of its electrons are the same distance from the nucleus as they orbit.

However, note that in lithium (^7_3Li), the three electrons are not all in the same orbit. Two circle the nucleus at an equal distance, but the third is farther out. Electrons that orbit a nucleus at the same distance are said to belong to the same **electron shell**. There is room for only two electrons in the innermost electron shell of lithium, so the third electron must enter a different shell, farther out from the nucleus.

The same pattern can be seen in elements of atomic numbers 4 (beryllium) through 10 (neon). The first two electrons form the innermost shell, and the fourth through tenth electrons of these elements are at approximately the same distance from the nucleus as the third electron, defining a second electron shell outside the first.

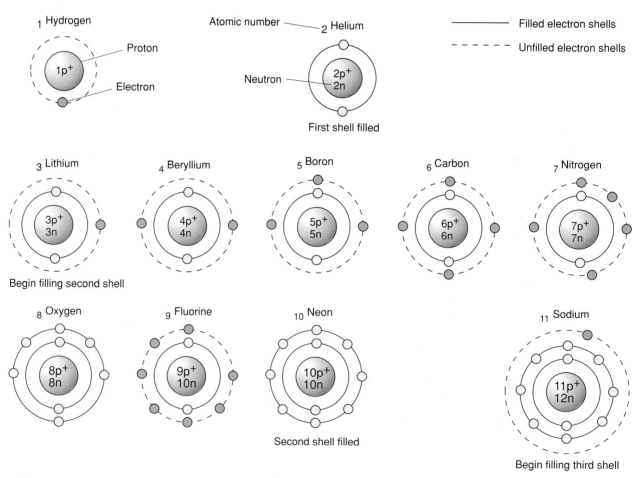

FIGURE 2–2
Electron structure of elements having atomic numbers 1 through 11.

The maximum number of electrons that can exist in the first shell is two; the maximum number that can exist in the second shell is eight. As shown in Figure 2–2, sodium ($^{23}_{11}$Na) has two electrons filling its inner shell and eight electrons filling its second shell. The eleventh electron must begin a third shell. There are strict laws governing the number of electrons that can fit into a given shell, but the nature of these laws need not concern us here.

Reactions Involving Atoms

Atoms undergo two kinds of transformations: **nuclear reactions,** which affect the nucleus, and **chemical reactions,** which involve changes only in the orbiting electrons. During nuclear reactions, a nucleus changes by gaining or losing nuclear particles. During chemical reactions, atoms become attached (*bonded*) to one another without altering their nuclei in any way.

Nuclear Reactions. In nuclear reactions, the nucleus of an atom of one element is converted to the nucleus of an atom of a different element. As an example, consider the following reaction involving one of the isotopes of uranium:

$$^{238}_{92}\text{U} \rightarrow\ ^{234}_{90}\text{Th} + ^{4}_{2}\alpha + \text{energy}$$

The uranium nucleus is converted to a thorium nucleus by losing a particle of nuclear material composed of two protons and two neutrons. This particle is called an **alpha (α) particle,** and this type of nuclear reaction is known as an **alpha decay.** In addition to losing a piece of the original nucleus, some of the energy that held the uranium nucleus together is also released. Nuclear reactions such as this, in which a nucleus splits, are called **fission** reactions. Nuclear reactions of the opposite kind involve *combining nuclei* of two atoms to make a different element and are called **fusion** reactions. Fission reactions are well known in humankind's nuclear reactors and weapons; fusion reactions occur in the Sun, where hydrogen nuclei join to form helium nuclei.

When nuclear reactions occur spontaneously, the energy released is called **radioactivity,** and the elements involved are said to be radioactive. We will examine nuclear reactions more closely in Chapter 8, where we will discuss their use in determining the age of Earth materials.

Chemical Reactions and Compounds. A few elements are called **inert elements** because their atoms always occur independently, not attached to any others. Examples are the inert gases helium, neon, argon, krypton, and radon. Most atoms, however, are found in combination with others in *compounds*. Some elements occur in the **native state,** in which their atoms form compounds only with one another and not with atoms of other elements. Common examples are gold, sulfur, carbon, and oxygen. In most compounds, however, atoms of one element combine with atoms of other elements. Two examples of simple compounds of this type are water, a combination of hydrogen and oxygen atoms (H_2O), and table salt, a combination of sodium and chlorine atoms (NaCl).

Bonding

The forces that hold atoms together in compounds are called **bonds.** Bonding has a profound effect on the atoms that it joins. When atoms of different elements are bonded together, they often lose their distinguishing characteristics and form a compound that displays properties of its own. Chlorine, for example, is a green poisonous gas, and sodium is a yellow solid that explodes when placed in water. The two combine to make salt, a safe, edible, colorless or white solid.

Bonding is controlled by the nature of the outermost electron shells of atoms. If an atom's outermost electron shell is completely filled, the element is inert and will not enter chemical reactions. Atoms of elements that have *unfilled* outer shells bond to one another by gaining, losing, or sharing their outermost electrons in a variety of chemical reactions.

The importance of outer electrons in governing chemical behavior is shown in the *periodic table* of the elements (Appendix A), where elements with similar outermost shells are aligned in columns. Elements in the same column tend to behave in very similar ways during chemical reactions, whereas those on opposite sides of the periodic table behave very differently. All the inert gases are in the right-hand column; they do not participate in chemical reactions. Fluorine and chlorine, in the next column to the left, both lack one electron in their outermost shells; they behave in much the same way chemically. At the opposite end of the table are lithium and potassium, each with only one electron in their outermost shells; they act like one another but behave very differently from either chlorine or fluorine.

Types of Bonding

Four types of bonding are important in geology: *ionic, covalent, metallic,* and *van der Waals'.* All Earth materials are held together by one of these bond types, or a combination of two or more.

Ionic Bonding. In **ionic bonding,** atoms gain or lose electrons to achieve filled outermost shells and thus attain stability. As an example, examine the bonding of sodium and chlorine in the compound sodium chloride, NaCl (Figure 2–3). An atom of sodium has a single electron in its outermost shell and can attain the same stable configuration as neon if it can *lose* that electron. Similarly, a chlorine atom can have the same electron structure as the inert gas argon if it can *add* one electron to its outer shell. The transfer of an electron from the outer shell of sodium to that of chlorine produces stable electron structures in both elements simultaneously, but also changes the electric balance in each atom. Sodium, having lost an electron, has an extra positive charge (11 protons in the nucleus, but now only 10 electrons), whereas chlorine has gained an electron and now has a net negative charge (17 protons, 18 electrons).

These charged particles are called **ions.** Positively charged ions such as sodium are called **cations,** and negatively charged ions such as chlorine are called **anions.** Ions are distinguished from atoms by a different type of symbol, Na^+ and Cl^- in this case. The oppositely charged sodium and chlorine ions attract one another and are bonded together. Bonds between ions are called ionic bonds and are very common in Earth materials. Notice that neither the electrons of the inner shells nor the neutrons are involved in this or any other bonding process.

Covalent Bonding. In **covalent bonding,** atoms *share* electrons rather than gaining or losing them. Electrons in the outermost shells of the atoms involved leave their positions around individual nuclei and form a **molecular shell,** a different kind of electron cloud that surrounds two or three nuclei. In this way, even atoms of the same element can be bonded to one another.

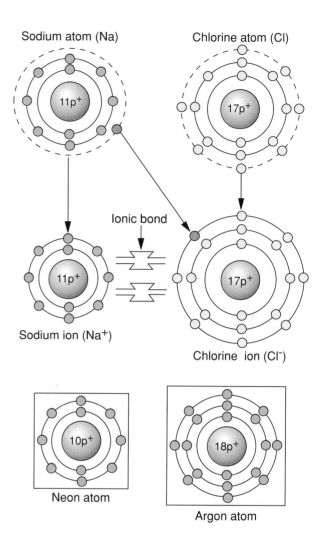

FIGURE 2–3
Ionic bonding between sodium and chlorine in the compound sodium chloride (NaCl—common salt).

Consider the bonds that hold oxygen atoms together in the compound O_2 (Figure 2–4). Each oxygen atom has two electrons filling its innermost electron shell and six of the possible eight in its outer shell. In the compound O_2, the six outer-shell electrons from each of the atoms join to form a molecular shell around the two nuclei. The attraction between the positively charged nuclei and negative molecular-shell electrons holds the two atoms together. As in ionic bonding, the inner electrons are not affected, but this time no ions are formed and no inert-gas electron structures are produced.

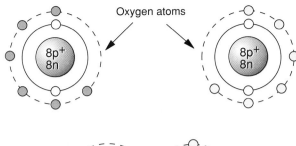

Oxygen atoms

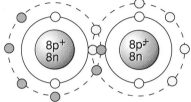

Outer shells overlap

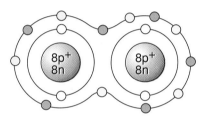

Molecular shell forms around
covalently bonded oxygen molecule

FIGURE 2–4
Covalent bonding between oxygen atoms in the compound O_2.

The smallest unit of a compound that can exist is called a **molecule**. Single oxygen atoms do not occur on Earth; the pair of covalently bonded oxygen atoms is thus called the *oxygen molecule*. Covalent bonding also occurs commonly in solids. For example, it holds the carbon atoms together in the mineral diamond and is found in many other minerals.

Metallic Bonding. Bonding between atoms of copper, gold, or silver occurs by a process similar to covalent bonding called **metallic bonding**. As in covalent bonding, the outer electrons are shared by different nuclei, but unlike covalent bonding, there are many nuclei involved. Molecular shells generally involve small numbers of atoms—two in the case of O_2—but in metallic bonding, *all* the outermost electrons of *all* the atoms present are shared by *all* the

nuclei present in the metal. These electrons are relatively free to travel from one nucleus to another throughout the metal, unlike electrons in the restricted positions involved in ionic or covalent bonds. This freedom of electron movement through the entire substance is what makes possible the flow of electricity in these metals.

Van der Waals' Bonds. Many Earth materials possess additional forces called **van der Waals' bonds** that assist ionic and covalent bonds in holding atoms and molecules together. These forces are different from other types of bonds because they are not strong enough to be the only bond type present and because they do not involve the transfer or sharing of electrons. They result from the positions of atoms in a molecule or of electrons in a single atom.

Strong van der Waals' Bonds. Strong van der Waals' bonds occur between molecules of water—Figure 2–5(a). Water is a covalent compound in which electrons from the hydrogen and oxygen atoms form a molecular shell. The molecule itself has no charge, but the two positively charged hydrogen nuclei are both located on one side of the compound. This creates what is called a **polar compound,** one in which there is a weakly positive side and a weakly negative side (in this case, the side opposite the hydrogen nuclei). These weak charges can attract the oppositely charged ends of other water molecules, or even ions present in other compounds. Water is the most important polar compound on Earth; we shall see how its polar nature makes it an effective contributor to internal processes (Chapter 4) and surface processes (Chapter 9).

Weak van der Waals' Bonds. Weak van der Waals' bonds occur in nonpolar substances, even between atoms of the inert gases—Figure 2–5(b). As the two orbiting electrons circle a helium nucleus, they may be positioned on the same side of the nucleus for an instant. Each helium atom is electrically neutral, but with both electrons on one side of the nucleus it temporarily has positive and negative sides. The atom then acts like a tiny magnet, attracting adjacent helium atoms. Graphite, the "lead" in pencils, and diamond are both made of pure carbon. The softness of graphite is due to the fact that it is held together partly by very weak van der Waals' forces.

FIGURE 2–5
Van der Waals' bonding (a) between water molecules and (b) between helium atoms.

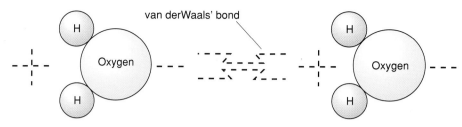

(a) "Strong" bonding in water, a polar compound

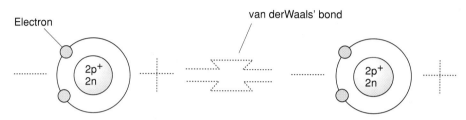

(b) Weak bond between two helium atoms

FIGURE 2–6
Major chemical subdivisions of the Earth.

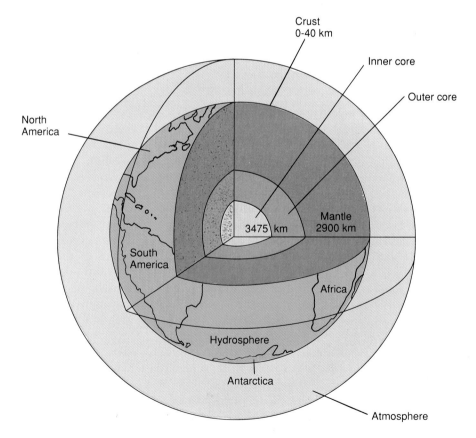

We shall look more closely at these two forms of carbon in Chapter 3.

Composition of the Earth

Our planet can be divided into three chemically distinct regions—the atmosphere, hydrosphere, and solid earth. Each has a unique combination of elements and a unique proportion of solids, liquids, and gases (Table 2–1; Figure 2–6).

The **atmosphere** is the gaseous envelope that surrounds the Earth. As is apparent in Table 2–1, it is composed mostly of nitrogen and oxygen but contains other gases (carbon dioxide, water vapor) and small amounts of solids such as dust particles, hailstones, and snowflakes.

The **hydrosphere** is the liquid outer covering of the Earth that includes the oceans, lakes, streams, and underground water. Most of our planet's surface is water—71% is covered by oceans alone. The hydrosphere is essentially water (Table 2–1), but a great deal of material is dissolved in both seawater and freshwater, including gases such as oxygen and ions such as Na^+ and Cl^-. Solids, like the sediment carried by the Mississippi River, are carried in suspension in the Earth's surface waters.

The Earth itself is nearly all solid and is divided into three regions, the **crust, mantle,** and **core,** as shown in Figure 2–6. Geologists have been able to drill only into the Earth's crust, but processes like volcanic eruptions occasionally bring samples of what we believe to be mantle material to the surface. The composition and very existence of a mantle and core are inferred from remote sensing methods described in Chapter 19.

Oxygen and silicon are by far the most abundant elements in the crust and mantle (see Focus 2–1), and most materials from these parts of the Earth are compounds that contain these two elements. The core is thought to be a combination of iron and nickel, and we now believe that the outer part of the core is liquid. We will discuss evidence for these hypotheses in Chapters 19 and 20.

Interfaces: Where the Action Is. Wherever atmosphere, hydrosphere, and solid earth interact, their differences bring about important physical and chemical processes. For example, shoreline erosion takes place where oceans and solid continents interact; rocks break down chemically to produce soils where atmosphere and solid earth come into contact; and waves form and move across oceans or lakes where atmosphere and hydrosphere meet. Interfaces between these chemically and physically distinct regions are among the most active and most rapidly changing geologic regions on the planet.

Chemical and Physical Boundaries. In Chapter 1 we divided the outer few hundred kilometers of the Earth into the lithosphere and asthenosphere on the

TABLE 2–1
Composition of the Earth (by weight).

Element	Crust Only	Entire Earth	Hydrosphere		Atmosphere	
Oxygen	46.3%	29.5%	Oxygen	86.3%	Nitrogen	75.5%
Silicon	28.2	15.2	Hydrogen	10.8	Oxygen	23.15
Aluminum	8.2	1.09	Chlorine	1.8	Carbon	trace
Iron	5.6	34.6	Sodium	1.05	Hydrogen	trace
Calcium	4.1	1.13	Potassium	0.04	Others	1.35
Sodium	2.4	0.57	Calcium	0.04		
Potassium	2.1	0.07				
Magnesium	2.3	12.7				
Titanium	0.5	0.05				
Nickel	trace	2.39				
All others	trace	2.70				

HOW RARE IS RARE?

Elemental abundances in Table 2–1 are given in *weight* percentages, so that oxygen is shown as comprising 46.3 grams and silicon 28.2 grams out of every 100 grams of the Earth's crust. In fact, oxygen atoms are even more abundant than those of silicon than is indicated in Table 2–1, because each oxygen atom weighs much less than each silicon atom: oxygen = 16 amu, silicon = 28.09 amu. When the relative weights of the elements are taken into account, the *atomic abundances* of the eight most common elements in the crust are actually:

Oxygen 62.55%	Calcium 1.94%
Silicon 21.22	Iron 1.92
Aluminum 6.47	Magnesium 1.84
Sodium 2.64	Potassium 1.42

Oxygen and silicon thus account for more than 83% of all the atoms in the crust!

Modern civilizations depend on many Earth resources, some of which, like aluminum and iron, also are among the most abundant atoms in that part of the Earth (the crust) which is accessible to us. What about some of the other, less common materials that are part of everyday life in a modern, industrialized nation? What about copper for the wires that carry our electricity? The zinc used in galvanizing metal to make it rust resistant? The vanadium, nickel, and chromium added to steel to make it hard, durable, and stainless? Table 1 gives an indication of the atomic abundance of these and some other industrially important materials.

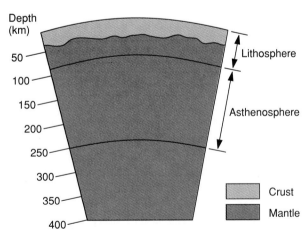

FIGURE 2–7
Physical and chemical divisions of the outermost part of Earth.

basis of *physical* criteria—the different rigidities of the materials in the two regions. The division here into crust and mantle is based on *chemical* criteria; the relationship among lithosphere, asthenosphere, crust, and mantle is shown in Figure 2–7. The lithosphere contains the entire crust and a part of the upper mantle, whereas the asthenosphere is a region 100 to 150 km thick in the upper mantle.

Plate tectonics has revolutionized our view of the planet because it requires that even the rock of the solid Earth be mobile. This was one of the reasons that the theory was not widely accepted immediately. We do not expect solids to be mobile, yet we saw in Chapter 1 that glacial ice can flow even though it seems to be a brittle solid. We shall see several examples in later chapters of this apparent paradox:

TABLE 1
Approximate atomic abundances of selected elements.

Metals and Other Industrial Materials		
Element	Atomic Percentage	Of every *million* atoms in crust there are this many atoms of this element:
Sulfur	0.035%	350
Chromium	0.008	80
Vanadium	0.004	40
Nickel	0.003	30
Zinc	0.002	20
Copper	0.0015	15
Lead	0.000165	1.65
Tin	0.000053	0.53
Precious Metals		Of every *billion* atoms in crust there are this many atoms of this element:
Silver	0.000002%	20
Gold	0.00000005	0.5

Clearly, these elements are rare in the crust. Compare silicon's abundance (over 210,000 atoms per million atoms in the crust) with that of any of these important industrial elements. It is 600 times more abundant than sulfur, and *400 million times* more abundant than gold. If it were not for geologic processes that concentrate the scarce elements for us, we would not be able to find them, much less mine them. The geologic processes are as varied as the melting and crystallization of rock in the crust and the transportation of sand grains by a stream: both were responsible for the California Gold Rush of 1849.

solid, seemingly permanent Earth materials are indeed capable of movement.

ENERGY: WHAT MAKES EARTH PROCESSES OPERATE?

We have just answered the first question asked at the beginning of this chapter, which was: What are Earth materials made of? We know that all Earth materials are composed of some combination of 88 naturally occurring elements whose atoms are bonded to one another by several kinds of forces. In Chapters 3, 4, 6, and 7 we will take a close look at some of these combinations—the minerals and rocks of the Earth's crust.

However, the second question still remains: What makes Earth processes happen? The answer is a single word: **energy**. Energy is defined as the capacity to do work or to cause activity to take place. Without energy, Earth would be a dead, motionless planet; with it, Earth is a dynamic, active world. There are many kinds of energy: *kinetic,* including heat and electric energy; *radiant,* including light and X-rays; *nuclear; gravitational;* and *chemical.* Most play roles in causing and sustaining Earth processes. To understand the complex factors that govern Earth behavior, we must now look at the sources and the interplay of different energy forms. Energy, of course, is needed by human beings for our own work and activity. The sources and uses of energy for human needs are discussed in Chapter 21.

Kinetic Energy

The ability of a moving object to induce activity in other objects is called **kinetic energy**. In most cases, the composition of the moving object is unimportant. It is the *movement* of the object that causes activity, not its composition or state of matter. A rolling billiard ball has kinetic energy (Figure 2–8). When it strikes a stationary ball, it causes the second ball to move.

Similarly, molecules of water flowing in the Mississippi River have kinetic energy and can therefore make sand grains move along the riverbed. The amount of kinetic energy depends on the mass m and the velocity v of the moving object:

$$\text{Kinetic energy} = \frac{mv^2}{2}$$

The faster an object moves, or the greater its mass, the more kinetic energy is involved. Large, fast-flowing rivers can move more sediment and do more geologic work than small, slow-moving streams because they have more kinetic energy. Imagine how much kinetic energy must be involved in moving an object the size of the North American plate! (The velocity is incredibly slow, but the mass is incredibly large.)

Heat. Atoms are constantly moving. Even in what appear to be extremely rigid solids, atoms and ions are vibrating back and forth around their fixed positions. Heat energy is the kinetic energy associated with this vibrational motion. The faster the atoms and ions move, the more heat energy they have.

Temperature and heat are not the same, although they are related. The **temperature** of a substance is the *average* kinetic energy of its atoms, whereas its **heat energy** is the *total* (sum) of the kinetic energy of all of its constituent atoms. The difference is illustrated by comparing a match and a roaring bonfire: both have about the same temperature, but there is little doubt about which has more heat energy. (Try barbecuing a chicken over a match if you are not sure!) In this book, temperatures will be given in degrees Celsius (°C). (For comparison with the Fahrenheit scale, water freezes at 0°C and boils at 100°C; a room temperature of 68°F is 20° on the Celsius scale—see conversion formulas in Appendix B.)

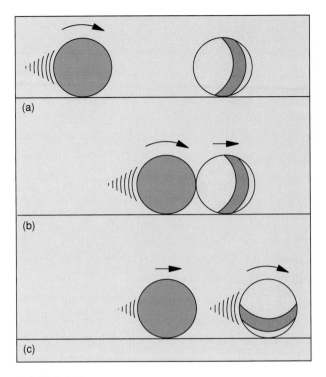

FIGURE 2–8
Kinetic energy. A rolling billiard ball has kinetic energy and can therefore impart some of that energy to a stationary ball, causing it to move.

Heat energy moves from one place to another, always from hot objects to cooler ones, by the processes of *conduction, convection,* and *radiation* (Figure 2–9).

Conduction transfers heat between materials that are in contact with one another. Particles vibrating rapidly in the hotter material bump into those in the colder one, making them vibrate faster, just as a moving billiard ball causes another to move. Figure 2–9(a) shows how heat energy from the heated end of a bar is conducted toward the cold end. Similarly, sunlight shining on rocks in a desert directly heats only their outer surfaces; the heat is then *conducted* inward to warm their interiors.

In **convection,** heated particles do not just vibrate in place but move much greater distances, carrying the heat with them. For this reason, convection is most effective in liquids and gases, where particles can move readily. Convection occurs in a container of water that is being heated. Water molecules at the bottom of the container become hot, rise upward,

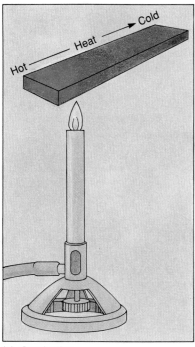

(a) Conduction

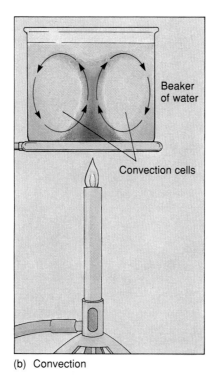

(b) Convection

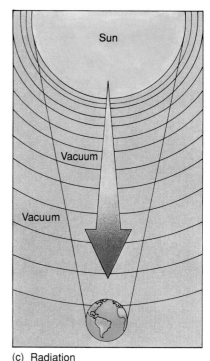

(c) Radiation

FIGURE 2–9
The three methods of heat-energy transfer.

and are replaced by colder molecules circulating from the top of the container. Rising molecules heat their surroundings by conduction as they move, and a circular motion called a **convection cell** is established. A very similar process occurs in the atmosphere as it is heated near the surface of the Earth, and is responsible for the major climate patterns on the planet.

In **radiation,** heat energy is converted into a different energy form called radiant energy (electromagnetic waves) and can be transmitted even through a vacuum where there are no particles to vibrate or be moved. Heat from the Sun is carried to the Earth by radiation through the vacuum of space.

Electric Energy. The special type of kinetic energy associated with the movement of electrons through a substance is called **electric energy**. We are most familiar with it in wires made of metallically bonded materials. The outer-shell electrons are free to move in these metals and their movement transmits the electric energy. Electric energy plays only a small part in Earth processes today, in the form of lightning,

but experiments dealing with the origin of life suggest that lightning in the atmosphere of the primitive Earth may have helped create organic molecules.

Radiant Energy

Light, infrared radiation, ultraviolet radiation, X-rays, and radio waves are all kinds of **radiant energy**. All are electromagnetic waves—energy forms that move in the wavelike motion shown in Figure 2–10. All forms of radiant energy can move through a vacuum, and all do so at the same velocity—the speed of light (3×10^8 meters per second).

The forms of radiant energy differ from one another in wavelength and frequency. **Wavelength** is the distance from a point on the wave to the next equivalent point on the wave (for example, the distance from wave crest to wave crest in Figure 2–10). **Frequency** is the number of wavelengths that pass an observer in a given interval of time.

Some radio waves have wavelengths of hundreds of kilometers, whereas X-ray wavelengths may be as

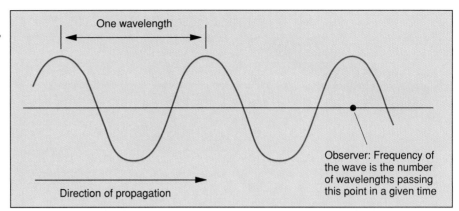

FIGURE 2-10
The form of electromagnetic waves, like ripples in a pond.

One wavelength

Observer: Frequency of the wave is the number of wavelengths passing this point in a given time

Direction of propagation

short as 10^{-8} (0.000000001) cm. The different wavelengths result in different properties of the radiant energy forms. For example, only radiant energy with wavelengths between 3900×10^{-8} and 7700×10^{-8} cm is visible (we call it light). We must use instruments to detect the others, although our skin reacts to the longer-wavelength infrared energy—in a way that we call sunburn.

Nuclear (Atomic) Energy

We saw earlier that some of the energy that holds the nucleus of ^{238}U together is released when the atom is converted to a thorium atom. This type of nuclear-binding energy is released during all nuclear reactions, be they fission or fusion, and is called **nuclear energy** or **atomic energy**.

Gravitational Energy (Gravity)

Gravitational energy, normally called **gravity,** is one of the most important causes of Earth processes. We tend to take gravity for granted because we are never without it and because it varies so little that we cannot detect its subtle variations without delicate instruments. Nevertheless, it controls nearly every aspect of our lives and of geologic processes.

Gravity is a force of attraction that objects have for one another due to their mass. All objects are affected by gravity regardless of their composition or state of matter. Sir Isaac Newton showed that the force of gravitational attraction between two bodies depends only on the masses of the two objects involved (M_1 and M_2), and on the distance between them (d):

$$\text{Gravity} = \frac{GM_1 \times M_2}{d^2}$$

G is a constant whose value depends on the units of mass and distance used. The larger the masses, or the closer the objects are, the greater the force of attraction will be between them. The smaller the masses or the farther apart the objects are, the smaller the force of attraction will be.

Earth's mass is almost incomprehensibly large—6.5×10^{21} metric tons—so it exerts a strong attractive force on any object on its surface or in its atmosphere. Indeed, its gravity is enough to hold the Moon in orbit. A mass dropped from a cliff "falls" because of gravitational attraction. Gravity thus controls rainfall, snowfall, landslides, the movement of water downhill in streams, and even the shapes of volcanoes.

Chemical Energy

During certain chemical reactions, some of the energy by which atoms and ions are bonded to one another is released. This type of energy is called **chemical energy**. When petroleum products are used as fuel for automobiles, the chemical energy that holds the complex organic molecules together is released and is used to turn the wheels.

Energy Conversions

Most kinds of energy can be converted easily into other kinds, resulting in the complex network of energy transfers that governs all Earth behavior. In nearly all Earth processes, energy may be converted from one form to another, but *energy cannot be cre-*

ated or destroyed. (The only exception is in nuclear reactions where some mass may be converted into energy.)

During all nonnuclear Earth processes, energy is transformed from one type to another, but does not disappear. A rolling ball has kinetic energy, but where does that energy go when the ball stops rolling? Some is used to heat atoms in the material the ball is rolling on; this conversion of the kinetic energy of large-scale motion to heat (small-scale vibration) is called *friction*. Some of the kinetic energy is used to push air molecules out of the path of the ball.

When energy is not actually being used to produce activity or do work, it can be stored in a passive state in materials. Stored energy, regardless of its type, is called **potential energy**. A car battery contains potential electric energy that is released as the ignition switch is turned to connect the battery to the starter motor. An atom of ^{238}U is a reservoir of potential nuclear energy that is released when a nuclear reaction occurs. Organic molecules in petroleum and coal are sources of potential chemical energy. A boulder poised at the top of a steep slope has potential kinetic energy that is released once it begins to move.

Earth's Energy Networks

Changes from one type of energy to another—from active to potential, potential to active, nuclear to heat, and so on—take place during all Earth processes. A brief look at one process reveals the complexity of Earth's energy network.

As an example, consider a process as "simple" as a flowing stream. This process actually begins in the Sun. Nuclear fusion reactions in the Sun release large amounts of nuclear energy. This is converted to radiant energy, some of which reaches the Earth. There it is converted to heat energy in the atmosphere and oceans. The heat energy counteracts the chemical bonding forces between water molecules in the ocean, causing a change of state from water to water vapor (evaporation). The heat is changed to kinetic energy, causing the vapor to rise into the atmosphere. Much of this kinetic energy is stored as potential energy when the vapor moves laterally, but it is released when gravity pulls raindrops back to Earth. It is this kinetic energy that causes water to flow in streams. Eventually, the water returns to the

oceans, completing the water cycle that we discussed in Chapter 1.

Sources of Energy for Earth Processes

Enormous amounts of energy are required each second to drive Earth processes, and the most important energy sources for these processes are heat and gravity. We shall see in the following chapters that many aspects of our planet's behavior depend on the interaction of these two agents.

All internal Earth processes require energy, including earthquakes, melting of rock to make lava, and movement of lithospheric plates. So do surface processes, such as the flow of ocean currents, the carving of landscapes by glaciers, and the formation of soil. We now know that most internal and surficial processes are driven by energy obtained from very different sources. For the most part, internal processes are driven by internal sources of energy—those that originate within the Earth—and surface processes are driven by external energy sources, mostly from the Sun.

Internal Energy Sources. Simple observations help us realize that there must be a source of *heat* within the Earth. As miners descend into the Earth, the temperature of the rock around them increases. In deep mines such as the diamond mines of South Africa and the copper mines at Butte, Montana, temperatures increase so rapidly that cool surface air must be circulated through the lower levels; otherwise the mines would be too hot for humans to work in. Temperatures in the Earth's crust and upper mantle increase at rates of 15–60°C per kilometer of depth. The mines in Butte are more than a mile deep; the temperature at their lowest levels would be anywhere from 24–96°C hotter than at the surface.

These vertical rates of change in the Earth's temperature are called **geothermal gradients**. Rocks are poor conductors of heat, so the increasing temperatures indicate that we are approaching the heat source. Therefore, there must be a source of heat energy inside the Earth.

The most important internal source of heat energy is radioactivity. Our ideas about the evolution of the Earth hold that the radioactive elements uranium, thorium, potassium, and rubidium are concentrated in the Earth's crust and upper mantle, where they

FIGURE 2–11
Tidal range along the St. Croix River at Calais, Maine/St. Stephen, New Brunswick. (a) High tide, (b) low tide. Photos by Allan Ludman.

(a)

(b)

supply the energy for processes such as volcanic eruptions. Part of the nuclear energy released during nuclear fission reactions is given off as radiant energy, and part is released as kinetic energy carried by particles such as alpha particles ejected from nuclei. Collisions with the ejected particles and interactions with the radiant energy cause atoms and ions in rocks to vibrate, thus producing heat energy.

Radioactivity is not the only source of internal heat. During the initial accretion of particles that formed the Earth, gravitational attraction is thought to have produced enough friction that the entire planet melted. Cooling of a planet, like cooling of a pie, begins at the outside and slowly works inward. The rocks that make up Earth's crust and upper mantle are such poor conductors of heat energy that some of that initial heat of accretion may still be trapped deep inside the Earth, and serve as a second internal source of heat energy. Additional but small amounts of heat are produced by chemical reactions and by friction from movement of Earth materials caused by the tides.

Gravity also plays a major role in internal processes. If you were buried several miles beneath the surface, you would be crushed by the pressure of the overlying rocks. Earth materials also are subjected to these pressures and must make adjustments as they are buried deeper and deeper. It is Earth's gravity pulling downward on those rocks that causes the pressure. The deeper we go, the greater grows the pressure from the overlying rocks.

The standard unit of internal Earth pressure is the **kilobar** (kb), equal to nearly 1000 times the atmospheric pressure at the Earth's surface. Just as temperature increases with depth, so too does pressure, in what is called the **geobaric gradient**. The geobaric gradient in the crust is approximately 1 kb for every 3.3 km of depth. If we know the Earth's geothermal and geobaric gradients, we can begin to estimate what conditions must be like deep within the planet. Humans could not survive even at moderate depths in the crust: at average geothermal and geobaric gradients, the pressure at a depth of 10 km would be 3000 times that at the surface and the temperature would be 300°C. We would be both flattened and roasted.

External Energy Sources. Surface processes are driven by heat and gravity originating in the Sun and are modified by Earth's own gravity. We saw in our discussion of Earth's energy network how nuclear reactions in the Sun provide heat for Earth processes. The gravitational pull of the Sun is enormous— enough to keep all nine planets and their satellites in their orbits. Even though the Sun is millions of kilometers from the Earth, its gravity is strong enough to affect the surface processes that we call *tides*. The Moon is smaller than the Earth and much smaller than the Sun, but it is so close that its gravitational attraction also is important in causing the tides.

The daily changes in sea level that occur along the world's coastlines are called the **tides**. Twice a day the oceans lap up onto the shores of the east coast of North America at high tide, and twice a day they recede into the ocean basins at low tide. Elsewhere, the tides are less regular, and some areas experience only one high and one low tide each day. The process is quiet and gradual at most shorelines. The difference between high and low water is only 1 or 2 m along the east and west coasts of North America and a little less in the Gulf Coast region.

In some places, however, shoreline topography controls the ebb and flow of the tides in such a way that the daily changes are drastic. Few who have seen the tides in the Bay of Fundy (between Nova Scotia and New Brunswick) can forget the spectacular changes in the shoreline at high and low tides (Figure 2–11). Tidal ranges in the Bay of Fundy and at a few other places on the Earth may be as much as 15 m.

Origin of the Tides. The tides are caused by a complex interaction among the gravity fields of the Earth, Sun, and Moon, coupled with Earth's rotation (Figures 2–12 and 2–13). The Moon exerts a strong gravitational pull on the Earth, particularly on

FIGURE 2–12
Causes of the tides. (a) Moon's gravity causes a tidal bulge in the ocean on the side of Earth facing Moon. (b) Centrifugal force due to Earth's rotation produces an equatorial bulge all around the Earth. On the side facing the Moon, the equatorial bulge adds to the Moon's pull but on the other side it counters the Moon's gravity, causing the bulge on that side. The Sun also significantly affects the tides—see Figure 2–13.

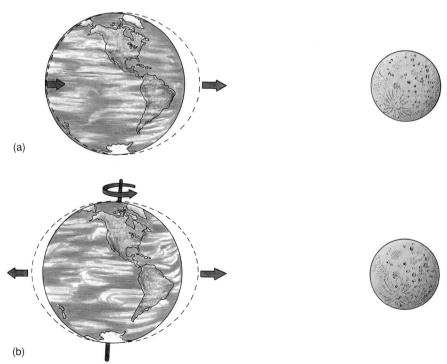

(a)

(b)

the side facing the Moon. This pull causes a slight bulge in the Earth and a much greater tidal bulge in the hydrosphere as the more mobile ocean waters are pulled toward the Moon—Figure 2–12(a).

Surprisingly, a second tidal bulge forms simultaneously on the opposite side of the Earth. This is due to the fact that Earth's rotation creates a centrifugal force that drives the water outward at the Equator —Figure 2–12(b). On the side of the Earth facing the Moon, this centrifugal force is added to the Moon's gravitational attraction; on the opposite side, it actually overcomes this attraction to produce the second bulge. High tides are thus felt twice a day, when the Moon is directly overhead and when the Moon is on the opposite side of the Earth. Low tide occurs at points on the Earth's surface 90° away from the high-tide locations—also twice a day.

As the Moon orbits the rotating Earth, the tidal bulges change position because different points on the planet's surface are brought into alignment with the Moon. The energy needed to move water onto the land and then back into the ocean must come from somewhere—essentially from the rotation of the Earth—and this leads to a surprising conclusion of how the Earth has changed throughout time (Focus 2–2).

Although the Sun is nearly 400 times farther from the Earth than the Moon, its mass is so great that it too affects the tides. Twice a month, when the Moon and Sun are aligned with the Earth—Figure 2–13(a), the gravitational pull of the Sun assists that of the Moon and enhances tidal effects. High tides are higher than usual and low tides are lower. These greater-than-normal tidal ranges are called *spring tides*. When Earth-Moon-Sun positions are as shown in Figure 2–13(b), the Sun's gravity partly counteracts the Moon's, resulting in lower than normal tidal ranges called *neap tides*.

FIGURE 2–13
Effect of the Sun and Moon on Earth tides. (a) Spring tides: at new moon and full moon, the Sun's gravity acts with the Moon's to produce greater than normal ranges of high and low tide. (b) Neap tides: At the one-quarter and three-quarter phases of the moon, the Sun's gravity partly counteracts the Moon's to produce less than normal tidal ranges. (The term "spring tide" has nothing to do with the season but refers to the "springing forth" of a larger tide.)

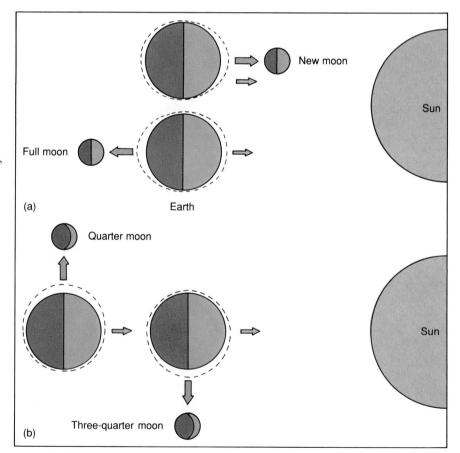

Imagine standing on the east coast of the United States as the Earth's rotation brings your position into alignment with the Moon. The ocean to your east has already bulged outward, and North America is driven into this mass of water, causing the tide to rise onto the land as you watch. Motion of any kind requires kinetic energy, and some of the kinetic energy associated with Earth's rotation is used in transporting the tides onto and off the landmasses.

During the transfer of kinetic energy from rotation to tidal motion, some of Earth's energy is expended in overcoming the friction of moving water against the solid Earth. The tides thus exert a frictionlike force on the Earth. They act as a brake that gradually slows rotation. We need not worry that our planet will stop rotating about its axis in the very near future. Estimates of tide-induced slowing of rotation are on the order of 0.001 second per century, so that since Europeans first visited North America the day has grown only five one-thousandths of a second longer.

On the other hand, we shall see in Chapter 8 that the Earth has a history measuring billions of years, and even very small changes may add to significant results over such long periods of time. A slowing of 0.001 second per century translates to 10 seconds per million years, or 10,000 seconds per billion years. This is about 2 hours and 46 minutes. Earth now takes about 24 hours to complete one rotation, even after billions of years of tidal braking. One billion years ago, it presumably rotated in 21 hours and 14 minutes, and one billion years in the future the day will be lengthened to almost 27 hours. 13 o'clock becomes a real possibility!

SUMMARY

1. All matter is composed of small particles called atoms, which are themselves made up of protons, neutrons, and electrons. A single atom consists of a nucleus that contains positively charged protons and uncharged neutrons surrounded by negatively charged orbiting electrons.

2. Atoms with the same number of protons are said to have the same atomic number and are atoms of the same element. They behave in the same way during chemical reactions. Atoms with the same atomic number but different atomic masses are called isotopes.

3. Atoms may enter into nuclear or chemical reactions. In nuclear reactions, atoms of one element are con-verted to atoms of another by changing the number of protons in the nucleus. In chemical reactions the nuclei are not affected; the orbiting electrons interact to bond atoms to one another, forming compounds.

4. Bonding forces may be ionic, covalent, metallic, or van der Waals' types, depending on the nature of the electron shells involved. Compounds often exhibit physical and chemical properties different from those of their constituent atoms.

5. The Earth can be divided into several internal regions distinguished by composition and physical conditions. The atmosphere consists of gases, dominantly nitrogen and oxygen. The hydrosphere includes all of the surface waters, is liquid, and is composed mostly of H_2O. The solid Earth can be divided into three parts—crust, mantle, and core. The crust is composed

of eight major elements: oxygen, silicon, aluminum, iron, calcium, sodium, magnesium, and potassium. The compositions of the mantle and core are less well known, but the core is thought to consist mostly of metallic iron and nickel.

6. Energy is the capacity to do work or to create activity and is required for all Earth processes. There are several kinds of energy, including kinetic, radiant, nuclear, gravitational, and chemical.

7. Kinetic energy is the energy associated with motion. Heat is a form of kinetic energy resulting from the vibrational motion of atoms in solids, liquids, and gases. Heat may be transmitted through materials by conduction or convection and through a vacuum by radiation.

8. Radiant energy forms are electromagnetic waves defined by their specific wavelengths. They include visible light, infrared and ultraviolet radiation, X-rays, gamma rays, radio waves, and others.

9. Nuclear energy is the energy used to hold the nuclei of atoms together. It is released when atoms undergo nuclear reactions (fission or fusion).

10. Gravity is a force of attraction between objects that increases as the masses of the objects increase, and decreases as the distance between objects decreases.

11. All energy forms may be stored as potential energy until they are actually used. Energy forms are interconvertible so that in most Earth processes energy is neither created nor destroyed.

12. Energy for Earth processes forms a complex network. Most energy for internal Earth processes comes from nuclear reactions in radioactive elements concentrated in Earth's crust and upper mantle, and from Earth's gravity.

13. Energy for surface processes comes largely from the nuclear energy of the Sun. Gravity plays an important role in all Earth processes. The interaction of lunar, solar, and Earth's gravity, plus the centrifugal force of the rotating Earth, produce the tides.

REVIEW QUESTIONS

1. How is it possible for atoms of two different elements to have the same atomic mass?
2. Describe the role of each of the three subatomic particles in chemical reactions.
3. Explain, with the aid of diagrams, the ionic bonding between lithium and fluorine in the compound LiF (lithium fluoride).
4. What are the differences between ionic and covalent bonding?
5. How are van der Waals' bonds different from the other bond types?
6. What are the differences between nuclear and chemical reactions?
7. Light and X-rays are both forms of radiant energy. How do they differ from one another?
8. What evidence do we have that demonstrates that solar energy is not a major factor in causing internal Earth processes?
9. How do we know that gravity alone cannot explain the tidal features that we observe on the Earth and that centrifugal force also must be involved?
10. The Sun is many thousands of times more massive than the Moon. Explain why the Sun's role in causing the tides is less important than that of the Moon.
11. Describe, in as much detail as you can, the energy network involved in the eruption of a volcano.

3
Minerals

Multicolor tourmaline crystals known as the "Rabbit Ears," grown on quartz crystals (right). Specimen collected at the Himalaya Mine, San Diego County, California. Tourmaline from the permanent collections of the Houston Museum of Natural Science.

The solid part of the Earth is composed of special types of compounds called minerals, which range from spectacularly beautiful rubies and emeralds to dull, earthy "rust." In this chapter we will discuss what makes minerals unique, examine the processes by which they form, and introduce those that are most common in the Earth's crust. We also will show how geologists use minerals to interpret Earth processes.

WHAT IS A MINERAL?

To some, anything that is neither animal nor vegetable must be a mineral. To others, minerals are related to vitamins and must be part of our diet if we are to stay healthy. To a *geologist,* however:

A **mineral** is a naturally occurring, inorganic solid with an ordered internal arrangement of atoms or ions and a specific chemical composition.

By definition, all minerals are solids, so neither water nor gases can be minerals. Some minerals are native elements such as gold, sulfur, and diamond (carbon). Some are simple compounds, such as halite (NaCl). Others have complex formulas that are longer than a line in this book.

All minerals form by chemical reactions and are held together by one of the bond types, or some combination of bond types, described in Chapter 2. When we apply the definition of a mineral to Earth materials, some gems and "mineral resources" prove not to be minerals at all. Amber (hardened sap of ancient trees), pearls (secretions of oysters), and ivory (the tusks of elephants and walruses) are not minerals because they are always the result of organic processes. Oil and natural gas are not solids and should not be called mineral resources. Opal is a naturally occurring inorganic solid, but it is only a *mineraloid* because it lacks the rigidly ordered internal structure and specific chemical composition required of minerals.

FIGURE 3–1
The complex crystal shapes of snowflakes. Note the characteristic sixfold symmetry. Photo by American Museum of Natural History.

Several common materials do satisfy the definition. Salt is a mineral, a natural inorganic solid with ordered internal arrangement of its specific ions, sodium and chlorine. A snowflake is a mineral, forming naturally by inorganic processes, with a constant chemical composition, and with its hydrogen and oxygen atoms in fixed positions within a geometrically ordered structure. It is this rigid structure that yields the complex but always symmetrical shapes of snowflakes (Figure 3–1). Their *ordered internal structure* is what sets minerals apart from other types of compounds, and we shall discuss that structure in detail in the following sections.

WHAT'S IN A NAME?

Minerals, as compounds, could be identified by their chemical formulas, but they are normally called by names, either scientific or common. For example, sodium chloride (NaCl) is the mineral halite, commonly called "rock salt." Silicon dioxide (SiO_2) is the mineral quartz.

Mineral names come from many sources. Some are named for where they were discovered: labradorite was named for Labrador in Canada. Other names describe a mineral's appearance or physical properties: azurite is deep blue (azure); orthoclase is named from the Greek words *orthos* (meaning straight) and *klastos* (meaning broken) because, when struck, it forms fragments bounded by 90° angles. Minerals also are named after famous people: sillimanite after Benjamin Silliman, an early American mineralogist, and armalcolite after the first *Apollo* astronaut crew to visit the Moon (Armstrong, Aldrin, and Collins).

THE INTERNAL STRUCTURE OF MINERALS

Under favorable growth conditions, minerals form regular geometric shapes called **crystals** (Figure 3–2). Some are simple shapes with only a few faces, but others, such as snowflakes, are more complex. In 1669, after years of studying crystals, the Danish scientist Nicolaus Steno proposed that all minerals possess an ordered internal structure. He found that each of the hundreds of quartz crystals in his collec-

FIGURE 3–2
The shapes of mineral crystals vary widely. Clockwise from upper left, examples shown are cubes of pyrite, six-sided prisms and pyramids of smoky quartz, tabular crystals of wulfenite, prisms of beryl, stubby prisms of apophyllite, and rhomb of rhodochrosite. Macmillan Publishing/Geoscience Resources photo.

tion had the same symmetrical arrangement of crystal faces and had identical angles between faces, regardless of their size, color, or degree of perfection. He also observed that crystals of galena were identical to one another, but had a different kind of symmetry than quartz. Steno reasoned that this could happen only if a very small, basic unit were arranged in the same way within every quartz crystal, and similarly within every galena crystal, although a different arrangement. Steno knew nothing of atoms, of course—they were discovered long after his death.

Two hundred fifty years later we learned that Steno was right. In 1912 German physicist Max von Laue irradiated minerals with X-rays, themselves a new discovery. He correctly deduced that the resulting geometric patterns (Figure 3–3) were caused by X-rays interacting with ordered planes of atoms.

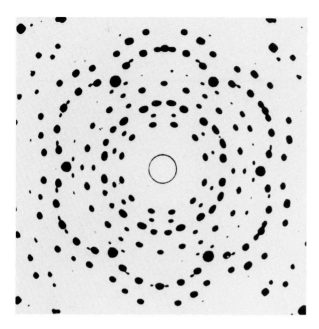

FIGURE 3–3
This regular, geometric pattern results when X-rays pass through the ordered atomic structure of a mineral. Photo by Eastman Kodak company.

FIGURE 3–4
The internal atomic structure of halite or sodium chloride, NaCl. This ordered array of alternating Na^+ and Cl^{2-} ions was interpreted in 1914 by W. H. Bragg and W. L. Bragg from X-ray images like Figure 3–3.

Within two years, the British father-and-son team of W. H. Bragg and W. L. Bragg successfully determined the geometrically perfect internal arrangement of sodium and chlorine ions in the mineral halite (Figure 3–4) from its X-ray pattern.

Today, mineralogists using sophisticated X-ray instruments have learned the structures of most common minerals. The ordered structure of minerals is called a **crystalline structure** because it was discovered first in well-formed crystals, but X-rays show that it is present in every sample of a mineral. Some mineral specimens grow too rapidly or in an environment too crowded with other specimens to form crystals, but even these apparently shapeless masses possess the internal structure characteristic of the specific mineral. Each mineral's structure is a unique three-dimensional arrangement of particular atoms or ions and can be used like fingerprints to help in mineral identification.

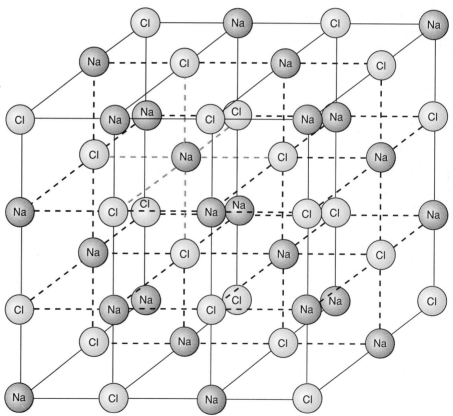

Why Do Minerals Have Crystalline Structures?

Atoms and molecules in gases and liquids can move freely. For example, once two oxygen atoms are joined by strong covalent bonds to form an O_2 molecule, only weak van der Waals' forces hold the molecule to other molecules. This allows freedom of movement, for each oxygen atom is bonded strongly to only one other atom. In the mineral kingdom, however, each atom or ion is firmly bonded to *several* others, and this locks them in place. In the mineral halite (NaCl), for example, each sodium ion is bonded to six chlorine ions, and each chlorine to six sodiums (Figure 3–4). The reason for this arrangement lies in the sizes and charges of the ions.

Ionic size and charge. The halite structure demonstrates the role of *ionic size* in controlling a mineral's internal arrangement of atoms or ions. Imagine an environment in which millions of sodium atoms lose electrons to millions of chlorine atoms, forming sodium cations (+) and chlorine anions (−). Each sodium ion then attracts as many chlorine ions as can fit around it—*not just the one involved in the electron transfer*. Similarly, each chlorine ion attracts as many sodium ions as can surround it.

The number of anions that can fit around a cation is called the **coordination number** and depends only on the relative sizes of the ions involved. Each anion barely touches the surface of the cation, and the anions repel one another so that none actually come into contact. In halite, six Cl^- ions surround every Na^+ ion; they lie at the corners of an eight-sided figure, with the sodium at the center (Figure 3–5). This geometrically perfect arrangement is found in every sample of halite and results in the same basic crystal shape every time halite grows.

In halite, the six chlorine ions surrounding sodium act to neutralize its +1 charge. Each Cl^- uses one-sixth of its total −1 charge to do so, leaving it with five-sixths of a negative charge. Every Cl^- ion must then find five more sodium ions with which to bond in order to reach electrical neutrality itself. Like all compounds, NaCl is electrically neutral. Its formula shows that equal numbers of Na^+ and Cl^- ions are present, but the *sizes and charges* of these ions dictate the physical geometry by which the ion-by-ion neutrality is achieved.

The manner in which each mineral manages its coordination and charge balance depends on the combination of ions involved. Ions vary considerably in size, as shown in Table 3–1. Those with the greater number of electron shells are the largest and those with fewest shells are the smallest. There are many possible combinations and several kinds of coordination, including the fourfold coordination of oxygen around silicon (Figure 3–5) for the two most common ions in the Earth's crust. As a result, there are many kinds of mineral structures.

Polymorphs: The Role of Temperature and Pressure. Diamond is the hardest mineral. It is translucent and may be colorless, yellow, blue, or green. Graphite, on the other hand, is one of the softest minerals. It is opaque, gray-black, and has a greasy feel. Although their appearances and physical properties are dramatically different, both minerals have exactly the same chemical composition: each is made entirely of carbon. Graphite and diamond are **polymorphs,** minerals that have identical compositions but different internal structures.

Diamond and graphite form under very different conditions, and their internal structures reflect this (Figure 3–6). Diamond's structure is much more compact than that of graphite because diamond

TABLE 3–1
Ionic size and coordination number.

Relative ion size	Approximate ionic radius X 10^{-8} cm	Coordination with oxygen	Possible substitutions
	Oxygen 1.40		
	Silicon 0.34	4	Si^{4+}, Al^{3+}
	Magnesium 0.66	6	Al^{3+}, Mg^{2+}, Fe^{2+} Mn^{2+}, Fe^{3+}, Cr^{3+}
	Calcium 0.99	8	Ca^{2+}, Na^+, Sr^{2+}
	Potassium 1.33	12	Na^+, K^+, Rb^+

FIGURE 3-5
The coordination principle. Every ion is surrounded by as many ions of the opposite charge as can be packed around it. The number of neighbors (the *coordination number*) depends only on the relative sizes of the cations and anions involved. In addition to the examples shown, coordination numbers of 2, 3, 8, and 12 are possible.

Halite (NaCl)
Coordination number 6

Quartz (SiO_2)
Coordination number 4

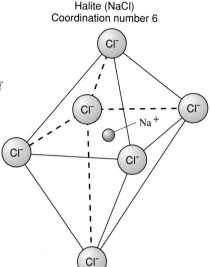
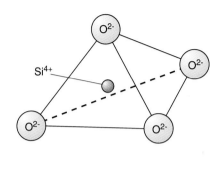

(a) Stick models showing positions of ions

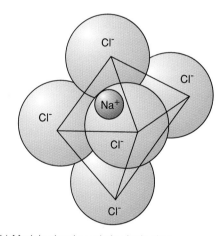

(b) Models showing relative ionic sizes

forms under extremely high pressures. Each carbon atom in diamond is bonded covalently to *four* others, forming a close-knit, three-dimensional framework. In graphite, each carbon is covalently bonded to *three* others, forming a series of two-dimensional sheets. These sheets are then held together by weak van der Waals' forces. As a result of their different structures, the density of diamond (the mass per unit volume) is much greater than that of graphite: 3.5 g/cm^3 (grams per cubic centimeter) for diamond versus 2.3 g/cm^3 for graphite.

Polymorphs such as diamond and graphite are useful for determining the physical conditions (pressure, temperature) of Earth processes and the locations in the Earth where minerals form. Experiments show that the pressures needed to form diamond are greater than those found anywhere in the Earth's crust, so rocks containing diamond must have originated in the mantle and somehow moved close enough to the surface that we could mine them. Such processes are rare, so it is no wonder that diamonds are scarce and therefore precious.

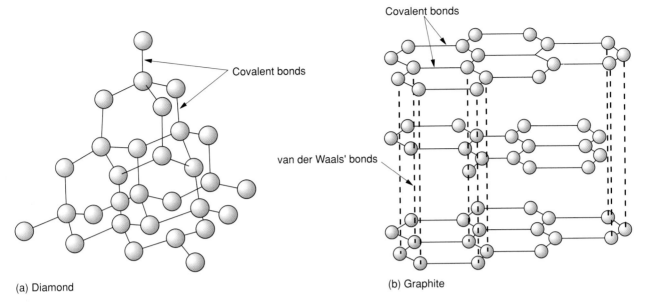

(a) Diamond

(b) Graphite

FIGURE 3-6

The internal structures of graphite and diamond. (a) In diamond, all carbon atoms are covalently bonded to one another in a dense, tightly packed structure. This accounts for diamond's tremendous hardness. (b) In graphite, the covalent bonds link carbon atoms in two-dimensional sheets; the sheets are held to one another by weak van der Waals' bonds. These bonds are so easily broken that the graphite sheets can slide against one another, making graphite a slippery lubricant.

VARIATIONS IN MINERAL COMPOSITION

Minerals such as graphite (C), quartz (SiO_2), and halite (NaCl) have fixed chemical compositions. However, nearly every mineral contains some impurities in its structure; most impurities are ions that were trapped during the mineral's growth. These impurities may bring about variations in a mineral's physical properties. Pure quartz, for example, is colorless, but various impurities turn it pink, purple, green, yellow, or other colors.

Olivine, the mineral known to jewelers as peridot, is different because it has a systematic variation in composition rather than a random inclusion of stray ions. In some mineral structures, several common ions can take the place of certain others, a process called **ionic substitution.**

Ionic Substitution

The olivine group of minerals, represented by the formula $(Fe,Mg)_2SiO_4$, is a common example of ionic substitution. The parentheses indicate ions that can substitute for one another (Mg and Fe) and indicate that the two can be present in any proportion. Pure Mg_2SiO_4 can exist as the mineral forsterite, or pure Fe_2SiO_4 as the mineral fayalite, and so can minerals with any composition intermediate between the two, such as $FeMgSiO_4$.

We saw earlier that the size and charge of ions determine mineral structure. Oxygen is the only anion in olivines and must surround all the cations present. The silicon ion, Si^{4+}, is small and is coordinated by only four oxygens, but both iron, Fe^{2+}, and magnesium, Mg^{2+}, ions are of the appropriate size to be surrounded by six oxygens. They also have the same +2 charge. As a result, either can fit into the same place in the olivine structure. Indeed, iron and magnesium are interchangeable in many minerals and are said to substitute for one another. In general, *two ions can substitute easily for one another if they have the same charge and size.* (Note that neither Fe^{2+} nor Mg^{2+} can substitute for silicon—they are too large to fit into its structural position.)

The olivine group is an example of a **solid-solution series,** a family of minerals that can have any composition between two extremes, called **end members,** and still maintain the same structure. Forsterite and fayalite are said to be end members of the olivine solid-solution series.

Coupled Ionic Substitution. Even if they do not have the same charge, ions can substitute for one another as long as they are the same size and some way can be found to maintain *electric neutrality.* The plagioclase feldspar solid-solution series, the most abundant mineral group in the Earth's crust, is an example of this kind of chemical variation.

The end members of the plagioclase feldspar series are albite, $NaAlSi_3O_8$, and anorthite, $CaAl_2Si_2O_8$. Ca^{2+} and Na^+ are so similar in size that they commonly substitute for each other in minerals, even though their charges are different. When a calcium ion substitutes for sodium in albite, an extra positive charge is added, but this problem is solved by simultaneously substituting an aluminum ion, Al^{3+}, for a silicon ion, Si^{4+}. Electric neutrality is maintained because the combined charges of Ca^{2+} + Al^{3+} —five—equal the charges of the Na^+ + Si^{4+} that they replace. This type of substitution is called **coupled ionic substitution** because one change requires (or is coupled to) the other.

IDENTIFICATION OF MINERALS

A mineral's unique combination of chemical composition and internal structure produces a diagnostic set of physical properties. These properties allow us to identify an unfamiliar mineral without using the sophisticated or expensive equipment needed for chemical or structural analyses. In most instances, the properties are due to the presence of a specific atom or ion in the mineral or to a particular type of bonding.

The physical properties useful in mineral identification include crystal shape, color, luster, hardness, the manner in which a specimen breaks, and specific gravity. Other helpful attributes are less familiar: streak (the color of a mineral's powder), magnetism, malleability, and ductility.

Crystal Shape (Habit)

We have seen that a mineral's internal structure results in a regular crystal form, if growth conditions permit. This crystal form is called a mineral's **habit** and is useful in identifying many minerals (Figure 3–2). Garnets, for example, form 12-sided or 24-sided crystals, but never the cubes favored by halite or the elongate hexagonal (six-sided) crystals of quartz. When unfavorable growth conditions pre-

TABLE 3–2
Mohs scale for mineral hardness.

Mohs Scale		Common Testing Materials	
Hardness Number[1]	Standard Mineral	Material	Hardness
10	Diamond		
9	Corundum		
8	Topaz, beryl		
7	Quartz	← Streak plate	7
		← Hard steel file	6.5
6	Orthoclase		
		← Window glass	5.5
5	Apatite	← Pocketknife	5.0
4	Fluorite		
3	Calcite	← Penny	3.0
		← Fingernail	2.5
2	Gypsum		
1	Talc		

[1]This is a scale of *relative* hardness. It shows that quartz is harder than anything lower than 7 and softer than anything higher than 7. In terms of *absolute* hardness, corundum is twice as hard as topaz and nearly 400 times harder than talc.

vent the formation of crystal faces, other properties must be used for identification.

Hardness

Hardness is the resistance of a mineral to being scratched and indicates the bond strength in a mineral. Diamond is harder than quartz because the bonds holding diamond's carbon atoms together are stronger than the forces that bond silicon to oxygen in quartz. We can determine hardness by seeing whether a mineral can scratch (or be scratched by) standard test minerals on a reference scale called **Mohs hardness scale.** This scale, devised by German mineralogist Friedrich Mohs in 1812, remains the standard for determining mineral hardness (Table 3–2). For example, a hardness of 5.5 on the Mohs scale means that a mineral is harder than apatite but softer than orthoclase.

As shown in the table, household materials such as a steel knife or a piece of glass also can be used, once their hardnesses are calibrated against the scale. A mineral will scratch any substance softer than itself, but will be scratched by any harder substance. For example, garnets have a hardness around 7. They will scratch a glass plate (hardness 5.5–6.0) but not a ruby (a gem-quality form of corundum with a hardness of 9).

In some minerals, bonds are stronger in certain directions, and care must be used in testing hardness. Kyanite is one of the best examples of such directional hardness. A pocketknife (hardness 5) can scratch kyanite if used parallel to the long dimension of its crystal (hardness 4–5), but cannot scratch it if used parallel to the short dimension (hardness 6–7).

Cleavage

Minerals break in distinctive ways. For example, minerals of the mica family can be peeled into smooth, thin sheets (Figure 3–7). Feldspars, important constituents of many rocks such as granite, also break along smooth surfaces when struck, but they do not break into sheets. The feldspar breakage surfaces are at nearly right angles to one another (like the halite in Figure 3–8). The tendency of these minerals to break along sets of parallel planes is called **cleavage,** and the surfaces are called **cleavage planes.**

Cleavage occurs in those minerals that have planar zones of weak bonding. In graphite, for example, cleavage occurs because weak van der Waals' bonds are easily broken. As a result, the cleavage planes are parallel to the sheets of carbon atoms shown in Figure 3–6.

To describe cleavage, it is necessary to note two things: the *number of different cleavage directions* (two in feldspars, one in micas) and the *angles between the cleavage directions.* Two minerals might each have two directions of cleavage, but the cleavage planes in one could be at right angles (as in halite), whereas the cleavage planes in the other could be nonperpendicular (as in calcite—Figure 3–8).

Fracture

Some minerals do not cleave at all because they have no planar zones of weak bonds in their structures. Such minerals break by what is called **fracture.** Fracture may occur along irregular, jagged, or splintery surfaces or along smooth, curved surfaces in what is called **conchoidal fracture** (Figure 3–9). Quartz is a common mineral that displays a striking conchoidal fracture. Ordinary glass also shows conchoidal fracture.

Luster

A mineral's **luster** is the manner in which light is reflected from its surface. Luster depends on the

FIGURE 3–7
Perfectly developed cleavage in one direction shown by pieces of the mica called muscovite. Macmillan Publishing/Geoscience Resources photo.

composition of a mineral, the presence of defects or chemical impurities in its structure, and pitting or chemical alteration of its outer surfaces. Pure, unflawed diamonds have a *brilliant* luster, but inclusions of tiny impurities or flaws in the internal structure can cause a cloudy appearance. Most terms used to describe luster are self descriptive. Galena, the principal ore of lead, has a *metallic* luster—its surface looks like shiny aluminum foil. Other minerals have lusters that are silky, waxy, earthy, pearly, or glassy.

Specific Gravity

The **specific gravity** of a mineral is a comparison of its density with the density of water, as shown in the relationship:

$$\text{Specific gravity} = \frac{\text{Density of a mineral}}{\text{Density of water}}$$

Density depends on the mass of the atoms in a mineral and the compactness with which they are bonded. Water at 25°C has a density of 1 g/cm^3, and graphite has a density of 2.4 g/cm^3. Graphite thus is 2.4 times as dense as water, so its specific gravity is 2.4—and if placed in water, graphite sinks.

If the specific gravity of a substance is drastically higher (or lower) than that of most materials, we

FIGURE 3–8
Angles between cleavage directions. Both halite and calcite have three directions of cleavage, but the angular relationships are different in the two minerals.

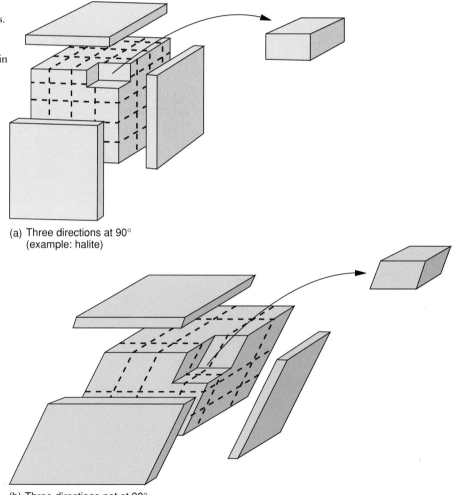

(a) Three directions at 90°
(example: halite)

(b) Three directions not at 90°
(example: calcite)

detect it when we hold the substance in our hands. If it seems much heavier (or lighter) than what we expect for the size of the sample, its specific gravity is abnormally high (or low). To determine the specific gravity of a material whose density feels more in the normal range, we use a simple procedure: weigh a sample of the substance and compare its weight with that of an equal volume of water. The specific gravity is the weight of the material divided by the weight of the equal volume of water.

Most of the Earth's crust is composed of minerals that have specific gravities between 2.40 and 4.50, but some minerals have much higher values. Gold (15.0 to 19.3) and platinum (14.0 to 19.0) are the highest. Common ore minerals like galena (7.5) also are much denser than the usual crustal materials.

Color

Color is a useful property in identifying some minerals, but is much less valuable for others. Minerals with metallic lusters generally have colors that are diagnostic. Galena, for instance, is always battleship gray. Pyrite, also known as "fool's gold," is brassy yellow. Unfortunately, impurities and crystal defects cause many nonmetallic minerals to vary widely in color (Figure 3–10). Thus, quartz can be colorless, white, green, blue, gray, yellow, black, rose, smoky, or purple. Color *can be* a useful property, but it must be used with care.

FIGURE 3–9
Conchoidal fracture exhibited by a quartz crystal. Macmillan Publishing/Geoscience Resources photo.

FIGURE 3–10
The color of a mineral may vary widely, as in these examples of quartz. Clockwise from upper left: colorless quartz, citrine (yellowish quartz), smoky quartz, and amethyst (purple quartz). Macmillan Publishing/Geoscience Resources photo.

IDENTIFICATION OF MINERALS | 55

FIGURE 3–11

Color versus streak. For some minerals, the streak is the same as the mineral color, but for other minerals the two colors are different. The yellow streak of orpiment (right) is the same as the color of the mineral. However, the brassy yellow pyrite (left) produces a black streak. Macmillan Publishing/Geoscience Resources photo.

Streak

The color of a mineral's powder is called its **streak**. It is so-called because it is determined by rubbing the mineral on a hard, white porcelain plate (hardness about 7), which grinds the mineral to a powdery streak on the plate. Streak is a more reliable property than surface color because pulverizing a mineral minimizes the effects of impurities, tarnish, or structural defects. For many minerals, color and streak are the same, but for many others they are surprisingly different, and this difference often can be a key to identification (Figure 3–11). (The few minerals that are harder than the streak plate leave no streak, of course—they scratch the plate.)

Unlike color, streak tends to stay constant. Thus, sphalerite, the major ore of zinc, has a creamy yellow streak regardless of whether the mineral is colorless, black, white, or green. If a brassy yellow mineral gives a black streak, chances are it is pyrite or another type of "fool's gold." However, if it streaks brassy or golden yellow, save the powder—it probably is real gold!

Other Properties

The preceding properties are easily determined and are sufficient for identification of many common minerals, but several other properties also may prove helpful. Geologists often taste, sniff, and rub their fingers across mineral samples because taste, odor, and feel are diagnostic properties for certain minerals. Halite obviously tastes salty; kaolinite, a common mineral in clay, sticks to the tongue. The streak of many sulfur-bearing minerals smells like rotten eggs; that of some arsenic minerals smells like garlic. Graphite and talc feel greasy. Apply these tests carefully; geologists sniff first, to test for arsenic, before we taste!

Some minerals have one color in normal light, but look very different in ultraviolet light, a property called **fluorescence** (Figure 3–12). Other useful properties include **magnetism** (the ability of a mineral to attract a magnet), **malleability** (the ability of a mineral to be pounded into thin sheets), and **ductility** (the ability to be pulled into thin wires).

One simple chemical test often is made because it requires no equipment and only a drop or two of dilute hydrochloric acid. Minerals such as calcite that contain the carbonate anion, $(CO_3)^{2-}$, effervesce ("fizz") when hydrochloric acid is put on them. A rapid chemical reaction takes place between the carbonate mineral and acid, releasing carbon dioxide and water. The carbon dioxide bubbles out through the liquid to produce the "fizz." In the case of calcite, the reaction is:

$CaCO_3$	$2HCl$		$CaCl_2$	H_2O	CO_2
Calcite	+ Hydrochloric	$\rightarrow$	Calcium	+ Water	+ Carbon
	acid		chloride		dioxide

MINERALS AND ROCKS

You may be wondering what the difference is between a mineral and a rock. A **rock** is an aggregate of several different minerals or of many grains of a single mineral. It is rare for a mineral to grow in the isolation needed to form a perfect crystal. Most minerals are intergrown with many other mineral grains in a wide variety of rocks.

How Do Minerals Form?

Minerals grow from small clusters of atoms or ions. At first, a small number of ions forms a tiny **seed crystal**—a group of ions in the appropriate proportions of elements and arranged in the correct internal structure. More ions are attracted to oppositely charged ions in the seed crystal. As the additional ions fit into their structural positions, the seed grows into a visible grain. If seed crystals are far apart, each can grow without interference from the others, and well-formed crystals are produced (see Focus 3–1). If several seed crystals are close to one another, they interfere with each other as they grow, resulting in an interlocking aggregate of grains that displays no crystal faces.

Minerals grow in a wide range of physical conditions, some from liquids and others from solids. Ice forms by the solidification of water at 0°C. At slightly higher temperatures, halite forms by precipitating from saltwater as the water evaporates. Other minerals, such as olivine and feldspar, grow deep within the crust from molten rock at temperatures of 550° to 1100°C and pressures up to 10 kb. Still others, such as garnet, form in solid rocks at moderate temperatures (350°–600°C) and pressures (2–8 kb) by the rearrangement of ions already present in other minerals. (You can even grow mineral-like crystals yourself—see Focus 3–1.)

Some minerals, such as quartz, can grow under a wide range of conditions, from those of the surface to those several kilometers deep in the crust. Others, such as halite, form only within a restricted range of conditions. Those that form in a narrow range of

(a)

(b)

FIGURE 3–12
Some minerals *fluoresce;* that is, they exhibit different colors when viewed under ultraviolet light. Calcite (white), willemite, and franklinite in this specimen from Franklin, New Jersey, are pale in normal light (a) but fluoresce brightly in ultraviolet light (b). (The willemite fluoresces green.) Macmillan Publishing/ Geoscience Resources photos.

GROWING CRYSTALS AT HOME

C rystals have become expensive collectors' items, but the color and geometric perfection that make them sought after can be duplicated cheaply in your own home. It is possible to mimic nature's crystal-making procedures by using household objects and some inexpensive chemicals that can be purchased at the local pharmacy or camera shop. We provide here a recipe and complete instructions for growing *colorless octahedral crystals of alum*. Other shapes and colors can be grown from other starting materials (see Holden and Singer, 1960 in Further Readings for more recipes).

Ingredients

1. Pint jar with a screw top
2. 6 ounces alum—potash alum, $KAl(SO_4)_2 \cdot 12H_2O$
3. White sewing thread
4. Hammer and a sharp-pointed nail
5. Water.

Instructions

1. *Make a saturated solution of alum in water.* This will be the environment in which the crystal will grow. To do so, place about 16 ounces (2 cups) of water in the jar, and slowly add about 3 ounces of alum powder, stirring continuously. When the alum no longer dissolves and begins to collect on the bottom of the jar, place the open jar in a pan of hot water and heat until the alum dissolves completely. Add another ounce of alum, stir until it dissolves, and then remove the jar from the hot water and let it cool. As the solution cools, alum crystals will precipitate because more alum can be dissolved in hot water than in cool water. When the jar has reached room temperature, carefully pour the liquid (now perfectly saturated at room temperature) into another container and discard the solid alum from the bottom of the jar. Return the solution to the original jar.

2. *Prepare a crystal seed.* Carefully pour about an ounce of the saturated solution into a clean shallow dish and allow it to stand for a few hours. As the water evaporates, the solution becomes supersaturated and small crystals begin to form. Using a pair of

FIGURE 1

Screw-top jar set up for crystal growing. The seed crystal is suspended in the center of the jar by a length of thread. It can be raised or lowered by rolling the pencil so that the growing crystal avoids contact with the sides or bottom of the jar.

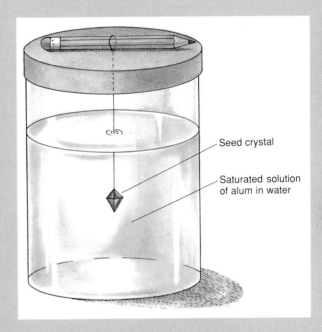

Seed crystal

Saturated solution of alum in water

tweezers, *carefully* remove individual crystals when they are about 1/8 to 1/4 in. long and place them on a dry napkin. These are the seed crystals.

3. *Place the seed in the saturated solution.* Using the thread, tie a simple knot around one of the seed crystals. Using the hammer and nail, punch a hole in the center of the jar lid. Suspend the crystal on the thread in the jar of saturated solution; after passing the end of the string through the hole in the lid, tightly fasten the lid and secure the free end around a pencil, as shown in Figure 1. This crystal now is isolated from others and can grow without interference.

4. *Watch your crystal grow.* Ions in the saturated solution will be attracted to the crystal seed, and as they are attached to it the crystal will grow. The result should look like the diamond crystal shown in Focus 3–2. If the seed dissolves, remake the saturated solution, adding a bit more alum powder, and repeat the seeding process.

conditions are used by geologists to estimate the conditions under which the rocks that contain them formed.

Rocks and Rock-Forming Processes

Rocks are divided into three types—*igneous, sedimentary,* and *metamorphic*—according to the nature of the processes that brought the minerals together. Minerals also can grow during each of the three processes and thus are igneous, sedimentary, or metamorphic minerals. We will look briefly at these processes here, and study them in more detail in Chapters 4, 5, 6, and 7.

Molten rock is called **magma,** and rocks that form by the crystallization of magma are known as **igneous rocks**. Seed crystals form in a magma as it cools, and the igneous minerals grow in the hot liquid. Eventually, all of the ions in the liquid fit into crystal structures and the magma is completely solidified as an igneous rock. Igneous minerals and rocks form deep in the crust and mantle and at the surface during volcanic eruptions. Olivine, feldspars, and quartz can crystallize from magmas and thus may be igneous minerals.

When rocks are exposed at the surface of the Earth, they eventually crumble or dissolve. This breakdown is caused partly by chemical reactions with the atmosphere and partly by physical processes such as wedging caused by ice and plant roots. Fragmented and dissolved rock material can be transported by streams, glaciers, and wind, then deposited or precipitated from solution, and finally cemented together to produce a new rock, such as sandstone. Rocks produced at the surface by these processes are called **sedimentary rocks**.

Sedimentary minerals are the new minerals that grow chemically during the deposition or cementation processes, usually by precipitation of dissolved ions from water. Halite, for example, is a sedimentary mineral formed by evaporation of seawater or saline lakes in arid areas. Some quartz also is sedimentary because it precipitates between fragments and acts as a cement.

Note that some sedimentary rocks may be composed of reworked fragments of any kind of previously existing rock, not just of new sedimentary minerals. Thus, old minerals—igneous, sedimentary, or metamorphic—may be found in sedimentary rocks.

When igneous and sedimentary rocks are subjected to temperatures and pressures greatly different from those under which they first formed, they often respond by forming new minerals. These minerals form without melting or dissolving, through the recombination of ions from previously existing minerals. The new minerals are **metamorphic minerals,** and the resulting rocks are called **metamorphic rocks**. Common examples of metamorphic minerals are those of the garnet group.

Many minerals, such as halite, can form only by one of the three types of rock-forming processes, but others, such as olivine, may form by two processes (igneous and metamorphic). A few minerals can form during all three kinds of processes. For example, quartz and some of the feldspars may be igneous, sedimentary, or metamorphic minerals.

The Rock Cycle

Any rock can be converted to an igneous, sedimentary, or metamorphic rock if it is acted on by the appropriate processes. Nature is thus the original recycler, using the same ions over and over again to make new rocks in another important Earth cycle called the **rock cycle** (Figure 3–13). The rock cycle is more complex than the water cycle because it uses deep internal processes as well as those that operate at the surface.

One way to show how the rock cycle works is to look at what happens when lithospheric plates collide. When sedimentary rocks formed on the ocean floor are subducted, heat and pressure first convert them to metamorphic rocks as they are buried deep in the crust. Eventually, they reach a level in the mantle where they melt to form magma. This magma rises and becomes igneous rock by solidifying deep in the crust or after erupting at the surface. Breakdown of lava at the surface converts the igneous rock back to sedimentary rocks, completing the cycle.

The rock cycle follows other paths as well. Sedimentary, metamorphic, and igneous rocks exposed at the surface will be converted to new sedimentary rocks, or they will be converted to metamorphic rocks if they are buried beneath new deposits. Metamorphic rocks may be brought to the surface during mountain building and converted to sediments, or

FIGURE 3–13
The rock cycle. Depending on the processes to which it is subjected, any rock may be converted to an igneous, sedimentary, or metamorphic rock. Arrows show the possible conversions.

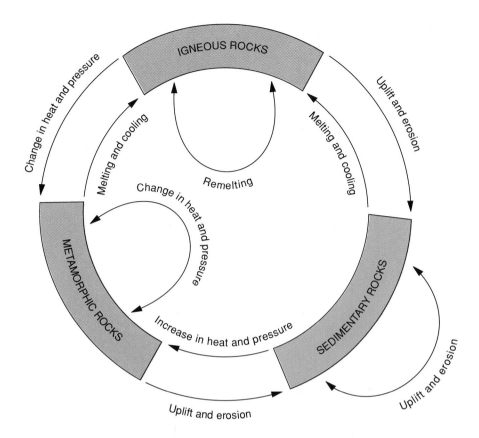

they may be melted during subduction. With deeper burial, they also may be remetamorphosed into a new metamorphic rock. An igneous rock also can remelt and become a different igneous rock! Figure 3–13 summarizes the many possible paths that the rock cycle can take. We will be looking at the processes that occur along those paths throughout the rest of this book.

MINERALS OF THE EARTH'S CRUST

There are nearly 3000 known minerals, and new ones are found every year. Fortunately for students, only a few of the 3000 are abundant; we need to know only about these few to understand the processes by which igneous, sedimentary, and metamorphic rocks form. These few are called the **rock-forming minerals.** We will examine some of the most important groups in the rest of this chapter.

Minerals are classified by the anion they contain. For example, those that contain the sulfur anion are

sulfides, and those containing chlorine are called chlorides. We saw in Chapter 2 that the crust is composed predominantly of oxygen and silicon. Thus, it is not surprising that the Earth's crust consists mostly of minerals containing *both elements*—minerals called **silicates.** The rock-forming minerals are generally divided into two major groups: silicates and nonsilicates.

Silicate Minerals

The relative sizes of the Si^{4+} and O^{2-} ions are such that four oxygens can fit around a silicon ion. They form a four-sided figure called a **silicon-oxygen tetrahedron,** with silicon at the center and the oxygens at the corners—Figure 3–5(b). The SiO_4 tetrahedron is not electrically neutral; the charge of the four oxygens adds to -8, and that of the silicon is $+4$. Consequently, the SiO_4 tetrahedron acts as an **anion complex** with a total charge of -4. Importantly, an SiO_4 tetrahedron *cannot exist by itself* because it is not neutral and must attract cations to achieve neu-

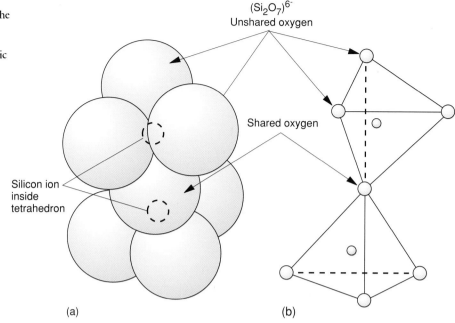

$(Si_2O_7)^{6-}$
Unshared oxygen

Shared oxygen

Silicon ion inside tetrahedron

(a)

(b)

trality. These cations hold the tetrahedra together and stabilize the silicate mineral structures.

The silicon-oxygen tetrahedron is the basic building block of all silicate minerals, but building blocks can be put together in many ways. A brief look at the silicate minerals will give more insight into the variety of mineral structures.

Types of Silicate Minerals. The simplest silicate structure is one in which SiO_4 tetrahedra are held together by cations. This type of structure is called an **independent tetrahedron structure** and is found in several minerals, including those of the olivine family—$(Fe,Mg)_2SiO_4$. Each oxygen gives some of its negative charge to a silicon ion and some to iron ions or magnesium ions. The tetrahedra must be arranged in exactly the right way so that their oxygens also fit into the correct coordination around the iron and magnesium.

For independent tetrahedra to form, there must be four oxygens for every silicon. Indeed, every mineral that has the independent tetrahedron structure has SiO_4 in its formula, reflecting this ratio. If there are not enough oxygens to go around during rock-forming processes, independent tetrahedra cannot exist and a different kind of structure must be

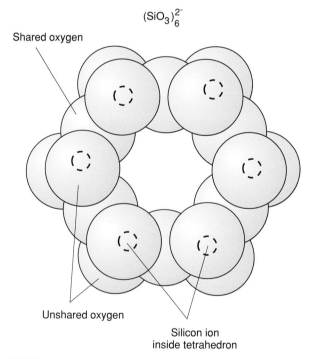

$(SiO_3)_6^{2-}$

Shared oxygen

Unshared oxygen

Silicon ion inside tetrahedron

FIGURE 3–15
Geometry of the ring silicate structure. Each tetrahedron shares two oxygens with neighboring tetrahedra. In this case, the $(Si_6O_{18})^{12-}$ anion complex is formed.

formed. In such a case, the available oxygens must be *shared* between tetrahedra.

In some minerals, two tetrahedra share an oxygen ion and are linked in a **twin-tetrahedron structure** (Figure 3–14). Notice that each Si^{4+} ion is surrounded by four oxygens, as required by the coordination principle, but that one of the oxygens (the *shared oxygen*) is part of *both* tetrahedra. The twin tetrahedron has a composition of Si_2O_7 and a net charge of -6; it acts as an anion complex and attracts cations to satisfy its remaining charge.

In other silicate minerals, each tetrahedron shares two of its four oxygens with adjacent tetrahedra. This can be accomplished in two ways. In the **ring silicates,** rings made of three, four, or six tetrahedra are formed, and *these rings* become the anion complexes that act as building blocks in the mineral structure (Figure 3–15). The six-tetrahedron ring shown in Figure 3–15 is typical of the minerals tourmaline and beryl. In other minerals, the tetrahedra line up to form long chains by sharing oxygens with their neighbors in what is called a **single-chain structure**

FIGURE 3–16
The single-chain silicate structure. (a) Geometry of the chain shows each tetrahedron sharing two oxygens with its neighbors along the chain. The shared oxygens are electrically satisfied by being bonded to two silicon ions. (b) Unshared oxygens retain a negative charge and attract the cations that hold the chains together.

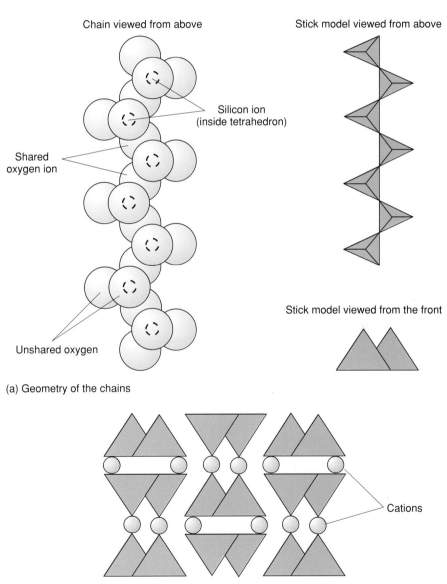

Chain viewed from above

Stick model viewed from above

Silicon ion (inside tetrahedron)

Shared oxygen ion

Unshared oxygen

(a) Geometry of the chains

Stick model viewed from the front

(b) Cations hold the chains together

Cations

(Figure 3–16). These chains act as large anion complexes and are held together by cations. In both the ring and single-chain structures, the Si:O ratio is 1:3.

With progressively smaller amounts of oxygen, even more sharing must take place, resulting in even more-elaborate anion complexes. **Double chains** form when some tetrahedra share two oxygens and others share three, resulting in an Si:O ratio of 4:11—Figure 3–17(a). **Sheet silicate structures** with a ratio of 2:5 occur when every tetrahedron shares three of its four oxygens with its neighbors—Figure 3–17(b). This sheet structure is reflected clearly in the sheetlike cleavage found in the mica minerals. When all four oxygens of every tetrahedron are shared with neighboring tetrahedra, a complex three-dimensional network called a **framework silicate structure** is formed (Figure 3–18).

Three of the most abundant minerals in the crust have framework structures—quartz, plagioclase feldspar, and potassic feldspar.

Rock-Forming Silicate Minerals. The **olivine group** consists of the solid-solution series forsterite-fayalite and has the independent tetrahedron structure. Olivine minerals are typically green and have high specific gravities for silicate minerals (3.27 to 4.37). Most olivine occurs in stubby, irregularly shaped grains. As a result, rocks composed entirely of olivine are compact masses lacking the elongate crystals common in rocks containing feldspars or amphiboles. Olivines form throughout the crust in igneous rocks, but the magnesium-rich varieties are found also in metamorphic rocks. Olivine is thought to be a major constituent of the mantle and has been found in

FIGURE 3–17
Geometry of (a) double-chain and (b) sheet silicate structures. In double chains, some tetrahedra share three oxygens with neighbors; others share four. In sheets, all tetrahedra share three oxygens.

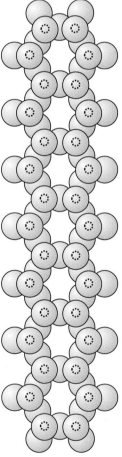

(a) Double chain (Si_4O_{11})

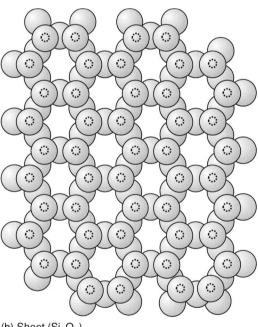

(b) Sheet (Si_2O_5)

many meteorites. Specimens with particularly brilliant luster are valued as the gemstone peridot.

The **garnet family** consists of two solid-solution series: $(Fe,Mg,Mn)_3Al_2(SiO_4)_3$ and $Ca_3(Al,Fe,Cr)_2(SiO_4)_3$. Each has three end members. With all of the substitutions that are possible, it is not surprising that garnets come in several colors. Dark red is the most common, but pink, brown, yellow, and bright green types also are known. Garnets are independent tetrahedron silicates; most form in metamorphic rocks, although a small proportion crystallize in igneous rocks.

The **pyroxene family** consists of several solid-solution series, all of which have the single-chain structure. The weakest bonds in the pyroxene structure lie between the chains and result in the characteristic two-directional cleavage at nearly right angles ($87°$ and $93°$). Pyroxenes are commonly medium-to-dark green or bronze-brown. One type, known as jadeite, has a color and unique luster that makes it valuable as the familiar gemstone "jade."

Pyroxenes are important constituents of igneous and metamorphic rocks and are thought to be important components of the mantle. The most common pyroxenes belong to one of two solid-solution series: the complex augite group, $(Ca,Na)(Mg,Fe,Al)(Si,Al)_2O_6$, and the simpler enstatite group, $(Mg,Fe)SiO_3$.

The **amphibole family** consists of several solid-solution series similar to those of the pyroxenes, but with a double-chain structure in which the hydroxyl anion complex (OH^-) is present. Amphiboles are difficult to distinguish from pyroxenes because both are chain silicates, contain the same elements, are the same color, and occur in the same kinds of rocks. In many instances, the distinction can be made only by the different cleavages of these two mineral families: amphiboles also cleave between their chains, but at angles of $57°$ and $123°$ rather than at right angles.

Amphiboles are important igneous and metamorphic minerals, the most common belong to the hornblende "series," actually a collection of over ten solid-solution series. Hornblendes form under a wide range of conditions, but other amphiboles are more restricted in their occurrence.

The **mica family** consists of several sheet silicate minerals in which large cations—usually K^+, Na^+, or Ca^{2+}—bond silicon-oxygen sheets together. The most-abundant micas are the colorless mineral muscovite and the dark brown or green mineral biotite. Both are important constituents of metamorphic rocks and form in many kinds of igneous rock as well. All micas have such excellent cleavage in one direction (between the silicon-oxygen sheets) that they can be peeled into thin, flexible sheets (Figure 3–7).

The **feldspars** include two groups of minerals, the plagioclase and potassic feldspar families. Each has a framework structure in which aluminum ions substitute for silicon in some tetrahedra, and large cations are added to the structure in a coupled ionic substitution.

The **plagioclase feldspars** are a solid-solution series of framework silicates varying in composition between end members albite, $NaAlSi_3O_8$, and anorthite, $CaAl_2Si_2O_8$. Plagioclase minerals are the most abundant in the Earth's crust and are found in nearly every igneous and metamorphic rock. In addition, sodium-rich plagioclase (albite) can form in some sedimentary environments.

Plagioclases vary widely in color, occurring in white, gray, black, and colorless varieties, but can be recognized by their good two-directional cleavage at

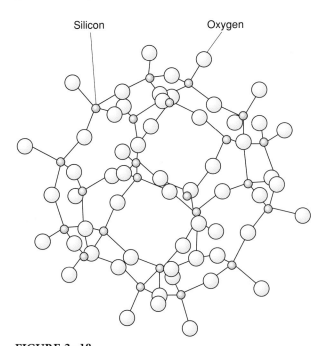

Silicon   Oxygen

FIGURE 3–18
Part of a framework silicate structure. All four oxygen ions of each tetrahedron are shared with neighboring tetrahedra in this complex three-dimensional network.

nearly right angles and their moderate hardness (6 on Mohs scale). Some plagioclase grains exhibit very fine grooves called *striations* on one of the two cleavage directions; these striations are a diagnostic property for the entire family.

The **potassic feldspars** also are framework silicates and have the composition $KAlSi_3O_8$. There are three polymorphs of $KAlSi_3O_8$, each of which forms in different environments and can help pinpoint the processes or conditions of formation. **Sanidine** crystallizes in volcanic igneous rocks and in some metamorphic rocks that have experienced very high temperatures and low pressures. **Orthoclase** forms under higher-pressure metamorphic conditions, in igneous rocks, and to a lesser extent in sedimentary rocks. The third polymorph, **microcline,** forms in igneous and metamorphic environments similar to those of orthoclase, but is not produced in sedimentary processes.

Quartz, SiO_2, is one of the commonest minerals and can form in igneous, metamorphic, and sedimentary rocks. It is a framework silicate, but unlike the feldspars there is no substitution of aluminum for silicon. Quartz is identified readily in rocks by its glassy luster, high hardness (7), and conchoidal fracture. Impurities and defects in its crystalline structure help quartz to occur in many colors. Its colorless variety is known as rock crystal, and each of its colored varieties has been given a different name: amethyst (purple), rose quartz (pink), smoky quartz (gray or black), citrine (yellow), and aventurine (green).

Nonsilicate Minerals. Although there are more *kinds* of nonsilicate minerals than silicates, the nonsilicates constitute only a very small part of the Earth's crust. Few nonsilicates are important rock-forming minerals; most occur as **accessory minerals** (minor minerals) in silicate rocks. On the other hand, nonsilicates are very important economically because they include most of the ore minerals from which we get metals and nonmetallic resources. A list of some of the more important nonsilicate minerals is given in Table 3–3. Our discussion here will focus on those few nonsilicates that are rock-forming minerals (highlighted in Table 3–3).

There are several **carbonate** minerals, each of which contains the carbonate $(CO_3)^{2-}$ anion complex. The two most common are **calcite,** $CaCO_3$, and **dolomite,** $CaMg(CO_3)_2$. These are important constituents of sedimentary rocks, but also are abundant in some metamorphic rocks. Both are relatively soft (around 3 on Mohs scale) and display excellent cleavage in three nonperpendicular directions. Both

TABLE 3–3
Important nonsilicate minerals. Rock-forming minerals are boldfaced.

Native Elements	Oxides (O^{2-})	Sulfides (S)
Gold (Au)	Hematite: Fe_2O_3	Galena: PbS
Silver (Ag)	Magnetite: $FeO \cdot Fe_2O_3$	Pyrite: FeS_2
Platinum (Pt)	Corundum: Al_2O_3	Bornite: Cu_5FeS_4
Diamond (C)	Ilmenite: $FeTiO_5$	Sphalerite: ZnS
Sulfur (S)	Chromite: $FeCr_2O_4$	

Carbonates (CO_3)	Halides (Cl^-, F^-)	Sulfates $(SO_4)^{2-}$
Calcite: $CaCO_3$	**Halite: NaCl**	**Gypsum: $CaSO_4 \cdot 2H_2O$**
Aragonite: $CaCO_3$	Sylvite: KCl	**Anhydrite: $CaSO_4$**
Dolomite: $CaMg(CO_3)_2$	Fluorite: CaF_2	Barite: $BaSO_4$
Malachite: $Cu_2CO_3(OH)_2$		
Azurite: $Cu_3(CO_3)_2(OH)_2$		

Hydroxides $(OH)^-$	Phosphates $(PO_4)^{3-}$	Others
Goethite: FeO(OH)	**Apatite: $Ca_5(F,Cl,OH)(PO_4)_3$**	Borax: $Na_2B_4O_7 \cdot 10H_2O$
Gibbsite: $Al_2(OH)_6$	Monazite: (Ce,La,Y,Th)PO_4	Carnotite: $K_2(UO_2)_2(VO_4)_2 \cdot 3H_2O$
Boehmite: AlO(OH)	Turquoise: $CuAl_6(PO_4)_4(OH)_8$	Wulfenite: $PbMoO_4$

HOW CAN A DIAMOND BE CUT?

Diamonds are among the most beautiful and prized of all gems, and their multifaceted shapes reflect light brilliantly from rings, necklaces, and other jewelry. Diamond, however, is not found in nature in such magnificent shapes. Instead, its crystal habit is an octahedron—Figure 1(a)—and most of the largest specimens occur as irregularly shaped masses—Figure 1(b). How can a diamond be "cut" into

FIGURE 1
(a) A diamond crystal (16.76 carats) showing its typical eight-sided habit. (b) Glass model of the Cullinan diamond (3106 carats). The pencil is the same size in both photographs. Photos courtesy of Dept. of Library Services, American Museum of Natural History.

(a)

(b)

the intricate shapes we have come to enjoy? After all, diamond is the hardest natural substance, so nothing can cut it.

A brief look at the history of diamond cutting shows that it has changed over the past 500 years as we have learned more about the internal structure and physical properties of minerals. Ancient jewelers could not cut or modify the shape of diamond at all. Baffled by its hardness, they simply mounted the stone in a setting that enhanced its shape. Today, three different diamond-shaping methods are used, each taking advantage of the physical properties of cleavage and hardness.

Diamond Cleaving

Diamond cleaves in four directions, parallel to the faces of the octahedron, and jewelers can take advantage of this to alter the shape of a specimen. A scratch is made on the surface of the diamond, using the only substance that can scratch it—another diamond. A sharp, flat blade is inserted in the scratch and oriented parallel to the cleavage plane. The blade is then tapped sharply with a mallet. If all is oriented correctly, the mineral splits smoothly along a cleavage plane; if not, a precious gem shatters into tiny pieces. Irregularly shaped diamonds can be shaped in this way once the cleavage directions are located by modern X-ray methods.

Diamond Polishing, or Grinding

During the Middle Ages, jewelers found that certain faces and directions in diamond crystals are softer than others. Powdered diamonds of nongem quality were mixed with olive oil to make a paste that was then used to grind away a gem-quality diamond along its soft planes. The locations of these zones of softness became jealously guarded professional secrets. Two of the most commonly used planes resulted in the so-called table cut, Figure 2(a), and lozenge cut, Figure 2(b). About the year 1520, more detailed knowledge of diamonds led to the more elaborate rose cut, Figure 2(c). Finally, near the end of the seventeenth century, a new motif was created by adding 57 more facets to the table cut. These facets led to even more spectacular light reflection, and the new shape was called the brilliant cut, Figure 2(d).

FIGURE 2
Diamond shapes produced by polishing and grinding.

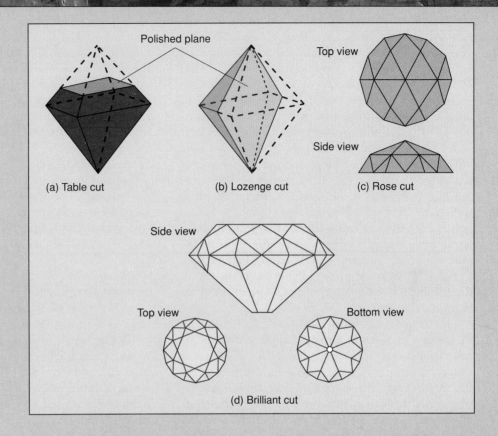

(a) Table cut

(b) Lozenge cut

(c) Rose cut

Polished plane

Top view

Side view

Side view

Top view

Bottom view

(d) Brilliant cut

Diamond Cutting

The so-called "cuts" described above were not originally made by cutting because diamond is very brittle and shatters easily. Eventually, diamond cutting became a reality when wires were impregnated with diamond dust and rubbed rapidly across a stone. Today, high-speed saw blades are impregnated with a diamond-dust paste and are used to cut diamonds into desired shapes.

react with hydrochloric acid, but differently: calcite effervesces vigorously and dolomite only weakly.

Gypsum, $CaSO_4 \cdot 2H_2O$, and anhydrite, $CaSO_4$, are the most common minerals containing the **sulfate** anion complex $(SO_4)^{2-}$. They are typically light gray to white and are among the softest minerals known (gypsum is the standard mineral for 2 on the Mohs scale). Both minerals occur in massive aggregates or well-developed crystals, some up to 1.5 m long. Gypsum and anhydrite are exclusively sedimentary in origin, and form by evaporation of seawater.

Elements next to the inert gases in the periodic table are called halogen elements, and their minerals are called **halides**. The only major halide rock-forming mineral is halite, NaCl, although fluorite, CaF_2, is a very common accessory mineral. Halite occurs both in clear cubic crystals that exhibit cleavage in three mutually perpendicular directions and in granular masses that form layers tens of meters thick ("rock salt") in many areas. Like gypsum and anhydrite, halite is a sedimentary mineral formed by evaporation of seawater.

One important group of minerals—the **clay minerals**—does not fit into our simple chemical classification scheme because some clays are silicates and others are nonsilicates. All form at the Earth's surface by **weathering,** which is the interaction between previously formed minerals and the gases of the atmosphere. These minerals contain aluminum and various amounts of water and/or the hydroxyl anion complex in their structures. Clays commonly occur in very fine-grained mixtures that are plastic and can be molded when mixed with a little water.

The most common clay minerals are the silicates kaolinite, montmorillonite, and illite (a mineral much like muscovite but with less potassium and more silicon). Their silicon:oxygen ratios indicate that these clay minerals have the sheet silicate structure. Gibbsite and boehmite are the most common nonsilicate clays. All the clays contain aluminum and various amounts of water or the hydroxyl anion complex.

USES OF MINERALS

Many minerals are useful, and we go to extraordinary lengths to discover and mine them. Today we tunnel thousands of meters below the surface for copper sulfides and diamonds, pump steam into the ground to melt and recover native sulfur, and even talk about dredging the deep ocean floors for manganese oxides. Minerals are valuable resources only if they are concentrated in such large amounts that recovery is easy. In Chapters 4, 6, and 7 we will look at some of the rock-forming processes that bring about these concentrations, and in Chapter 21 will look at all of our natural resources in some detail. Here, we will focus on those attributes of minerals that make them useful to humans. A list of a few of the important mineral resources, both metals and nonmetals, is given in Table 3–4.

Some minerals are valuable because of the *elements they contain*. The three metals we use most—iron, aluminum, and copper—come from mineral oxides (magnetite and hematite for iron), mineral hydroxides (bauxite for aluminum), and mineral sulfides (bornite for copper). Minerals containing uranium, such as carnotite and uraninite, are important sources of nuclear fuels.

Other minerals are useful because of their *physical properties*. Industrial abrasives used for smoothing, polishing, and grinding are chosen from minerals having great hardness—quartz (hardness 7), corundum (9), and diamond (10). The drills used to penetrate kilometers of rock in the search for oil and natural gas are coated with diamonds to help grind through the softer rock-forming minerals of the crust. Graphite, on the other hand, is so soft that it rubs off on paper; it is graphite that you write with when you use a common black pencil. (Actually, graphite is too soft to be used alone in pencils because it smudges so easily. It must be mixed with clay minerals to reach the desired hardness.)

Talc and graphite often are used as lubricants in industry because of their softness and greasiness. The greasy feel of these minerals is due to the presence of very weak van der Waals' bonds in their structures. Finger pressure alone can break these bonds, causing sheets of silicon-oxygen tetrahedra (in talc) or of carbon atoms (in graphite) to slide past one another. It is this sliding that we perceive as greasiness. In contrast, extremely strong bonds give very high melting points to feldspars and to the Al_2SiO_5 polymorphs named andalusite, sillimanite, and kyanite. These minerals are used in making porcelain and the refractory bricks that line the insides of steelmaking blast furnaces.

TABLE 3-4

Minerals as natural resources (excluding gemstones).

Metal	Mineral Ores of Metal
Iron	Magnetite, hematite, siderite, goethite, limonite
Copper	Cuprite, bornite, chalcopyrite, chalcocite
Zinc	Sphalerite
Lead	Galena
Aluminum	Bauxite (a mixture of gibbsite, boehmite, diaspore)
Chromium	Chromite
Molybdenum	Molybdenite
Tin	Cassiterite
Titanium	Rutile, ilmenite

Use	Minerals Having Nonmetallic Uses
Abrasives	Diamond, corundum, garnet, quartz
Construction	Gypsum, calcite, dolomite, asbestos
Ceramics	Feldspars, clays, kyanite, wollastonite
Insulators	Muscovite
Optical lenses	Quartz, fluorite
Flux in metallurgy	Fluorite
Drilling muds	Barite, clay minerals
Medicinal	Sulfur (sulfa drugs), kaolinite, orpiment
Acid manufacture	Sulfur, pyrite, marcasite for H_2SO_4; fluorite for HF; halite for HCl
Fertilizers	Apatite, potassic feldspar
Cleansers	Borax
Lubricants	Graphite, talc
Energy sources	Carnotite, uraninite, thorianite
Water purification	Zeolite minerals

The metallic bonding in gold, silver, platinum, and copper enables them to conduct electric currents easily. Quartz also responds to electricity, but in a different way. Thin slices of quartz deform slightly and vibrate when subjected to an alternating electric current; this allows their use in tuning to specific radio frequencies in "crystal-controlled" transmitters and receivers.

The sheet structures of some clay minerals permit them to absorb large amounts of water; when this happens, they swell to many times their normal volume. Such clays are used for a wide range of purposes, from thickening "thickshakes" in fast-food restaurants to stopping leaks in reservoirs or artificial lakes by swelling and plugging tiny cracks. Changes in clays brought about by heat make them important in construction (to make bricks) and manufacturing (clay pots, tiles, and ceramics).

The open-framework structure of the zeolite minerals permits small ions to pass through while in solution, but not large ones. Zeolites thus are used as the active ingredients in many water purifiers and softeners because they can selectively remove ions of specific sizes from the water.

A combination of hardness, bright color, and brilliant luster is responsible for the use of minerals as gems (Figure 3-19; also see Focus 3-2). Very few mineral specimens, even of diamond, have the right combination of properties to be gemstones, and that is why gems are so rare and expensive. Many gems are uncommonly spectacular samples of common minerals. For example, ruby and sapphire are varieties of corundum, the same mineral used in emery cloth that we can buy cheaply at hardware stores. Moonstone is either orthoclase or albite. Agate, amethyst, and carnelian are varieties of quartz.

There is literally no end to our uses for minerals, and new applications continually are found. Rubies, long used to make jewelry, now are used to generate laser beams, something unimaginable 25 years ago. The mixture of clay minerals that we call bauxite was relatively unimportant until we learned how to extract aluminum from it. The more we learn about minerals, the more useful they become.

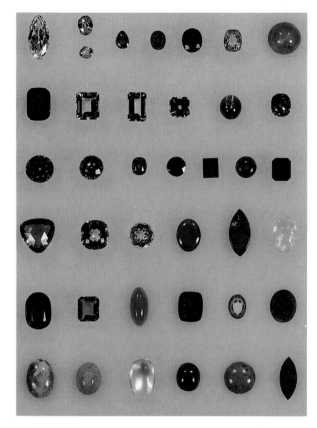

FIGURE 3–19
The spectacular color and luster of these minerals make them valuable gems. All have been "cut" and/or polished; none appear in their untouched out-of-the-ground state. Photo by American Museum of Natural History.

SUMMARY

1. Minerals are naturally occurring, inorganic solid compounds that make up most of the Earth. Each possesses an ordered internal arrangement of atoms or ions and a specific chemical composition.
2. A mineral's structure is determined by the size and charge of the ions or atoms present and by the conditions of formation.
3. Each ion in a mineral is coordinated with as many oppositely charged ions as can fit around it, and the total electric charge of a mineral must be zero.
4. Polymorphs are minerals with identical compositions but different structures resulting from different conditions of formation.
5. Ions may substitute for one another in a mineral if they are of approximately the same size and charge.
6. Minerals grow during igneous, sedimentary, and metamorphic processes by the addition of ions to seed crystals. Igneous minerals form from molten silicate materials at high temperatures. Sedimentary minerals form at the Earth's surface by weathering or precipitation from water. Metamorphic minerals grow within solid rocks at conditions of temperature and pressure intermediate between those of the surface and melting.
7. A mineral's physical properties are the result of its unique combination of composition and structure. Bond strength and orientation determine hardness, melting point, feel, and breakage (cleavage or fracture), whereas composition determines color, luster, streak, magnetism, reaction to acids, taste, and odor. Internal structure controls crystal form and, along with composition, determines a mineral's specific gravity.
8. Most of the Earth's crust is made up of silicate minerals in which the tetrahedral silicon-oxygen anion complex, $(SiO_4)^{4-}$, is the basic building block. By sharing oxygen ions, tetrahedra may be linked in pairs, rings, chains, sheets, and three-dimensional frameworks. The ratio of silicon and oxygen ions at the time of mineral formation determines which structure will form.
9. A few silicate mineral families—olivine, plagioclase and potassic feldspars, pyroxene, amphibole, mica—and the mineral quartz comprise most of the crust. There are many nonsilicates, but few are important as rock-forming minerals; these include calcite, dolomite, and gypsum.
10. Many minerals are useful to humans because of their physical or chemical properties, particularly the nonsilicates. Many oxides and sulfides are ores from which metals such as iron, copper, and aluminum are produced. Minerals also are used as abrasives, lubricants, electrical conductors, electronic tuning devices, water softeners, jewelry, and for many other purposes.

REVIEW QUESTIONS

1. What are the differences between liquid compounds (such as water) and minerals?
2. What factors control the internal structure of a mineral?

3. Lithium is in the same column of the periodic table as are sodium and potassium. All three form cations with a +1 charge and behave the same way during chemical reactions. Sodium and potassium commonly substitute for one another, but lithium is not interchangeable with either of them. Why not?

4. Andalusite, sillimanite, and kyanite are polymorphs of Al_2SiO_5. Their specific gravities are 3.16, 3.23, and 3.62, respectively. From this information, what can you infer about the conditions under which they form?

5. Discuss the controlling factor(s) for each of the following mineral properties: hardness, cleavage, specific gravity, electrical conductivity.

6. What are the factors that control the structure of a silicate mineral?

7. All silicate minerals whose tetrahedra share three oxygens have sheet structures. However, not all of those that have two shared oxygens per tetrahedra have identical structures. Explain.

8. Diamond is much scarcer than quartz, but it is far more useful in determining the conditions of formation of the rocks in which it is found than is quartz. Why is this so?

9. In olivine, each oxygen belongs to a single SiO_4 tetrahedron and must be bonded to some number of iron or magnesium ions. Using Table 3–1, discuss how many of these ions each oxygen must be bonded to, and describe the way in which each ion present in olivine balances its electric charge.

10. The mineral idocrase has a particularly complex structure, as indicated by its formula: $Ca_{10}Mg_2Al_4(Si_2O_7)_2(SiO_4)_5(OH)_4$. In what way is this structure different from those of all the other silicate minerals discussed in this chapter?

FURTHER READINGS

Desautels, Paul. 1974. *Rocks and minerals*. New York: Grossett & Dunlap. (A brief, clear description of the major mineral families, followed by superb color photographs of many minerals and mineral groups.)

Holden, A., and Singer, P. 1960. *Crystals and crystal growing*. Garden City, NY: Doubleday & Co., 320 p. (A paperback for the layperson interested in mineral structures and crystals. Includes detailed instructions and "recipes" for growing several kinds of crystals in the home.)

Hurlbut, C., and Klein, C. 1978. *Dana's manual of mineralogy*. 19th ed. New York: John Wiley & Sons, 532 p. (A sophomore-level mineralogy text, with details of internal structure, common crystal shapes, composition, and the origin of physical properties. Includes descriptions of the commonest rock-forming and ore minerals and excellent determinative tables for identification of unknown minerals from their physical properties.)

Mott, N. 1967. The solid state. *Scientific American* (September). (An introduction to the ordered, geometric world of solids, written for those having no scientific background.)

Vanders, I., and Kerr, P. F. 1967. *Mineral recognition*. New York: John Wiley & Sons. (For the amateur mineral collector, useful in field-identification of mineral specimens.)

4

Igneous
Processes and
Rocks

Photomicrograph of high-titanium basalt collected by *Apollo* astronauts on the Moon. Gray, elongate crystals are plagioclase feldspar; bright colored grains are pyroxene and olivine; black grains are ilmenite. Photo by Allan Ludman.

Deep beneath the Japanese islands, a part of a subducting plate melts, and the resulting magma erupts at the surface to build volcanoes such as Mount Fuji. Beneath Yellowstone National Park, a magma body is cooling slowly, giving off enough heat to generate the steam for Old Faithful geyser. During the formation of the Appalachian Mountains millions of years ago, intense heat and pressure caused melting within the crust; large bodies of magma formed and cooled without ever reaching the surface.

These are just a few of the igneous processes that we will examine in this chapter. We will focus on how rock melts to form magma and magma crystallizes to form igneous rock, show how to recognize igneous rocks, and explain how to read the geologic history that each preserves.

IGNEOUS ROCKS

Some igneous rocks form at the Earth's surface during eruptions of volcanoes, and are called **volcanic** rocks (Figure 4–1). They are also called **extrusive** rocks because the magmas extrude (flow out) onto the surface from within the Earth. When magma reaches the Earth's surface it is called **lava**. Extrusive rocks are forming today at the world's mid-ocean ridges at the estimated rate of 16 km^3 per year. They are forming on the slopes of the Hawaiian, Aleutian, and Cascade mountains in the United States and on hundreds of other volcanoes worldwide. Older volcanic rocks underlie much of the area drained by the Columbia and Snake Rivers in Idaho, Washington, and Oregon. Ancient lavas form prominent ridges in Virginia, New Jersey, Connecticut, and Massachusetts.

Some magmas solidify below the surface where we cannot directly observe the rock-forming process. The resulting rocks are called **plutonic** (after Pluto, god of the underworld) or **intrusive** because they commonly cut across or intrude their way into the surrounding rock. We see intrusive rocks at the surface today because uplift and erosion have exposed formerly deep-seated parts of the Earth's crust. Intrusive rocks, for example, make up most of the Sierra Nevada mountain range of California and underlie both ends of the Appalachian Trail—Mount Katahdin in Maine and Stone Mountain in Georgia.

Other intrusive rocks form major parts of mountain ranges throughout the world.

Texture

It is easy to tell the difference between a rock that formed quickly during a violent eruption and one that formed by slow cooling many kilometers beneath the surface—once you know what to look for. The clues are contained in the *texture* of every igneous rock. A rock is an aggregate of grains, each an individual crystal or group of crystals. The **texture** of an igneous rock includes such aspects as grain size, shape, and the way in which grains are intergrown. Some textures immediately identify a rock as igneous, help to distinguish intrusive rocks from extru-

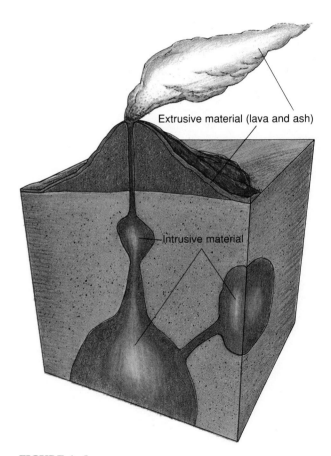

Extrusive material (lava and ash)

Intrusive material

FIGURE 4–1
Intrusive vs. extrusive igneous rocks. Intrusive igneous rocks solidify beneath the surface. Extrusive rocks made of volcanic ash and lava solidify at the surface.

sive ones, and aid in understanding how magmas cool.

Some textural features are so small that they cannot be seen, even with the aid of a magnifying glass. We must study them with a microscope, but to do so it is necessary to prepare a slice of rock so thin that it is transparent. First, a small chip is cut from the rock with a diamond saw. The chip is cemented to a glass slide (Figure 4–2) and ground with abrasive powder until it is so thin that light can pass through it. The resulting **thin section** can be studied with a micro-scope. Figure 4–3 is a photograph of a thin section of a fine-grained volcanic rock called basalt; it shows how rocks with grains smaller than a millimeter can be studied in detail.

Interlocking Grains. Minerals in most igneous rocks interlock like pieces of a three-dimensional jig-saw puzzle because of the way in which the minerals grow (Figure 4–4). Each crystal expands outward from its seed until it encounters its growing neigh-bors, and the grains become intimately intergrown as

FIGURE 4–2
Making a thin section. (a) A chip is cut from a hand sample using a diamond saw. (b) Chip is cemented to a glass slide and is then ground sufficiently thin (0.03 mm) for light to pass through it. (c) Completed thin section is studied with a microscope that can reveal small-scale mineralogic and textural features.

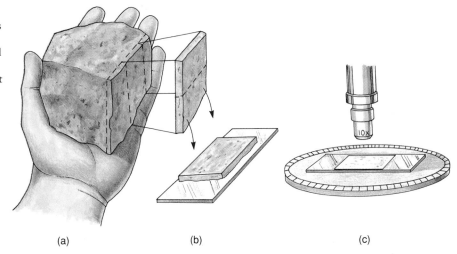

(a)   (b)   (c)

FIGURE 4–3
Photomicrograph of a thin section of basalt, a fine-grained igneous rock. Gray crystals are plagioclase feldspar; brightly colored minerals are pyroxenes. Note how the crystals are tightly interlocked. Macmillan Publishing/Geoscience Resources photo.

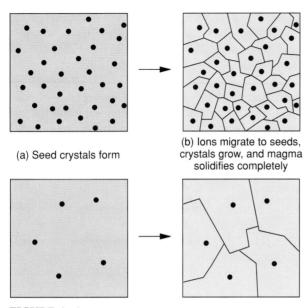

(a) Seed crystals form

(b) Ions migrate to seeds, crystals grow, and magma solidifies completely

FIGURE 4−4
Formation of interlocking grains. With many seeds (upper diagrams), crystals interfere with one another early in their growth, resulting in many fine-grained crystals. With fewer seeds (lower diagrams), much coarser crystals form.

FIGURE 4−5
Pegmatitic texture. The coarse pink grains of potassic feldspar in this pegmatitic granite are *smaller* than those in many pegmatites. Macmillan Publishing/Geoscience Resources photo.

they expand to fill all the available space. Interlocking grains are not restricted to igneous rocks; numerous seed crystals also form during some sedimentary and metamorphic processes. However, the combination of interlocking grains and the types of minerals present are usually enough to identify a rock as igneous.

Grain Size. There are igneous rocks in the Black Hills of South Dakota with crystals up to 10 m long and others in Hawaii with grains so small that they can barely be seen with a microscope. Most igneous rocks have readily visible grains that lie between these extremes. Grain size in an igneous rock depends on the balance between the rate at which seed crystals *form* and the rate at which they *grow*.

As a magma cools, a small number of seed crystals form. The crystals grow because ions in the magma move toward them, attracted by the positive and negative charges of the ions in the seeds. The ultimate size of the crystals depends on how fluid the magma is and how fast it cools.

Just as it would be more difficult for you to swim through molasses than through water, ions find it more difficult to move through thick magmas than through those that are highly fluid. Ions in a highly fluid magma can migrate easily to the initial seed crystals. The crystals can grow easily to a large size and produce a coarse-grained rock. In contrast, ions in a thick magma cannot migrate easily; instead, many new seeds form as the temperature decreases and ions need travel only short distances to them. The result is a fine-grained rock.

The **rate of cooling** must also be considered. If a magma cools rapidly, its ions will have little opportunity to migrate, regardless of its fluidity. In general, *the slower a magma cools and the more fluid it is, the coarser the resulting rock will be. The faster it cools and the thicker it is, the finer grained the rock will be.*

Volcanic rocks cool far more rapidly than plutonic rocks, just as pies cool more rapidly on a breezy window sill than in a warm oven. As a result, volcanic rocks are generally fine grained and plutonic rocks are normally coarser grained. Igneous rocks so fine-grained that their minerals can be seen only with a microscope are said to have an **aphanitic** texture (from the Greek word for invisible). Those having minerals visible to the naked eye are said to be

phaneritic (from the Greek for visible). Rocks with grains coarser than about 5 cm are called **pegmatitic** (Figure 4–5).

A magma need not cool at a constant rate. Consider a magma at great depth within the Earth, cooling slowly and developing large crystals. If it is driven upward toward the surface before solidifying completely, the late-formed crystals will be much finer grained than the early-formed ones. The resulting texture, called **porphyritic** (Figure 4–6), indicates two cooling rates. The coarser, slowly cooled grains are called **phenocrysts;** the finer, rapidly cooled grains are referred to as the rock's **groundmass.**

Degree of Crystallinity. Some lavas cool so quickly that not even seed crystals have a chance to form—ions are frozen in place before they can become part of mineral structures. There are thus no minerals in the resulting rocks, and the rocks are said to be *noncrystalline.* Such rocks are generally dark in color, glassy, and are called **volcanic glasses.** Obsidian, the dark glassy rock used by the Aztecs and other early cultures for cutting implements and weapons, is a typical volcanic glass (Figure 4–7).

Textures Produced by Gas. Gases dissolved in magma play important roles in forming textures in

FIGURE 4–7
Obsidian, a volcanic glass. Note the glassy luster and the conchoidal fracture characteristic of broken glass. Macmillan Publishing/Geoscience Resources photo.

volcanic rocks. Magmatic gases escape to the atmosphere during eruptions and the result is similar to what happens when a can of carbonated beverage is shaken vigorously and opened. A froth composed of gas and magma forms and—if it cools quickly enough—is preserved as a porous rock that looks much like a sponge (Figure 4–8). Such a rock is said to have a **vesicular** texture.

Gases are released explosively in some eruptions, breaking lava into tiny spatters of liquid and hurling them into the air. The airborne lava solidifies to small glass fragments that accumulate to form a rock when they fall to the ground. When the hot fragments pile up on the ground, they may be welded together by their own heat. Some rocks, called **welded tuffs,** are composed of millions of welded glass fragments (Figure 4–9). Although each fragment is a piece of glass, tuffs look very different from the homogeneous rock obsidian; their texture is said to be **pyroclastic,** from the Greek roots *pyro* (fire) and *klastos* (broken).

Not all fragments that have been erupted into the air are tiny. During the 1963 eruption of the volcano Surtsey south of Iceland, geologists were bombarded by blocks the size of *automobiles*—and survived to

FIGURE 4–6
Porphyritic texture. Phenocrysts of plagioclase feldspar are much larger than the fine-grained mafic minerals in the rock's groundmass. Macmillan Publishing/Geoscience Resources photo.

tell about them. All material ejected into the air by a volcano is called **tephra**.

You now see what we meant in Chapter 1 when we stated that Earth's history is written in its rocks—a magma's cooling history is clearly recorded in the texture of igneous rocks. Table 4–1 summarizes the textures and interpretations we make from them.

Classification of the Igneous Rocks

There are many different kinds of igneous rocks, and almost as many different ways of naming them. We will use a simple classification scheme based on a

FIGURE 4–8
Scoria, showing the porous texture produced by gas escaping from a magma. Macmillan Publishing/Geoscience Resources photo.

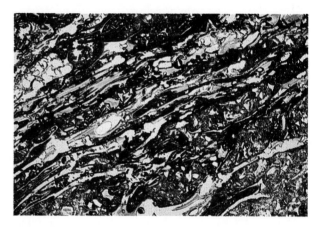

FIGURE 4–9
Welded tuff. Photomicrograph of a rhyolitic tuff showing a few mineral grains embedded in a matrix of welded glass shards. Many of the shards show the typical narrow, curved shape. Macmillan Publishing/Geoscience Resources photo.

TABLE 4–1
Summary of igneous rock textures and their interpretations.

Texture	Description	Crystallization Location	Interpretation
Pegmatitic	Very coarse grained	Intrusive	Crystallization from extremely fluid magma
Phaneritic	Grains visible to the naked eye	Intrusive and/or extrusive	Relatively slow cooling; usually intrusive but may form in slowly cooled centers of thick lava flows
Aphanitic	Grains not visible to the naked eye	Intrusive and/or extrusive	Relatively fast cooling; usually extrusive but may form in shallow intrusives or at edges of large intrusives
Porphyritic	Some grains coarse; most fine	Intrusive and/or extrusive	Two cooling rates; phenocrysts cooled more slowly (deeper?) than finer-grained groundmass
Glassy	Amorphous; no minerals formed	Extrusive	Extremely rapid cooling at surface
Vesicular	Porous, spongy	Extrusive	Rapid surface cooling with release of gases
Pyroclastic	Fragmental	Extrusive	Explosive eruption

rock's *mineralogy* and *texture* because both can be identified easily from hand samples (Figure 4–10).

Igneous rocks are often described by the proportions of feldspar and ferromagnesian minerals they contain:

Felsic rocks (named for *fel*dspar and *si*lica) have large amounts of potassic feldspar and sodium-rich plagioclase feldspar; they often contain quartz as well.

Mafic rocks (named for *ma*gnesium and *ferric*, Latin for iron) have abundant ferromagnesian minerals such as olivine and pyroxenes. Mafic rocks have considerable calcium-rich plagioclase, but contain little quartz and almost no potassic feldspar.

Intermediate rocks, as their name implies, have mineral contents between those of mafic and felsic varieties.

Ultramafic rocks consist almost entirely of ferromagnesian minerals, usually olivine and pyroxenes.

Feldspars and quartz have lower specific gravities than ferromagnesian minerals and are generally much lighter in color. As a result, felsic rocks are lighter colored and less dense than intermediate, mafic, and ultramafic rocks. Felsic rocks also have the highest concentration of *silica*—the combination of silicon and oxygen ions—of all the rock types. We will discuss why silica content is an important factor

FIGURE 4–10

Classification of igneous rocks, based on both mineral content (upper part of the diagram) and texture (lower part). A given rock sample's *composition* is shown in the upper part; its *name* is determined by its texture in the lower part. For example, the rock shown by the sample line consists of about 20% plagioclase feldspar, 65% pyroxene, and 15% olivine. The rock is named where the line crosses the lower diagram, depending upon its texture: if phaneritic, it is a gabbro; if aphanitic, a basalt; and so on.

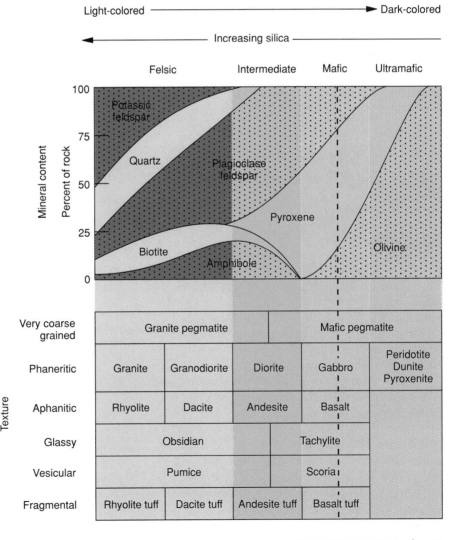

(a)

(b)

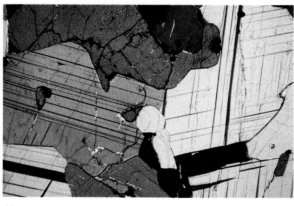

(c)

FIGURE 4–11
Basalt and gabbro. (a) Hand sample of basalt. (b) Hand sample of gabbro. Pale gray crystals are plagioclase feldspar; dark grains are pyroxene. (c) Thin section of gabbro. Plagioclase grains are gray and striped; pyroxenes are brightly colored. Macmillan Publishing/Geoscience Resources photos.

in igneous processes later in this chapter and will show in Chapter 5 how it affects the formation of volcanoes.

Figure 4–10 shows how igneous rocks are classified. The upper part of the diagram shows the combinations of minerals present in igneous rocks. The lower part focuses on texture, showing how rock names are applied. Rocks in the same vertical column have exactly the same mineral composition, but differ in texture. For example, the line on the right side of Figure 4–10 indicates a dark-colored rock composed mostly of pyroxene and plagioclase feldspar with some olivine. Depending on its texture, this rock may be a gabbro (coarse grained), basalt (fine grained), scoria (porous), and so on.

The Most Abundant Igneous Rocks

The three most common igneous rock types are basaltic, granitic, and andesitic. We will describe them briefly here and show later in the chapter how they can be used to construct hypotheses about tectonic processes and the structure of continents and oceans.

Basaltic Rocks. Basaltic rocks include **basalt,** which is volcanic in origin, and gabbro, which contains the same minerals but is plutonic. Together, they account for nearly 75% of the igneous rocks in the crust. They are mafic rocks composed mainly of calcium-rich plagioclase feldspar and one or two kinds of pyroxene, with minor amounts of olivine or

(a)

(b)

(c)

(d)

FIGURE 4–12

Granite and rhyolite. Both contain the same minerals but look different because of their different textures. (a) Hand sample of granite. (b) Thin section of granite. White and gray grains are quartz; plaid grains are potassic feldspar (microcline); striped grains are plagioclase feldspar; brightly colored grains are biotite and hornblende. (c) Hand sample of rhyolite. (d) Rhyolite porphyry in thin section. Large quartz phenocrysts are set in a very fine-grained groundmass. Macmillan Publishing/Geoscience Resources photo.

quartz. Basalts are generally dark gray to black and are fine grained—Figure 4–11(a), but thin sections reveal elongate plagioclase crystals intergrown with irregularly shaped pyroxenes (see Figure 4–3).

Gabbros are much coarser grained—Figure 4–11(b)—and vary in color, depending on the amount and color of the plagioclase. Porous basaltic rocks (scoria), glasses (tachylite), and basaltic tuffs are common volcanic varieties (Figure 4–10).

Granitic Rocks. Granites (intrusive) and rhyolites (extrusive) are the next most abundant igneous rocks. They are felsic, so they contain more silica than basaltic rocks and are less dense. Granites and rhyolites are generally light-colored rocks (see Figure 4–12), occurring in shades of gray, pink, or white depending on the color of the potassic feldspar. Granitic rocks contain potassic feldspar, sodium-rich plagioclase, and quartz, with lesser amounts of a ferromagnesian mineral—usually biotite or hornblende. Some also contain the colorless mica, muscovite.

Individual crystals in granites do not show the well-developed shapes common in many mafic rocks. This does not indicate different cooling processes, but rather the tendency of some minerals to form good crystals more readily than others. Rocks with the chemical composition of granite display the widest range of textural variation among the igneous rocks; obsidian is the glassy variety, pumice the frothy type. Rhyolite tuffs are very common, and the coarsest grained of all igneous rocks, the pegmatites, are generally of granitic composition.

Andesitic Rocks. Andesites and diorites are intermediate rocks with mineral content, color, and density between those of granitic and basaltic rocks. Most rocks of this composition occur as the extrusive variety andesite. Intrusive masses of diorite are far less common than those composed of granite or gabbro. The most abundant minerals are a plagioclase feldspar midway between calcic and sodic end members of the solid-solution series, an amphibole (usually hornblende), and pyroxene.

Ultramafic Rocks. Ultramafic rocks are rare in the crust, but are believed to compose most of the mantle. They are by far the densest and darkest-colored igneous rocks. Peridotite, a rock made of olivine and pyroxene, is the most abundant. Other types include dunite, a rock made entirely of olivine, and pyroxenite, one composed entirely of pyroxene.

WHAT IS MAGMA LIKE?

Certain properties of magma—*temperature, composition,* and *fluidity*—play important roles in the final nature of igneous rocks that form from it. We will look briefly at each of the three.

Temperature

Magma is very hot. Direct measurements of erupting lava and melting experiments in the laboratory suggest that most magmas have temperatures in the range of 650–1350°C. Granitic magmas have the lowest temperatures, andesitic magmas intermediate

temperatures, and basaltic magmas are hottest. The basalts of Hawaii, for example, exhibit a consistent temperature range of 1000–1200°C. It is important to note here that magmas have very different melting points: granites are lowest, basalts the highest. At 750°C, most granites are completely molten, but a basaltic magma would be totally solid.

Composition

Nearly all magmas are composed mainly of ions that produce silicate minerals, and silicon and oxygen are by far the most common elements. Silica (SiO_2) amounts to as much as 80% by weight of felsic rocks like granite, leaving only 20% for all the other cations that must hold the SiO_4 tetrahedra together. Basaltic magmas, on the other hand, contain only about 48% silica by weight. Geologists argued for years whether there could be nonsilicate magmas in nature. The answer came when the volcano Ol Doinyo Lengai, near the Olduvai Gorge in Tanzania, East Africa, erupted *carbonate magma*. Such magmas are extremely rare, however, and produce less than 1% of the total volume of igneous rock.

Some components of magma do not wind up in igneous rocks. Magmas contain dissolved gases, most of which escape during cooling and are not incorporated in minerals. Water vapor and carbon dioxide are by far the most abundant gases, but researchers have collected significant amounts of sulfur dioxide (SO_2), pure sulfur gas, nitrogen, chlorine, hydrogen, methane (CH_4), and ammonia (NH_3) at volcanic vents. Our atmosphere probably evolved by the escape of volcanic gases from Earth's interior during the early history of the planet.

Fluidity and Viscosity

The **viscosity** of a fluid is a measure of the sluggishness with which it flows; it is essentially the opposite of *fluidity*. For example, water, molasses, and petroleum jelly are all fluids, listed in order of increasing viscosity. Some lavas have very low viscosity and flow like molten rivers at many kilometers per hour, but others are highly viscous and move at only a few meters per hour.

A magma's viscosity is determined by its temperature and composition. High temperature often leads to highly fluid magmas because, at high temperatures, ions have high kinetic energy. They therefore move rapidly and freely—the two requirements for a highly fluid substance. Two aspects of composition control viscosity: the amounts of silica and water. High silica content causes high viscosity because the attraction between Si^{4+} and O^{2-} ions is so strong that many temporary bonds form in the magma. These bonds restrict the freedom of ions to move and make magma more sluggish. Basaltic magmas, with their low SiO_2 contents and high melting points, are typically the most fluid; granitic magmas are the most viscous.

The role of water in controlling viscosity is due to the fact that it is a polar compound (see Chapter 2). The polar water molecules interfere with the formation of even the short-range bonding between ions that characterizes magmas, and it breaks many of the bonds that do form. The more bonds that are broken, the more freely ions may move—and the more fluid a magma will be.

In summary, a magma is an extremely hot liquid composed of enormous numbers of ions. Most are oxygen and silicon ions, although many other cations and a few anions are also present. These ions attract or repel one another, depending on their

charges, and often form temporary bonds until the heat energy present breaks them apart.

WHERE DO MAGMAS FORM?

Melting takes place within the Earth wherever the internal energy sources, predominantly radioactive elements, supply enough heat to break the bonds that hold minerals together. The most abundant radioactive elements—uranium, thorium, potassium, and rubidium—are concentrated in the crust and upper mantle, and most magmas form in these regions. Concentrations of heat great enough to melt rocks are widely scattered so that neither the crust nor the mantle is entirely molten.

Melting is not a simple process because a rock's melting point depends on its composition, the pressure it is under, and the amount of water present (Figure 4–13). Increased pressures raise melting points because higher pressures favor the most densely packed arrangement of ions, and liquids are less dense than the rocks from which they melt. In contrast, water lowers melting points because of its polar nature. The weakly charged sides of the water

FIGURE 4–13
Estimating minimum depths of melting. Melting is possible where geothermal gradient curves intersect the melting curves of granite and basalt. At an "average" geothermal gradient of 30°C/km granite magma can form at depths of 20 km or more and basalt magma at 25 km or more if appropriate rocks are available for melting.

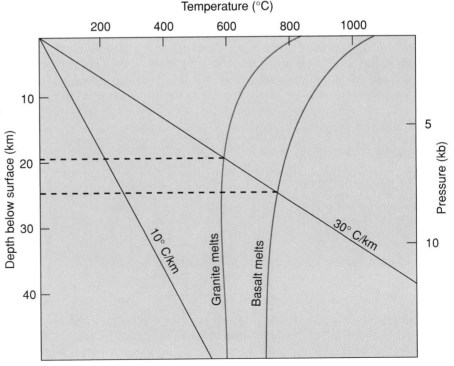

molecule exert a pull on ions in minerals, weakening the bonds that lock the ions in place. This lowers the amount of heat needed to break the bonds and thus lowers the melting point.

We can estimate the *shallowest* depth at which melting occurs by comparing the melting temperatures of rocks with the Earth's geothermal and geobaric gradients (Figure 4–13). Granitic rocks melt at lower temperatures than basaltic rocks at all pressures (compare the melting curves in Figure 4–13). The minimum depth of melting is estimated by drawing geothermal gradients (the straight lines) on the melting curve diagram. Melting begins where the melting and geothermal gradient curves intersect. It is clear that granite magmas can form closer to the surface than basalt magmas, but no closer than 20 km at the average geothermal gradient of 30°/km. Most granitic magmas form in the crust. Basalt magma cannot form any closer than 25 km below the surface, and all other indications (see Chapter 20) point to a much deeper source: the asthenosphere. Andesites, as we shall see later in this chapter, also originate in the mantle.

MIGRATION OF MAGMA

Once a magma has formed, several factors cause it to rise toward the surface. Pressure exerted on a tube of toothpaste forces the fluid paste into an area where pressure is lower—the atmosphere. Similarly, the great pressure exerted on magma because of its depth forces it upward toward the lower pressures of the surface or shallow crust. Gas dissolved in a magma tends to rise even more readily, pushing the magma upward the way carbon dioxide causes soda to erupt from a vigorously shaken bottle. Finally, magmas have lower densities than the rocks from which they are formed. This causes them to rise buoyantly above their denser surroundings, just as oil floats above vinegar in salad dressing.

A magma faces a more difficult task than salad oil because it must rise through solid rock. It moves slowly by squeezing through any available cracks and by melting some of the surrounding rock in a process called **stoping** (Figure 4–14). In stoping, projections of magma extend along fractures until they engulf blocks of the host rock. The blocks then settle into the magma as it fills the spaces they had occupied. These blocks, which stand out sharply in composition and texture from the igneous rock formed when the magma solidifies (Figure 4–15), are called **xenoliths** from the Greek words *xenos* (stranger) and *lithos* (rock).

Earthquake studies of Mount Saint Helens in 1980 enabled us to track the upward movement of lava beneath the volcano. There, and at other volcanoes we have studied, magma rises slowly, at a few tens of meters per day. A few magmas move a lot faster. This happens when gases trapped in an ultramafic magma in the upper mantle encounter a crack that extends upward. The gases surge toward the surface, acting like an air drill, punching a hole as they go. A pipelike passageway is thus created, and a magma may travel from the mantle to the surface in a matter of hours. The magma cools to form a cy-

FIGURE 4–14
Upward movement of magma by stoping. (a) Magma extends projections into host rock along fractures. The projections penetrate upward and as they grow they join to engulf blocks of the host rock. (b) Blocks of host rock are dislodged and the magma rises to occupy the space from which they came.

(a)

(b)

FIGURE 4–15
Xenoliths (dark fragments) of basalt enclosed in lighter-colored Baring Granite, Calais, Maine. Photo by Allan Ludman.

lindrical igneous body called a **diatreme** that preserves the shape of the passageway. Pieces of the mantle surrounding the "pipe" are carried upward as xenoliths by the rising magma, providing us with very rare samples of what the mantle is like. These xenoliths are the source of the world's diamonds.

CRYSTALLIZATION OF MAGMA

Ideally, the crystallization of a magma should be exactly the opposite of the process by which it melted. For example, the **crystallization point** of a mineral is exactly the same as its melting point. The analogy of melting and freezing ice comes to mind, but the process is not as simple because ions freed from several different minerals must somehow find one another and reassemble their complex internal structures. In addition, few minerals crystallize in the same way that ice freezes. To understand what goes on in a cooling magma, we will first look at the three ways by which the most important rock-forming minerals crystallize: *simple, continuous, and discontinuous crystallization.*

Simple Crystallization

Some common minerals, such as quartz and forsterite, behave much the same way that ice does. At sea-level pressures (about 1000 millibars or 15 pounds/in^2), the crystallization point of molten quartz is 1713°C. A magma made of pure quartz remains completely molten until it cools to 1713°C. At that temperature, there is no longer enough heat energy to keep broken the temporary bonds that form in magma between the silicon and oxygen ions. All of the bonds become permanent and the magma becomes completely solid at 1713°C.

Continuous Crystallization

Of the plagioclase feldspars, only the pure end members—albite and anorthite—crystallize simply, at 1118°C and 1553°C respectively. Anorthite has a higher melting point than albite because the bonds between Ca^{2+} and O^{2-} in anorthite are stronger than those between Na^+ and O^{2-} in albite. All other plagioclase feldspars are solid-solution mixtures of these end members, and they behave in a more complex manner.

Consider a magma formed of exactly equal amounts of molten albite, $NaAlSi_3O_8$, and anorthite, $CaAl_2Si_2O_8$, as shown in Figure 4–16. Crystallization begins at 1444°C, below the crystallization point of anorthite and above that of albite. Even though there are equal numbers of Ca^{2+} and Na^+

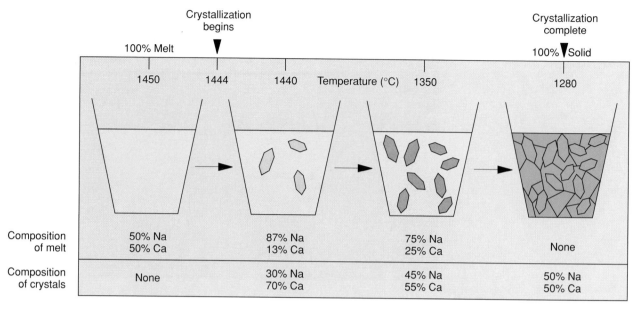

	Crystallization begins			Crystallization complete
100% Melt	▼			100% ▼ Solid
1450	1444	1440	Temperature (°C) 1350	1280

| Composition of melt | 50% Na 50% Ca | 87% Na 13% Ca | 75% Na 25% Ca | None |
| Composition of crystals | None | 30% Na 70% Ca | 45% Na 55% Ca | 50% Na 50% Ca |

FIGURE 4–16

Continuous crystallization of plagioclase feldspar. The first crystals that form are richer in calcium than the starting magma. The magma gradually crystallizes over an interval of temperature; as it does, more and more sodium enters the feldspar crystals until the crystals have the composition of the initial magma.

ions in the magma, the first crystals that form contain many more calcium ions than sodium ions. This reflects the greater strength of calcium-oxygen bonds—they can survive at high temperatures. A few sodium ions also enter the crystals, but at this temperature there is enough heat to break most sodium-oxygen bonds. Removal of more calcium than sodium from the magma enriches the remaining magma in sodium.

As cooling proceeds, more sodium ions can be added to the crystals because more sodium-oxygen bonds are stable at lower temperatures. The compositions of feldspar grains and residual magma change *continuously* over the crystallization interval, and only when the magma is completely solidified does the feldspar attain the original 50:50 composition. The *progressive* change in composition caused by interaction between magma and early-formed seeds is called a **continuous reaction**.

Thus, solid-solution minerals such as the plagioclase feldspars do not have a single crystallization

point like quartz, but rather a temperature *interval* over which the solid forms continuously and during which both the growing mineral and remaining magma change compositions continuously. Many plagioclase feldspar crystals exhibit a compositional zoning that reflects this change in composition (Figure 4–17). They are richer in calcium in their centers and become progressively richer in sodium toward their edges, reflecting the change in the ions that were added during growth.

Discontinuous Crystallization

Most ferromagnesian minerals crystallize from magma in yet a third way, called **discontinuous crystallization**. As an example, we will follow the formation of the pyroxene enstatite, $MgSiO_3$, as shown in Figure 4–18. When magma with the composition $MgSiO_3$ crystallizes, the first crystals to form are not enstatite, or even a different pyroxene, but are instead crystals of the olivine mineral forster-

FIGURE 4-17

Photomicrograph of zoned plagioclase feldspar. The concentric compositional zoning reflects the progressive change in mineral composition produced during continuous crystallization described in Figure 4-16. Macmillan Publishing/Geoscience Resources photo.

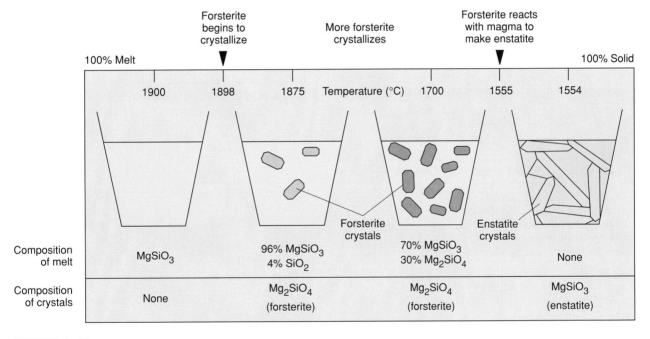

FIGURE 4-18

Discontinuous crystallization of the pyroxene enstatite. The first crystals that form from an $MgSiO_3$ magma are of the olivine mineral forsterite. Up until 1555°C, only forsterite crystallizes. At 1555°C, all the forsterite reacts with the remaining magma and is converted to enstatite.

ite, Mg_2SiO_4. This happens because, even though the ions are present in exactly the right proportion to make the single-chain pyroxene structure, the most stable ion grouping at high temperature is to form olivine with some of the ions and leave the remainder in the magma. At this stage neither the magma nor the crystals have the composition of pyroxene.

With further cooling, forsterite continues to crystallize until a reaction takes place that distinguishes this kind of crystallization from all others. Ions from the remaining magma are added to the independent tetrahedron structure of the forsterite crystals, converting them to enstatite single chains. This is called a **discontinuous reaction** because the magma abruptly stops crystallizing one kind of mineral and forms another.

In continuous crystallization of plagioclase feldspar, some plagioclase is always present once seed crystals have formed. In discontinuous crystallization, the final mineral appears only at the last stage of the process.

What Happens When More than One Mineral Crystallizes?

So far we have looked at simple magmas, those made by melting a single mineral. However, most igneous rocks contain several minerals, and the ions from each mineral affect the crystallization of the others. The result is similar to what happens in the radiators of our cars during the winter. We add antifreeze to water, analogous to mixing mineral components in a magma. Pure water freezes at 0°C, but the molecules of antifreeze get between the water molecules, preventing them from bonding to one another to form ice. The crystallization (freezing) point of the water is lowered by the presence of the antifreeze.

In a similar way, the presence of several minerals' ions in a magma lowers the temperature at which crystallization would proceed if each mineral were alone. Thus, granite, a combination of quartz, potassic feldspar, and sodium plagioclase, can melt completely at temperatures as low as 650°C, more than 1000°C lower than the melting point of quartz and 500°C lower than the melting point of sodic plagioclase.

Crystallization of Magmas in Nature

The crystallization of millions of tons of magma in the Earth's crust is far more complex than that of a few grams of melt in a laboratory. There are two basic reasons for the added complexities: (1) natural magmas crystallize several minerals, some simulta-

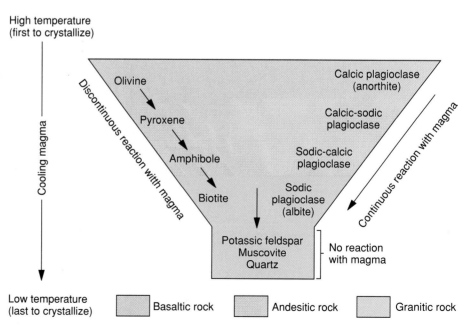

FIGURE 4–19
Bowen's Reaction Series. The order of crystallization of minerals from magma is shown by the arrows. Minerals that crystallize together are found together as shown by the areas representing the three most common igneous rock types.

neously and others in a sequence determined by their melting points, and (2) the process of crystallization may be altered during the tens of thousands of years it takes for magmas to cool.

Sequence of Crystallization.

Every granite contains minerals that display all three types of crystallization behavior: simple (quartz), continuous (sodic plagioclase), and discontinuous (hornblende or biotite). Basaltic and andesitic rocks also contain minerals that form through different processes. N. L. Bowen, a pioneer in melting and crystallization experiments, reduced many of these complexities to a simple diagram that shows the sequence in which minerals generally form in magma (Figure 4–19). In his honor, the scheme is called **Bowen's Reaction Series.**

The right side of Figure 4–19 illustrates the continuous reactions by which plagioclase feldspars crystallize. On the left are the discontinuous reactions that produce the ferromagnesian minerals. Minerals on the same horizontal line can crystallize simultaneously; those that are far apart vertically generally crystallize at very different temperatures. For the most part, igneous rocks are composed of minerals that crystallize together or within a narrow range of temperatures. Thus, minerals that are close to each other in Bowen's Reaction Series are commonly found together in igneous rocks, as shown by the shading.

Notice that basaltic rocks contain the minerals located at the top of the diagram and granite's minerals are located at the bottom. This is exactly what we would expect, considering the melting temperatures of these rocks. Minerals that are widely separated in the diagram such as quartz and calcic plagioclase are rarely found in the same igneous rock.

Bowen's Reaction Series does *not* indicate that olivine and calcic plagioclase are always the first minerals that form in a magma, or that quartz is always the last mineral to crystallize. What crystallizes first depends on a magma's composition, temperature, and pressure. In some magmas, an amphibole such as hornblende is the first and only ferromagnesian mineral to form; in others, the first crystals may be of potassic feldspar. The diagram merely shows the sequence of minerals most commonly observed in nature.

Changes in Cooling History.

We said earlier that crystallization is ideally the reverse of melting. Magmas do not necessarily follow ideal behavior because there are several processes, called processes of **magmatic differentiation,** by which minerals can be removed from a magma as it cools. A single magma may produce several different rock types, depending on the extent of differentiation.

One of the most effective magmatic differentiation mechanisms is called **crystal settling.** Nearly all crystals are denser than their parent magmas and therefore tend to sink toward the floors of their magma chambers. As they sink, early-formed minerals are removed from possible reactions with residual magma (Figure 4–20). This is particularly effective

FIGURE 4–20
Magmatic differentiation by crystal settling. (a) Early-formed crystals have higher density than magma and sink toward the floor of the magma chamber. (b) Early crystals accumulate to form one rock type at floor of magma chamber. Remainder of magma crystallizes to form a second type of rock.

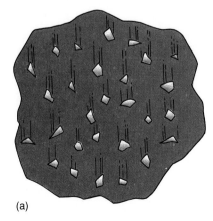

(a)

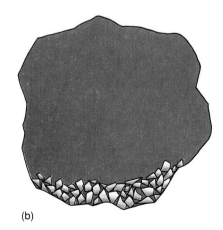

(b)

in magmas with low viscosity because crystals can sink more easily than in viscous magmas. The magma behaves as if the settled crystals had never been present and crystallizes from that point as if nothing had solidified earlier. Imagine the effect on discontinuous crystallization if the early olivine crystals were removed—enstatite might never form. The settled minerals form one rock, and the residual magma eventually crystallizes to form another.

One of the classic examples of crystal settling is the Palisades Sill, a body of basaltic rock that forms cliffs nearly 300 m high on the west shore of the Hudson River in New Jersey and New York. Early-formed olivine crystals settled toward the bottom of the sill and formed a rock composed largely of olivine. The remaining magma, depleted significantly in $(Fe,Mg)_2SiO_4$, then crystallized to produce rocks in which there is very little olivine. The settling of mafic minerals such as olivine and pyroxene may thus lead to the accumulation of an *ultramafic* rock from a magma of *mafic* composition.

Magmas often form in or intrude unstable areas of the crust where they are squeezed by intense mountain-building forces. The liquid is forced through cracks in the host rock, but large crystals cannot pass through these cracks and are left behind. This process of magmatic differentiation is called **filter pressing** because the crystals are filtered from the liquid.

SHAPES OF INTRUSIVE ROCK BODIES

The general term **pluton** is applied to any mass of intrusive rock regardless of its size or shape, but some plutons have been given special names based mainly on their shape (Figure 4–21). Flat, tabular (tabletop shaped) plutons are called **dikes** if they cut across layers that existed in their host rocks before intrusion, or **sills** if they are parallel to these layers. Many dikes and sills are small, only a few centimeters across, but some, such as the Palisades Sill mentioned earlier, are immense.

The Palisades Sill exhibits a feature common to many tabular plutons and lava flows—regularly

FIGURE 4–21
Shapes of plutons.

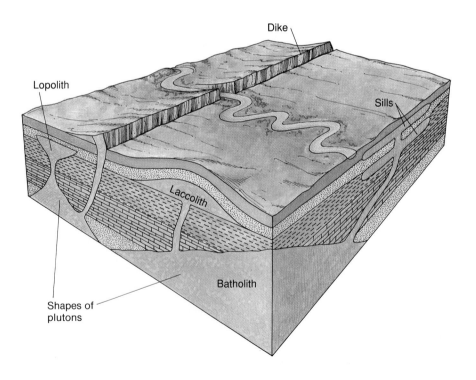

Dike

Lopolith

Sills

Laccolith

Shapes of plutons

Batholith

spaced fractures called **columnar joints** (Figure 4–22). Columnar joints form during cooling when the already solidified rock contracts as heat leaves it. The columns generally form at right angles to the flat surfaces of the plutons or flows.

Many plutons are much less regular in shape than sills and dikes. **Laccoliths** are flat bottomed but have dome-shaped tops that push the overlying rocks up into an arch. **Lopoliths** are the opposite shape—flat on top, curved at the bottom—and are much larger than laccoliths. Irregularly shaped plutons are called **stocks** or **plugs** if they are small, and **batholiths** if they underlie an area of over 75 km².

Most batholiths are enormous accumulations of granitic magma. For example, the Sierra Nevada Batholith (Figure 4–23) is composed of many plutons and underlies nearly 40,000 km² of eastern California. Despite its great size, it is "tiny" compared to its neighbor to the north, the Coast Range Batholith (Figure 4–23).

FIGURE 4–22
Columnar jointing in basalt at Devils Postpile National Monument in California. Thousands of these vertical columns, many almost 50 meters in length, are found along the top edge of a 900,000-year-old (Pleistocene) lava flow about 275 m long. During the past 20,000 years, glaciers polished and striated the tops of the columns. Photo by Bruce F. Molnia.

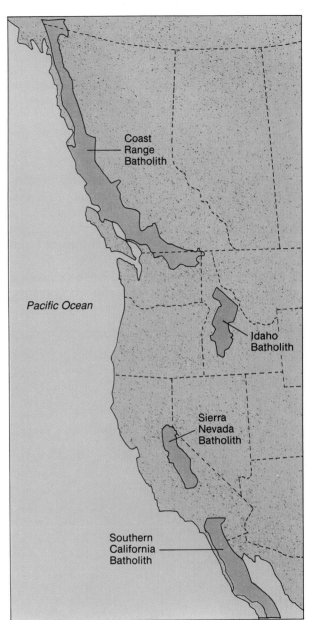

FIGURE 4–23
Large batholiths of the western United States.

IGNEOUS ROCKS: A WINDOW INTO THE EARTH

We have not yet succeeded in drilling entirely through the Earth's crust, so ideas about the composition of the mantle are hypotheses based on melting experiments and studies of earthquake waves, rather than on examination of actual rocks. Igneous rocks provide us with our only direct samples of mantle material; they also help us understand the structure of the crust beneath continents and oceans and the nature of many plate tectonics processes.

The Crust Is Different Beneath Continents and Oceans

At one time, many scientists thought that oceans were just parts of continents that had been flooded by seawater. However, the worldwide distribution of the three major igneous rock types (see Figure 4–24) shows that this hypothesis is incorrect and gives some indication of the differences.

Basalts are found everywhere. They have erupted on every continent, are found in subduction zones, and form the floors of every ocean. Gabbroic intrusive rocks are part of nearly every mountain range. Granitic plutons and rhyolitic volcanic rocks are widely distributed *on the continents,* but are almost totally absent from the oceans. The rare rhyolites found in the oceans have differentiated from basaltic magmas, rather than crystallized directly from a granitic magma. It appears that granitic magma is not present in any ocean basin, but is widespread on the continents.

This simple observation leads to profound statements about differences between the crust underlying the oceans and the crust beneath the continents (Figure 4–25). We saw earlier in this chapter that granitic magma melts at lower temperatures than basaltic magma. Therefore, if there is enough heat energy to form basaltic magma in the oceans, there must also be enough to form granitic magma in the

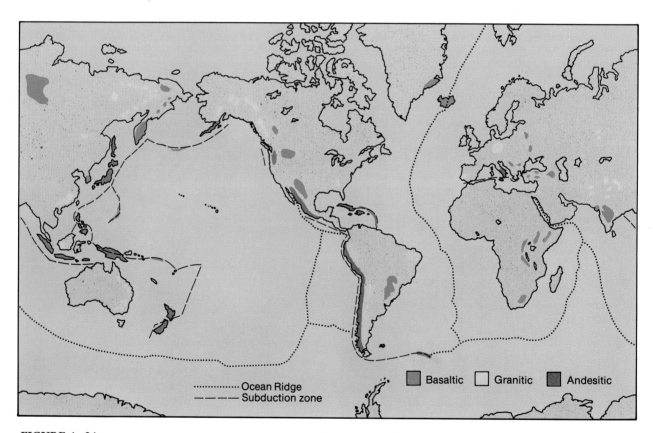

FIGURE 4–24
Distribution of the major igneous rock types. For simplicity, only those rocks younger than 225 million years are shown. After Hyndman, *Petrology of igneous and metamorphic rocks,* 2d ed., 1985, McGraw-Hill Book Co.

same place. Thus, the absence of granitic rocks from the oceans cannot be due to a lack of heat energy, but must instead be due to the absence of rocks that can melt to produce granitic magma.

The abundance of granite and rhyolite on the continents indicates that rocks of appropriate source compositions are present there, but not in the oceans. Basaltic rocks are found both on the continents and in the oceans, indicating that rocks capable of melting to form basaltic magma underlie both regions.

Composition of the Mantle

The range of geothermal gradients suggests that temperatures needed to form basaltic magmas cannot be attained in the crust, so basaltic magma must form in the mantle—probably in the asthenosphere. Every basalt and gabbro is therefore a direct sample of the upper mantle, but this does not mean that the mantle is made of basalt. Just as magmatic differentiation may yield several very different rock types from a single magma, the process of **partial melting** can produce several different melts from a single rock.

Remember that melting and crystallization are opposite processes. In continuous or discontinuous crystallization, the magma changes composition constantly. In continuous and discontinuous *melting* the initial liquid is thus different from the 30% melt, the 50% melt, and so on. Melting experiments suggest that basalt can be formed by partially melting an

ultramafic rock in the upper mantle rather than by totally melting gabbroic rocks. Xenoliths found in many basalts support this hypothesis. These inclusions were picked up by the rising magma, and the presence of diamonds in some indicates that the xenoliths must be representative of the rock types present in the upper mantle, as we discussed in Chapter 3. The xenoliths are of ultramafic rocks and are probably fragments of the mantle material that was left behind when partial melting took place to produce the basaltic magma.

Igneous rocks can thus tell us something about the large-scale structure of the Earth. If we apply our knowledge of melting and magmatic differentiation to what we suspect our planet was like in its early stages of formation, we may be able to explain how this structure came about. The reasoning involved is explained in Focus 4–1.

Igneous Rocks and Plate Tectonics

Igneous activity provides us with some of the best evidence for the plate tectonics theory. We have learned that modern convergent plate boundaries (subduction zones) and divergent plate boundaries (mid-ocean ridges) generate different kinds of igneous rock. Chemical studies reveal that the basalt which has erupted at ocean ridges is chemically identical to the basalt that forms the floors of the ocean basins. This supports our concept of seafloor spreading: basalt erupting at the ridges is moved away from the ridge crests, subsides, and eventually becomes the ocean floor.

FIGURE 4–25
Schematic of differences between continents and oceans based on distribution of igneous rocks. Rocks that melt to form granite are found only beneath the continents, but those that melt to basalt are found beneath both.

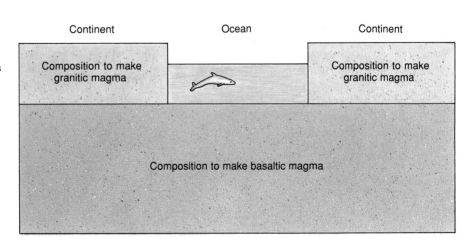

Continent   Ocean   Continent

Composition to make granitic magma

Composition to make granitic magma

Composition to make basaltic magma

FORMATION OF THE CRUST, MANTLE, AND CORE—AN IGNEOUS PERSPECTIVE

The hypothesis that the Earth is not homogeneous, but rather has a crust, mantle, and core, has come from many different kinds of observation, some of which will be discussed later in this book. Magmatic processes help us understand how this structure might have evolved.

In Chapter 1 we saw that the Earth probably accumulated from a cloud of dust and gas, but went through a stage early in its history during which it was completely molten. At that time, the Earth was a gigantic magma chamber, subject to all of the possible differentiation processes. Many scientists believe that gravity caused denser components to settle toward the center of the molten mass, in a way comparable to crystal settling, whereas less-dense material floated upward. The densest material became the core, and the lightest became the crust.

Ocean-ridge basalts can be distinguished chemically from subduction-zone basalts, so that ancient basalts now found on the continents can be related to their original tectonic settings. As an example, a group of basalts about 500 million years old found along the east coast of Maine and adjacent Canada probably erupted near a mid-ocean ridge; other basalts nearby probably erupted above a subduction zone.

Figure 4–24 shows that andesites and diorites are found only in narrow belts at the borders of oceans and continents next to subduction zones. Recall that subduction zones are places where two lithospheric plates collide and one is driven down into the mantle. Something about the subduction process must be responsible for the formation of these rocks (Figure 4–26).

The subducted plate consists of basaltic oceanic crust and sediments deposited upon it. Partial melting of the subducted rocks and assimilation of some material from the overlying plate by the rising magma combine to make andesite. Some andesites are found today in old mountain ranges far removed from any subduction zone. Andesites are formed today only in subduction zones, so the principle of

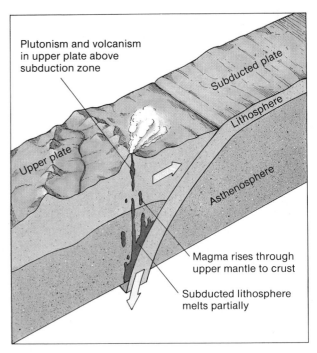

FIGURE 4–26

Origin of andesite in subduction zones. Subducted oceanic lithosphere melts when it reaches depths at which there is sufficient heat and pressure. Resulting andesitic magma rises, erupting above the subduction zones or forming diorite plutons.

uniformitarianism suggests that subduction was involved in the ancient mountain-building process.

How Do Granites Form?

Thus far we have seen that basaltic and andesitic rocks form from magmas generated in the mantle. Granitic magmas are different because they form within the crust. The heat needed to melt crustal materials is provided by one of two sources: (1) some is generated during the folding of rocks in mountain-building processes, and (2) some is provided by large volumes of basaltic magma that have risen into the crust. The basaltic magma forms in the asthenosphere and rises buoyantly because it is less dense than its surroundings. Many such magmas are intruded into the lower crust of continents where their heat causes rocks of the appropriate composition to melt and form granitic magma.

IGNEOUS ROCKS AS NATURAL RESOURCES

Many igneous rocks and minerals have become useful to modern civilization. A few of the igneous rock and mineral resources are described here as examples of the economic importance of igneous processes. Others will be discussed in more detail in Chapter 21.

Uses of Igneous Rock

Some igneous rocks are used because they are beautiful, strong, and durable. Granite, for example, has been used as a building stone for thousands of years. Gabbro, particularly a variety with large, iridescent plagioclase crystals, makes a spectacular ornamental facing stone when polished. Most uses of igneous rock are less spectacular. Fine-grained basalts are crushed to make stable road foundations. Pieces of scoria (sold as "lava rock") are placed in gas barbecue grills to help distribute heat evenly.

Pumice is used as an abrasive to smooth wood or remove calluses from hands and feet; finely crushed and added to soap, it is used in hand cleaners to remove grease. In the past, pulverized pumice was also used as a tooth powder because it abraded stains and dental plaque very effectively from teeth. Unfortunately, the pumice, with a Mohs scale hardness between 6 and 7, also abraded the material of which our teeth are made (essentially the calcium-phosphate mineral apatite, with a hardness of 5), so other materials are now used instead.

Igneous Minerals as Resources

We saw in Focus 2–1 that many useful elements are present in very small amounts in the Earth's crust, and that some natural processes must concentrate these elements in rocks if we are to recover them economically. Crystal settling is particularly effective in doing so because many ore minerals are igneous, and these generally have very high densities. Present in many igneous rocks as "accessory minerals" are oxide and sulfide minerals such as hematite (Fe_2O_3), magnetite ($FeO \cdot Fe_2O_3$), chromite ($FeCr_2O_4$), and bornite (Cu_5FeS_4). These are major ores of iron, chromium, and copper. Settling of early-formed ore minerals in a magma concentrates them in abundant proportions and can make mining feasible. A single example will illustrate the principle.

Chromite has a specific gravity of 4.62, far greater than that of the mafic magmas from which it crystallizes. Settling of chromite crystals during magmatic differentiation has produced layers of nearly pure chromite in many of the world's large mafic plutons. The Bushveld igneous complex in South Africa contains several chromite layers, and one zone called the Merensky Reef is a major source of platinum that has also settled out from the parent magma.

Hydrothermal Solutions

Water released from magma during the last stages of cooling often contains dissolved ions such as silicon, oxygen, sulfur, copper, lead, and zinc. These solutions easily penetrate fractures in host rock and precipitate ore minerals in the fractures. Underground water not derived directly from the magma may be heated when it passes through pores and fractures near a magma that is cooling. This heated water may then dissolve ions from the host rock, transport them to other locations, and precipitate them as ore minerals.

The heated water, whether from magma itself or from the local groundwater supply, is called a **hydrothermal solution,** from the Greek *hydros* (water) and *therme* (heat). Hydrothermal solutions were responsible for copper deposits at Butte, Montana, and Chuquicamata, Chile, and for extensive lead and zinc deposits of the midcontinental United States near Joplin, Missouri.

SUMMARY

1. Igneous rocks result from the solidification of molten material called magma that is formed by concentrations of radioactively generated heat in the crust and upper mantle.
2. Most magmas are of silicate composition, although their chemical compositions vary considerably.
3. Crystallization of silicate minerals from magma is complex, reflecting three different ways in which minerals enter and leave the liquid state. Minerals such as quartz crystallize simply from magma at their melting points; solid-solution series minerals like plagioclase and olivine crystallize over a range of temperature and change compositions continuously; ferromagnesian minerals like pyroxenes undergo discontinuous reactions with magma in which early-formed minerals react with residual magma to make new minerals.
4. Crystallization of a magma generally involves simultaneous crystallization of two or more minerals. Bowen's Reaction Series indicates the sequence of crystallization and the compatibility of minerals crystallizing from a single magma.
5. The final products of crystallization are variable because of magmatic differentiation. Settling of early-formed crystals and separation of crystals from magma by filter pressing may considerably alter magma composition and hence final rock types.
6. Some magma moves to the surface where it solidifies to form extrusive (volcanic) rocks. Some rises through the crust but solidifies beneath the surface to form intrusive (plutonic) rocks.
7. The textures of igneous rocks reveal their cooling histories and enable us to determine whether they were intrusive or extrusive. Rapid cooling at the surface leads to fine-grained rocks; slower cooling at depth produces coarse-grained rocks. Glassy rocks result from extremely rapid cooling.
8. A rock may be identified as igneous by its texture, mineral content, and intrusive or extrusive relationships with surrounding rocks. Igneous rocks are classified and named by their mineralogy and texture.
9. Basalts and gabbros are the most abundant igneous rocks and are found throughout the oceans and on the continents. Granites and rhyolites are restricted to the continents and continental margins. Oceans and continents are underlain by rocks of different compositions.
10. Andesites are found in subduction zones and form by the melting of the subducted plate and assimilation of some materials of the upper plate by the rising magma.
11. Earth's crust, mantle, and core formed largely by gravity separation during an early stage in the planet's history when it was completely molten.

REVIEW QUESTIONS

1. What information can the texture of an igneous rock provide about its crystallization history?
2. Why do most magmas rise toward the surface from the places where they form?
3. What factors determine the physical properties of a magma?
4. Contrast and compare the crystallization behaviors of enstatite, plagioclase feldspar, and quartz.
5. How can an igneous rock be produced from a magma that does not have the same composition as the rock?
6. Explain, using the mineral enstatite as an example, how different degrees of partial melting can produce different magmas from the same starting material.
7. Why is it unlikely to find pyroxene and potassic feldspar together in an igneous rock?
8. What properties of accessory minerals make it possible for them to be concentrated during the crystallization of a magma? Explain.
9. Basaltic magmas are much less viscous than granitic magmas, despite the fact that they contain far less water. Explain how this is possible.
10. Some pegmatites contain crystals as much as 10 m long, and yet evidence suggests that they cooled relatively quickly. How can such large crystals form during rapid cooling?

FURTHER READINGS

Bowen, N. L. 1928. *Evolution of the igneous rocks*. New York: Dover Publications, reprinted from Princeton University Press original. (A classic discussion of melting and crystallization by a pioneer in the field, with detailed examples of magmatic differentiation.)

Ehlers, E. G., and Blatt, H. 1982. *Petrology: Igneous, sedimentary, and metamorphic*. San Francisco: W. H. Freeman & Company, 732 p. (An excellent in-depth explanation of all three rock types and the processes that form them.)

Hyndman, D. W. 1985. *Petrology of igneous and metamorphic rocks*. 2d ed. New York: McGraw-Hill Book Co. (An excellent treatment of igneous rocks and their relationship to plate tectonics processes.)

Williams, H., Turner, F. J., and Gilbert, C. M. 1982. *Petrography: An introduction to the study of rocks in thin section*. 2d ed. San Francisco: W. H. Freeman & Company. (Descriptions of the three rock types, superbly illustrated with drawings of textures as seen in thin section.)

5
Volcanoes

The 1980 eruption of Mount Saint Helens in the Cascade
Mountain Range, Washington State. Photo courtesy of U.S.
Geological Survey/Cascades Volcano Observatory.

The eruption of a volcano can be a terrifying event. Thunderous explosions, hot choking clouds of glowing ash, and rivers of molten lava have affected Alaska, Hawaii, and Washington State in this century alone. Few geologic processes are so violent or so awesome in the amount of energy expended in a short period of time—or so feared.

Not all volcanic activity is violent, however, and not all eruptions build the cone-shaped mountain that we call a volcano. Why not? How many different kinds of volcano are there? Why are there volcanoes in Alaska, Washington, Oregon, California, and Hawaii, but not in Kansas, Florida, or Maine? In this chapter, we will answer these questions as we focus on the surface expression of igneous activity. We also will examine the harmful effects of volcanoes, ways in which volcanoes may be beneficial to human beings, and the ways in which geologists use volcanism to interpret tectonic processes.

WHAT ARE VOLCANOES MADE OF?

In Chapter 4 we saw that magma may erupt onto the surface in one of two forms. It may remain fluid and flow along the ground as lava, or it may be shot into the air where it solidifies into tiny fragments that rain down as tephra. Both types of material—lava and tephra—build volcanoes and other landforms.

Lava

Lava contains the same ions, has the same properties, and is at the same range of high temperatures as intrusive magmas. (The only exception is the loss of some of the gas that drives the magma to the surface.) Thus, the hotter a magma is and the less silica it has, the more fluid the erupting lava will be. There are two major types of lava, pahoehoe and aa, distinguished by their fluidity.

Pahoehoe Lava. Highly fluid lava behaves like the water in a stream, except that it is much hotter—about 1200°C. It flows rapidly down steep slopes, meanders more slowly across flats, and sometimes even creates "lavafalls" analogous to waterfalls (Figure 5–1). Much lava in the Hawaiian Islands is of this type, and the Hawaiian word **pahoehoe** (pronounced pa-'hoy-hoy) is applied to highly fluid lava throughout the world. Pahoehoe forms a glassy crust as it chills in contact with the air, and this crust commonly retains a "ropy" appearance that records the lava's high fluidity and flow directions (Figure 5–2). The escape of gases from nearly solidified pahoehoe lava produces a highly vesicular texture.

The fluidity exhibited by pahoehoe lavas requires either a very high temperature, a low silica content, or a combination. All pahoehoe is basaltic because only basalt has the right composition; rhyolitic and andesitic magmas have too much silica and are too "cool" to be so fluid.

Aa Lava. Thicker, more viscous lava is known by the Hawaiian word **aa** (pronounced 'ah-ah). The surface of an aa flow consists of sharp, angular chunks of solidified but still hot material; underneath lies a core of highly viscous lava (Figure 5–3). Aa advances across the countryside much more slowly than pahoehoe, more like partly solidified hot tar than like water. An aa flow advances somewhat like the movement of a bulldozer tread (Figure 5–4). Viscous lava within the flow pushes slowly but doggedly onward, dragging the brittle crust with it and plowing chunks under as it moves.

Aa lava may be andesitic or rhyolitic, but much is basaltic, just like pahoehoe. Basaltic aa differs from pahoehoe in that it has cooled more and partly solidified before erupting, so that it is more viscous.

Features Associated with Lava Flows. Several unique geologic features are produced by advancing pahoehoe flows. Two of the most common are lava tubes and pillows.

Lava Tubes. Many pahoehoe flows contain long, tubelike cavities called **lava tubes**. These form when the top, bottom, and sides of a flow solidify, leaving a central channel through which lava continues to flow. During waning stages of an eruption, the amount of lava coming from the source may decrease so that this channel is no longer completely filled. When final cooling occurs, a tunnel is left within the flow (Figure 5–5). Lava tunnels may be thousands of meters long and 5 to 10 m high.

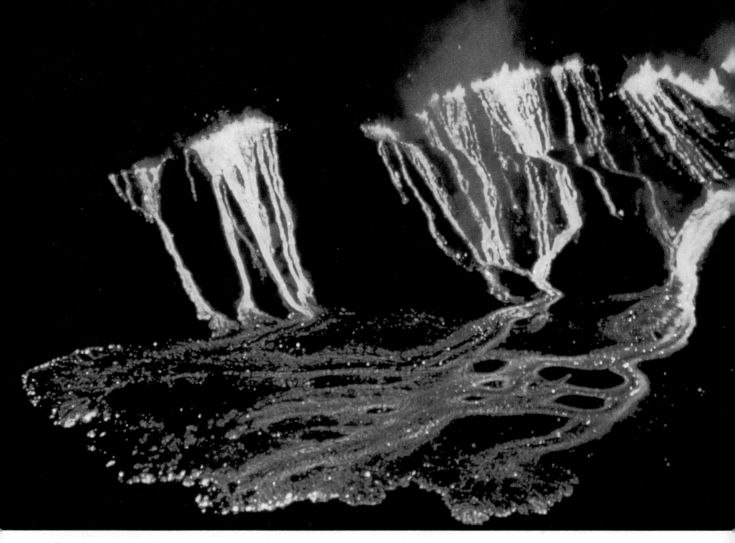

FIGURE 5–1
1959 lava flow along the east rift of Kilauea, Hawaii Volcanoes National Park. Flow is being
fed by lavafalls from Byron Ledge overlook. Photo courtesy of U.S. Geological Survey/
Hawaiian Volcano Observatory.

The longest lava tubes known are on the Moon. Long, sinuous lunar valleys called **rilles** had long puzzled astronomers, and the Hadley Rille was one of the first features visited by *Apollo* astronauts. Rilles are similar to lava tunnels on Earth, so they are thought to be lunar lava tunnels whose roofs have collapsed. They were among the first extraterrestrial volcanic features ever identified.

Pillows. When pahoehoe flows into a river, lake, or ocean, there is often an explosion as magma meets water, and billowing clouds of steam are created. If the eruption occurs deep under water, however, the pressure of the overlying water prevents the escape of steam, and the confrontation between lava and water is less violent. As lava erupts onto the seafloor, small tongues of pahoehoe break through the already solidified front of the flow; the pahoehoe chills on contact with the cold seawater to form globular masses called **pillows** (Figure 5–6).

Focus 5–1 describes an extraordinary feature of some pahoehoe flows.

Tephra

During many eruptions, particles of lava are explosively ejected into the air. These particles chill rapidly

FIGURE 5–2
Ropy surface of a solidified 1986
pahoehoe lava flow, Kalapana,
Hawaii. Photo by Allan Ludman.

FIGURE 5–3
Blocky surface of an aa lava flow on the flanks of Kilauea, Hawaii Volcanoes National Park.
Compare with the pahoehoe in Figure 5–2. Photo by Allan Ludman.

FIGURE 5–4
Mechanism by which an aa flow advances.

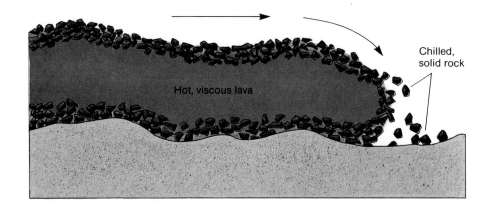

Chilled, solid rock

Hot, viscous lava

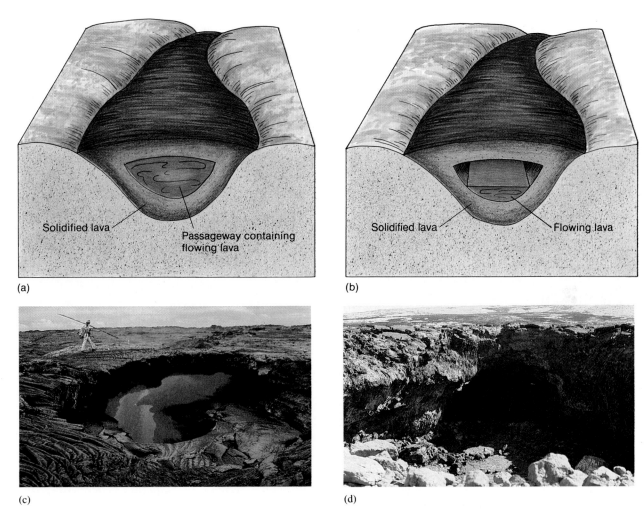

Solidified lava

Passageway containing flowing lava

(a)

Solidified lava

Flowing lava

(b)

(c)

(d)

FIGURE 5–5
Lava tube. (a) Surface of flow solidifies but lava continues to flow through the opening inside. (b) Level of lava drops in opening. (c) Partially collapsed roof of a lava tube along east rift of Kilauea showing red-hot lava within the tube. (d) Small lava tube north of Lihue on the island of Hawaii. Photo (c) courtesy of U.S. Geological Survey/Hawaiian Volcano Observatory; photo (d) by Allan Ludman.

A FOREST OF LAVA TREES

Figure 1 shows one of the most eerie and fascinating features associated with pahoehoe flows: the remains of a forest preserved in the same lava that killed its trees. This particular photograph is from Lava Trees State Park on the island of Hawaii, but similar "lava trees" occur in several other places throughout the world that are affected by highly fluid lavas.

FIGURE 1
Lava trees in Lava Trees State Park, Hawaii County, Hawaii. Photo by Allan Ludman.

Figure 2 shows how the living forest is engulfed and eventually preserved by the lava. Hot pahoehoe flows through the forest, and the lava chills as it comes into contact with the water-laden tree trunks. It solidifies around the trunks, perfectly preserving their shapes—and in some instances details of their bark and inner tissue. Lava that does not come into contact with the trees does not harden in the forest, but flows onward, leaving "stumps" of chilled lava to indicate where individual trees had been.

FIGURE 2
Origin of lava trees. Highly fluid lava flows through a forest, chilling against tree trunks but remaining fluid elsewhere. Lava drains from the forest, leaving chilled molds of the trunks.

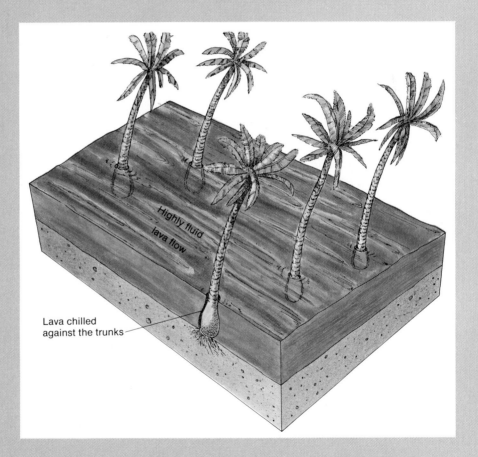

FIGURE 5–6
Bulbous and cylindrical pillow lavas at Hole-in-the-Wall, Waterton-Glacier International Peace Park, Glacier County, Montana. Photo courtesy of R. G. McGimsey/U.S. Geological Survey.

and fall as solidified or partly solidified tephra particles. Tephra ranges in size from blocks the size of trucks to dust so fine that it can pass through the air filters of cars—as we found out in 1980 when Mount Saint Helens erupted in the state of Washington. One of the best ways to visualize the force involved in a volcanic eruption is to imagine how much energy is needed to propel thousands of truck-sized blocks and tons of dust into the atmosphere. The large particles fall back to Earth and are added to the slopes of the volcano that discharged them, but the finest particles may be blasted into the upper atmosphere and travel great distances around the world (Figure 5–7).

Tephra may be rhyolitic, andesitic, or basaltic in composition. Some tephra is made of volcanic glass—lava that chilled almost instantly in the air so that minerals did not have a chance to form. Some of the most spectacular examples of this glass are translucent, golden fibers found on many volcanoes. In Hawaii, these are called *Pele's hair*, after the Hawaiian goddess of fire.

Crystals also are found in many tephra deposits. These formed when magma cooled within the Earth, but were carried upward and erupted into the air along with the unsolidified lava. Rising magma often picks up xenoliths, and these also may be ejected during an eruption. They are then incorporated with the tephra in a complex igneous rock.

The shapes of tephra particles often tell of their history. Large blocks tend to be irregular, but some blobs of lava hurled into the air cool during flight and have a streamlining that indicates aerodynamic control of their shape (Figure 5–8). Many of the smallest fragments of volcanic glass, called **shards,** have curved surfaces and sharp, angular corners. Shards form from nearly solidified frothy lava, a mixture of gas and lava similar to the foam on a glass of beer (Figure 5–9). When the bubbles burst at the Earth's surface, the fragments of the bubble walls become the shards. If hot shards stick to each other when they land, they form the **welded tuff** described in Chapter 4.

Lava or Tephra: Which Will Erupt?

The temperature and the amounts of silica and gas in a magma determine which kind of material will erupt. The hotter a magma is when it reaches the surface, the more freely it can move. Very hot magmas can produce pahoehoe lavas; cooler magmas can

FIGURE 5–7
Satellite image of western United States and Canada taken on May 18, 1980 showing spread of ash cloud from Mount Saint Helens. Photo by National Oceanic and Atmospheric Administration.

FIGURE 5–8
Volcanic bomb showing aerodynamically streamlined shape. Macmillan Publishing/Geoscience Resources photo.

produce aa. The more silicon and oxygen a magma contains, the more viscous it is. Highly siliceous magmas tend to produce very viscous (aa) lavas.

A magma's gas content determines the explosiveness of its eruption. The more gas there is, the more explosive an eruption can be and the more particles will be ejected into the air to form tephra. Remember that *the most abundant gas in any magma is water vapor*.

Table 5–1 summarizes these three attributes (silica content, water content, temperature) for the three most common magma compositions. It comes as no surprise that basalt can erupt as pahoehoe, aa, or tephra. Andesite, with lower temperature and higher contents of silica and water vapor, erupts as aa or tephra. Rhyolite, with the lowest temperature and highest silica and water contents, erupts only as tephra and exceptionally viscous forms of lava.

FIGURE 5–9
Origin of volcanic glass shards. (a) Magnified view of frothy magma. Bubbles are formed in nearly solidified magma as gases rise toward surface. (b) Bubbles burst, gases escape, and walls of bubbles are shattered. These pieces fall back to Earth to become the shards in tuffaceous rocks.

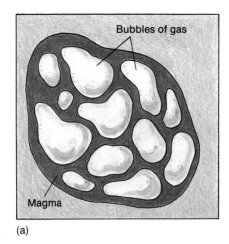

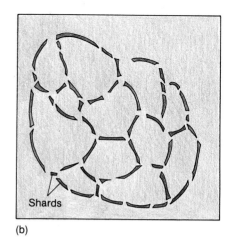

(a)

(b)

TABLE 5–1
Nature of erupted materials.

Magma Type	Temperature	Silica Content	Water Content	Material Erupted
Basaltic	Very high—high	Low	Low	Pahoehoe, aa, tephra
Andesitic	High—moderate	Intermediate	Moderate	Aa, tephra
Rhyolitic	Low	High	High	Tephra; rarely very viscous lava

VOLCANOES

A **volcano** is a mountain produced by the accumulation of volcanic materials. A volcano grows when magma leaves its subterranean chamber and rises to the surface through a roughly cylindrical passageway called a **feeder conduit**. The opening of the conduit at the surface is called the volcanic **vent,** and in most cases the vent is located at the center of the volcano, at least initially. In later stages, secondary feeders may branch off the main conduit, and thus eruptions can occur on the flanks of a volcano as well as at its center.

There are three types of volcanoes, each characterized by a unique shape and by the materials from which it is constructed: cinder cones, shield volcanoes, and composite cones.

Cinder Cones

A **cinder cone** is a volcano made of tephra. It is shaped like a cone with moderately steep slopes and generally has a bowl-shaped depression called a volcanic crater at its apex, with the vent at the center of the crater (Figure 5–10). After tephra is ejected from the vent, it falls back to the ground, with most landing near the vent itself. This produces a mound of debris that grows higher as more tephra erupts. The slope of the volcano's flanks is determined by how steeply particles can be piled before gravity causes the pile to slump; thus the volcano's shape is controlled by gravity. (The same phenomenon—called the "angle of repose"—controls the steepness of an anthill or a pile of sand.) The crater is kept free of tephra by explosions that accompany eruptions.

Cinder cones are the smallest volcanoes. They vary widely in size from small hills only a few tens of meters high to mountains several hundred meters tall; most are only a few hundred meters in diameter at ground level. Cinder cones can build very quickly, and geologists were fortunate to be able to watch one of the biggest, Paricutín, almost from the day it was born.

Volcano Paricutín, 1943–1952. Paricutín, in west central Mexico, is one of the largest cinder cones in the world, rising approximately 400 m from the plain that surrounds it. One day in February 1943, a farmer plowing a cornfield noticed hot fumes rising from a small crack in the ground. The fumes were sulfurous and were soon joined by cinders, steam, and small soil fragments hurled upward from the growing vent. The farmer was soon chased from his field by nauseating gases and tephra that fell from the sky, and by severe shaking as the ground beneath his feet began to rumble. When he returned to the cornfield the next day, he found a small cinder cone nearly 10 m high where the fumes had emerged. Within a week it had grown to nearly 130 m. Within a month, the formerly flat upland fields of the area were the home of a 300 m high, extremely active cinder cone. Eventually, Paricutín also erupted lava, not from the central crater, but from vents along its flanks.

Although cinder cones eject tephra explosively, their eruptions are not as violent as those of the much larger composite cones, the other type of volcano in which tephra plays an important role. This is due to the smaller amount of both energy and material involved. It was thus possible for the farmer—and the hundreds of visiting geologists—to stay close by and monitor the eruption of Paricutín. The volcano last erupted in 1952 and today sits quietly in the midst of the fields that it destroyed.

Shield Volcanoes

Shield volcanoes are made of lava, most of which is basaltic pahoehoe. In contrast to the moderate slopes of cinder cones, shield volcanoes have very gentle slopes (Figure 5–11) because of the action of gravity on the highly fluid pahoehoe. As it is so fluid, the lava that erupts from the conduit cannot be piled into the steep slopes that typify cinder cones. A simple experiment at home demonstrates this principle:

FIGURE 5–10
Haleakala National Park, Hawaii. (a) Group of small cinder cones in the caldera of Haleakala. (b) Cross-section through a cinder cone. Photo (a) by Allan Ludman.

(a)

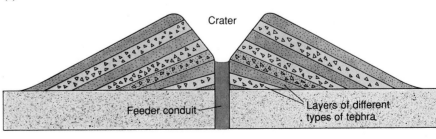

(b)

FIGURE 5–11
Hawaii Volcanoes National Park.
(a) Profile of Mauna Loa, a classic
shield volcano. (b) A shield volcano
is built of many overlapping lava
flows. (c) Cross-section through a
shield volcano showing its internal
plumbing. Photo (a) by Allan
Ludman.

(a)

(b)

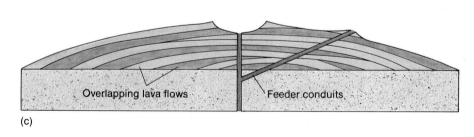

Overlapping lava flows   Feeder conduits

(c)

pour pancake syrup onto a flat surface and see how
steep a cone you can produce. Successive flows increase the size of the volcano but cannot steepen its
slopes.

Shield volcanoes are the largest ones on Earth,
and are best illustrated by the Hawaiian Islands. The
island of Hawaii alone contains five famous shield
volcanoes, including Mauna Loa, Mauna Kea, and
Kilauea (see Focus 5–2). The summit of Mauna Loa
is more than 4000 m above sea level, but most of the
volcano is under water because its base sits on the

floor of the Pacific Ocean nearly 6000 m below the
surface. The total height of the pile of lava thus is
more than 10,000 m, making Mauna Loa far taller
than Mount Everest. The diameter of shield volcanoes is much greater than that of cinder cones, reaching several kilometers. The volume of lava erupted to
form Mauna Loa makes it the largest single mountain on the Earth. Shield volcanoes are found only in
the oceans, for reasons that we shall discuss later.

Shield volcanoes also take much longer to grow
than cinder cones. In early stages, most eruptions are

at the central vent, and a large crater—called a **caldera**—forms at the summit (Figure 5–12). This happens because the lava formerly occupied a chamber directly beneath the vent. When the chamber is drained during an eruption, it is no longer able to support the cone above it. The summit then collapses, forming the caldera. During later central eruptions, the crater is filled with lava, forming a lava lake, and beautiful fiery fountains several tens of meters high are created as eruptions continue through the lake (Figure 5–13). Eventually, the lake may overflow, sending tongues of lava tens of meters wide down the flanks of the volcano.

In later stages, many eruptions take place on the flanks rather than at the summit caldera. Flank eruptions occur when secondary conduits form and lava reaches the surface at elevations below the summit (see Figure 5–11). Other eruptions may be from elongate fissures hundreds or even thousands of meters long (Figure 5–14); these produce broad sheets of lava rather than the tongues that emanate from vents. Some of the erupted lava may be aa rather than pahoehoe, depending on its preeruptive cooling history and its water content. In addition, just as some lava may erupt from a cinder cone, small amounts of tephra may be found in shield volcanoes. Indeed, the cinder cones shown in Figure 5–10 formed within the summit caldera of Haleakala on the Hawaiian island of Maui.

Composite Cones

Most of the best known volcanoes of history and mythology are composed of layers of tephra and lava

FIGURE 5–12
Aerial view of the summit caldera of Kilauea volcano, Hawaii Volcanoes National Park. The caldera measures 4.5 × 3 km; the Halemaumau crater near the far side of the caldera is itself 1 km across. Photo courtesy of J. D. Griggs/U.S. Geological Survey.

LOIHI: A NEW HAWAIIAN ISLAND GROWS IN THE PACIFIC

The appearance of a new volcanic island like Surtsey near Iceland is an exciting event because it gives us a rare glimpse of volcanic activity as it changes from submarine to subaerial eruptions. However, in some cases, we can monitor a volcano's growth even before it emerges from the sea as a new island.

Just 25 km southeast of the island of Hawaii, another volcano is disturbing the peace on the seafloor (Figure 1) and may be in the process of producing a new vacation retreat. Although its summit is still nearly 1000 m below the surface of the Pacific Ocean, the volcano already has a name:

FIGURE 1

Location of the submarine volcano Loihi, southeast of the island of Hawaii. Elevations in meters. After Figure 6.3 in Malahoff, "Geology of the Summit of Loihi Submarine Volcano," Chapter 6 in Decker, Wright, and Stauffer, eds., *Volcanism in Hawaii*, U.S. Geological Survey Professional Paper 1350, 1987.

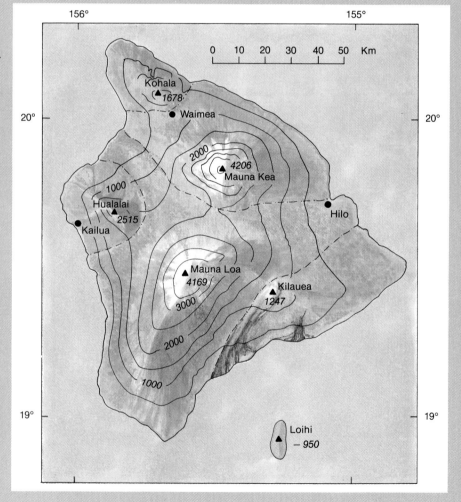

Loihi. Earthquake activity helped geologists locate the volcanic center, but bubbling water caused by the release of gases from the vent made location easy. Geologists have studied Loihi with underwater photography and submarine surveying methods.

Figure 2 shows the elongate shape of Loihi's summit and gives an idea of its size. Detailed photographic surveys of the volcano show that it is composed of tongues and sheets of chilled pahoehoe lava, with some aa flows as well. The two craters at the summit are called **pit craters,** and appear to be the result of subsidence—sort of miniature calderas. At this stage in its history, Loihi seems to be a small-scale version of nearby Kilauea, but if Loihi continues its present pattern it can be expected to produce a volcano much like its larger and older relative.

When should you start thinking of buying a condominium on the shoreline of Loihi? The best available estimate is that Loihi is a young volcano—one that began erupting perhaps half a million years ago. In that time it has grown to more than 3000 m above the seafloor, but it still has another 1000 m to go. Do not plan your Loihi vacation for the very near future!

FIGURE 2
Artist's concept of the summit of Loihi, based on submarine photography and surveying. After Figure 6.4 in Malahoff, "Geology of the Summit of Loihi Submarine Volcano," Chapter 6 in Decker, Wright, and Stauffer, eds.,

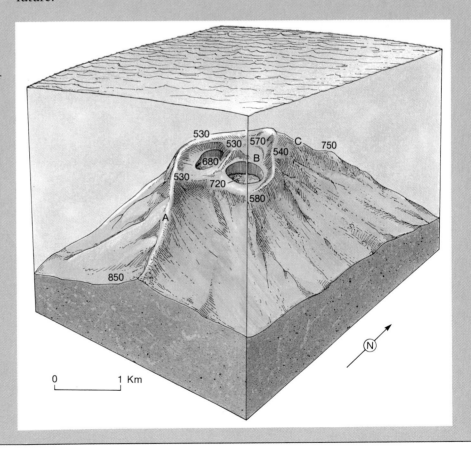

in roughly equal amounts and are called **composite cones**. Classic composite cones are Etna (Sicily), Vesuvius (Italy), Mount Fuji (Japan), Mount Rainier and Mount Saint Helens (Washington State), and most of the Aleutian Islands off the coast of Alaska (Figure 5–15).

Composite cones are intermediate in size between cinder cones and shield volcanoes. Alternating eruptions of lava and tephra produce the classic profile of a composite cone. The steepness of the profile is due to welding of tephra pieces that enables the slope to exceed the limits imposed by gravity on unwelded particles such as those of cinder cones. As these volcanoes are composed of strata (layers) of different types of extrusive rock, they are known also as **stratovolcanoes**.

The growth of a composite cone involves the two very different kinds of eruptions that yield lava and tephra. A single composite cone may erupt basaltic, andesitic, and rhyolitic lavas at different times, so pahoehoe and aa types are commonly present. The eruption of tephra from a stratovolcano is the most violent of all volcanic activity. Indeed, the 1883 eruption of Krakatau in Indonesia produced what is said to have been the loudest noise ever heard by humans; it was heard more than 3000 kilometers away!

The three case histories that follow illustrate some of the processes unique to composite cones.

Mount Pelée, Martinique, 1902. In late March 1902, hikers noticed sulfurous fumes in a small crater near Mount Pelée's summit on the Caribbean island of Martinique. Activity increased gradually for a month, with minor explosive eruptions of tephra and formation of a small cone within the crater. Fleeing the ominous mountain, residents of the surrounding countryside flocked to Saint Pierre, the largest town on the island, located 6 km from the crater, and the population of Saint Pierre swelled from 20,000 to nearly 30,000. Just before 8 A.M. on May 8, 1902, virtually all were killed as Saint Pierre was destroyed almost instantaneously by the explosive eruption of Mount Pelée.

There is some dispute about how many survived, but most accounts describe one or two people who lived to tell about the eruption. After a loud series of explosions, a large black cloud rose upward from the crater and then sank toward the ground and rolled downslope toward Saint Pierre. The cloud was a hot, glowing mass of tephra and poisonous gases—several hundred degrees Celsius. One survivor was a prisoner in the local dungeon. He was spared because his cell was partly below ground, out of the direct path of the glowing cloud, but even he was injured seriously. Gases and ash reached his cell through ventilation ducts and burned him severely in seconds.

The sudden and violent onslaught of the glowing cloud of gas and tephra is nature's version of the surge of material outward from the base of a nuclear mushroom cloud. The fiery cloud, called a **nuée ardente** ("glowing cloud" in French), is a feature

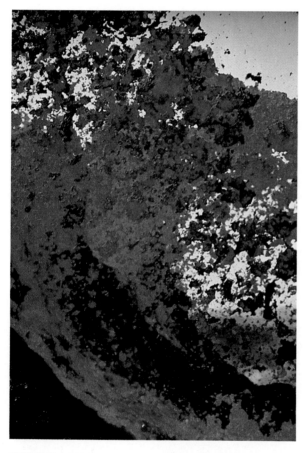

FIGURE 5–13
Arching lava fountain formed during the February 1983 eruption of Kilauea, Hawaii Volcanoes National Park. Photo courtesy of U.S. Geological Survey/Hawaiian Volcano Observatory.

unique to stratovolcanoes because of their high gas content and the viscosity of their magmas. Once discharged from the volcano, the cloud sinks to the ground because it is denser than the surrounding air, and moves downslope under the influence of gravity. On Martinique, a nuée ardente from Mount Pelée knocked down stone walls, set Saint Pierre on fire, twisted steel beams into knots, and as we said took nearly 30,000 human lives in seconds. Figure 5–16 shows a nuée ardente erupting from Mount Saint Helens in 1980.

Mount Saint Helens, Washington State, 1980. Throughout the spring of 1980, Americans watched television nightly to follow the eruptive activity of Mount Saint Helens, a stratovolcano that is part of the Cascade Range. After nearly 100 years of inactivity, a new eruptive cycle began. Sensitive instruments detected a series of small earthquakes beneath the mountain, caused by the rise of magma from deep magma chambers toward the surface. The tremors began on March 20, 1980; within a week steam and ash erupted and a new crater began to form near the summit. On March 28, ash and steam erupted explosively, rising more than 1.5 km above the volcano and generating avalanches of ash-laden snow down the east flank. By March 30, the huge tephra cloud had deposited ash as far as 250 km away.

This proved to be just a preliminary event to the major eruption. Seismologists detected continuous tremors beneath the mountain, indicating the ascent of large volumes of magma. These stopped on May 8, but not because the activity was over. Magma had arrived near the surface and the build-up of magma and gases under high pressure caused the north face

FIGURE 5–14
Curtain of fire caused by a fissure eruption along the northeast rift zone of Mauna Loa volcano, Hawaii Volcanoes National Park. Fountains are approximately 25 m high. Photo courtesy of R. B. Moore/U.S. Geological Survey.

of the mountain to bulge outward by about 100 m. It was apparent that a large-scale eruption was imminent, and evacuation warnings were issued.

On May 18, Mount Saint Helens blew its top with a roar heard 300 km away as the pent-up gas blasted into the air (Figure 5–17). The eruption was expected, but its direction was a surprise. Instead of erupting upward as anticipated, the initial blast was directed laterally, outward from where the bulge had weakened the north face of the mountain. The full force of the eruption was directed at a nearby hill where a U.S. Geological Survey scientist was monitoring the volcano from what was thought to be a safe vantage point. He was the first to report the eruption and became its first casualty as well. The shock wave toppled trees like matchsticks several ki-

FIGURE 5–15
Composite cone stratovolcano. (a) Mount Shishaldin in Alaska's Aleutian Islands shows the classic profile. (b) Profile through a stratovolcano. Photo (a) courtesy of Dr. Hannes Brueckner.

(a)

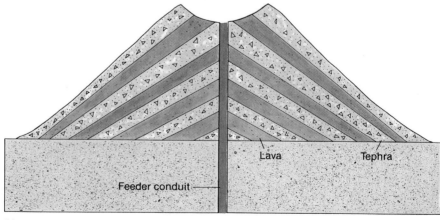

(b)

FIGURE 5−16
1968 nuée ardente rushing down the slope of Mayon volcano, Philippines. Note that tephra is ejected vertically before settling into the nuée ardente flow. Photo courtesy of R. G. Moore/ U.S. Geological Survey.

lometers from the mountain. Nuées ardentes, mudflows, and debris flows rolled down the north slope, filling the Toutle River with as much as 60 m of steaming mud. A tephra flow followed, damming the stream and creating an instant flood hazard.

Subsequent activity was controlled by the buildup and release of gas within the mountain. After the major explosion released the enormous pressure that had accumulated, a relatively quiet period ensued. Further eruptions occurred whenever sufficient pressure built up again, but these were minor compared with the blast of May 18. Eventually, viscous rhyolitic lava flowed slowly from the conduit and formed a domelike cap that effectively plugged the vent. This marked the end of the eruptive cycle. The future of Mount Saint Helens depends on a new supply of magma and the accumulation of enough gas to blast off the lava cap.

Mount Mazama, Oregon, 4500 B.C. Crater Lake National Park in Oregon may reveal some of that future. Crater Lake was originally the site of another

(a)

(b)

(c)

(d)

FIGURE 5–17
Series depicting the explosive eruption of Mount Saint Helens, August 7, 1980. Photo
courtesy of P. Lipman/U.S. Geological Survey.

composite cone in the Cascade Range, named
Mount Mazama by geologists. An explosive eruption
more powerful than that of Mount Saint Helens
blew off a large part of the volcano, and collapse of
the crater due to loss of support following the eruption created a large caldera. Renewed eruptive activity built a subsidiary cone within the caldera (Figure
5–18). Eventually, Mount Mazama ceased activity

FIGURE 5–18

Crater Lake, Oregon, fills the crater of an extinct stratovolcano called Mount Mazama. Wizard Island is a young cone built after the caldera formed. Photo courtesy of U.S. Geological Survey/Cascades Volcano Observatory.

and water gradually filled the caldera, giving Crater Lake its name. The secondary cone was isolated in the lake and is today called Wizard Island.

OTHER LANDFORMS PRODUCED BY EXTRUSIVE IGNEOUS ACTIVITY

Volcanoes are the most familiar extrusive igneous landform, but are not the only result of such activity. Small dome-shaped landforms or extensive plateaus may form instead.

Lava Domes

The solidified lava cap over the vent in Mount Saint Helens is a small example of a volcanic dome. Exceptionally viscous lavas, generally rhyolitic, cannot flow far from their vents. Instead, they ooze slowly upward and build small hills called **lava domes**. Mono Craters of eastern California, for example, are lava domes made almost entirely of obsidian (Figure 5–19). This kind of eruption can be simulated by slowly squeezing an upright tube of the glue used to hold plastic models together. The bulbous mass of glue emerges in the shape characteristic of lava domes. Many lava domes, like that in Mount Saint Helens, form late in the history of a volcano and fill earlier craters.

In very rare instances, the rising magma is so viscous that it cannot punch through the surface rocks to erupt; instead it arches the surface rocks upward. The growth of such lava domes is slow and relatively safe to watch from a respectful distance. The Santiaguito dome in Guatemala, for example, has been growing for decades, but so peacefully that geologists have climbed on it to measure its growth.

Flood Basalts

When highly fluid pahoehoe erupts in the oceans, the result is sometimes a shield volcano. When pahoehoe erupts on the continents, the result is often an accumulation of hundreds or thousands of meters of basalt that form extensive plateaus—large upland areas—without ever building a volcano. Such basalts are called **flood basalts** because of the amount of lava involved. They are found throughout the world, on every continent. The Columbia Plateau of Oregon, Washington, and Idaho consists of flood basalts that are thought to have once covered an area of 300,000 km² (Figure 5–20).

Flood basalts form by eruption of pahoehoe from fissures several kilometers long rather than from pipelike conduits. The lavas spread quietly outward and, because they erupt along an entire fissure, never pile up to form a volcano. Individual sheetlike fissure flows may cover thousands of square kilometers, and successive flows from the fissure or from others nearby may build extremely thick masses of lava. The Columbia Plateau, for example, contains hundreds of flows that total more than 600 m in thickness (Figure 5–21). Individual flows, however, may only be a few centimeters to tens of meters thick.

Ash Falls and Ash Flows

Deposits of fine-grained tephra from eruptions such as that of Mount Saint Helens sometimes form blankets that cover the countryside for great distances around the vent. The deposits that simply fall from the air are called **ashfalls;** those from nuées ardentes are called **ashflows**. Ashflows are deposited rapidly, but an ashfall may not be complete for many days.

Fine dust from the Krakatau eruption circled the world in the upper atmosphere for over a year before falling back to Earth; during that year, sunsets were spectacular because of diffraction and absorption of light through the dust, and temperatures were lower because some of the Sun's energy was blocked by the dust.

Some ash deposits are only a few centimeters thick, but cover thousands of square kilometers. Others, such as that produced by the 1912 eruption of Mount Katmai in Alaska, are over 10 m thick. Ashflow deposits from nuée ardente eruptions surround many major stratovolcanoes. Plateaus underlain by alternating lava flows and ashflows may cover thousands of square kilometers, as in the San Juan Mountains of Colorado.

FIGURE 5–19
Aerial view of Mono Craters, California (Mono Lake in foreground). Well-developed lava dome fills the crater on the right. Photo courtesy of C. D. Miller/U.S. Geological Survey.

FIGURE 5−20
Geographic extent of the flood basalts of the Columbia Plateau.

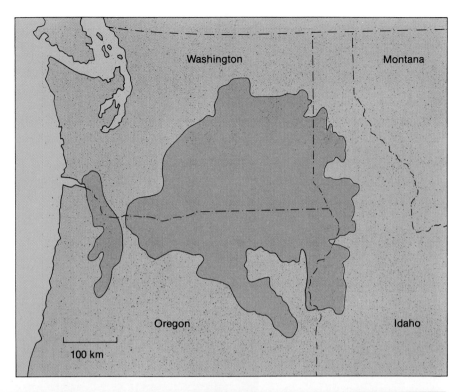

FIGURE 5−21
A sequence of basaltic lava flows on the east wall of Grand Coulee, Columbia Plateau, Washington. Photo by John Shelton.

FIGURE 5–22
Devils Tower, Wyoming, is a volcanic neck. The remainder of the volcano has been eroded. Photo courtesy of D. E. Trimble/U.S. Geological Survey.

Erosional Remnants of Volcanic Landforms

Solidified lava flows are often more resistant to erosion than are sedimentary rocks with which they may be interlayered. As a result, the igneous rocks form a protective cap and eventually become high areas overlooking lowlands where the less-resistant rocks were unprotected. Basalts of the Holyoke Range of central Massachusetts, for example, stand high above the red sedimentary rocks of the Connecticut River valley. In contrast, rock formed from blocky, poorly consolidated aa lava or from unwelded tephra particles may be more easily eroded than adjacent rocks and can be rapidly removed by erosion.

In some instances, the conduits that brought the magma to the surface are preserved, rather than the erupted material. Volcanic **feeders** are commonly preserved as dense, crystalline igneous rocks that stand up as isolated remnants when the rest of the volcano or flow has been eroded away (Figure 5–22). Some of these remnants, called **volcanic necks,** are well-known landscape features: Devils Tower in Wyoming, Ship Rock in New Mexico, Arthur's Seat overlooking Edinburgh in Scotland, and many steep-sided pinnacles in the Auvergne region of France.

VOLCANOES AND HUMAN BEINGS

Many of the effects of volcanic eruption are clearly hazardous, but others, perhaps surprisingly, are actually beneficial. We shall look at a few examples of each.

Hazards of Volcanoes

We have already seen how explosive eruptions (Mount Pelée) can destroy a modern city with nuée ardente. Heavy ashfalls can be every bit as deadly. In A.D. 79, Vesuvius erupted in Italy, burying the nearby town of Pompeii beneath a thick pile of ash and killing hundreds. A disaster for the Roman citizens of Pompeii, the event was a gift to archaeologists because the hardened ash preserved minute details of Pompeiian homes, utensils, and even the bodies and clothing of its inhabitants.

Heavy rains often accompany volcanic eruptions because the dust particles act as raindrop seeds for water released during eruption and other water already in the air. When these rains fall on unwelded tephra on the steep slopes of a volcano, an unstable mass forms that can rush downhill as a volcanic mudslide. Other volcanic mudslides occur when the hot ash melts glaciers high on the flanks of volcanoes. Such a mudslide destroyed Colombian villages in 1985 (Figure 5–23).

Lava flows themselves can be destructive. Pahoehoe flows from Kilauea have advanced through populated areas several times on the island of Hawaii. From 1987 through 1989, a housing subdivision was destroyed as pahoehoe flows first blocked roads (Figure 5–24) and then engulfed homes. Aa flows are slower, but no less dangerous. The 1973 eruption of Helgafell in Iceland produced aa that moved slowly toward the fishing port of Heimaey, giving residents a chance to evacuate before engulfing part of the town.

FIGURE 5–23
This volcanic mudslide, caused by the November 1985 eruption of Nevado del Ruiz, buried the city of Armero, Colombia. Rescuers carry a young woman through the mud and debris. Photo by Wide World Photos.

An eruption can cause damage hundreds of kilometers away from the volcano:

☐ Submarine and coastal eruptions may trigger shock waves called **tsunamis** that travel at hundreds of kilometers per hour across the ocean. When these waves (mistakenly called tidal waves, for they have nothing to do with the daily tides) strike a shoreline, they can cause

FIGURE 5–24
1987 Pahoehoe flow cutting a highway on the island of Hawaii. Photo courtesy of J. D. Griggs/U.S. Geological Survey.

tremendous damage. Most of the damage in Indonesia caused by the eruption of Krakatau in 1883 was done by tsunamis far from the volcano, where fishing villages had no knowledge of the eruption and therefore no warning.

☐ Dust and ash from Mount Saint Helens were inhaled by thousands of people. What effect will this have on their lungs and health in the future? (This hazard of course exists with any volcanic eruption that generates voluminous dust.)

☐ Hard volcanic ash in Iceland coated grass in pastures used by grazing sheep, resulting in excessive wear of the animals' teeth, the death of many, and a serious disruption of Iceland's economy.

The eruption of a single large stratovolcano may have worldwide effects by altering Earth's climate. Most volcanic dust returns to Earth within a year of its eruption, but in that year each particle reflects a tiny amount of solar radiation back into space. The amount of solar heat lost can be significant. The eruption of Krakatau cooled the atmosphere so much that it created "the year without a summer." The 1982 eruption of El Chichon in Mexico is estimated to have lowered surface temperatures worldwide by as much as 0.5°C. If this change seems trivial, imagine what would happen if several large volcanoes erupted in a single year. The effect on land plants and on plankton living at the surface of the ocean would have significant impact on the world's food chains. Loss of a single year's crops in many areas would mean starvation for millions.

Do We Have Any Defense? Most deaths due to volcanic eruptions occur because people are caught by surprise. Loss of life could thus be lessened if we were able to predict eruptions so that the danger areas could be evacuated, but how predictable are eruptions? Careful monitoring of Mount Saint Helens, Kilauea, and volcanoes in Iceland and Japan make us better able to estimate the moment of eruption.

The shift of magma toward the surface just before an eruption can be detected in several ways. Ground temperatures rise, the land surface tilts, the frequency of small earthquakes increases, and more gas is released from vents surrounding the volcano. If all of these parameters are carefully monitored, major eruptions can be predicted. However, geologists cannot usually pinpoint the exact time or the precise nature of an eruption—hence the tragic events on Mount Saint Helens.

Once an eruption has begun, can we deflect lava or nuées ardentes away from our cities? At this time, such methods are more science fiction than fact, but some encouraging experiments have been made. In Hawaii, the movement of lava toward inhabited areas has been slowed by bombing the active lava tubes to block the flows. During the 1973 eruption of Helgafell, Icelanders used high-pressure hoses to pour tons of cold North Atlantic water onto the front of the aa flow advancing toward Heimaey. They hoped to speed up cooling to the point that the lava would solidify and stop moving. The flow did indeed slow and eventually stop, but we do not know if this was due to the human effort or simply to the waning of a natural process.

Benefits of Volcanoes

Lava and tephra brought to the surface help create new land area in the oceans. After all, Iceland and the Hawaiian Islands are made entirely of volcanic deposits. As an example of creation of new land, in 1963 fortunate scientists had a front-row seat at the birth of a new volcanic island. Geologists watched the island of Surtsey appear, caused by a submarine volcano off the south coast of Iceland building its cone above sea level. A more recent example is the infancy of a volcano that will eventually become a new Hawaiian island (Focus 5–2).

Volcanoes create landforms that can be beneficial. The same lava that nearly destroyed Heimaey flowed into the Atlantic Ocean and actually improved Heimaey harbor by building a natural breakwater.

New land not only means new places for people to live. New volcanic rock is often a valuable resource because it contains many of the nutrients need for plant growth. When exposed to the air, volcanic glass and porous tephra weather rapidly and yield these nutrients to form soils that can support abundant plant life. The luxuriant plants of Hawaii reflect a fertile soil in an ideal tropical environment, but even in the inhospitable North Atlantic, plants successfully rooted on the newborn island of Surtsey within a year of its emergence from the sea. Coffee in

FIGURE 5-25
Sulfurous steam erupted from fumarole in Kilauea volcano, Hawaii Volcanoes National Park. Crystallized sulfur forms a light-colored coating on the otherwise black basalt. Photo by Allan Ludman.

Central America, rice in Indonesia, and vegetables in the Philippines all grow on soil derived from volcanic rock.

Volcanoes also create mineral resources. They often erupt gases when no lava or tephra is being extruded; such gas-producing vents are called **fumaroles**. Sulfurous gases are released by many fumaroles, and pure sulfur deposited from these vapors coats many cracks and fissures (Figure 5-25). Volcanoes produce only a small percentage of the world's supply of sulfur, but these deposits are important resources in Chile and Bolivia. Copper in the Chilean Andes and the Keweenaw Peninsula of Michigan is mined from volcanic rock, as are iron in Argentina and gold in Colorado.

Volcanic rock itself also can be useful. Pumice is light in weight but quite strong; as a result it is used to strengthen plaster, stucco, cement, and some cinder blocks. Volcanic ash sometimes alters to **bentonite,** a group of clay minerals that swell enormously when water is absorbed into their sheet structures.

Bentonite thus is used as a sealer for artificial lakes and reservoirs. Mixing bentonite with water produces a mud with just the right density, viscosity, and lubricating ability to be valuable in drilling for oil. This "drilling mud" is injected down the well to lubricate the drill and carry some of the drilled particles away, just as one lubricates a drill hole when boring through hard ceramics or bricks at home. (On the other hand, the ability of bentonite to swell when wet creates problems in land stability and may cause landslides. We will discuss these in chapter 10.)

VOLCANOES AND PLATE TECTONICS

The appearance of new volcanic islands such as Surtsey and the disappearance of others (Krakatau, Santorini) is clear evidence that the Earth is a dynamic planet and proves that the static Earth model mentioned in Focus 1-2 must be incorrect. One of the reasons that plate tectonics has been so widely accepted is that the theory explains several aspects of volcanicity. Prior to plate tectonics, no theory adequately explained both the distribution and composition of erupted materials.

Geographic Distribution of Volcanoes

Active volcanoes are not scattered randomly on the Earth's surface; they occur in belts in some areas and are absent from others. For example, the western coasts of North and South America contain several active volcanoes, whereas there are none on the eastern coast of either continent. The entire Pacific Ocean basin is surrounded by chains of active volcanoes, often called the "Ring of Fire" (Figure 5–26). By contrast, the margins of the Atlantic are nonvolcanic. The ocean ridges found near the middle of every ocean basin are actually enormous chains of submarine volcanoes. At the very least, the locations of the world's volcanoes indicate areas of high heat flow and relatively easy access to the surface by magma.

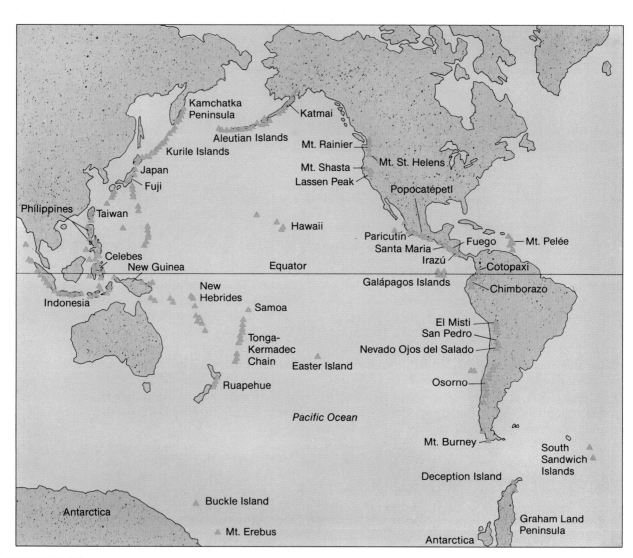

FIGURE 5–26
Location of active volcanoes in the Pacific Ring of Fire.

The plate tectonics theory explains the distribution of volcanoes very well. Compare Figure 5–26 with the map of Earth's plates (Figure 1–6). Chains of volcanoes coincide with both convergent and divergent boundaries, as predicted by the hypothesis. Friction as one plate is subducted beneath another at a convergent boundary generates enough heat to cause volcanism. This explains the Pacific Ring of Fire and several other chains of volcanoes. Seafloor spreading at divergent boundaries provides fractures that ease the rise of magma at ocean ridges. Plate tectonics equally explains the absence of volcanoes: most areas lacking volcanic activity turn out to be the centers of plates, places far removed from either subduction or seafloor spreading.

Distribution of Volcanic Rock Types

The plate tectonics model also explains the compositions of lavas and tephra erupted in different tectonic settings. In chapter 4 we showed that andesitic rocks are the result of partially melting a subducted plate; we would expect to find andesitic volcanism only at convergent boundaries, and that is exactly the area to which it is restricted. Andesites comprise a large proportion of the rocks formed above subduction zones, along with varying amounts of basalt and rhyolite. In contrast, volcanism at the ocean-ridge spreading centers is almost entirely basaltic. Andesites are not found at any ocean ridge. Volcanoes on the continents can be attributed to either subduction of a plate beneath the continent, as in the case of the Andes Mountains of South America, or to rifting leading to formation of a new ocean basin, as in East Africa.

A few volcanic islands contradict the generalization that volcanoes form at plate boundaries. The Hawaiian chain is far from the nearest ridge or subduction zone. It lacks the andesites characteristic of subduction zones, and its overwhelmingly basaltic composition is different enough from the basalt of ocean ridges to suggest a different tectonic setting. For reasons that we do not fully understand, there appear to be narrowly focused sources of heat in the mantle called **hotspots** that cause localized melting of basalt in the asthenosphere. The basalt rises through narrow conduits to erupt *anywhere* on a plate, even in the center as in the Hawaiian islands.

We shall discuss such hotspot-produced volcanoes in more detail in Chapters 16, 18, and 20.

SUMMARY

1. Landforms produced by extrusive igneous activity include volcanoes, plateaus underlain by lava or tephra, and erosional remnants such as volcanic necks. These landforms are made of lava, tephra, or a combination.

2. The nature of an eruption depends on the composition, temperature, and water content of the magma involved. The more water there is, the greater is the probability of an explosive eruption. Highly siliceous rhyolitic magmas produce either explosive tephra eruptions or peaceful, extremely viscous lavas. Andesitic magmas yield explosive tephra or aa lavas. Basaltic magmas can generate pahoehoe, aa, or tephra in either violent or "quiet" eruptions.

3. Volcanoes are mountains built of lava and/or tephra, and their shapes are controlled by the action of gravity on the erupted materials. Cinder cones are moderately steep-sided volcanoes composed mostly of tephra. Shield volcanoes are broad, gently sloping structures made mostly of basalt. Composite cones (stratovolcanoes) are a composite of lava and tephra, often in alternating layers.

4. Certain features and processes are associated with particular kinds of eruptions. Pillows and lava tubes form during pahoehoe eruptions, lava domes during eruption of highly viscous rhyolite, and nuées ardentes during violent stratovolcano eruptions.

5. Lava plateaus and ashfall blankets result from eruptions at linear fissures rather than from the pipelike vents of volcanoes. Individual flows or falls may be thin, but hundreds of such deposits may lead to accumulations of igneous rock.

6. Volcanoes threaten humans in several ways. Direct effects, such as lava flows, heavy ashfalls, and nuées ardentes, are a hazard near the vent, but tsunamis and climatic changes can affect a much larger area.

7. Volcanism can be beneficial by creating new land in the oceans, providing the raw materials from which fertile soil is formed, providing mineral resources, and providing useful volcanic rock such as pumice.

8. Most volcanic activity takes place at ocean ridges and subduction zones, and chains of volcanoes help define boundaries between the major plates. Some midplate volcanism occurs above mantle hotspots.

REVIEW QUESTIONS

1. What are the different kinds of lava? Describe their properties and the reasons for the differences.
2. Why is it more likely that lava tubes and pillows will form during eruption of pahoehoe than of aa lava?
3. Basaltic magma erupts as pahoehoe, aa, or tephra, but rhyolite generally erupts as tephra. Explain this difference.
4. Explain how gravity ultimately controls the shapes of volcanoes.
5. Describe the several roles played by water in creating and controlling volcanic eruptions.
6. Why are some volcanic eruptions more violent than others?
7. What are the dangers inherent in living near a volcano on a continent? What additional problems arise if you live near an island volcano?
8. Discuss the processes by which a caldera forms.
9. What differences would you expect between the volcanic activity on Iceland and that in the Cascade Range of the northwestern United States?
10. Draw profiles of each of the three types of volcano to scale, and explain why the three shapes are so different.

FURTHER READINGS

Decker, R., and Decker, B. 1981. *Volcanoes*. San Francisco: W. H. Freeman & Co. (A general treatment of volcanism and its resulting landforms.)

Francis, Peter. 1976. *Volcanoes*. Baltimore: Penguin Books, 368 p. (A well-written explanation of volcanic activity with case histories of some classic eruptions and their effects on people.)

Green, N., and Short, N. M. 1971. *Volcanic landforms and surface features: A photographic atlas and glossary*. New York: Springer-Verlag. (A collection of photographs showing volcanoes and volcanic features from around the world.)

Oakshott, G. B. 1976. *Volcanoes and earthquakes: Geologic violence*. New York: McGraw-Hill Book Co., 143 p. (A short paperback presenting types of eruptions and the plate-tectonics reasons for volcanicity.)

Takahashi, T. J., and Griggs, J. D. 1987. Hawaiian volcanic features: A photoglossary. In *Volcanism in Hawaii*, U.S. Geological Survey Professional Paper 1350, p. 845–902. (A series of stunning color photographs that illustrate many features of Hawaiian volcanoes.)

Thorarinsson, S. 1967. *The new island of Surtsey*. New York: Viking Press. (An interesting view of the evolution of the volcanic island Surtsey by a geologist who watched it grow and who was among the first to set foot on it.)

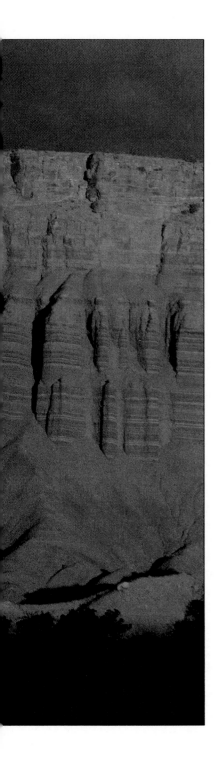

6

Sedimentation and Sedimentary Rocks

Moenkopi Formation near Capitol Reef National Park, Utah. Photo by Stephen Trimble.

Sedimentary rocks provide a fascinating picture of past surface conditions on the Earth. They do so because they form at or near the surface and have not been melted or severely altered by the high pressures and temperatures that produce igneous and metamorphic rocks. Sedimentary rocks may record the erosion of a mountain range, the drying up of an ocean basin, or the spread of glaciers into temperate latitudes millions—or even billions—of years ago. The remains of ancient plants and animals are incorporated commonly into the sediments that become sedimentary rocks. Their remains are thus preserved, and provide a record of the life that existed on the Earth's surface in the past.

The formation of a sedimentary rock involves a number of processes. Rocks exposed at the surface react with the atmosphere and hydrosphere to produce loose particles called **sediments**. These sediments are eroded by water, wind, and ice and usually are transported away from their point of origin. Sediments accumulate layer by layer, along with minerals deposited from solutions and the remains of any organisms or plants that lived in the depositional area. Subsequent physical and chemical changes transform the sediment layers into sedimentary rock. **Sedimentary rocks** thus are formed from sediments at or near the Earth's surface by a combination of physical, chemical, and biological processes.

SEDIMENT-FORMING PROCESSES

Sediments are composed of three major types of materials: rock and mineral particles derived from the breakdown of older materials, particles produced through the life activities of plants and animals, and crystals that precipitated from concentrated solutions at or near the Earth's surface. The sediment at any one place on the Earth's surface may be composed of any combination of these three components (Figure 6–1).

Weathering of Older Materials

When older materials (such as rocks, sediments, or shells of organisms) are exposed to the atmosphere, surface waters, and activities of plants and animals,

they eventually disintegrate into fragments. Some of the original minerals in the rock particles also may decompose to form new minerals. They may free soluble ions as well. All of these changes in exposed rock are called **weathering** (described in more detail in Chapter 9).

A major agent of fragmentation is the alternating freeze-and-thaw of water in a crevice, which can break solid rock into pieces—ice occupies 9% more volume than water and is a powerful wedge. Another agent is plant roots; once they get a toehold in a crack, they act as wedges as they grow, and they too will break apart a rock.

Sediments derived primarily from the physical breakdown of other materials are called **clastic** sediments (Figure 6–1). The word *clastic* is from the Greek *klastós*, meaning "broken in pieces." Most clastic particles are derived from weathering that occurs on land. These particles are composed in part of silica, and thus are referred to as **siliciclastic**. Other types of clastic particles are neither silicic nor originated on land; an example is the shell fragments found along many shorelines, particularly in tropical areas.

At the same time, chemical processes may alter some of the minerals present. For example, when the feldspars in granite are exposed to acid-bearing water, they change to clay minerals, which crumble easily. The combination of physical fragmentation and chemical change slowly reduces rock to a mixture of clastic particles that may range in size from boulders to the minute particles comprising mud. These particles can be removed by water, wind, or ice, to be transported elsewhere and accumulate as layers of clastic sediment. The soluble ions produced by weathering are washed from the soil and transported by streams to the oceans and into desert basins on the continent, such as California's Death Valley, where they are deposited by chemical processes.

Biologic Activity

Sediments manufactured by plants and animals are called **biogenic** sediments. The most abundant biogenic particles are the shells of marine organisms. These frequently make up a large portion of the sediments accumulating along shorelines and on the deep ocean floors. In tropical waters, such as along

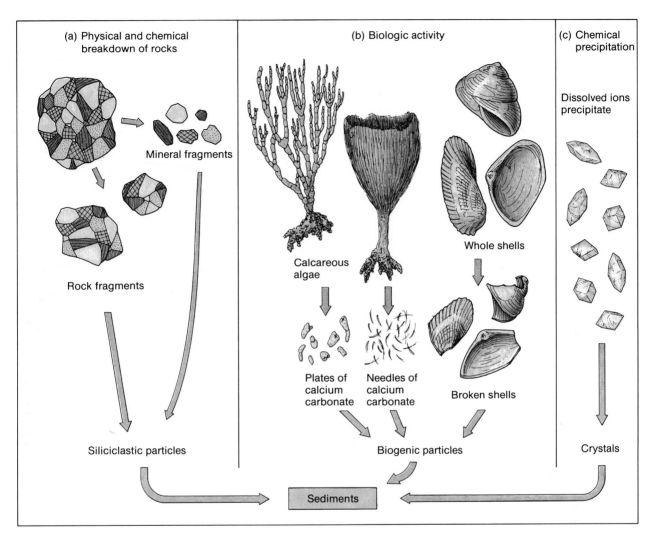

FIGURE 6–1
Sediments are composed of various mixtures of (a) siliciclastic particles from the physical and chemical breakdown of older materials, (b) debris from biologic activity, and (c) crystals chemically precipitated from concentrated solutions. These are the components that form sedimentary rocks.

the shore in southern Florida and the Bahamas, most sediment is of biogenic origin. The warm, clear, shallow waters in these areas provide optimum conditions for the growth of many organisms that secrete calcium carbonate.

Calcium carbonate may be in the form of calcite or its polymorph, aragonite. Muds composed of aragonite crystals accumulate in the quieter areas. These

small crystals were once thought to have crystallized directly from seawater, but recent studies have shown that calcareous algae remove calcium carbonate from seawater and incorporate it into their skeletons. When the algae die, the needlelike aragonite crystals that made up the strongest parts are deposited on the ocean floor (Figure 6–2). Other types of calcareous algae produce plates of aragonite, which

FIGURE 6–2
Calcareous algae and mud on floor of Florida Bay in the Florida Keys. These algae produce the calcareous silt and sand-sized particles that cover the bottom. Water depth about 3 m. Photo by Nicholas K. Coch.

form coarser-grained biogenic sediments—Figure 6–1(b).

Coral reefs are another common source of biogenic sediments in tropical areas. Waves smashing against the reef erode pieces of the coral and break up shells of other animals that live within the reef to produce a coarse-grained biogenic sediment.

Biogenic sediments also form and accumulate on

FIGURE 6–3
Evaporite deposits at the Devil's Golf Course on the floor of Death Valley National Monument, California. Evaporites have been etched into irregular shapes by rain and wind. Photo by Nicholas K. Coch.

DEVILS GOLF COURSE

INTERBEDDED SALT AND WATER-BEARING GRAVELS ARE MORE THAN 1,000 FEET THICK BENEATH THE DEVILS GOLF COURSE. GREAT HORIZONTAL FORCES EXERTED BY CRYSTALIZING SALT, PUSH THESE COLUMNS UPWARD WIND AND RAIN CARVE THEM INTO FANTASTIC SHAPES.

FIGURE 6–4
Geologic section showing deposition of evaporites in a basin partially separated from the ocean by a barrier.

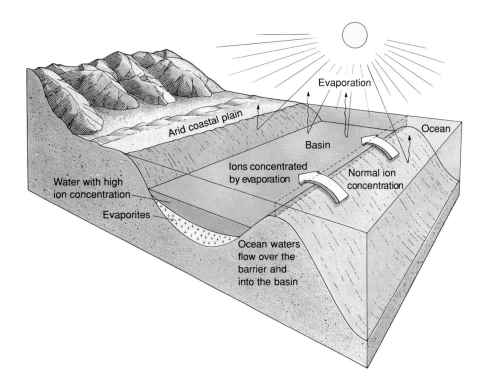

land. **Peat** is a biogenic sediment formed from plant materials that accumulate in swamps. After peat is buried, it begins a slow transformation into coal—a process to which we will return later in this chapter.

Chemical Precipitation

Chemical sediments are composed of crystals that have precipitated out of concentrated solutions. Ions freed from minerals during weathering are carried from their source by water. These dissolved ions may be deposited when either chemical changes or a physical change such as evaporation make certain ions less soluble. The ions then precipitate as crystals that accumulate as a layer of *chemical sediment*. Precipitation continues until the concentration of the remaining ions becomes low enough for them to remain in solution.

Chemical sediments deposited from concentrated solutions are called **evaporites**. Evaporites form in a wide variety of environments. Some form in hot, dry desert valleys such as Death Valley in California, where evaporites are deposited around the edges of temporary lakes as the lakes undergo evaporation. Eventually, an evaporite deposit covers the area formerly occupied by the lake (Figure 6–3). At White Sands National Monument, New Mexico, underground water reaches the surface and evaporates, precipitating crystals of gypsum ($CaSO_4 \cdot 2H_2O$). Wind subsequently erodes the gypsum crystals and builds them into the white sand dunes for which this area is famous. Evaporites also are forming in salt flats and ponds along dry, hot coastlines such as those of the Red Sea.

Some thick deposits of ancient evaporites are believed to have accumulated on the slopes and floors of deep marine basins that were separated from the open ocean by a submerged barrier. This barrier allowed only a minimal exchange of water from the ocean into the basin (Figure 6–4). Ocean waters carrying a normal concentration of ions entered through and over the submerged barrier. As evaporation occurred within the basin, the dissolved ions reached such a concentration that crystals began to grow from solution and accumulate on the seafloor. This process is producing modern evaporite deposits in the Red Sea and the Persian Gulf, both of which

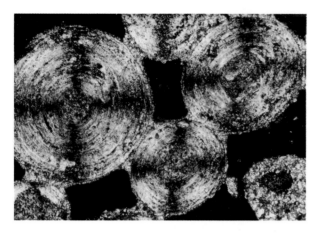

FIGURE 6–5
Photomicrograph in polarized light of ooid cross section, Bahamas. The concentric layers of calcium carbonate result from deposition of aragonite on all sides of the ooid as it rotates in turbulent tropical water. Photo courtesy of Peter A. Scholle.

have only limited connection with the ocean. Laboratory studies show that as evaporation proceeds, different compounds are deposited as their ion concentrations increase. Theoretically, calcium carbonate precipitates first, followed by gypsum and halite.

Another type of chemical deposit in some tropical marine sediments is the **ooid,** a concentrically layered spherical particle of aragonite. Ooids form in shallow, warm, agitated seawaters where calcium carbonate is deposited around a nucleus of a shell fragment, quartz grain, or another "seed" particle. This calcium carbonate is a chemical sediment because it forms from the direct precipitation of ions out of seawater rather than through the activities of living organisms. Concentric layering in ooids (Figure 6–5) results from the equal deposition of calcium carbonate on all sides of the ooid as it is tossed around in the agitated water.

SEDIMENT CHARACTERISTICS

Now that we have briefly considered the different ways in which sediments form, we can look at their distinguishing characteristics. These different properties provide the geologist with clues to the environmental conditions in which the sediments accumulated.

Composition

The most important control of siliciclastic sediment composition is the *mineralogy* of the source rocks. Sediment mineralogy is quite variable. For example, weathering of a granite yields a sediment containing quartz grains, but weathering of a basalt yields a sediment lacking in quartz, because quartz is not present in basalts. Further, the duration and intensity of chemical weathering in the source area can significantly change the mineralogic composition of a sediment.

Even after the sediments have been transported from their source rocks and deposited, weathering processes continue to modify their mineralogical composition until they are buried under later deposits, shielded from the atmosphere and hydrosphere. If the rate of burial is fast, it is possible to preserve more of the easily weatherable minerals. If the rate of burial is very slow, weathering may destroy some minerals and thus increase the concentration in the sediment of more stable minerals, such as quartz. (The mineral quartz is very resistant to weathering and abrasion during transport, under most conditions; this accounts for its great abundance in siliciclastic sediments.) In addition, new minerals such as clays appear due to the chemical alteration of feldspars.

Thus, just as magmatic differentiation can produce many different types of igneous rocks from one magma (see Chapter 4), differences in weathering intensity and duration, changes during sediment transport, and the differing rates of burial can produce several different types of sediments from the same source rock. Whereas there are a great number of minerals in metamorphic and igneous rocks, there are far fewer in sedimentary rocks. This is because many of the minerals in igneous and metamorphic rocks are unstable under surface conditions and are transformed into more stable minerals during weathering. Consequently, the great majority of particles in sedimentary rocks are composed of comparatively stable quartz, feldspar, calcite, iron oxides, clay minerals, and rock fragments.

The mineralogic composition of a sediment does not always remain fixed, even after burial. Later in this chapter we will discuss additional mineralogical changes which can occur as buried sediment is transformed into sedimentary rock.

TABLE 6-1
The Wentworth grade scale.

Particle Size Range (mm)	Name of Particle	Name for Sediment Composed of that Particle Size
>256	Boulder	Boulder gravel
256–64	Cobble	Cobble gravel
64–4	Pebble	Pebble gravel
4–2	Granule	Granule gravel
2–0.063	Sand	Sand
0.063–0.004	Silt	Silt
<0.004	Clay	Clay

Grain Size

The sand beneath your feet as you walk across a beach can be quite coarse, whereas that in adjacent dunes is much finer grained. Similarly, the boulders at the base of a cliff along a lake are very different from the mud scooped out of the center of the lake. Obviously, sediment particle size is quite variable.

The **Wentworth grade scale** (Table 6–1) uses the common terms **boulder, cobble, pebble, granule, sand, silt,** and **clay** to describe the sizes of sediment particles. Terms such as fine, medium, and coarse are used to subdivide the sand, silt, and clay classes.

The terms of the Wentworth grade scale refer only to *particle size;* they imply nothing about the composition of the sediment. The Wentworth grade scale, when used with mineralogic information from other sources, allows us to describe particles with precision. For example, the composition of sand-sized particles on beaches in different areas can be described using terms such as *quartz sands* (Virginia), *calcium carbonate sands* (South Florida), and *olivine sands* (Hawaii). In a similar fashion, the size term "clay" refers only to particles smaller than 0.004 mm.

To avoid confusion, it is important to note that a clay-*sized* particle is not necessarily a clay-*mineral* particle (see Chapter 3). A clay-*sized* particle could be composed of any mineral. For example, a *calcareous clay* would be composed of calcium carbonate particles smaller than 0.004 mm.

Sorting

Why is there such a range of particle size in sediments from different areas? One factor is the type of source rock. For example, weathering of a coarse-grained granite will yield larger quartz particles than weathering of a fine-grained granite (assuming that weathering and transport conditions are the same for each granite). The type of transporting agent also exerts a major effect, as we shall show later in this chapter.

The range of particle size in a deposit is called the **sorting**. *Well-sorted sediments* have particles of the same size, whereas *poorly sorted sediments* have a wide range of particle sizes.

Roundness

The particles in many sedimentary rocks have rounded edges. This is because the originally angular sedimentary particles have been rounded by abrasion during transport. The degree of development of rounded edges on sedimentary particles is called **roundness**. Softer minerals, such as calcite and gypsum, become rounded at a faster rate than harder quartz particles do. Furthermore, roundness increases with the number of collisions that sedimentary particles have as they are being transported.

Color

A dramatic display of colored sedimentary rocks can be seen from the edge of the Grand Canyon. Some are green, gray, or black; others are brilliant white, tan, orange, or red.

Why is there such a diversity of color in sediments and sedimentary rocks? In some rocks, color reflects the color of the original minerals. For example, a feldspar-rich rock derived from a granite may have a pink color because of the salmon-pink color of the orthoclase feldspar in the rock. Similarly, a rock composed entirely of quartz may have a white color.

In many rocks, however, the color is not caused by the original minerals, but rather by the *weathering products* of those minerals. For example, when black iron-bearing minerals such as biotite and hornblende are deposited in a sedimentary layer, they react with oxygen and water in the sediment to form brightly colored iron oxides such as hematite (red) and limonite (yellow and orange). Most sedimentary rocks contain less than 5% iron, but weathering of only a portion of even this small amount of iron is sufficient to impart a brilliant red color to the rock. Sedimentary rocks with high percentages of organic material tend to be gray, bluish gray, or black. In general, finer-grained rocks have more intense colors than coarser-grained rocks of the same composition.

TRANSPORT AND DEPOSITION OF SEDIMENT

Water rushing down a mountain stream, a mudflow moving down a desert canyon after a rainfall, wind blowing sand into dunes, and sediment-laden glaciers moving slowly across the land—each can transport large quantities of sediment. However, each of these transporting agents moves sediment particles in different ways.

Types of Transport

In wind, glacial, and mudflow transport, most of the sediment is carried as discrete particles suspended in the flow. In glaciers, particles may be carried on top of the glacier as well as within it. Water transport, however, involves a greater variety of physical and chemical mechanisms.

The mechanisms of water transport are visible in any shallow stream. Larger particles, such as gravel, may slide or roll across the bottom. In contrast, sand-sized particles may bounce along the bottom. Silt- and clay-sized particles mostly are suspended within the flow; it is these suspended particles that make water cloudy. At the same time, ions are being carried invisibly in solution within the flow. (The presence of the ions can be proved if you boil away a volume of water. When all the water has evaporated, a crust of minerals made up of formerly dissolved

ions forms a ring or coating on the inside of the receptacle.)

Water- and wind-transported sediments are deposited as the kinetic energy of the flow decreases. When the particles settle out, they accumulate on the surface as a layer. Particles in mudflows are deposited when water is lost through evaporation or infiltration into the ground. Glacially transported particles are deposited when the ice melts and the material carried within and upon the glacier settles to the ground or is washed beyond the glacier by meltwater.

Effects of Sediment Transport on Sediment Characteristics

The type of sediment transport has a great influence on the size and sorting of the sediment:

☐ Glaciers are solid and have a strength that enables them simultaneously to carry boulders as well as finer material. Consequently, glacial deposits have some of the largest particle sizes and poorest sorting.

☐ Stream deposits can be quite variable in grain size, depending on the energy of the stream, but generally are much better sorted than glacial deposits. A slow-moving stream deposits only sand and finer material, whereas a stream in flood may deposit gravel as well.

☐ Wind can carry only sand and finer-grained sediment because it has only 1/40 the density of water. Consequently, wind-deposited sediments usually are fine-grained sands with very good sorting.

The roundness of sedimentary particles also is related very closely to the type of sediment transport. We mentioned that, as particles collide with each other, their rough edges are rounded. Wind-transported particles have the greatest opportunity for colliding with each other because air offers little resistance to movement. Consequently, particles deposited by wind have the greatest roundness—Figure 6–6(a). Stream-transported particles have intermediate roundness because particles must push water aside in order to collide; this lessens the force of impact. Grains carried in mudflows or frozen in glacial ice have little opportunity to abrade each

(a)   (b)

FIGURE 6–6
Photomicrographs of quartz grains from (a) a dune and (b) glacial till. The differences in
rounding reflect the greater incidence of grain collisions in wind deposits compared to those in
glaciers. Macmillan Publishing/Geoscience Resources photos.

other, and consequently glacial particles often are
quite angular—Figure 6–6(b).

SEDIMENTARY ENVIRONMENTS AND FACIES

Sediments are deposited in many different places.
Figure 6–7 illustrates some common environments
in which sediments accumulate: a stream, a delta
where the stream enters a lake, the lake itself, a
marsh, a beach, and dunes. Each of these areas is a
sedimentary environment—an area with identifying physical, chemical, and biological characteristics,
and containing a distinctive sediment type.

The type of sediment in any particular environment reflects the conditions in that area. Fast-
flowing stream water carries sand and gravel along its
bed, whereas slower-moving water may be cloudy
with suspended sediment. When the stream reaches
the quiet water of the lake, its velocity decreases and
it deposits sand and coarser sediment in a delta. Finer
silt and clay particles, along with organic particles,
are carried in suspension out into the lake. These fine

particles eventually settle through the lake waters to
form dark muds on the lake bottom. Occasional
stream floods may transport fine sand particles in
suspension far into the lake, leaving thin sand layers
between the dark mud layers normally deposited
there. Some of the coarse material supplied by the
stream is moved along the shore of the lake to form
sandy beaches. Winds blowing across these beaches
remove the finer sand, building it into dunes behind
the shore. On the opposite shore, dense growths of
vegetation trap fine-grained sediments and organic
material in the marshes.

In this example, we have considered only a few of
the many possible sedimentary environments. More
examples of sedimentary environments and their associated deposits are shown in Figure 6–8. We will
describe all of these environments in greater detail in
subsequent chapters.

The distinctive type of sediment deposited in each
sedimentary environment shown in Figures 6–7 and
6–8 is a **facies**. Each of the sedimentary facies grades
laterally into other facies which are being deposited
simultaneously in adjacent environments. For example, the *deltaic facies* in Figure 6–7 grades into the

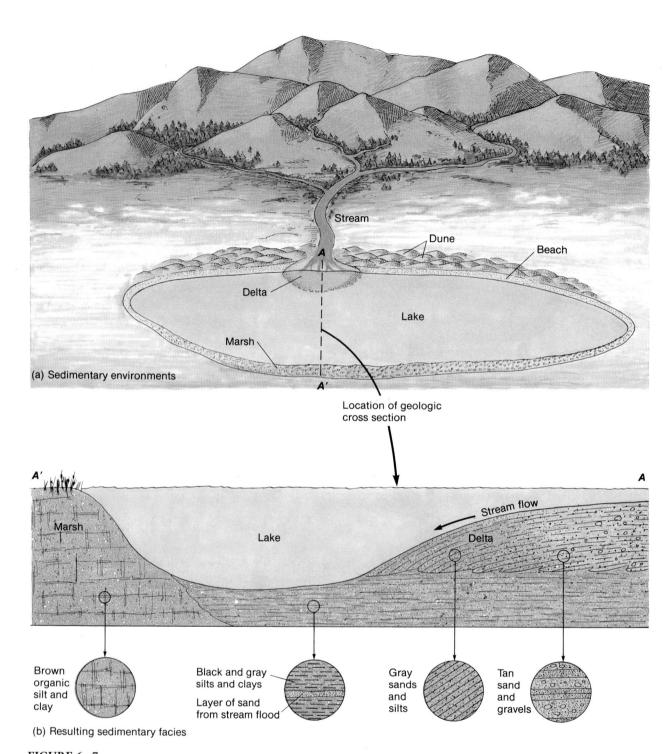

Stream

Dune

Beach

A

Delta

Lake

Marsh

A'

(a) Sedimentary environments

Location of geologic
cross section

A'

Marsh

Lake

Stream flow

Delta

A

Brown
organic
silt and
clay

Black and gray
silts and clays

Layer of sand
from stream flood

Gray
sands
and
silts

Tan
sand
and
gravels

(b) Resulting sedimentary facies

FIGURE 6–7

Relationship of (a) sedimentary environments and (b) resulting sedimentary facies (viewed in
cross section along line *A –A'*).

lake facies. Eventually, these may be buried and transformed into distinctive *rock facies*. In a similar fashion, ancient rock facies that grade laterally into each other must have been deposited at the same time. This fact is an extremely important tool used in establishing the geologic history of an area. We shall return to it in Chapter 8, when we describe how the ages of rocks are determined.

Sediments deposited in one environment may be eroded and reworked again in a different environment. For example, glacial deposits in cliffs may be eroded by coastal waves, and streams may erode banks composed of older lake deposits. Thus, individual grains of sediment may bear the imprint of several previous sedimentary environments. As these cycles of erosion, transportation, and deposition repeat over a long period of time, the original sediment changes more and more from its original character.

LITHIFICATION

Eventually, sediment reaches an area where it is protected from further reworking and accumulates as a layer. As the layer becomes buried under successively younger deposits, it is subjected to higher temperatures and pressures. The sediment then undergoes a series of transformations that convert it into a sedimentary rock. The process of turning a sediment into a rock is called **lithification**. The major physical and chemical changes that occur during lithification are shown in Figure 6–9. All of these changes lithify the sediment, but some change the mineralogy and physical characteristics of its component grains. The most important mechanisms of lithification are compaction, cementation, and matrix formation.

Compaction

The most important process in lithification is the squeezing together of particles by the pressure of more recently deposited sediments on those below. This process is called **compaction**—Figure 6–9(a). Most sediments initially have significant amounts of open space between their particles. The volume of these open spaces, or **pores,** compared to the total volume of the rock is called the **porosity**. Initially, the sediment may have a high porosity, and its pores

may be filled with air or water. As the sediment is buried progressively by younger sediments, the pressure on its grains increases. The particles then rearrange more compactly, resulting in a decrease in porosity and expulsion of most of the fluids that originally filled the spaces. Compaction thus results both in a decrease in porosity and a decrease in thickness of the layers—Figure 6–9(a).

Cementation

Sediments can be lithified by the growth of crystalline materials between the particles in a process called **cementation**—Figure 6–9(b). Common cementing agents include silica (SiO_2), calcite ($CaCO_3$), and iron oxide (Fe_2O_3). Cementing agents in sedimentary rocks can have a number of different origins. They can be derived from the liquids trapped within the pores at burial, from fluids circulating through the pores after deposition, or from the solution and reprecipitation of sedimentary particles themselves.

Matrix Formation

Particles in some sedimentary rocks are not lithified by cementation but by the solidification of a fine-grained material called the **matrix**—Figure 6–9(c). In some sedimentary rocks, the matrix originates from fine-grained particles that were part of the sediment when it was deposited. In other rocks, at least part of the matrix is believed to form after burial by the chemical alteration of some of the smaller particles to form new matrix grains.

Matrix formation is a mineralogic change that lithifies some sediments, but mineralogic changes that do not result in lithification may occur as well. Some of the original sedimentary particles may even be removed by solution during lithification. For example, acidic waters circulating through sedimentary pores may dissolve shell material composed of calcium carbonate. Another nonlithifying mineralogic change is the conversion of a mineral to a more stable form. Many present-day aquatic organisms secrete calcium carbonate shells composed of the mineral aragonite. Shells found in sedimentary rocks are not aragonite but calcite, because aragonite commonly changes to its more stable polymorph, calcite, during lithification.

(a)

(b)

(c)

(d)

FIGURE 6–8

Some common sedimentary environments: (a) Glacial environment at the terminus of the Athabaska Glacier, Canada. (b) Desert alluvial fan and playa lake—Death Valley, California. (c) River—Brazos River, Texas. (d) Estuary, tidal channel, flat, and marsh—southwest coast of France.

Changes in Organic Materials

The transformation of organic-rich swamp sediments into coal is a good illustration of the changes that occur in organic materials during lithification. As the organic sediment is buried under the stagnant swamp waters where it accumulated, increased pressure and temperature begin to break down the vegetal debris, releasing gases in the process. The formation of coal involves a series of intermediate steps with progressive release of more and more gaseous components, enriching the residue in carbon.

The first stage of the process of transformation is **peat,** a soft, wet, spongy, fibrous, organic material with a low carbon content and heating value. In many areas of the world, such as Ireland and Great Britain, peat is cut from bogs, stacked to dry, and used as a low-grade fuel. Ideal conditions for peat formation are an abundance of plant debris, stagnant water to prevent oxidation of the vegetal debris, and the absence of siliciclastic sediments that would dilute the organic content of the accumulating material.

As temperature and pressure increase, peat is transformed successively into **lignite, bituminous coal** ("soft" coal), and finally **anthracite coal** ("hard" coal). The highest heat value is in bituminous coal. Anthracite has somewhat less heat value and a higher

FIGURE 6–8 *continued*
(e) Dunes—Padre Island National Seashore, Texas. (f) Beach—Southampton, New York. (g) Reef—Florida Keys. (h) Fine-grained bottom sediments at a depth of approximately 350 m on the upper continental slope near the Georgia-South Carolina border. The two red dots in the center of the photograph are laser light points 10 cm apart that provide a scale. (a) photo by John Shelton; (b) courtesy of Robert Q. Oaks, Jr.; (h) courtesy of Paul Gayes; others by Nicholas K. Coch.

carbon content, but it is a premium fuel because it burns more cleanly than bituminous coal and causes less air pollution.

SEDIMENTARY ROCK TYPES

Sedimentary rocks are classified as clastic, biogenic, or chemical rocks, depending on the *origin of their major component*. Texture (the size and arrangement of the grains) and mineralogic composition are used to further subdivide the three major rock groups.

Siliciclastic Rocks

Siliciclastic rocks are classified by using a combination of terms denoting the texture and composition of the component particles. The upper half of Table 6–2 lists the textures by which siliciclastic rocks are classified and Figure 6–10 shows what these textures look like.

☐ Rocks having particles that are boulder, cobble, or pebble sized (Table 6–1) are called **conglomerates** if the particles are rounded—Figure 6–10(a) and **breccias** if the particles are angular—Figure 6–10(b).

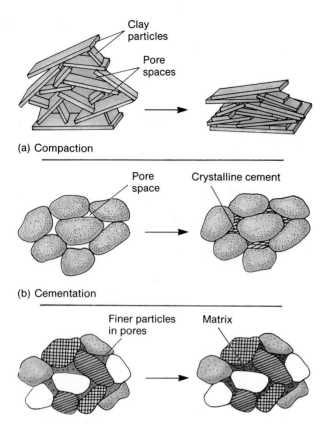

Clay particles

Pore spaces

(a) Compaction

Pore space

Crystalline cement

(b) Cementation

Finer particles in pores

Matrix

(c) Formation of a matrix by recrystallization of fine material in pores

FIGURE 6−9

Processes in the lithification of sedimentary rocks: (a) compaction, (b) cementation, (c) matrix formation.

□ Rocks composed of sand-sized particles are called **sandstones**—Figure 6−10(c). If the sandstone has less than 15% matrix it is called an **arenite**. If it has more than 15% matrix, it is called a **wacke** (pronounced "'whack-ee").

□ Rocks composed of silt-sized particles are **silt-stones;** those composed of clay-sized particles are **shales**—Figure 6−10(d).

The composition of the particles is identified by an adjective that precedes the textural term (see lower half of Table 6−2). For example, a siliciclastic rock composed of sand-sized quartz grains cemented together is called a **quartzose sandstone**. A silici-clastic rock composed of quartz grains and more than 15% matrix is called a **quartz wacke**. Sand-

stones composed of 10−25% feldspar are called **feldspathic sandstones;** those with higher percentages of feldspar are referred to as **arkosic sandstones** or **arkoses**. If the sandstone is composed of rock fragments and matrix, it is referred to as a **lithic wacke**.

Biogenic Rocks

Biogenic rocks are classified according to their composition, texture, and in the case of crystalline limestones, crystal size. The two commonest biogenic rocks are **limestone**—composed of calcite, $CaCO_3$, and **dolostone**—composed of the mineral dolomite, $CaMg(CO_3)_2$. Another example is **coquina,** a clastic rock of biogenic origin. Coquina is composed of shell fragments of calcareous marine organisms—Figure 6−11(b).

Limestones can have a wide variety of components, including shells, shell fragments, and ooids, along with sand, silt, and clay-sized particles derived from several different sources (Figure 6−11). Some limestones are composed largely of shells and shell

TABLE 6−2

Terms used to classify siliciclastic sedimentary rocks.

Texture	
Particle Size	Rock Name
Gravel	
Rounded particles	Conglomerate
Angular particles	Breccia
Sand	Sandstone
Less than 15% matrix	Arenite
More than 15% matrix	Wacke
Silt	Siltstone
Clay	Shale

Composition	
Component	Rock Name Modifier
Quartz	Quartzose
Feldspar	
10−25% feldspar	Feldspathic
More than 25% feldspar	Arkosic
Iron oxide	Ferruginous
Calcite	Calcareous
Rock fragments	Lithic

(a)

(b)

(c)

(d)

FIGURE 6–10
Types of siliciclastic rocks: (a) conglomerate, (b) breccia, (c) sandstone, (d) shale. Macmillan
Publishing/Geoscience Resources photos.

fragments. An example is **chalk,** a soft, white rock composed mainly of the skeletons of microscopic marine organisms—Figure 6–11(a). Another example is coquina.

Some very fine-grained limestones are formed from the lithification of calcium carbonate muds derived from the breakdown of calcareous algae (Figure 6–2). These rocks are called **lithographic lime-**

(a)

(b)

(c)

FIGURE 6–11
Types of biogenic rocks: (a) chalk, (b) coquina, (c) lithographic limestone. Macmillan Publishing/Geoscience Resources photos.

stones because of their use in the lithography ("stone printing") process—Figure 6–11(c). However, most limestones are composed of a mixture of textural elements and are classified by using terms that describe their major component. For example, a rock containing ooids of calcium carbonate cemented by calcite is referred to as an *oolitic limestone*.

Most dolostones are produced by the chemical alteration of limestone. Though dolostones could be considered chemical rocks, we classify them here as biogenic rocks because the *original* rock was produced by biogenic processes and because the chemical change involved is *replacement* rather than the deposition of crystals from solution as in chemical rocks. The exact mechanism of dolostone formation is controversial. Most geologists agree that it involves increasing the magnesium content in the rock, but the controversy is over *how* this is accomplished.

Field studies in hot, shallow intertidal areas suggest one mechanism for producing dolostone. As calcium carbonate sediments are deposited in these areas, calcium is removed from the local seawater, enriching it in magnesium. Evaporation of both the overlying waters and the waters circulating through

FIGURE 6–12
Types of chemical rocks: (a) halite, (b) gypsum, (c) chert, (d) phosphorite. Macmillan Publishing/
Geoscience Resources photos.

the pores of the calcium carbonate sediments below enrich the waters further in magnesium. Magnesium-enriched waters circulating through the pores of the limestone replace some of the calcium in the rock and convert the limestone to a dolostone. Commonly, many of the original details, such as shell fragments, are altered or obliterated in the dolomitization process.

Chemical Rocks

Chemical rocks are produced by precipitation of crystals from concentrated solutions. The texture of these rocks is a mosaic of interlocking crystals, similar to the texture in igneous rocks. Crystal size can vary from too small to be visible without magnification to coarse-grained crystalline aggregates. Chemical rocks are classified by the composition of their component crystals, and the rock name is commonly that of the dominant mineral. The commonest types of chemical rocks are **halite** (NaCl), **gypsum** ($CaSO_4 \cdot 2H_2O$), **chert** (SiO_2), and **phosphorite** (complex phosphates of calcium). These rocks are shown in Figure 6–12.

FOSSILS

Sedimentary rocks are unique in that they provide a rich record of the changes of life through geologic time. The traces, imprints, or parts of ancient plant or animal life that have been preserved through natural processes are called **fossils**. Although the great majority of fossils are found in sedimentary rocks, they also have been preserved, in rare cases, in some types of igneous and metamorphic rocks.

What parts of these ancient life forms become fossilized? In general, only the hard parts of the organisms, such as shells, teeth, and bones are preserved, because the soft fleshy parts of the organisms decompose quickly after death. However, there are several exceptions. For example, organisms having only soft body parts have been preserved in sediments that were deposited slowly in quiet waters with low oxygen levels in the Ediacara Hills of Australia (Figure 6–13). The slow decomposition under these conditions allows the animal to keep its shape as the sediment is being lithified by compaction. In some cases, both soft and hard parts are preserved; an example is the Ice Age mammoths, several of which have been found in frozen soil in Siberia. The mammoths apparently fell into cracks in the ice, and their carcasses froze before the flesh could decompose. Remains of Arctic vegetation even have been found in their stomachs!

In some cases, very delicate organic structures can be preserved in fossilization. Examples include numerous complete insect fossils that have been preserved in the hardened tree sap called **amber** (Figure 6–14). The insects were trapped by the sticky sap as they walked around the trees and became incorporated into the tree sap as it spread over them. The sap solidified into amber with time, protecting the in-

FIGURE 6–13
Late Precambrian Ediacara fauna of Australia. Reconstruction of living conditions. Present interpretation is that the stalked forms are sea pens and the floating disk and bell-shaped organisms are jellyfish. Photo by Smithsonian Institution.

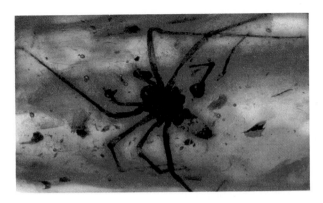

FIGURE 6–14
Spider preserved in amber. Such fossils retain many delicate structural details. Photo © American Museum of Natural History.

sects from decomposition and preventing them from being crushed during subsequent compaction of the sediment.

Fossils found in sedimentary rocks form in different ways. Water that passes through dead plants and animals provides an opportunity for minerals to precipitate from solution; the minerals may replace or-ganic tissues and still preserve delicate structures. Such a mineral transformation is called **replacement**. Fossilized wood at the Petrified Forest National Park in Arizona (Figure 6–15) formed when silica replaced the original woody tissues of the tree trunks, forming a hard, resistant fossil. A similar type of fossilization occurs when the hard parts of former animals are replaced by different minerals. For example, in some rocks the original calcium carbonate in marine shells has been replaced by silica.

Leaves and other plant materials that fall into the muds of stagnant lakes and swamps are preserved from oxidation and decay. In the transformation of the muds to shale, the plant material is transformed to carbon imprints, preserving the form of the original fossil material (Figure 6–16). The process of forming such imprints is called **carbonization** and is part of the process of coal formation. Thus, carbonized plant remains are especially common in coal beds and the shales associated with them.

A buried organism may be dissolved after burial (Figure 6–17). This leaves a cavity, called a **mold,** that can perfectly preserve the organism's shape. The cavity may later be filled by material that forms a three-dimensional **cast** of the original organism.

FIGURE 6–15
Petrified wood at Petrified Forest National Park in Arizona. Shown are cross sections of logs and trunk sections that have been replaced by silica. Photo © Norman R. Thompson, 1990.

FIGURE 6–16
Carbonized plant remains in shale. Macmillan Publishing/
Geoscience Resources photo.

Many of the fossils collected from rocks are casts of the original organisms.

Some types of fossils do not represent the remains of the organisms themselves but rather their *effects* on the sediment during deposition. Organisms live on and within the sediment and leave a record of their activities in burrows, tracks, and trails as they move around their environment. Sedimentary structures resulting from the life activities of organisms are called **trace fossils**. Trace fossils provide important information on the behavior of ancient organisms.

SEDIMENTARY STRUCTURES

When you examine newly deposited sediments, you may see ripple marks on sandy beaches or mud cracks formed on a mud flat as the surface dries out. If you dig a trench in the sand at a beach, or along a river shore, you will probably see layers of sand of different size and perhaps of different composition. Such features, formed by physical, chemical, or biological mechanisms at the time of sediment formation, are referred to as **sedimentary structures**. There are many different kinds of sedimentary structures, and because they form at the time of sediment deposition, they provide a detailed picture of past changes in surface conditions.

Sedimentary structures can be used to answer questions as diverse as:

Which way were the currents flowing (air, water, or ice)?

How fast were the currents moving?

What was the climate like?

Are these rock layers still in the same order in which they were deposited, or have they been overturned since deposition?

We will consider some common examples of sedimentary structures that tell us the most about conditions during deposition.

Stratification

One of the distinctive properties of sedimentary rocks is that they are layered, with the layers stacked upon each other like a deck of cards. The accumulation of sedimentary particles in layers is called **stratification**. The layers in sedimentary rocks are called **beds** if they are thicker than 1 cm and **laminae** if they are thinner than 1 cm. Each layer is separated from the one above it by a physical break in the rock called a **bedding plane**.

When environmental conditions change, causing a different sediment to be laid down, or when deposition resumes after ceasing for a while, another bed will begin to form. Many beds in chemical rocks record changes in the chemical conditions during deposition (temperature, water composition, and so forth).

Beds can be distinguished because of differences in particle (or crystal) size, composition, sorting, color, and shell content. As each distinctly different bed represents a particular set of environmental conditions, study of a vertical sequence of sedimentary rocks (such as that shown on the opening page for this chapter) allows a geologist to determine the changes in an area over a long period of time.

Thicker beds generally indicate a longer time of sediment accumulation. However, some thick beds can be deposited very quickly and some thin ones

FIGURE 6–17
Formation of a mold and cast of a fossil clam.

(a) Shell is buried in sediment which lithifies into sedimentary rock.

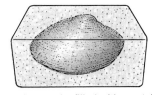

(c) The cavity may be filled with crystals precipitated from pore waters to form a cast.

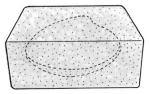

(b) Shell is dissolved by circulating waters, forming a cavity called a mold.

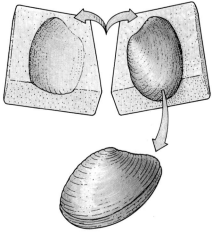

(d) If the rock is broken open, the cast is freed. The cast has a shape similar to the original shell.

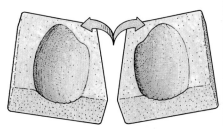

If the rock is broken open, a mold with the imprint of the outer surface of the shell is visible.

may take a very long period to accumulate. For example, a desert mudflow may deposit a meter of poorly sorted sediment in a few minutes, whereas sedimentation of an oceanic clay layer 1 cm thick may take thousands of years.

Graded Bedding. Beds in which the grain size decreases steadily from the bottom to the top of the bed are called **graded beds** (Figure 6–18). Graded beds are produced where sediment-laden fluids—water or wind—lose velocity and undergo a fairly rapid decrease in kinetic energy. As the energy decreases, successively finer sediments are deposited on the bottom. For example, where a flooding stream spills over its banks, there is an abrupt decrease in the stream's kinetic energy and a graded bed may begin to be deposited over the adjacent flood area. As the velocity begins to drop, the largest particles are de-

posited first, with smaller particles deposited on top as the velocity decreases with time.

Rhythmic Bedding. Stratification composed of the regular alternation of two different types of layers is called **rhythmic bedding**. This type of bedding implies a sequential alternation of two different depositional conditions.

One environment in which rhythmic bedding occurs is in the lakes that develop at the borders of melting glaciers. In the summer, streams from the melting glaciers have high energy and deposit their sediment load into the lake as a series of graded beds (Figure 6–19). In the winter, both the stream and the lake freeze over and the coarse particles no longer can be deposited. At this time, the fine sediment and organic particles that have remained suspended in the lake waters are deposited to form a dark, fine-

FIGURE 6–18
Graded bed. Particle size grades
upward from gravel to mud.

Graded bed

FIGURE 6–19
Rhythmic beds of laminated silt and
fine sand.

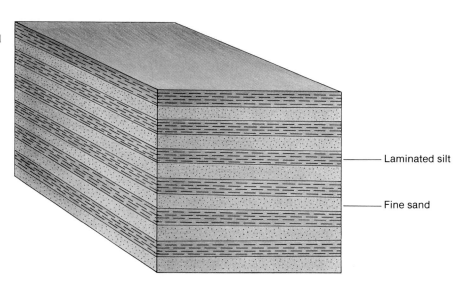

Laminated silt

Fine sand

grained layer on the bottom of the lake. As the ice melts the following spring, coarse river sediments again reach the lake, and another series of graded beds is deposited until the lake and stream freeze over once again. Such *yearly* couplets of a coarse, graded bed (summer) and a fine lamina (winter) constitute a special type of rhythmic bed called a **varve**.

Cross-bedding. A type of stratification in which the layers *within* the bed are not horizontal but are inclined is called **cross-bedding**. Cross-bedding forms

as sand is moved in ripples or dunes across a surface by water or air. A section through a ripple or dune reveals sand layers that usually are inclined in the down-current direction—Figure 6–20(a). The cross-stratification within any one bed either makes a small angle with the bottom of the bed or is tangential to it—see the close-up view in Figure 6–20(b). In contrast, the cross-stratification is cut off (truncated) at the top of the bed. A geologist can therefore tell if a cross-bedded rock is right-side-up or overturned because the truncated beds mark the original top surface. Sedimentary structures—such as mud cracks, oscillation ripple marks, and cross-beds—that can be used to determine the tops and bottoms of layers are called **top-and-bottom structures**.

Ripple Marks

You may have noticed regular undulations in the surface of sandy material near the shore of a lake, river, or ocean, or even on the sides of a sand dune. These undulations form as flowing wind or water currents move over the surface and pile up the grains in a series of wavelike forms called **ripple marks**. Two kinds are recognized, each of which forms under different conditions. **Oscillation ripple marks**—Figure 6–21(a)—are symmetrical and are shaped by the back-and-forth bottom motion associated with waves. Their pointed crests are always directed upward. **Current ripple marks** are asymmetrical ripple marks formed as the current moves along the bottom—Figure 6–21(b). The steeper side of the ripple is always inclined down current, making it a useful structure for determining ancient current directions, or **paleocurrents**.

Mud Cracks

Clayey sediment is notable for swelling when moistened and shrinking when dried. When a wet, clayey sediment dries, it contracts and forms a series of polygonal **mud cracks** at its surface. Subsequently deposited sediments fill the mud cracks and preserve them. Long after the deposits have been lithified,

FIGURE 6–20
Formation of cross-laminated sand beds. (a) Migration of sandy ripples. (b) Detailed view of a bed with cross-lamination showing features mentioned in the text.

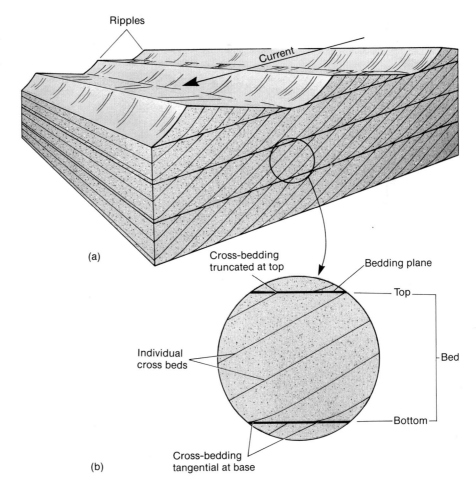

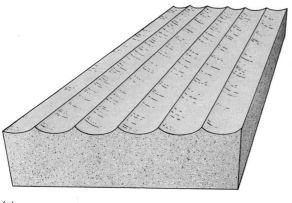

(a)

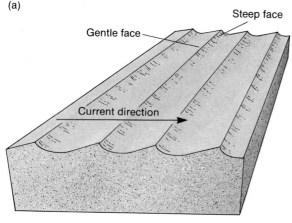

Gentle face

Steep face

Current direction

(b)

FIGURE 6–21
Types of ripple marks. (a) Oscillation (symmetrical) ripple marks. (b) Current (asymmetrical) ripple marks.

enormous forces, such as those along convergent plate boundaries, may heave the area up and turn the rock over. Mud cracks indicate the original top, because they open upward when they form.

Sedimentary Structures and Geologic History

The geologic history of an area is determined by studying the sequence of exposed rocks. This task can be quite complicated if postdepositional forces have tilted, or even overturned, the sedimentary rock layers. How can we determine if this has occurred?

It is possible to determine the original tops of the beds if the sequence contains top-and-bottom sedimentary structures such as oscillation ripple marks,

graded beds, mud cracks, and cross-bedding. The technique is illustrated in Figure 6–22.

Geologists examining the rock exposure shown in Figure 6–22 might first think that they were dealing with ten different layers. However, after more careful examination of the top-and-bottom sedimentary structures in those beds, it would be apparent that the upper half of the exposure is an overturned version of the lower half. Specifically, the cross-bedding in layer C_2 is tangential to what appears to be the top of the bed, and it is truncated at what appears to be the bottom of the bed. In addition, unit B_2 has oscillation ripple marks with crests that point downward, and unit D_2 has mud cracks that open downward. These relationships are not normal; they indicate that A_2 through E_2 were originally extensions of A_1 to E_1 that curled up and turned over on top of A_1 to E_1. The geologic history of the area is then reconstructed from the characteristics and ages of the lower five units (A_1 to E_1). This is an example of how geologists work in the field.

SUMMARY

1. Sediments are produced by the breakdown of older rocks, by organic activity, or by precipitation of crystals from solutions.
2. Sedimentary materials are carried physically, as particles moving along the bottom or suspended in the water or air, and chemically, as ions in solution. Sediment carried physically in the flow of air or water is deposited as the kinetic energy of the flow decreases. Dissolved load is deposited when chemical conditions change or when evaporation concentrates the dissolved ions so that they crystallize as evaporite minerals.
3. Mineralogic composition of sediments is a function of the mineralogy of the source rock, weathering, changes during transport, and mineralogic changes during burial.
4. Particle size is greatest in most glacial deposits and mudflows, finer in stream deposits, and very fine in wind deposits.
5. Particle sorting is best in windblown sands because air has a low density and can carry only a narrow range of particle sizes. The greater density of glacial ice enables

FIGURE 6–22

Use of top-and-bottom sedimentary structures to determine the depositional sequence of overturned rocks. (a) Five sedimentary rock layers are deposited and (b) folded into overturned beds. (c) The final rock sequence is: A—sandstone with oscillation ripple marks. B—shale. C—sandstone with cross-bedding and pieces of underlying shale. D—sandy shale with mud cracks filled in with material from layer E. E—sandstone with graded bedding and chips of layer D at base. Dashed lines show inferred reconstruction of bed sequence.

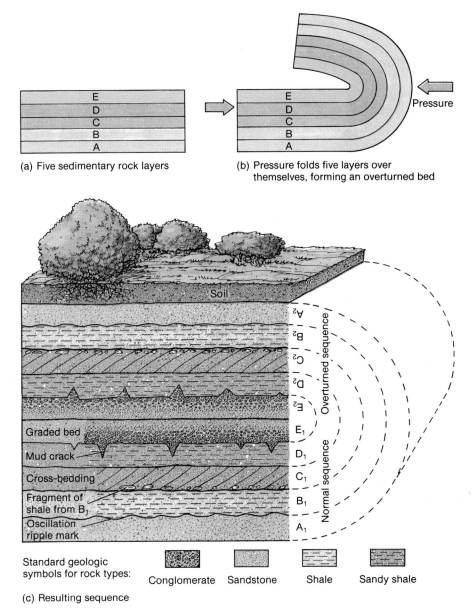

(a) Five sedimentary rock layers

(b) Pressure folds five layers over themselves, forming an overturned bed

Graded bed
Mud crack
Cross-bedding
Fragment of shale from B₁
Oscillation ripple mark

Standard geologic symbols for rock types:

Conglomerate   Sandstone   Shale   Sandy shale

(c) Resulting sequence

it to carry the broadest range of particle sizes. Consequently, glacial deposits have the poorest sorting of all sediments.

6. Once sediment is deposited, it is commonly eroded and transported again, passing through a number of different sedimentary environments. This constant recycling of the sediment, along with its dilution with other sediments, can change sediment greatly from its original composition, size, roundness, and sorting.

7. Eventually the sediment reaches an area on the continent or in the ocean basin where it is protected from further erosion and accumulates layer by layer. The sediments become lithified by compaction, cementation, and/or matrix formation after burial by younger sediments. The sediment also undergoes a series of postdepositional changes that include solution, formation of new minerals, and decomposition of organic material.

8. Sedimentary rocks are classified into three types, based on the dominant origin of their component crystals or particles. They are subdivided further on the basis of particle composition and texture.
9. The three major types of sedimentary rocks are clastic, biogenic, and chemical. However, many sedimentary rocks are composed of a mixture of these three components.
10. Stratification and sedimentary structures provide information about the environmental conditions that existed during accumulation of sediment. Some of these sedimentary structures enable geologists to determine the tops and bottoms of beds. This is important in determining whether the rock layers have been overturned since their deposition.
11. Fossils are the traces, imprints, or parts of ancient plant or animal life that have been preserved through natural processes. The fossils in sedimentary rocks enable geologists to determine the successive changes in life forms with time.

REVIEW QUESTIONS

1. Why do sedimentary rocks have far fewer minerals than igneous or metamorphic rocks?
2. Oscillation ripple marks are good top-and-bottom indicators, but current ripples are not. Why is this so?
3. What inferences could you make about ancient rock-forming environments, based on the following rocks?
   a. A white sandstone with well-sorted and rounded particles.
   b. A sandy conglomerate with poor sorting, angular particles, and a wide range of mineral types.
   c. Beds of gypsum.
4. Describe the formation of the following sedimentary structures:
   a. Cross-bedding.
   b. Graded bedding.
   c. Rhythmic bedding.
5. In what way are the textures of chemical sedimentary rocks similar to those of igneous rocks? Why is this so?

6. How can sedimentary structures be used to deduce:
   a. The tops and bottoms of sedimentary rock layers?
   b. The directions of ancient currents?
7. What are facies? If two facies interfinger, what is the geological significance of this relationship?
8. How are size and sorting related to the type of sediment-transporting agent? Give two examples.
9. Under what conditions can soft parts of ancient animals be fossilized? Give two different examples.
10. What conditions favor deposition of evaporites?
11. Evaporites are accumulating on the continents and along arid coastlines today. How could you distinguish rock samples from each of these areas?

FURTHER READINGS

Blatt, H., Middleton, G. V., and Murray, R. C. 1980. *Origin of sedimentary rocks*. Englewood Cliffs, NJ: Prentice-Hall, 782 p. (Detailed treatment of a wide variety of sedimentary rocks.)

Collinson J. D., and Thompson, D. B. 1982. *Sedimentary structures*. London: Allen & Unwin, 194 p. (A detailed and excellently illustrated treatment of sedimentary structures.)

Davis, R. A. 1983. *Depositional systems*. Englewood Cliffs, NJ: Prentice-Hall, 669 p. (An excellent overall treatment of both modern sedimentary environments and their rock equivalents.)

Miall, A. D. 1984. *Principles of sedimentary basin analysis*. New York: Springer-Verlag, 490 p. (Detailed treatment of the methods used by geologists to determine the change in the depositional conditions recorded by sedimentary rocks.)

Scholle, P. A., Bebout, D. C., and Moore, C. H. 1983. *Carbonate depositional environments*. Tulsa, OK: American Association of Petroleum Geologists Memoir 33, 708 p. (A detailed treatment of a wide variety of carbonate sedimentary environments. Profusely illustrated with aerial photographs, geologic sections, and rock photomicrographs, many in color.)

7

Metamorphism and Metamorphic Rocks

Banded calc-silicate from the Skowhegan area, central Maine. This rock formed by metamorphism of a gray, clay-rich limestone. White bands are composed of calcite; pale green bands contain diopside, $CaMgSi_2O_6$; reddish areas are rich in grossular garnet, $Ca_3Al_2(SiO_4)_3$. Photo by Allan Ludman.

A balloon rising into the upper atmosphere expands and bursts as it adjusts to the gradually decreasing pressure around it. Rocks within the Earth also change when they are subjected to conditions that are greatly different from those under which they first formed. When one rock is transformed into another kind of rock in response to changing conditions within the Earth, it is said to have undergone **metamorphism,** from the Greek roots *meta* (change) and *morph* (form). In this chapter we will examine the principles of metamorphism and then look at metamorphic rocks and the types of information they convey about the interior of the Earth.

METAMORPHISM

What Is Metamorphism?

Metamorphism involves *solid-state* changes that occur *within* the Earth. Both parts of this definition are important. "Solid-state" means that changes occur without melting or dissolving, and "within the Earth" means not by the surface processes of weathering and soil formation. Other changes that rocks undergo occur within the Earth but involve melting and are therefore igneous.

Some changes are more difficult to categorize. For example, as sedimentary rocks are buried beneath younger deposits, increased pressure closes pore spaces and ions often are added or removed by circulating water, changing the rock's mineralogy. These changes generally are considered to be part of a rock's sedimentary history. The distinction between metamorphism and these sedimentary processes is hazy, reminding us that rigid definitions do not always fit the gradual changes that take place in nature.

Causes of Metamorphism

Metamorphism is caused by three activities in a rock's environment: change in heat, change in pressure, and circulation through the rock of solutions that are rich in dissolved ions. These are called the **agents of metamorphism** (Figure 7–1). Early in the scientific investigation of metamorphism, other factors were thought to be equally important, such as the age of a rock. We now know that a rock's age has nothing to do with its metamorphism, and only the three factors listed here play important roles.

Heat. Most substances, including minerals, expand when heated. Their ions vibrate more widely, stretching and weakening the bonds that hold them in place—Figure 7–1(a). If enough heat is added, some of the bonds may break (although not enough for the entire structure to collapse and thus cause

FIGURE 7–1
Agents of metamorphism. (a) Heat causes minerals to expand, increasing interionic distances. This weakens bonds and makes some minerals unstable. (b) Pressure-compression forces ions closer to one another, breaking some bonds and creating new mineral structures. (c) In chemically active fluids, new ions are transported to a rock, changing its composition and creating new minerals.

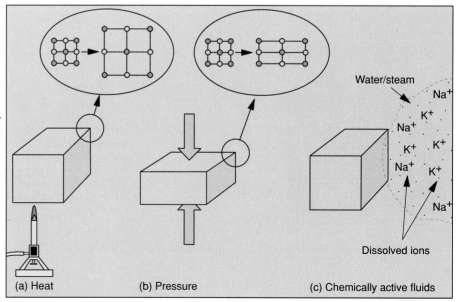

(a) Heat   (b) Pressure   (c) Chemically active fluids

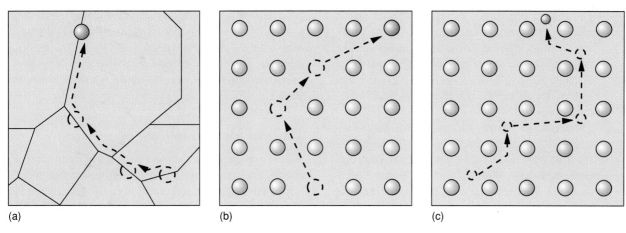

(a)   (b)   (c)

FIGURE 7-2
Migration of ions through solid rock during metamorphism. (a) Ions may move along the surfaces of grains, aided by whatever fluid is present. (b) Defects in crystal structures often result in some vacant ionic positions, allowing ions to migrate from one vacancy to another. (c) In a relatively open mineral structure, small ions may move among the mineral's ions.

melting). The ions that are freed can migrate through the solid rock to form seeds of new minerals (Figure 7-2). Most ions probably move along boundaries between grains, aided by whatever fluid is present, but some can actually pass through mineral structures—Figure 7-2(b) and (c). They move much more slowly than ions in magma or water, but over time they migrate enough to create new minerals.

Pressure. The effect of increased pressure is directly opposite to that of increased heat. Pressure compresses rocks and minerals, eventually forcing their ions closer together than their structures permit—Figure 7-1(b). The resulting strain on bonds causes some to break, and the ions become rearranged in a more compact structure. This is how graphite is converted to diamond, both in nature and in the laboratory. Geologists recognize two different kinds of pressure, each of which causes a different kind of change in rocks:

1. **Lithostatic pressure** (from the Greek *lithos*, meaning rock, and *status*, meaning position) is the pressure exerted on a rock as it is buried deeper and deeper below the surface—Figure 7-3(a). The greater the thickness of the overlying rocks, the greater the lithostatic pressure.

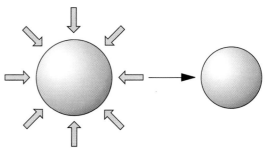

(a) Lithostatic pressures

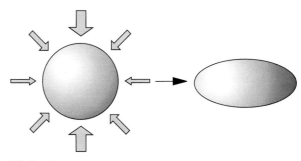

(b) Directed pressures

FIGURE 7-3
Types of pressure. (a) Lithostatic pressures are equal in all directions. They compress grains, but do not greatly deform them. (b) Directed pressures are greater in some directions than in others. They compress and change the shapes of grains in rocks.

Lithostatic pressure is of equal intensity from all directions; as a result it can compress a rock into a small volume but cannot flatten it. The ultimate cause of this kind of pressure is Earth's gravity, and the geobaric gradient of 3.3 km/kb (Chapter 2) is the gradient of lithostatic pressure.

2. **Directed pressure** differs from lithostatic pressure in that it is greater in some directions than in others and can therefore flatten and deform rocks—Figure 7–3(b). Directed pressures are not the result of gravity alone, but are produced by the tectonic forces that form mountains like the Alps or Himalayas.

Chemically Active Fluids. Solutions containing large quantities of dissolved ions are called **chemically active fluids**. Some are hydrothermal solutions that emanate from cooling magmas, and others are fluids released during metamorphism deep within the crust. Changes in temperature and pressure alter a rock's *physical* environment; the chemically active fluids change its *chemical* environment. The dissolved ions can be added to rocks, significantly changing their chemical composition and producing new minerals that could not otherwise have formed under any range of physical conditions. Sometimes it is the water itself that interacts with the minerals. Olivine, for example, can resist extensive changes in heat and pressure, but reacts rapidly to form talc if water is added.

Types of Metamorphism

All metamorphism involves heat, pressure, and chemically active fluids, but one agent may be more important than the others in some areas. In many instances, we can reconstruct the roles played by each agent, because the way in which a rock adjusts to metamorphism depends on which agent was dominant. Several different types of metamorphism are recognized, depending on the manner in which the agents acted (Table 7–1) and on which ones were most important.

Contact (Thermal) Metamorphism. The intrusion of a magma provides a source of heat for metamorphism, without altering pressure. Heat released from the cooling magma is conducted to the host rock, where it causes changes in texture and mineralogy.

TABLE 7–1
Types of metamorphism.

Agent	Type of Metamorphism	Description
Heat	Contact (thermal) metamorphism	Areas around intrusive igneous rocks; bases of lava flows
Pressure:		
Lithostatic	Burial metamorphism	Base of a thick pile of sedimentary or volcanic rocks
Directed	Dynamic (cataclastic) metamorphism	Zones of faulting and intense fracturing
Chemically active fluids	Metasomatism	Addition of ions to solid rock by fluids from cooling magmas
Combinations:		
All of the above	Regional metamorphism	Large areas subjected to intense compression at depth; mountain ranges such as the Alps; convergent plate boundaries
Heat and directed pressure	Impact (shock) metamorphism	Sites of meteorite impact (meteorite craters)

This is called **contact metamorphism** because its effects are most intense at the contact between pluton and host rock. Contact metamorphic intensity decreases with distance from the pluton and reaches zero at locations far enough from the intrusive body that the heat has no effect on the local rocks. The area within which contact metamorphism occurs is known as the pluton's **contact aureole**. Contact aureoles around small dikes may be only a few centimeters wide, but those bordering batholiths may extend a few kilometers from the pluton.

Burial Metamorphism. Metamorphism produced almost entirely by lithostatic pressure occurs during burial of sedimentary or volcanic rocks beneath later deposits or flows. Some heat is involved because the overlying rocks act as an insulating blanket and prevent heat caused by radioactive decay from escaping, but it is the pressure that dominates in this process, called **burial metamorphism**. The intensity of burial metamorphism increases both with the depth and the density of the overlying rocks.

Dynamic Metamorphism. Fractures in the Earth called **faults** separate blocks of rock that have moved relative to one another because of strong directed pressures. Neutral plate boundaries, the transform faults shown in Figure 1–6, are a large-scale example. The directed pressures along the fault bring about metamorphic changes (Figure 7–4). This kind of metamorphism is called **dynamic metamorphism** because it is caused by movement. Some frictional heat aids in the metamorphism, but most changes are controlled by the strong directed pressures generated as the two blocks of rock grind past one another along the fault.

Dynamic metamorphic intensity decreases rapidly away from fault zones. The extent of the effects depends on the duration and intensity of the faulting. Along major transform faults, such as the San Andreas fault of California, the zone of dynamic metamorphism is hundreds or even thousands of meters wide; in smaller faults it may be only a few centimeters across.

Metasomatism. When chemically active fluids— Figure 7–1(c)—change a rock's composition markedly, the process is known as **metasomatism**. This term is generally applied to the *addition* of large

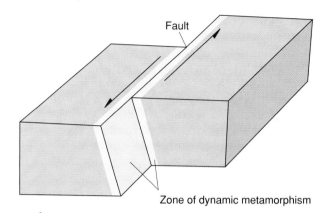

FIGURE 7–4
Dynamic metamorphism occurs during faulting. Arrows indicate movement on opposite sides of the fault. Crushing and grinding during dynamic metamorphism are concentrated in the region of faulting.

amounts of cations such as K^+ and Na^+ to rocks, and we will follow this standard usage. The *removal* of water or carbon dioxide from rocks also occurs during many metamorphic reactions, but even though the rock compositions change, most geologists do not consider this to be metasomatism.

Regional Metamorphism. A combination of all three agents acting on a very large scale is called **regional metamorphism**. Deeply buried (lithostatic pressure) sedimentary and volcanic rocks are squeezed and folded (directed pressure), intruded by igneous rocks (contact metamorphism), and permeated by chemically active fluids (metasomatism). Regional metamorphism occurs at convergent plate boundaries, as we shall see later in this chapter. Metamorphic effects are greatest in the deepest and most intensely folded parts of the affected area.

Regional metamorphism is so named because of the broad areas that it affects. For example, about 400 million years ago rocks from easternmost Canada to Alabama were subjected to regional metamorphism that accompanied the formation of the Appalachian Mountains. Some rocks were only slightly affected, but others were heated to the extent that they nearly melted.

Impact Metamorphism. The rarest form of metamorphism occurs when a meteorite collides with the Earth. During the collision, the enormous kinetic energy of the meteorite is converted to heat, and tremendous directed pressures are built up at the moment of impact. This process, called **impact metamorphism,** is rare on Earth because most meteors burn up in our protective atmosphere. A few do strike the surface, such as the one that created Meteor Crater in northern Arizona. On Mercury, Mars, and the Moon, however, where there is no shielding atmosphere, this is the most common type of metamorphism.

How Much Can a Rock Change?

Nearly every aspect of an igneous or sedimentary rock can change during metamorphism, including its texture, mineral content, and chemical composition. In some instances the changes are so great that the metamorphic rock bears little resemblance to its parent (Figure 7–5).

The amount of change depends on (1) the *intensity* of the metamorphic agents, (2) the *duration* of metamorphism, and (3) the *type of rock* involved. The roles of the first two, intensity and duration, are obvious: we would expect the greatest changes to ac-

company the greatest heat, pressure, and chemical alterations, and the longer these agents operate, the more sweeping the changes ought to be. The type of rock, however, turns out to be equally important. Rocks such as granite, whose minerals are stable over a wide range of physical and chemical conditions, change very little during metamorphism. In contrast, rocks whose minerals are stable only under a narrow range of conditions tend to change most readily and most completely. Shales and limestones belong in this category.

Low-grade metamorphic rocks are those subjected to low-intensity metamorphism. They retain many of their original features, such as bedding, texture, fossils, and minerals, and may not look much different from their initial states. **High-grade metamorphic rocks** have lost most of their primary features and look very different from their unmetamorphosed parent rock. Extreme metamorphism culminates in partial melting, resulting in an intimate mixture of igneous rock formed from the melt and high grade metamorphic minerals in the unmelted rock. These very high grade mixed rocks are called **migmatites**.

(a)

(b)

FIGURE 7–5
Scope of metamorphic change. Metamorphism of the siltstone with a fossil leaf shown in (a) can produce the schist with large garnet crystals shown in (b). Macmillan Publishing/Geoscience Resources photos.

FIGURE 7–6

Recrystallization may destroy original rock textures. (a) Fine-grained limestone containing a few large fossils. (b) The small calcite grains recrystallize until nearly all traces of the original grain size and organic content have been destroyed. However, note that traces of the fossils remain.

(a)

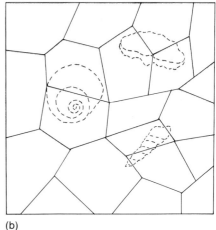

(b)

Metamorphic intensity gradually increases over a period of hundreds or thousands of years during contact and dynamic metamorphism, and over millions of years in the case of regional and burial metamorphism. Rocks adjust gradually as conditions change, and may pass through several different stages between parent rock and high-grade metamorphic product. Original minerals are converted to low-grade minerals in the early stages, and these metamorphic products are themselves transformed to higher-grade minerals as intensity increases. This step-by-step adjustment is called **progressive metamorphism,** and the changes that take place as the grade of metamorphism increases are called **prograde** metamorphic changes. In contrast, impact metamorphism occurs almost instantaneously, within tiny fractions of a second.

METAMORPHIC ROCKS

Each metamorphic rock contains a record of all the processes that have affected it. Thus, many serve as *geothermometers* or *geobarometers*—indicators of the temperatures or pressures at which metamorphism occurred. The clues to a rock's metamorphic history lie in its texture and mineralogy—the same features that reveal sedimentary and igneous rock histories.

Metamorphic Textures

The texture of a metamorphic rock may prove more difficult to interpret than that of an igneous or sed-

imentary rock because it can contain features produced both during metamorphism and during the process by which the unmetamorphosed rock first formed. Textural features inherited from the original rock are called **relict features**. They often are modified during metamorphism, but even such delicate features as fossils and vesicles may be recognizable in metamorphic rocks. In high-grade rocks, however, most original textures (and minerals) are destroyed and are completely replaced by metamorphic features.

All textural features of a rock can be altered during metamorphism. Initial low-grade textural changes often are simply the closing of pores and vesicles in sedimentary and volcanic rocks. As intensity of metamorphism increases, however, changes in grain size, shape, and orientation are brought about by the processes of *recrystallization* and *reorientation*.

During **recrystallization,** grains in the original rock change in size and shape as heat causes ions to migrate from their original positions to new crystal seeds. The result is often the complete destruction of the original texture. For example, recrystallization of clastic particles, matrix grains, and fossils in a sedimentary rock may yield interlocking grains of nearly uniform size (Figure 7–6). During **reorientation,** minerals change their positions, generally realigning themselves in response to directed pressures.

Grain Size and Shape. Grain size increases during most types of metamorphism. A typical example is the transformation of fine-grained siltstone to the coarse-grained metamorphic rock called **schist,**

FIGURE 7–7
Porphyroblastic texture with large staurolite crystals in finer-grained schist matrix, central Massachusetts. Photo courtesy of Charles V. Guidotti.

shown in Figure 7–5. As in igneous rocks, the amount of ionic migration determines the size of mineral grains. We would expect the coarsest-grained rocks to be those that were subjected to the most heat energy and remained hottest for the longest time. In general, this is exactly what we find.

Some minerals in a metamorphic rock may grow much larger than others, either because their ions did not have far to travel or because they migrate more easily than others. The result is called a **porphyroblastic texture** (Figure 7–7), and the large grains are called **porphyroblasts**. Many garnet porphyroblasts in high-grade rocks near Gore Mountain, New York, are larger than softballs, attesting to the ability of ions to migrate even in solid rocks. (In contrast to the igneous *porphyritic* texture, the different grain sizes do not indicate different rates of heating or cooling.)

FIGURE 7–8
Flattened pebbles produced by faulting of the Bygdin Conglomerate, Norway. The knife rests on a surface that shows extreme flattening and elongation during dynamic metamorphism. Pebbles showing more normal shapes are visible on the other face. Photo by Robert D. Hatcher, Jr.

Grain size often *decreases* during dynamic metamorphism because rocks along a fault are subjected to intense grinding and crushing as the blocks of material on opposite sides move past one another. This grinding forms *smaller* grains, rather than coarser ones.

The shapes of particles in a rock also change as the result of grain growth and breakage, as shown in Figures 7–5 and 7–6. Round clastic particles in sedimentary rocks also may change shape drastically when strong directed pressures are applied (Figure 7–8). In more extreme cases in faults, grains are sometimes stretched and flattened, resulting in a characteristic **mylonitic texture** (Figure 7–9) that uniquely identifies dynamic metamorphism.

Preferred Orientation. Directed pressures cause many grains be to reoriented in one of two ways, depending on their original shapes (Figure 7–10). **Lineation** is the parallel alignment of rodlike grains such as amphibole or plagioclase feldspar crystals. **Foliation** is the name given to a parallel alignment of sheetlike minerals such as the micas or of flattened grains such as those shown in Figure 7–8.

Directed pressures cause minerals or clasts to rotate from their original positions until they are aligned. Sheetlike and tabular minerals change position so that their flat surfaces are at right angles to the greatest directed pressures. This enables geolo-

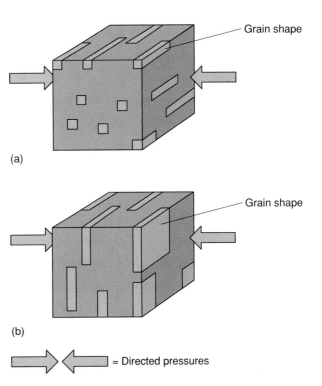

(a)

(b)

= Directed pressures

FIGURE 7–10
Preferred orientation of grains. Directed pressures cause grains to display (a) lineation (parallel alignment of rodlike minerals) and (b) foliation (parallel alignment of sheetlike grains).

gists to determine the directions of greatest pressure merely by studying the orientation of the foliation.

Foliation defined by parallelism of sheetlike minerals such as muscovite or biotite is given a special name—**schistosity**. Other foliations are formed by alignment of flattened clasts or of stretched grains in a mylonite. Any well-developed foliation indicates the operation of strong directed pressures, suggesting regional or dynamic metamorphism. Weakly developed alignments are characteristic of burial metamorphism.

Foliation and lineation do not always accompany metamorphism because directed pressures are not always involved; neither would be expected, for example, during contact metamorphism. Even in regional metamorphism, however, aligned grains may not be found if the affected rocks do not contain platy or rodlike minerals.

Random Orientation. Metamorphic rocks formed without directed pressures, particularly during con-

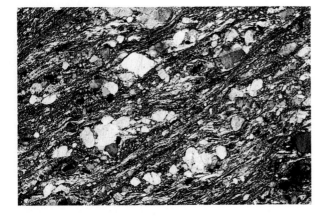

FIGURE 7–9
Mylonitic texture. Photomicrograph shows the strongly foliated, smeared-out appearance characteristic of mylonites. Macmillan Publishing/Geoscience Resources photo.

FIGURE 7–11
Granoblastic texture. Photomicrograph of a quartzite showing random orientation of tightly interlocked grains characteristic of a granoblastic-textured rock. Macmillan Publishing/Geoscience Resources photo.

tact metamorphism, exhibit neither foliation nor lineation. Their random, unaligned mineral orientation is called a **granoblastic texture,** and a rock with this texture is called a **granofels** (Figure 7–11).

Granoblastic-textured rocks must be interpreted with extreme caution. If platy and elongate minerals such as micas and amphiboles are part of the random texture, directed pressures were certainly absent during metamorphism because such minerals are readily aligned. We saw earlier, however, that if no platy or elongate minerals are present, a rock cannot display foliation or lineation even if strong directed pressures had been active.

Gneissosity. Some metamorphic rocks consist of alternating layers of foliated and granoblastic-textured materials (Figure 7–12). The layering is indicated by alternation of light and dark bands, with foliation restricted to one of the two, usually the darker. This combination of layering and foliation/nonfoliation is called **gneissosity** (pronounced "nice-ossity"), and rocks displaying it are called **gneisses** ("nices"). Most gneisses are high-grade rocks and their layering is the result of metamorphic ion migration rather than relict sedimentary bedding.

Relationships Between Grains. All metamorphic rocks have an interlocking texture and can be distinguished from igneous and chemical sedimentary

rocks having similar textures by their unique minerals. Foliation and lineation also help identify a rock as metamorphic, because few igneous or sedimentary processes cause a strong alignment of grains.

Metamorphic Minerals

All of the textural changes just discussed can occur without the formation of new minerals or without the disappearance of old ones. Most metamorphism, however, involves formation of new minerals. Whether minerals form by igneous, sedimentary, or metamorphic processes, they represent the most stable grouping of ions for a particular environment. For example, clay minerals such as kaolinite have open, sheetlike structures that are stable at the Earth's surface, where they form during weathering. However, these minerals cannot survive either the lithostatic pressures or temperatures found deep in the crust, and they must be replaced by more compact structures (hence new minerals) as burial and/or heating occur.

The simplest mineral changes are transformations from one polymorph of a mineral to another, be-

FIGURE 7–12
A gneiss. Gneissosity is a texture in which a rock is segregated into alternating bands of foliated (dark layers) and granoblastic (light) material. Macmillan Publishing/Geoscience Resources photo.

cause ions are only rearranged; none are added or removed, and little migration is necessary. Most mineralogic changes are more complex, however, because they involve combinations of ions from two or more of the original minerals. We shall examine some of these reactions soon, when we study the metamorphism of the most common rock types.

Pressure is an important agent of metamorphism, so most new metamorphic minerals tend to be denser than the original ones. Mineral reactions in limestones release carbon dioxide (decarbonation reactions), and those in shales release water (dehydration). The loss of these fluids decreases the volume of the rock, thus increasing its density. If directed pressures are active during these reactions, the new metamorphic minerals are generally foliated or lineated.

Factors Controlling the Mineralogy of Metamorphic Rocks. The mineral content of a metamorphic rock is determined by two equally important factors: *the parent rock* and *the intensity of the agents* of metamorphism. The parent rock, known as the **protolith,** determines what ions will be available for formation of minerals. The intensity of metamorphism controls how the ions will be arranged.

Role of the Protolith. The ions present in the protolith are the only ones available to make new minerals (except in cases where there has been extensive addition of materials during metasomatism). For example, a limestone that has no aluminum cannot form aluminous minerals, and iron-rich minerals cannot be produced from a rock that contains no iron.

Four broad compositional groups of rocks are recognized on the basis of their mineral content:

1. **Pelitic rocks** are those that originally contained a large proportion of clay minerals— shales, mudstones, or sandstones with clay-rich matrixes. Clay minerals contain large amounts of aluminum, and so do the metamorphosed pelitic rocks. ("Pelitic" comes from the Greek *pelos,* "clay mud.")
2. **Calcareous rocks** have high calcium contents; most were originally limestones and dolostones.

3. **Mafic rocks** were originally basalts or gabbros rich in ferromagnesian minerals and calcic plagioclase feldspars.
4. **Quartzo-feldspathic rocks** were either felsic igneous rocks or sedimentary rocks that contained large amounts of quartz and feldspar.

Table 7–2 lists some of the common metamorphic minerals found in these four types of rock. Note that some minerals are found only in one kind of rock, whereas others may occur in several. This happens because the ions needed to make a particular mineral may be found in more than one kind of rock. The mere presence of certain minerals in a metamorphic rock is enough to identify its protolith, whereas other metamorphic rocks cannot be interpreted as easily.

Role of Metamorphic Intensity. The mineralogy of a shale changes many times as it is progressively metamorphosed from low-grade conditions through high-grade conditions. Low-grade pelitic rocks thus have very different minerals than high-grade pelitic rocks. All compositional types behave the same way. At each stage of metamorphism, the minerals present are the most stable arrangement of the available ions.

Classification of Metamorphic Rocks

The classification of a metamorphic rock is based on its two most readily identifiable properties: *texture* and *mineral content* (Table 7–3). In many instances, the names also indicate the grade or type of metamorphism. For example, the name **migmatite** is applied only to very high-grade rocks that are mixtures of partial melt and residual metamorphic rock. **Mylonites** are rocks with mylonitic textures, regardless of their composition; they form only during dynamic metamorphism. **Hornfels** is a rock produced during contact metamorphism. Hornfelses are dense, fine-grained granoblastic rocks regardless of their composition or grade. The term hornfels is applied only when it is absolutely certain that contact metamorphism was involved. If there is any doubt, the descriptive term **granofels** is preferred because it does not specify the type of metamorphism.

During progressive metamorphism, each of the four chemically distinct types of rock undergoes a

TABLE 7–2

Metamorphic minerals and original rock composition.

Compositional Type	Original Rocks	Metamorphic Minerals[1]
Pelitic	Rocks rich in clay minerals (mostly shales and some siltstones and sandstones with clay-rich matrixes	Aluminum-rich minerals Andalusite Cordierite Kyanite Muscovite Sillimanite Staurolite Biotite[2] Chlorite[2] Garnet (almandine)[2]
Calcareous	Limestones and dolostones	Calcium and calcium-magnesium silicates Garnet (grossularite) Idocrase Wollastonite Actinolite[2] Diopside[2] Epidote[2] Talc[2]
Mafic	Basalts and gabbros	Garnet (almandine)[2] Hornblende Pyroxene (calcium-poor) Actinolite[2] Chlorite[2] Diopside[2] Epidote[2]
Quartzo-feldspathic	Felsic igneous rocks (granite, rhyolite) and sedimentary rocks (arkose, sandstone)	Quartz Plagioclase feldspar Potassic feldspar

[1]In addition to the minerals listed, quartz, plagioclase, and potassic feldspar may be present.
[2]These minerals can be formed in rocks of different initial composition. In such cases, the entire suite of minerals present must be used to identify the original rock type.

series of changes that is different from that of the others. We will now take a brief look at the metamorphism of each type to show how the rocks listed in Table 7–3 are related to each other and to further clarify the nature of metamorphic changes.

Metamorphism of Pelitic Rocks. The changes that shales experience depend on the type of metamorphism, as summarized in Figure 7–13. During early stages of regional and burial metamorphism, pressure completes the compaction begun during sedimentation. Platy clay minerals are aligned and pores are closed, resulting in the compact metamorphic rock called **slate**—Figure 7–14(a). With increased intensity, the clays undergo **dehydration** reactions, which remove water from the rock and produce small flakes of muscovite (mica) and chlorite. Foliation of these sheet silicates results in a shimmery surface in which grains are too small to be seen with the unaided eye. Rocks with this texture and mineralogy are called **phyllites**—Figure 7–14(b).

At still higher grades, grain size increases and biotite replaces chlorite by mineralogic reactions, resulting in a coarse-grained, strongly foliated, mica-rich rock known as **schist**—Figure 7–14(c). Continued metamorphism produces aluminous porphyroblasts in the schist, and these minerals should be mentioned as part of a rock's complete name. (For example, a coarse-grained, well-foliated rock composed of muscovite and biotite with a few porphyroblasts of garnet should be called a *garnetiferous muscovite-biotite schist.*)

TABLE 7–3
Classification of metamorphic rocks.

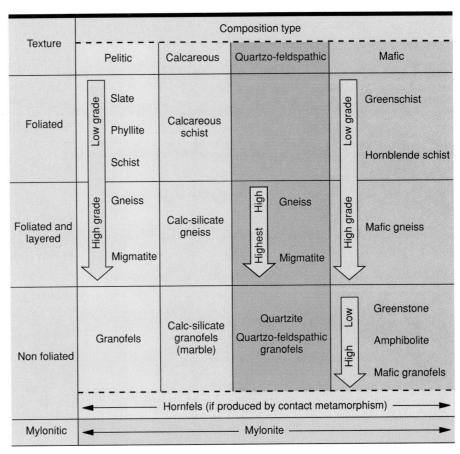

Texture	Composition type			
	Pelitic	Calcareous	Quartzo-feldspathic	Mafic
Foliated	Slate / Phyllite / Schist (Low grade)	Calcareous schist		Greenschist / Hornblende schist (Low grade)
Foliated and layered	Gneiss / Migmatite (High grade)	Calc-silicate gneiss	Gneiss / Migmatite (High / Highest)	Mafic gneiss (High grade)
Non foliated	Granofels	Calc-silicate granofels (marble)	Quartzite Quartzo-feldspathic granofels	Greenstone / Amphibolite / Mafic granofels (Low / High)
	Hornfels (if produced by contact metamorphism)			
Mylonitic	Mylonite			

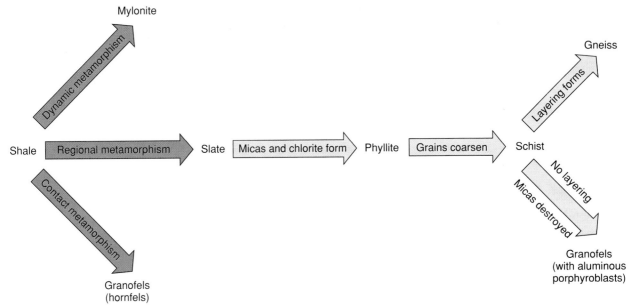

FIGURE 7–13
Metamorphism of pelitic rocks.

(a)

(b)

(c)

(d)

FIGURE 7–14
Metamorphosed pelitic rocks. From lowest grade to highest grade: (a) slate, (b) phyllite, (c) schist, (d) migmatite. Photos (a)–(c) Macmillan Publishing/Geoscience Resources; (d) by Allan Ludman.

Sometimes metamorphic layering develops at very high grades as the result of ionic migration, creating a **gneiss** by segregation of quartz and feldspars into light-colored layers, and micas into darker layers. If layering does not occur, the final product is a coarse, granoblastic rock composed of quartz, plagioclase, and potassic feldspars, with aluminous minerals such as garnet and sillimanite. At the highest grade, partial melting produces migmatite—Figure 7–14(d).

An example of a dehydration reaction that occurs commonly in high-grade pelitic rocks involves an interaction between muscovite and quartz:

$$\underset{KAl_2(AlSi_3O_{10})(OH)_2}{Muscovite} + \underset{SiO_2}{quartz} \rightarrow \underset{AlSi_2O_5}{sillimanite} + \underset{KAlSi_3O_8}{\underset{feldspar}{\underset{potassic}{}}} + \underset{H_2O}{water}$$

Metamorphism of Calcareous Rocks. Metamorphism of a pure calcite limestone or pure dolostone generally results in formation of the calcitic or dolomitic granofels called **marble**. Small amounts of other minerals give marble the wide variety of colors that makes it a favorite stone for sculpture, furniture, and buildings. If the limestone or dolostone contains abundant quartz and clay mineral grains, **decarbonation** reactions take place between these minerals and the carbonates, and released carbon dioxide. A common decarbonation reaction in limestones is

$$
\begin{aligned}
&\text{Calcite} + \text{quartz} \rightarrow \text{wollastonite} + \text{carbon dioxide} \\
&\text{CaCO}_3 \;+\; \text{SiO}_2 \;\rightarrow\; \text{CaSiO}_3 \;+\; \text{CO}_2
\end{aligned}
$$

Calcium-bearing silicate minerals that are produced during metamorphism, such as wollastonite, are called calc-silicate minerals. Some are elongate or tabular and are lineated or foliated if directed pressures were present when they formed. Metamorphic layering develops in many instances, and a **calc-silicate gneiss** is produced. The modifier *calc-silicate* is used to distinguish this kind of gneiss from aluminous quartz-feldspar gneisses formed from pelitic rocks.

Metamorphism of Mafic Rocks. Near volcanic islands such as Japan and the Aleutians, basalts are commonly interlayered with sedimentary rocks and can be metamorphosed along with them. Stages in the complex metamorphism of basalt are shown in Figure 7–15.

In the early stages, water driven from sedimentary rocks by dehydration reactions combines with minerals in the basalt (pyroxene, olivine, plagioclase) to form chlorite, epidote, and actinolite. All of these minerals except the feldspar are green, so low-grade metamorphosed basalts are called **greenstones** if granoblastic and **greenschists** if foliated.

At higher grades, greenstone and greenschist undergo dehydration reactions and progressively lose the water that had been added in the initial low-grade reactions. Hornblende forms from epidote, chlorite, and actinolite, and rocks composed mostly of hornblende and plagioclase result. They are called **amphibolites** if granoblastic and **hornblende schists** if foliated. Further metamorphism converts the hornblende to pyroxene, and a granoblastic **pyroxene granofels** is produced that may have nearly the same mineralogy as the basaltic protolith. A

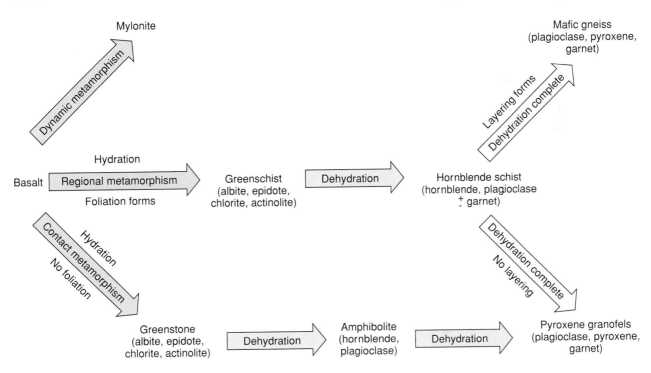

FIGURE 7–15
Metamorphism of mafic rocks.

mafic gneiss can be produced if layering develops by separation of pyroxene from the plagioclase.

Metamorphism of Quartzose and Quartzo-feldspathic Rocks. A quartz sandstone with a silica matrix undergoes little change during metamorphism. Eventually, matrix silica and siliciclastic particles recrystallize until there is no distinction between them. The resulting dense, tough granofels is called **quartzite**. Rocks such as granite and arkose experience similar textural changes and produce a quartzo-feldspathic granofels. There are no mineralogic changes in this case, because quartz and feldspars are stable over wide ranges of temperature and pressure.

INTERPRETATION OF METAMORPHIC ROCK HISTORY

A metamorphic rock has the potential to preserve records of both metamorphic and premetamorphic events. Our goal is to read this record in as much detail as possible, to learn three things about every metamorphic rock:

☐ *The type of metamorphism* involved, which may be indicated by texture and by field relationships. A mylonitic texture indicates dynamic metamorphism; metamorphism that decreases away from a pluton indicates contact metamorphism.

☐ *The nature of the protolith,* which can be deduced from the minerals present, as shown in Table 7–2. Minerals such as biotite can form during metamorphism of shales, basalts, or dolostones, so their presence cannot uniquely identify the protolith. We must examine all minerals present in the rock to accurately identify it. For example, the assemblage biotite–muscovite–chlorite–quartz–plagioclase is so aluminous that it could have come only from a pelitic protolith.

☐ *The intensity of metamorphism,* which is discussed next.

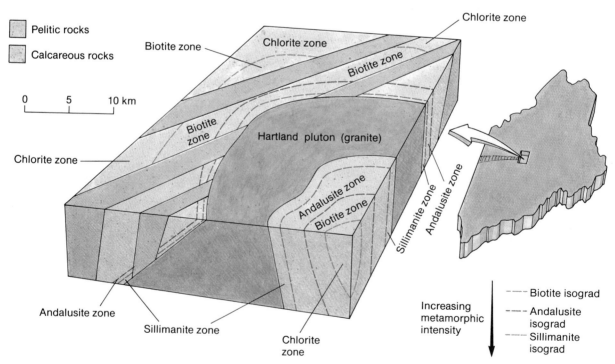

FIGURE 7–16
Contact aureole of the Hartland pluton, Maine.

Determining the Intensity of Metamorphism

Ideally, we want to know the *absolute intensity* of metamorphism: the exact temperature and pressure to which each rock has been subjected. Methods for determining absolute intensity have been developed only in the past few decades, and even now it is impossible to pinpoint conditions for every kind of rock. It is easy, however, to establish a scale of *relative intensity*—a scheme in which one rock can be described as having been more or less intensely metamorphosed than another. We already have used the terms "high grade" and "low grade" for relative intensity, but will now look at a more precise system.

Relative Intensity: Index Minerals and Metamorphic Zones. The contact aureole surrounding a pluton is an ideal place to study relative metamorphic intensity, because metamorphic grade decreases away from the pluton in all directions. We will study an aureole produced by a small granitic body to show how individual metamorphic minerals can be used to describe relative intensity. (Intensity of burial and regional metamorphism also can be described using the same principles.)

Figure 7–16 shows the relationship between the Hartland pluton, a granite that intruded into central Maine nearly 365 million years ago, and its host sedimentary rocks. The pluton has produced an aureole in surrounding pelitic and calcareous rocks that extends as much as 10 km from the contact. We will concentrate first on the pelitic rocks and then look at the calcareous types.

As one follows any single band of pelitic rock toward the granite, striking mineralogic changes are noticed. Green, chlorite-rich sandstones and shales far from the pluton are replaced by purplish-gray, biotite-rich slates and granofels closer to the contact. These are in turn followed by rocks containing porphyroblasts of andalusite and, within a few hundred meters of the contact, sillimanite crystals appear in the metamorphosed pelitic rocks. Enough relict textures are preserved to show that the rock within a single band in Figure 7–16 was originally all of the same type. Therefore, appearance of the different minerals must be due to different intensities of contact metamorphism. Sillimanite represents the highest grade of metamorphism, because it is found only

adjacent to the pluton where heat was most intense. Chlorite, found only in areas far from the heat source, represents the lowest grade. Similar reasoning explains the geographic distribution of andalusite and biotite.

The aureole can be divided into **metamorphic zones,** each identified by the presence of a mineral called a metamorphic **index mineral** that indicates a particular grade of metamorphism. Thus, chlorite is an index mineral that represents the lowest-grade metamorphism associated with the Hartland pluton. The *chlorite zone* encompasses the area where metamorphism was intense enough to produce chlorite but not intense enough to make biotite. Biotite, andalusite, and sillimanite also are index minerals, each recording an intensity greater than that of the adjacent index mineral farther from the granite, but less than that indicated by the adjacent index mineral closer to the contact.

Boundaries between mineral zones are called **isograds** (lines of *equal grade*) and are named after the higher-grade index mineral. Thus, the boundary between the chlorite and biotite zones is the biotite isograd; that between the biotite and andalusite zones is the andalusite isograd; and so forth.

Mineral zones also are found in the calcareous rocks surrounding the Hartland pluton, but are defined by different index minerals because of the different rock composition. If basalts had been present when the Hartland pluton intruded, a series of *mafic* index minerals also would have been produced.

Metamorphic Facies

The use of index minerals and mineral zones to define metamorphic intensity has a major drawback: it is restricted to rocks of a single compositional type. "Andalusite zone metamorphism" has no meaning to a geologist studying calcareous rocks because limestones do not have the aluminum needed to make andalusite, even if the correct temperature and pressure conditions were attained. Conversely, someone studying aluminous rocks has little use for calcareous or mafic index minerals. What is needed is a system that applies to all rocks.

The concept of **metamorphic facies** solves this problem. Just as a sedimentary facies contains all rocks deposited in a single environment, a metamorphic facies includes all rocks subjected to the same general conditions of temperature and pressure, regardless of their composition. Furthermore, rocks of one metamorphic facies pass into those of another at the point where intensity changes, just as sedimentary facies merge at boundaries between depositional environments. Facies are named after minerals present in one of the compositional types, but the name extends to all rocks subjected to the particular conditions.

Figure 7–17 illustrates the metamorphic facies and shows their relationships to location in the crust. Facies at the bottom of the diagram, occurring deep in the crust, have undergone greater pressure than those at the top; those at the right have experienced greater heat than those at the left. Boundaries between facies represent important mineral reactions. For example, the boundary between the amphibolite and granulite facies coincides with reactions marking the final dehydration of mafic rocks by conversion of hornblende to pyroxene.

A rock's mineral assemblage determines the facies to which it belongs. At a particular set of conditions, calcareous rocks would have one assemblage, pelitic rocks another, and mafic rocks a third, but all three assemblages would belong to the same facies. The most intensely metamorphosed rocks of the Hartland pluton aureole belong to the hornblende-hornfels facies, because the pelitic and calcareous rocks at the contact are typical of that facies throughout the world. Studies elsewhere of this facies enable us to predict what mafic assemblage would have been present if basaltic rocks had been present in the aureole. Indeed, the facies is named after the occurrence of hornblende in basalts metamorphosed to conditions comparable to those caused by the Hartland pluton. Rocks of the chlorite, biotite, and andalusite zones of the Hartland pluton's contact aureole belong to the albite-epidote-hornfels facies (also named for mafic mineral assemblages).

Absolute Metamorphic Intensity. Experiments that recreate metamorphic reactions are conducted to learn what temperatures and pressures exist when these reactions occur in nature. Powdered minerals are placed in an inert container and are heated and squeezed to the desired conditions. The contents of the container are then examined to see whether a reaction has taken place. By trial-and-error experiments such as this, the minimum conditions needed for a particular reaction can be estimated. Experiments such as this helped construct the stability diagram for the polymorphs of Al_2SiO_5 shown in Figure 7–18. The experiments show that sillimanite forms at higher temperatures than andalusite—a fact shown by the positions of these minerals around the Hartland pluton—and also indicate the temperatures and pressures at which the transition takes place. The absence of kyanite from the Hartland pluton contact aureole indicates that conditions of its formation were not found in the area, significantly narrowing the possible range of conditions.

FIGURE 7–17
Metamorphic facies and their positions in the crust.

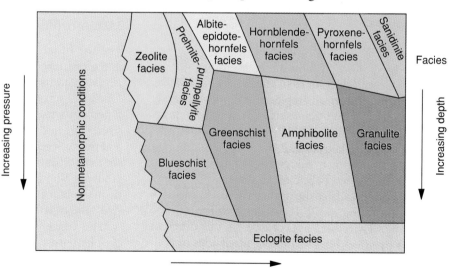

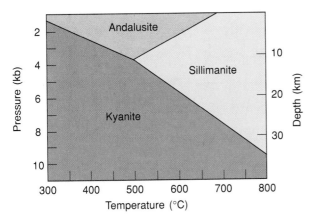

FIGURE 7–18
Stability diagram showing conditions under which kyanite, andalusite, and sillimanite can exist. With changes in temperature and/or pressure, one of the polymorphs can be converted to another.

METAMORPHISM AND PLATE TECTONICS

Most types of metamorphism can be found anywhere in the world. Contact metamorphism can happen at the margins of plutons or at the bases of lava flows on any continent or in the oceans. Similarly, burial, dynamic, and impact metamorphism are not restricted to any particular location.

Regional metamorphism is an entirely different matter. The tremendous scale of regional metamorphic events requires that large-scale forces be involved—forces comparable to those expected in plate collisions. The distribution of regional metamorphism in space and time provides vital clues to major Earth processes.

Geographic Distribution of Regionally Metamorphosed Rocks

Regionally metamorphosed rocks are found on every continent. Some occur in elongate belts that form the cores of mountain ranges such as the Himalayas and Alps and were formed during mountain building. Others are exposed in broad, flat terrains that cover hundreds of thousands of square kilometers at the centers of continents. These also formed during mountain building, but the mountains, once rivals of the Himalayas in height, have been leveled by hun-

dreds of millions of years of erosion. Extensive tracts of unmetamorphosed sedimentary and volcanic rock separate these two types of occurrence. For example, in North America, the broad, flat metamorphic terrain is called the Canadian Shield (Figure 7–19); it is separated from the metamorphic rocks of the Rocky, Appalachian, and Ouachita Mountains by the sedimentary strata of the Great Plains, the mid-continent platform, and the Appalachian plateau.

We have not sampled the oceans as thoroughly as we have the continents, but it appears that regionally metamorphosed rocks are rare in the oceans. They are restricted to narrow zones in island arcs such as the Japanese and Aleutian islands, but are not found on mid-ocean islands like Hawaii or Iceland, on ocean ridges, or on the seafloor. We saw in Chapter 4 that igneous activity in the oceans differs from that on the continents, and now we find that metamorphism also is drastically different. Compositional differences between continents and oceans were called upon in Chapter 4 to explain the distribution of different igneous rock types, but metamorphism affects rocks of all compositions. The scarcity of regionally metamorphosed rocks in the oceans indicates that the *processes* acting in the oceans also are different from those on the continents.

We saw in Chapter 5 that island arcs form above subduction zones. Now we see that such plate collisions also can explain regional metamorphism. In the plate tectonics model, subduction zones are places where two plates collide and one is forced down into the mantle beneath the other. Enormous directed pressures are produced in these collisions. Thick piles of sediment on the plates generate great lithostatic pressure, and heat is built up by the insulating effect of the overlying strata and by the upward movement of magmas. All the ingredients for regional metamorphism are present. This combination of metamorphic agents is not found anywhere else in the oceans. Thus, the plate tectonics model satisfactorily explains the restricted locations of regional metamorphism in the oceans.

Identifying Ancient Subduction Zones

When we carefully examine the types of metamorphism that occur in subduction zones, we recognize a "metamorphic signature" that enables us to identify sites of ancient plate collisions. Figure 7–20 shows

FIGURE 7–19
Distribution of regionally
metamorphosed rocks exposed at
the surface in North America.

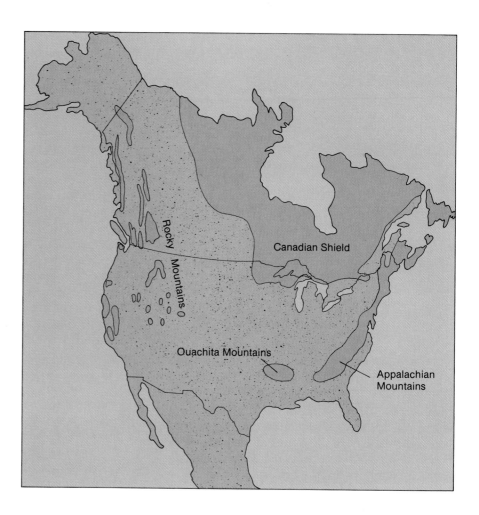

that the upper plate is subjected to both lithostatic and intense directed pressures close to the actual subduction zone, whereas areas farther from the zone experience less directed pressure and greater heat flow from the subduction-generated magmas. The effect of metamorphic agents thus differs in different parts of the upper plate.

Pressure dominates over heat as the most important agent in the subduction zone, resulting in what is called **high-pressure/low-temperature metamorphism** (A in Figure 7–20). Farther from the actual collision, rising heat becomes the dominant agent, and **high-temperature/low-pressure metamorphism** takes place (B in Figure 7–20). Another way of describing this is to note that temperatures increase less rapidly than pressures in the subduction zone and more rapidly than pressures far from the subduction zone.

When we examine rocks of ancient mountain belts, we often find that similar kinds of metamorphism have occurred. The evidence lies in the manner in which rocks of one metamorphic facies pass into those of another in a particular area (Figure 7–21). These paths are called **metamorphic facies series**. In New Zealand, parts of California, and areas of Japan closest to the Japanese trench, the zeolite facies is followed with increasing grade by the prehnite-pumpellyite and blueschist facies. This is the high-pressure/low-temperature type of metamorphism that we would expect in a subduction zone. In the Abukuma plateau area of Japan, the sequence is from the albite-epidote-hornfels facies through the pyroxene-hornfels facies. This is a high-temperature/low-pressure type of metamorphism, one that we expect to find in the upper plate far from the subduction zone.

FIGURE 7–20

Plate tectonics and metamorphism. Most of the ocean floor is devoid of metamorphism. Regional metamorphism occurs in the upper plate of subduction zones; dynamic metamorphism is restricted to the transform faults.

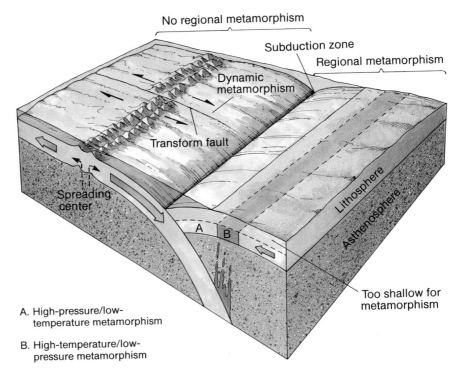

A. High-pressure/low-temperature metamorphism

B. High-temperature/low-pressure metamorphism

FIGURE 7–21

Metamorphic facies series and ancient subduction zones. Sequence of facies in California and New Zealand is that expected in an area close to a subduction zone, whereas that found in the Abukuma area of Japan is characteristic of the upper plate far from the subduction zone.

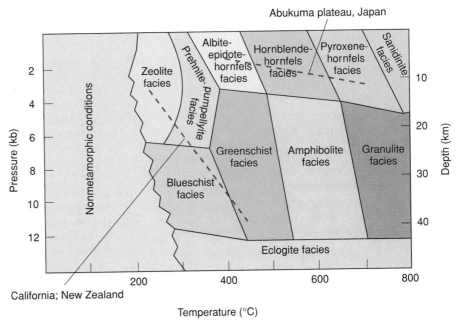

In some mountain belts, a pair of metamorphic facies series can be identified. One member of the pair is generally a high-pressure/low-temperature type, the other a high-temperature/low-pressure variety. By comparing the distribution of the facies series with that in Figure 7–21, geologists can infer not only that subduction has taken place, but in what direction the subducted slab was thrust downward.

METAMORPHISM AND PLATE TECTONICS | **181**

METAMORPHISM AND NATURAL RESOURCES

Metamorphism has created several materials that are even more important to industry than they are to geologists. These include both rocks and individual minerals.

Metamorphic Rock Resources

The two most widely used metamorphic rocks are probably marble and slate. Marble has been used for thousands of years as a decorative facing stone because of its almost endless color varieties (pink, tan, white, green, black, brown, and so on), texture, and layering. Marble is also a favorite medium for sculptors because it is relatively soft—it is, after all, composed mostly of calcite with a hardness of 3 on the Mohs scale.

Slate also comes in a wide range of colors, but is put to very different uses. Its dense, impermeable nature (caused by foliation of clay and mica flakes) and the ease with which it can be split into flat slabs make it an excellent roofing shingle. Many of the most expensive fish tanks and billiard tables use slate as a long-lasting base. We walk on it (as flagstones), set hot cups and pots on it (as trivets), and even write on it (until recently, all "blackboards" were made of slate).

Metamorphic Mineral Resources

Metamorphic minerals derived from nearly all grades and all types of rock have proven useful in our modern society. The graphite that we use as a lubricant or as the "lead" in pencils forms by metamorphism of carbon contained in shale, coal, or limestone. Corundum, used as an abrasive in its normal variety or as a gemstone in its ruby and sapphire varieties, is formed by metamorphism of highly aluminous rocks. Garnet also is used as an abrasive and a gemstone and is a common mineral in medium-grade rocks of several compositions.

Low-grade metamorphism of ultramafic rocks or impure dolostones yields the talc from which talcum powder is made. Low-grade ultramafic rocks also produce serpentine, best known in its fibrous variety, asbestos. Asbestos once was used extensively in construction, household items, and automobile brake linings, but discovery of its carcinogenic properties has led to a drastic reduction in its use.

High-grade minerals such as andalusite, sillimanite, and kyanite are used to make high-temperature porcelains used in spark plugs and in the refractory bricks that line steel blast furnaces.

Metasomatism is important in the formation of many commercially valuable ore deposits, because it often concentrates ions of the ore minerals to many times their normal abundances. This process is most common at the contacts between intrusive igneous rocks and their sedimentary hosts. Such deposits are called **contact metasomatic ores** and include tungsten minerals in Nevada and California, iron minerals in Utah and Pennsylvania, lead in New Mexico, and gold in Montana.

SUMMARY

1. Metamorphism consists of the solid-state changes that a rock undergoes when subjected to physical and chemical conditions different from those under which it first formed.
2. Changes in a rock's texture, mineralogy, and composition take place in response to heat, lithostatic and directed pressures, and chemically active fluids. Different types of metamorphism occur when these agents are applied in different proportions.
3. Textural adjustments during metamorphism include reorientation of grains, changes in grain shape, and recrystallization that changes grain size. Mineralogic changes result in new minerals which are more stable under the metamorphic conditions than the original ones. Many mineralogic changes involve the release of water vapor or carbon dioxide.
4. Metamorphism is a series of progressive changes. Rocks constantly adjust as metamorphic intensity increases, and early-formed metamorphic minerals often are converted to yet different minerals at later stages. Low-grade metamorphic rocks are those that have changed only slightly because of the low intensity of the metamorphic agents. High-grade rocks are produced by intense metamorphism and are generally very different from their low-grade or unmetamorphosed parents.
5. The mineralogy and texture of a rock depend on its original mineralogy; the intensity, duration, and type of metamorphism; and the addition or removal of ions.
6. Textures reveal the type of metamorphism to which a rock has been subjected. Foliation and lineation result

from directed pressures and hence form during regional or dynamic metamorphism. Granoblastic textures form in the absence of directed pressure (e.g., contact metamorphism) or in the absence of rodlike or platy minerals.

7. Metamorphic mineral assemblages indicate the intensity of the metamorphic agents. Either index minerals (and mineral zones) or complete assemblages (facies) can be used to determine relative metamorphic intensity. The absolute intensity is estimated by comparing laboratory-produced mineral assemblages with natural ones, assuming that the experimental reactions occur under the same conditions as those in nature.

8. Regional metamorphic rocks form the cores of continental interiors. They also occur in linear belts in mountain ranges but are absent from most of the ocean basins.

9. Regional metamorphism probably results from plate collision during subduction. Ancient thermal gradients comparable to those found in modern subduction settings can be estimated from the metamorphic facies series in eroded mountain ranges.

REVIEW QUESTIONS

1. Compare and contrast the roles of heat and pressure in causing metamorphism.
2. Discuss the different ways in which a rock may adjust to changes in its physical and chemical environment.
3. Each type of metamorphism usually can be identified by the effects that it has on rocks. How would you distinguish between rocks subjected to thermal, regional, and burial metamorphism?
4. Explain how minerals may be used to indicate the intensity of metamorphic agents.
5. Why are some minerals useful as index minerals whereas others are not?

6. Why are mineral assemblages more valuable than index minerals for assigning relative metamorphic intensities to rocks?
7. Why are metamorphic facies more useful than metamorphic zones as indicators of metamorphic intensity?
8. What factors determine the mineral content of a metamorphic rock?
9. A sheet of basalt sandwiched between layers of shale may be either a lava flow or a sill. Suggest a method that can distinguish between these possibilities by using the principles of metamorphism.
10. A sequence of interbedded shales and arkosic sandstones is subjected to intense regional metamorphism. Although these rocks experience exactly the same conditions, they will respond differently. Describe the differences and explain how they come about.

FURTHER READINGS

Ehlers, E. G., and Blatt, H. 1980. *Petrology: Igneous, sedimentary, and metamorphic.* San Francisco: W. H. Freeman & Company, 732 p. (An excellent in-depth explanation of all three rock types and the processes that form them.)

Hyndman, D. W. 1985. *Petrology of igneous and metamorphic rocks.* 2d ed. New York: McGraw-Hill Book Company. (An excellent treatment of the two rock types; relates them to large-scale plate tectonics processes.)

Spry, A. 1969. *Metamorphic textures.* New York: Pergamon Press, 350 p. (A classic treatment of metamorphic rocks; relates textures to the processes that produce them.)

Williams, H., Turner, F., and Gilbert, C. 1982. *Petrography.* 2d ed. San Francisco: W. H. Freeman & Company, 626 p. (A comprehensive overview of all three rock types, with drawings illustrating textures of common metamorphic rocks.)

8
Geologic Time

Grand Canyon sequence, Grand Canyon National Park, Arizona. Photo by Wiley/Wates/Profiles West.

ost people think of time in terms of recent events, lifetimes, or even generations. This is *historical* time. Geologists, however, deal routinely with much greater intervals of time. **Geologic time** is measured in relation to the origin of the Earth—an event that scientists believe occurred about 4.6 *billion* years ago. The human concept of time is so limited that it is very difficult to grasp the vastness of geologic time. For perspective, consider that humanlike creatures began to appear only in the *last one-tenth of one percent* of geologic time! To understand and appreciate the material in subsequent chapters of this book, you must think from now on in geologic time, rather than human time.

Geologists describe the age of Earth materials and events in two different ways, which we will refer to as relative age and absolute age:

☐ **Relative age** defines the age of a feature compared to that of others. For example, a sedimentary rock containing the bones of dinosaurs must be relatively older than a rock containing the bones of humans.

☐ **Absolute age** is another way to express the age of Earth materials and events, determining their age in actual number of years before the present by using the radioactive isotopes and other dating methods discussed later in this chapter.

We will start our discussion of geologic time by considering the sequence of discoveries that allowed early geologists to determine the relative ages of rocks. We will then consider how absolute ages were determined for these rocks, enabling geologists to construct a geologic time scale for the Earth.

RELATIVE TIME

By the end of the nineteenth century, geologists had used a combination of physical and biological methods to establish the relative ages of rocks in local areas. They then expanded their work until they were able to determine the relative ages of rocks on a regional or even continental scale. The concepts that were developed to determine the relative ages of rocks may seem obvious to us in hindsight, but many of them were revolutionary at the time.

Physical Methods for Determining Relative Ages

Several of the basic principles that geologists use to determine the relative ages of rocks came from observations made by Nicolaus Steno, a Danish geologist and anatomist, in the latter half of the seventeenth century. Steno noticed that many sediments accumulate by the deposition of individual particles from air or water. This results in a series of horizontal layers, with the oldest at the bottom and youngest at the top. Steno's observations led him to propose three basic principles in 1669:

1. The principle of **superposition,** perhaps the most important, states that in any sequence of rocks, the oldest is at the bottom and the youngest is at the top (assuming that the rocks have not been overturned) (Figure 8–1).

2. His principle of **original horizontality** states that sedimentary rocks normally are deposited in horizontal layers. If rock layers are inclined, we can attribute that to some postdepositional

FIGURE 8–1
Superimposed forests buried in volcanic ash, Yellowstone National Park, Wyoming, illustrating Steno's principle of superposition. Photo courtesy of National Park Service.

FIGURE 8–2

Tilted Precambrian sedimentary and volcanic rocks exposed along the Colorado River in Grand Canyon National Park, Arizona. The angular attitude of these beds indicates that they were tilted after deposition, according to the principle of original horizontality first proposed by Steno. Note the angular contact with overlying horizontal Paleozoic sedimentary rocks. Photo by Nicholas K. Coch.

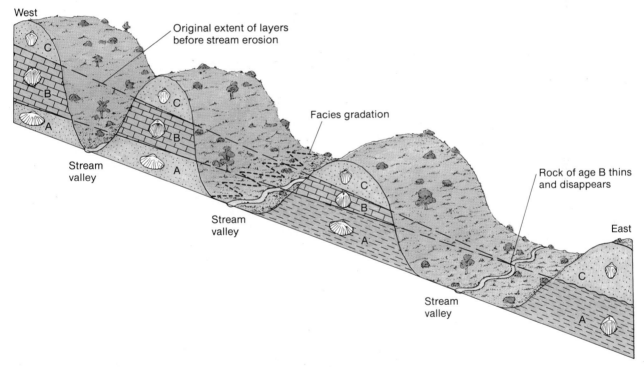

FIGURE 8–3

Even when rock exposures are widely separated geographically, their individual layers can be traced laterally by observing distinctive physical characteristics and fossils. This illustrates Steno's principle of lateral continuity. Rocks of ages A, B, C were deposited in the western part of the area, but rocks of age B were not deposited in the eastern part, creating a disconformity (wavy line) in the rock section there.

force that tilted or folded the originally horizontal layers (Figure 8–2).

3. The principle of **lateral continuity** states that, as a sedimentary layer is deposited, it extends outward horizontally until it either thins and disappears, or until it terminates against the boundaries of the basin in which it is accumulating. This last principle suggests that if a layer appears in a local area, the same layer (or its facies equivalent—see Chapter 6) probably would appear in adjacent areas as well (Figure 8–3).

Some of Steno's principles are shown in the geologic section in Figure 8–3. Layer A underlies the whole area, but it undergoes a facies change from fossiliferous sandstone in the west to fossiliferous shale in the east. Such a facies change can be seen today where modern beach and nearshore sands grade offshore into silt and clay. In contrast, rock layer B thins and disappears eastwardly. We will discuss the other aspects of this diagram later in this chapter.

In 1788 James Hutton, the Scottish geologist who had proposed the principle of uniformitarianism, devised two methods for determining relative age:

1. His principle of **cross-cutting relationships** states that a rock that cuts across another layer must be *younger* than that layer. For example, the granite dike shown in Figure 8–4 must be younger than the metamorphic rock it intruded, because that rock had to already exist to be cut across.

2. His principle of **inclusions** states that a rock including fragments of another rock must be *younger* than the rock whose fragments it contains. This principle is illustrated in Figure 8–5, in which the metamorphic rock in the lower part of the picture was exposed and

FIGURE 8–4
Zoroaster Granite dike (pink) cutting across the Vishnu Schist (dark grayish green) in Grand Canyon National Park, Arizona. The dike must be younger than the schist, based on the principle of cross-cutting relationships. Photo by Nicholas K. Coch.

FIGURE 8–5

Contact between Precambrian Vishnu Schist (dark rock in lower third of photo) and Cambrian Tapeats Sandstone (white and gray rock in upper portion), Grand Canyon National Park, Arizona. The sandstone contains pebble gravel derived from the underlying schist; it was incorporated while the sandstone was being deposited. Thus the Tapeats Formation is the younger, based on the principle of inclusions. Photo by Nicholas K. Coch.

weathered into fragments on an ancient land surface. The encroaching seas that deposited the overlying sandstone incorporated these fragments as inclusions in the base of the layer. The sandstone is younger than the metamorphic rock because the metamorphic rock had to be there beforehand to supply the fragments.

How does a geologist use these physical criteria to determine the relative age of a rock layer in a section? The techniques used are described in the following analysis of rock units shown in Figure 8–6.

First we use top-and-bottom sedimentary structures (Chapter 6) to determine whether the rock layers are *in the same order* in which they were originally deposited or whether they have been overturned since deposition. The mud cracks at the top of shale A open upward, and the oscillation ripple marks at the top of the sandstone B point upward; both indicate that the rock layers have the same orientation as when they were deposited. Therefore, the principle of superposition indicates that the oldest layer is shale A at the bottom of the section and that layers B, C, and D are progressively younger.

Igneous intrusion E cuts across layers A, B, C, and D, so it must be younger than all of them, according to the principle of cross-cutting relationships. E does not cut layers F and G, so it must have intruded *before* those layers were deposited.

Sandstone F contains fragments of both layer D (below) and the igneous intrusion E, so it must be younger than both, according to the principle of inclusions. Finally, the principle of superposition indicates that the uppermost layer G is younger than layer F.

Thus, we have used the principles of Steno and Hutton to determine that the relative ages of the rocks in this section are A (oldest), B, C, D, E, F, and G (youngest).

Biological Methods for Determining Relative Ages

As soon as the relative ages of rock units in local areas were determined from the principles of Steno and Hutton, it became apparent that rocks of the same relative age had the same kinds of fossils. Once that association was made, it became possible to determine the relative age of a sedimentary rock in another area by its fossils alone.

William Smith, a British engineer who supervised the digging of canals in England in the late 1700s and early 1800s, made the first observations linking rock strata and their fossil content. After examining many exposures in isolated mines and canal excavations, he was able to show the following:

1. Rock layers with distinct physical characteristics and the same overlying and underlying rock layers occurred over a wide area.

FIGURE 8–6
Using the principles of Steno and
Hutton to determine the relative
ages of rock layers in a section.

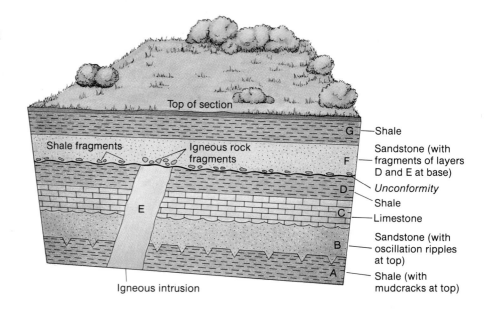

Top of section

G — Shale

Shale fragments — Igneous rock fragments

F — Sandstone (with fragments of layers D and E at base)

D — *Unconformity*

— Shale

C — Limestone

E

B — Sandstone (with oscillation ripples at top)

A — Shale (with mudcracks at top)

Igneous intrusion

2. Each of these sedimentary rock layers contained at least one type of fossil that was not found in the other layers.

3. The rock layers in geographically isolated exposures could be identified on the basis of their fossil content alone.

Smith's observations were very significant because they provided the means by which rock layers in widely separated areas could be related to one another. To see how these principles are applied in field studies, refer back to Figure 8–3. Note that each layer of rock has a distinct group of fossils associated with it. Layer A, a sandstone in the west, undergoes a facies change to become a shale in the east. However, the critical part of the layer, showing the facies transition, has been removed by stream erosion. A geologist studying this area would see layer A as two different rock units with an unknown time relation between them. However, examination of the sandstone and the shale shows that both of them have some of the *same fossils* and thus are of the same age, even though they have different physical characteristics. The geologist would therefore be able to deduce the actual facies relationship.

At about the same time that Smith was doing his work in Britain, across the English Channel Georges Cuvier was making important discoveries in the Paris basin of France. In using the principle of superposi-

tion to determine the relative ages of the rock layers, he showed in 1812 that each layer in the Paris basin not only had a distinct group of fossils, but that the fossils in the lowest layers (oldest by superposition) were most dissimilar to present-day animals, and that fossils closer in appearance to present-day forms occurred in the upper layers. Such an upward sequential change in fossil types is referred to as **faunal succession**.

Cuvier tried to explain the differences in fossils from layer to layer as the result of the catastrophic extinction of one organism and the appearance of a somewhat different form by the time that the next rock layer was deposited. Cuvier's contribution to relative dating was very important, because up to that time superposition could be used only for relative dating on a local scale. With the principles of superposition and faunal succession as tools, geologists were now able to apply relative dating on a regional scale, and eventually on a global scale.

The explanation for faunal succession was provided half a century later by Charles Darwin in his 1859 classic, *On the Origin of Species by Means of Natural Selection*. Darwin showed that fossil forms evolved slowly over a long period of time, resulting in progressive changes in appearance. He explained the marked differences between fossil assemblages in some adjacent beds not by catastrophes, but by the absence *at that place* of rock layers that would have recorded the transitional evolutionary stages of the fossils. Darwin's worldwide travels led him to see that fossils missing from a rock layer in one area could be found somewhere else. Thus, although the evolution of fossil forms continued through time, the *evidence* for this change might be missing at any one locality.

Gaps in the Geologic Record—Unconformities

As geologic studies continued in the latter half of the nineteenth century, it became more and more apparent that the rock record *at any one locality* was incomplete. Why were rocks deposited in some areas, but none in other areas at the same time? We can see the answer around us today: broad areas of the Earth, such as the oceans and the plains spreading away from great mountain chains, are major areas of sediment deposition; while at the same time, broad areas of the continents are dominated by erosion and have little if any sedimentation (such as the Colorado plateau, where the Grand Canyon is located). Throughout geologic time, such widespread areas of deposition or predominant erosion have coexisted. When deposition resumes in an area that has been subject to erosion for a long time, the overlying rocks may be much younger than the rocks below.

A break in the rock record, where some layers are missing, is an **unconformity**. The reason that the layers are missing may be because they were never deposited, or because they were eroded after deposition. The formation of an unconformity is shown in Figure 8–6. After deposition of layer D and the subsequent intrusion of E, there was a break in sedimentary deposition, during which portions of D and E were eroded. (Other layers that may have been above D and E—if any were deposited—also were completely eroded.) The information these units might have contained was irretrievably destroyed, creating a "gap" in the geologic record.

Later, when deposition resumed with layer F, pieces of D and E were incorporated into the base of F. These pieces were fragments that weathered from D and E but had not been transported out of the area. The actual surface of the unconformity is symbolized by the wavy line cutting across the tops of D

and E. Three basic types of unconformities are recognized:

1. A **disconformity** exists when sedimentary rocks of two significantly different ages are in contact and the beds above and below the unconformity are *parallel*—Figure 8–7(a). The loss of rock record may have resulted either from a lack of deposition (Figure 8–3) or from an actual removal of rock before the overlying layer was deposited—Figure 8–7(a). Disconformities are the commonest type of unconformity.

2. An **angular unconformity** occurs where the beds above and below the unconformity are *not parallel* with each other—Figure 8–7(b). An angular unconformity develops when the underlying beds are tilted or folded and then eroded before the deposition of the overlying beds. A good example of an angular unconformity is the contact between the tilted Precambrian sedimentary volcanic rocks and the overlying horizontal Paleozoic rocks in Grand Canyon National Park.

3. A **nonconformity** exists where intrusive igneous or metamorphic rocks have been eroded and then overlain by sedimentary rocks—Figure 8–7(c). The igneous and metamorphic rocks form at a great depth below the surface. Therefore, a great deal of time is required for the overlying rocks to be eroded away, permitting the igneous or metamorphic rocks to be exposed at the surface, eroded, and then overlain by the sedimentary rocks which are deposited only at the Earth's surface.

   An outstanding example of a nonconformity is that between the Precambrian Vishnu Schist and the Cambrian Tapeats Sandstone in the Inner Gorge of the Grand Canyon (see Focus 8–1, Figure 1; Chapter 1, Figure 1–1). The gap in the rock record at this particular location on the unconformity spans a period of about one *billion* years.

The Relative Geologic Time Scale

Even though the rock and fossil record is incomplete in any one place, geologists are still able to piece together a nearly complete record by examining a

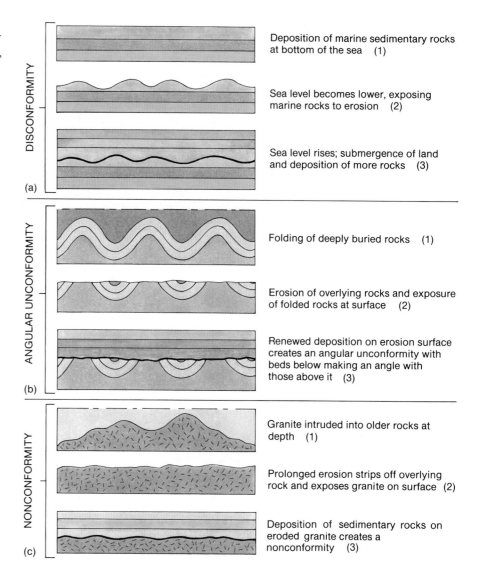

FIGURE 8-7
Development of different types of unconformities: (a) disconformity, (b) angular unconformity, (c) nonconformity.

DISCONFORMITY

Deposition of marine sedimentary rocks at bottom of the sea   (1)

Sea level becomes lower, exposing marine rocks to erosion   (2)

Sea level rises; submergence of land and deposition of more rocks   (3)

(a)

ANGULAR UNCONFORMITY

Folding of deeply buried rocks   (1)

Erosion of overlying rocks and exposure of folded rocks at surface   (2)

Renewed deposition on erosion surface creates an angular unconformity with beds below making an angle with those above it   (3)

(b)

NONCONFORMITY

Granite intruded into older rocks at depth   (1)

Prolonged erosion strips off overlying rock and exposes granite on surface   (2)

Deposition of sedimentary rocks on eroded granite creates a nonconformity   (3)

(c)

great number of rock exposures in different areas. By the latter part of the nineteenth century, the geologic column had been subdivided on the basis of differences in fossils in the sedimentary rock record, and a **relative geologic time scale** was being established. It is shown on the left side of Table 8–1.

The geologic time scale is divided into eons, eras, periods, and epochs:

☐ **Eons** include great spans of geologic time and are defined by very major events. For example, the Phanerozoic Eon includes the time during which advanced life has flourished on Earth. In contrast, the earlier Proterozoic Eon includes the abundance of primitive life, such as bacteria and algae, in the latter part of the Proterozoic.

☐ **Eras** are defined by major differences in dominant life forms. For example, reptiles dominated the Mesozoic Era, whereas mammals dominated the later Cenozoic Era.

☐ **Periods** are subdivisions of eras, based on evolutionary changes that are less dramatic than those used to distinguish eras.

☐ **Epochs** are subdivisions of periods that are defined by even subtler life changes than those used for periods.

TABLE 8-1
Relative and absolute geologic time scales.

Relative Time				Isotopic Time	
Eon	Era	Period	Epoch	MYBP[1]	Some Changes in Life Forms
Phanerozoic	Cenozoic	Quaternary	Holocene	0.01	● First modern humans
			Pleistocene	1.6	
		Tertiary	Pliocene	5	● First hominids
			Miocene	24	
			Oligocene	37	● First grasses
			Eocene	58	● First large mammals
			Paleocene	66	
	Mesozoic	Cretaceous	(Numerous epochs recognized)	144	● Flowering plants dominant, dinosaurs at peak
		Jurassic		208	● First birds and mammals
		Triassic		245	● Beginning of age of large reptiles (dinosaurs)
	Paleozoic	Permian	(Numerous epochs recognized)	286	● First reptiles
		Pennsylvanian[2]		320	● First winged insects
		Mississippian[2]		360	● First trees and land vertebrates (amphibians)
		Devonian		408	● First land plants and wingless insects
		Silurian		438	● First corals
		Ordovician		505	● First fish and shellfish
		Cambrian		570	
Proterozoic[3]					● Complex, soft-bodied, multicellular life in latest part
					● Colonial algae
				2500	
Archean[3]					● Blue-green algae appear
					● Primitive bacteria appear
					● No life known
				4000	
Hadean[3]					● No rocks of this eon are known on Earth
					● Solid Earth forms
				4600	

Precambrian

[1]Beginning of the time interval, in MYBP (millions of years before the present). Geologic dates based on Palmer, A. R., 1983, Decade of North American geology time scale (DNAG). *Geology* 11:503–04.

[2]In Europe, the Pennsylvanian and Mississippian Periods are known collectively as the Carboniferous Period, with upper and lower portions. In North America geologists identify the specific subdivisions Pennsylvanian and Mississippian.

[3]The eons of Precambrian time have been subdivided into eras, but are not shown. The Archean has three life stages (Early, Middle, Late); the Proterozoic has two.

THE GRAND CANYON—STAIRSTEPS THROUGH GEOLOGIC TIME

Grand Canyon National Park in Arizona contains one of the most exceptional series of exposed rock formations found anywhere in the world, in terms of color, thickness of rock section exposed, variety, and the span of geologic time represented (Figure 1). This sequence extends from the Colorado River upward through a vertical distance of 1370 m (South Rim) and 1680 m (North Rim) to the plateau above.

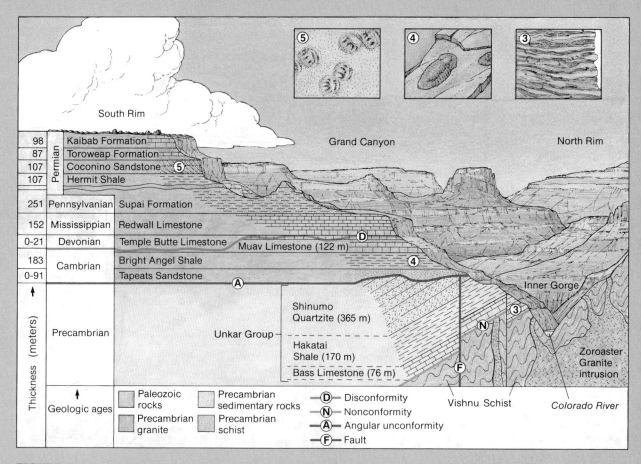

FIGURE 1

The three diagrams (5, 4, 3) at the top of this figure show representative fossils found at different levels in the canyon. They illustrate the founal succession that occurs in the Grand Canyon rock sequence. Photos of the actual fossils are shown in Figure 3–5. After Nations and Beus, 1976, "Teaching Historical Geology in the Grand Canyon," *Journal of Geological Education* 24:77–81. Used with permission.

Only a fraction of the thickness of rocks originally deposited in the Grand Canyon region has been preserved; several types of unconformities occur within the sequence. For example, the contact between the most ancient rocks—the Vishnu Schist—and the overlying Precambrian sedimentary and volcanic rocks is a *nonconformity* (N on Figure 1). The tilting and erosion of the Precambrian sedimentary and volcanic rock sequence and the subsequent marine deposition of the Tapeats Sandstone over the eroded land surface created an *angular unconformity* (A on Figure 1). The Paleozoic Tapeats Sandstone, Bright Angel Shale, and Muav Limestone were deposited during this submergence by the sea. The sea then retreated, leaving the area high and dry. Part of the rock record was removed by erosion before the sea rose over the land once again to deposit the Temple Butte Limestone. This created a *disconformity* (D on Figure 1). These unconformities point out how incomplete even an extensive sedimentary record can be.

The Grand Canyon sequence represents a wide variety of Earth processes. The lowermost rock, the Vishnu Schist, and the granitic dikes of the Zoroaster Granite that intrude it, represent a mountain-building event that occurred over a billion years ago. The lower Paleozoic formations (Cambrian through Mississippian) are composed of different marine facies. In contrast, the Mississippian-to-Middle Permian deposits are all continental. They represent stream deposits in the lower part (Supai Formation and Hermit Shale) and desert conditions (Coconino Sandstone) in the upper part. The Coconino Sandstone is characterized by large-scale cross-bedding (Figure 2) which was formed in ancient dune fields. The uppermost forma-

FIGURE 2
Cross-bedding in the Coconino Sandstone. Photo courtesy of E. D. McKee/U.S. Geological Survey.

tions, the Toroweap and Kaibab, were deposited as marine waters once again covered the Grand Canyon region at the close of the Paleozoic Era.

The Grand Canyon sequence also provides an exceptional example of faunal succession. Precambrian sedimentary rocks contain only the most primitive life forms such as algae (Figure 3). Lower Paleozoic marine rocks show abundant traces of invertebrates such as trilobites (Figure 4), whereas upper Paleozoic rocks show fossils of vertebrates such as reptiles (Figure 5).

Visitors to the Grand Canyon all agree that it is one of the most spectacular sights on Earth because of its size, the cliffs and terraces developed on its sides, and the strikingly different colors of its formations. Geologists savor these features too, but with the added knowledge that the canyon's layers are a unique geologic exposure because they provide a rock record of the greatest portion of geologic time visible anywhere.

FIGURE 3
Algal bedding in the Precambrian Bass Limestone, Grand Canyon National Park, Arizona. Photo by Nicholas K. Coch.

FIGURE 4
Fossil of primitive invertebrate (trilobite) in Cambrian Bright Angel Shale, Grand Canyon National Park, Arizona. Photo courtesy of E. D. McKee/U.S. Geological Survey.

FIGURE 5
Footprints of primitive reptile (lizard) in Permian Coconino Sandstone, Grand Canyon National Park, Arizona. Photo courtesy of E. D. McKee/U.S. Geological Survey.

The greatest portion of geologic time (3.9 billion years out of 4.6 billion) is the Precambrian, which extends from the formation of the Earth to the start of the Phanerozoic Eon. Although the Precambrian has long been considered an "era," it includes 85% of geologic time. We now know enough of events during the Precambrian to divide it into three eons (Table 8–1). These eons have been divided further into eras (not shown in the table).

The Paleozoic Era (from Greek words meaning *ancient life*) was marked by the first appearance of animals with hard shells, such as clams. This era was dominated by invertebrates (animals without backbones); fish, land plants, and amphibians made their first appearance.

The Mesozoic Era (meaning *middle life*) was dominated by reptiles such as the dinosaurs.

The Cenozoic Era (meaning *recent life*), the most recent, represents the time when mammals and flowering plants became dominant. Scientists believe that the oldest humanlike fossils are only about 3.5 million years old. For perspective, note that *modern* humans have existed for less than one-tenth of one percent of the Earth's history.

Establishment of the relative geologic time scale made it possible for geologists to establish the time-equivalence, or **correlation,** of rock layers in widely separated areas. We can see how correlation is accomplished by returning to Figure 8–3. The similarities of fossils in sandstone bed A in the west and shale bed A in the east establish that these beds were deposited at about the same time. We can therefore say that the sandstone and the shale *correlate*. Note how fossils help us to understand the geologic history of the eastern area. The absence of fossil B in the eastern part of the area suggests that rocks of age B were never deposited there, or if they were deposited, they must have been eroded before the deposition of layer C. The contact between A and C is therefore a disconformity; its surface is symbolized by the wavy line on Figure 8–3.

ABSOLUTE TIME

The physical and biological methods used in the eighteenth and nineteenth centuries to determine the *relative* age of rocks finally allowed us to place geologic materials and events in a relative time frame-

work. However, no one had an idea of the *actual* age in years of the rocks or events. For example, some people believed the age of the Earth to be less than 10,000 years, whereas others suspected it was much older. No one could yet *scientifically* answer basic questions: How old is the Earth? When did life begin? At what time did humans appear? Answers to these basic questions had to wait until accurate measurements of geologic time became possible early in the twentieth century. We shall examine these time-measurement methods, but first we should ask: How *do* we measure time?

We use a number of different devices to measure time, but all follow the same principles. Whether we use a sand-filled hourglass, a clock with a swinging pendulum, or some other device, all good timepieces *must operate at a fixed and known rate*. Thus time that passes can be measured by the amount of sand that falls through the hourglass or the number of pendulum swings. In addition, a good timepiece *must operate continually* and *must not be affected by any external factors*. All of these criteria must be satisfied by "geologic clocks" used by geologists to determine the ages of rocks.

Early Attempts

Geologists of the nineteenth century tried to use Earth processes as clocks to obtain absolute dates for past events.

In 1897, British physicist William Thomson (better known as Lord Kelvin) calculated that the Earth took 20 to 40 million years to cool from an assumed initial molten state to its present temperature. Kelvin's calculations now are known to be wrong because he assumed that all the Earth's heat was produced when it formed, and that the heat loss since then could be calculated. We now know that the breakdown of radioactive elements in the Earth's crust produces new heat, thus slowing the rate of cooling. Therefore the age of the Earth must be much greater than that proposed by Kelvin.

John Joly, another British physicist, estimated the age of the Earth in 1899 from the salt content of the oceans. He assumed an initial freshwater ocean on a primitive Earth into which streams were pouring salts (their dissolved load) at a fixed rate. He arrived at an age of 80 to 90 million years by measuring the average salt content of the present rivers and comparing that with the total salt content of the oceans. However, one of the problems with Joly's estimate was that he addressed only the salt being put *into* the oceans. We have seen that salt also is constantly being *removed* from the oceans to form evaporites (Chapter 6). In addition, fossil animals and other evidence indicate that the primitive oceans reached a salt content similar to today's as long ago as the Precambrian Era.

Radioactivity as a Dating Tool

The discovery of radioactivity by French physicist A. H. Becquerel in 1896 finally gave geologists a fixed-rate process that could be used to obtain the *actual* ages of rocks. Radioactivity is an excellent dating tool because nuclear reactions proceed at a constant rate that is unaffected by changes in temperature, pressure, or chemical reactions. British physicist Lord Ernest Rutherford first suggested in 1905 that radioactivity could be used to determine the age of the Earth. Shortly afterward, B. B. Boltwood in America used the radioactive breakdown of uranium to date some minerals that ranged up to about 2 billion years in age. With Boltwood's first analyses, **geochronology** was born—the science of obtaining isotopic dates for geologic materials and events.

Types of Nuclear Reactions Used in Geochronology

A product of a nuclear reaction is called a **radiogenic** product. In each reaction, the starting isotope is called the **parent,** and the isotope produced by decay of the parent is called the **daughter.** Each radioactive dating method used by geologists employs one of three different types of nuclear reactions: alpha decay, beta decay, or electron capture.

Alpha Decay. Alpha decay occurs naturally in several elements, including the isotopes of uranium. Recall from Chapter 2 that all isotopes of an element have the same atomic number. This means that they have the same number of protons. What differs from isotope to isotope is the number of neutrons. Some isotopes of certain elements decay; these are called **radioactive isotopes** or **unstable isotopes.** Uranium has atomic number 92 (its nucleus contains 92 protons). One of its more common isotopes is $^{238}_{92}U$,

which contains 146 neutrons and thus has an atomic mass of $92 + 146 = 238$. Another isotope is $^{235}_{92}U$, which has 143 neutrons and thus an atomic mass of 235. Both uranium-238 and uranium-235 are radioactive. Both break down into isotopes of the element thorium:

$$^{238}_{92}U \rightarrow {}^{234}_{90}Th + {}^{4}_{2}\alpha + energy$$
$$^{235}_{92}U \rightarrow {}^{231}_{90}Th + {}^{4}_{2}\alpha + energy$$

Notice that the uranium has lost two neutrons and two protons, which are ejected from the nucleus as a single particle called an **alpha particle** (α). The energy released was what originally held the alpha particle in place. A nuclear reaction that releases an alpha particle is called **alpha decay.**

Beta Decay. The type of nuclear reaction in which the parent-to-daughter transition is accompanied by release of a small negatively charged particle is called **beta decay.** The particle is called a **beta particle** (β^-).

An example of beta decay is the transition of the radioactive isotope of rubidium, $^{87}_{37}Rb$ to strontium:

$$^{87}_{37}Rb \rightarrow {}^{87}_{38}Sr + \beta^- + energy$$

Notice that the daughter has gained a proton, but there is essentially no change in atomic mass. Where does the additional proton come from? How can it be added without changing the mass of the daughter from that of the parent? Furthermore, the beta particle has the physical properties of an electron, but how can an electron be ejected from the nucleus, which is composed only of protons and neutrons?

These questions can be answered if we think of a neutron as consisting of a proton joined to an electron. A review of Chapter 2 shows that the mass of a neutron is nearly exactly the mass of a proton plus an electron, and the neutral charge also fits this model. If one neutron in the $^{87}_{37}Rb$ nucleus is converted to a proton and an electron, the atomic mass remains virtually unchanged, but the atomic number is increased by 1. The electron is then expelled as a beta particle.

Electron Capture. Electron capture is involved in the radioactive decay of potassium-40, $^{40}_{19}K$ to the gas argon-40, $^{40}_{18}Ar$:

$$^{40}_{19}K \rightarrow {}^{40}_{18}Ar + energy$$

In this case, the atomic number decreases by 1 and the atomic mass remains the same. This decay scheme differs from the other types of radioactivity in that it involves an electron from the electron shells as well as a nuclear particle. One electron combines with one of the protons in the nucleus to form a neutron. Hence the name **electron capture**. No particle is ejected in this process, but energy is emitted.

Series Decay. In some cases the daughters, such as $^{234}_{90}Th$ and $^{231}_{90}Th$, *also* are radioactive. The radioactive daughter continues to break down into a succession of radioactive isotopes until a nonradioactive daughter is produced. In the case of $^{234}_{90}Th$, the final stable form is an isotope of lead, $^{206}_{82}Pb$. This type of decay, in which there are a series of steps between the parent isotope and a stable daughter isotope, is called **series decay**.

Obtaining Isotopic Dates

Isotopic dating depends upon the radioactive decay of isotopes at known rates, or half-lives. This is how it works: as minerals crystallize, they may incorporate atoms of some radioactive isotopes into their structure. These atoms decay—that is, their radioactivity converts them to daughter isotopes—at a constant rate, called the **decay rate**. Decay rates for all radioactive isotopes used in geochronology have been determined directly from laboratory experiments. **Half-life** is the time required for one-half of the total number of parent atoms to decay into the daughter isotope. This process is shown schematically in Figure 8–8 for the potassium-argon method, where the parent $^{40}_{19}K$ has a half-life of 1.3 billion years. (For illustrative purposes, only a few atoms are shown, but in reality *great numbers are involved*.)

The mineral shown in Figure 8–8 incorporated some radioactive isotopes of $^{40}_{19}K$ into its structure when it crystallized 3.9 billion years ago. One half-life later (2.6 billion years ago), 50% of the parent atoms, $^{40}_{19}K$, had been converted into daughter atoms, $^{40}_{18}Ar$. After two half-lives, only 25% of the original parent atoms were left. Three half-lives from the time of mineral formation, only 12.5% of the original parent atoms are left; and so on. Note that as the percentage of parent decreases, the percentage of daughter increases (Figure 8–8). The sum of the parent and daughter atoms at any time remains the same. A date obtained by such use of radioactive isotopes is called a **radiometric date**.

Isotopic ages for minerals are obtained by crushing the rock and separating the minerals that contain the parent and daughter products. These minerals are then analyzed to determine the percentage of parent and daughter atoms. As the half-life is known, the age of the mineral can be obtained. For example, if the percentages of parent and daughter atoms is the same—i.e., their ratio is 1:1—the time elapsed since formation of the mineral is one half-life. Consider the example in Figure 8–8. A laboratory analyzes the minerals and reports that the percentage of parent atoms is 12.5% (parent:daughter ratio = 1:7). This means that three half-lives (3 × 1.3 billion years) have elapsed since the crystal formed. The rock containing the crystal is now known to be 3.9 billion years old.

Not all kinds of rocks are useful for isotopic dating. Igneous and metamorphic rocks are best for determining dates because the minerals in them crystallized (or recrystallized) at the time of rock formation. In contrast, most minerals in sedimentary rocks are derived from the breakdown of older rocks and thus cannot indicate an accurate date for the time of formation of the sedimentary rock. An exception is glauconite, one of the few minerals that form as sediments accumulate; thus this potassium-rich mineral can provide a date for the rock. Some success has been obtained in dating glauconite in sedimentary rocks.

Accuracy of Isotopic Dating: Resetting the Geologic Clocks

In any of the dating schemes involving a parent-to-daughter breakdown, three criteria must be satisfied to obtain an accurate date for the crystallization of a mineral:

1. The mineral must have measurable contents of both parent and daughter. For example, with present technology, isotopic dating methods can be used to determine ages as great as 12 half-lives. (After 12 half-lives have passed, only 1/4000 of the parent survives.)

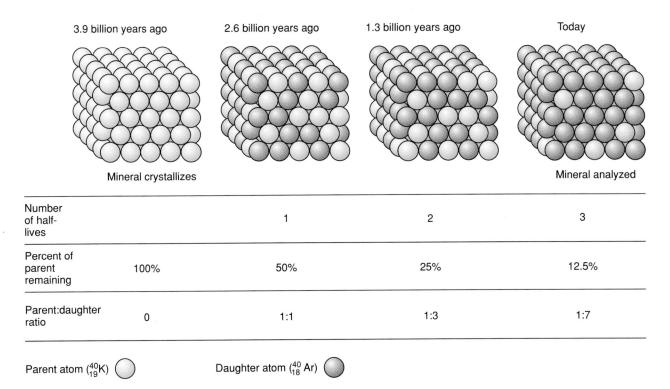

	3.9 billion years ago	2.6 billion years ago	1.3 billion years ago	Today
	Mineral crystallizes			Mineral analyzed
Number of half-lives		1	2	3
Percent of parent remaining	100%	50%	25%	12.5%
Parent:daughter ratio	0	1:1	1:3	1:7

Parent atom ($^{40}_{19}$K) ◯   Daughter atom ($^{40}_{18}$Ar) ◯

FIGURE 8–8
Changes over time in the proportion of parent and daughter atoms in a crystal. Example shows potassium-argon decay. (Only a few atoms can be shown in this illustration; in reality, great numbers are involved.)

2. Conditions must have precluded any gain or loss of parent or daughter.
3. No daughter isotope can have been trapped in the mineral at the time of formation, or if some was, it must be possible to determine the amount and adjust the isotopic date accordingly. (The trapped daughter product is called **nonradiogenic.** If a correction for the trapped daughter product is not made, the calculated isotopic age will be greater than the true one. The technique for this correction is described later in the chapter.)

Additional errors may result from sampling and analytical procedures. *Every* isotopic date involves some uncertainty, but under ideal conditions, the error in the isotopic date can be as small as ±2%. If the same age is determined from the same material using several methods, the age is said to be **concordant,** and a high degree of confidence can be placed in the date.

If a rock is heated during metamorphism, some products of radioactive decay may be driven from the crystals. If these rocks are then dated, the reduced daughter content will give an apparent date *younger than* the true date. When the daughter product is a gas, as in the potassium-argon method, the daughter can be lost more readily than in dating methods where parent and daughter are nongases. In this case, the date obtained cannot be the original time of crystallization of the mineral but rather that of the later metamorphism.

Where the daughter product is not a gas, metamorphic changes may drive the daughter from the crystals in which it formed, but it still may be retained in the *rock*. In this case, an age can be obtained from the rock by crushing it and analyzing the whole sample, instead of just analyzing the mineral. An isotopic date obtained in this manner is called a **whole-rock date**.

Isotopic Dating Methods

Major isotopic dating schemes are outlined in Table 8–2. The wide range of half-lives and source materials enables geologists to date a wide variety of geologic materials. In many cases, two different methods can be used to obtain a concordant date on a sample.

Uranium-Lead Dating. This method was the first used to date minerals. It is based on the radioactive decay of two uranium isotopes—uranium-235, $^{235}_{92}U$, and uranium-238, $^{238}_{92}U$. The $^{235}_{92}U$ ultimately breaks down to form lead-207, $^{207}_{82}Pb$, whereas the $^{238}_{92}U$ breaks down to form lead-206, $^{206}_{82}Pb$. All naturally occurring uranium deposits contain both uranium-235 and -238 isotopes; therefore, a rock can be dated using *both* isotopes. The date determined from the second isotope serves as a check on the first.

At first, only uranium ores could be dated by this method. It is now possible to analyze minute amounts of uranium and lead in the mineral zircon, $ZrSiO_4$, and in other minerals that are common in many igneous rocks. Recent developments in uranium-lead dating enable geochronologists to determine (1) whether lead or uranium has been lost or gained since the rock was formed, (2) the time when the change occurred, and (3) the correct age of the rock, in spite of the alteration.

Rubidium-Strontium Dating. Rubidium-87 undergoes beta decay, forming strontium-87. Metamorphic rocks and felsic plutonic rocks are commonly dated using the rubidium-strontium method.

Potassium-Argon Dating. The radioactive isotope potassium-40 decays to form two radiogenic products:

$$^{40}_{19}K \left\langle \begin{array}{l} \longrightarrow 89\% \text{ beta decay to } ^{40}_{20}Ca \\ \longrightarrow 11\% \text{ electron capture to } ^{40}_{18}Ar \end{array} \right.$$

Most of the potassium (89%) undergoes beta decay to form radiogenic calcium-40. This decay branch is not useful for isotopic dating because the radiogenic calcium-40 usually cannot be distinguished from the large amounts of nonradiogenic calcium-40 originally present in the crystal. About 11% of the potassium-40 undergoes decay by electron capture, forming argon-40, a chemically nonreactive gas that accumulates within the crystal. This decay branch is useful in isotopic dating because no radiogenic argon-40 is incorporated into minerals at their formation. However, argon-40 is a gas, so subsequent heating of the rocks during metamorphism may drive off part or all of the argon-40. Consequently, the isotopic age of the rock after metamorphism would be *younger* than its true age. The date is important, however, because it indicates the time of *metamorphism*.

The half-life of potassium-40 permits it to be used in dating materials as young as 100,000 years. It can thus be used to date rocks that are much younger than those which can be dated by the uranium-lead and the rubidium-strontium methods (Table 8–2). The potassium-argon method has been used extensively to date rocks from the last few million years, when major human evolutionary changes occurred. For example, potassium-argon dating of volcanic ash beds interlayered with sedimentary beds containing human fossils has allowed anthropologists to date successive stages of human evolution.

Samarium-Neodymium Dating. The samarium-neodymium technique is the newest major isotopic dating method and is based on the breakdown of $^{147}_{62}Sm$ to $^{143}_{60}Nd$, with a half-life of 106 billion years. It has proved very useful for obtaining the ages of previously undatable rocks, because samarium and neodymium both are rare-earth elements that are relatively immobile in solution. Therefore, these two elements are not easily removed from a rock even when it has been weathered, metamorphosed, or altered by hydrothermal solutions. This method's major use has been for dating very old mafic rocks and rocks derived from the mantle of the Earth.

Carbon-14 Dating: Window on the Earth's Recent Past. The great majority of radiometric dates used in geology are obtained from radioactive isotopes other than carbon-14. However, the carbon-14 method is probably the best-known isotopic dating method because of its extensive application in the dating of

TABLE 8–2
Major radioactive isotopes used to obtain absolute dates.

Method	Parent	Daughter	Decay Scheme	Half-life of Parent (yr)	Effective Age Range (yr)	Materials Commonly Dated	Comments
Uranium-lead	$^{238}_{92}U$	$^{206}_{82}Pb$	Chain of alpha and beta decays	4.5 billion	100 million[1] to 4.6 billion	Zircon, sphene Uraninite[2] Whole rocks[2]	Uranium-238 and uranium-235 are found together in all uranium-bearing deposits. A radiometric date obtained from one isotope can be cross-checked against that obtained from the other.
Uranium-lead	$^{235}_{92}U$	$^{207}_{82}Pb$	Chain of alpha and beta decays	0.71 billion	100 million[1] to 4.6 billion	Zircon, sphene Uraninite[2] Whole rocks[2]	
Rubidium-strontium	$^{87}_{37}Rb$	$^{87}_{38}Sr$	Beta decay	48.8 billion	100 million[1] to 4.6 billion	Muscovite Biotite Potassic feldspar Whole igneous rock Metamorphic rock	Very useful in studies of metamorphic and plutonic rock. Oldest Earth-rock dates obtained by this method.
Potassium-argon	$^{40}_{19}K$	$^{40}_{18}Ar$	Electron capture	1.3 billion	100,000[1] to 4.6 billion	Muscovite Biotite Sanidine Hornblende Glauconite[2] Whole rock (basalts and clean glasses)	Potassium-argon dating of volcanic deposits interlayered with sediments containing fossil human bones has provided dates for successive stages of human evolution. If rock is heated subsequently, radiogenic argon may be driven off, giving an age younger than the true one.
Carbon-14	$^{14}_{6}C$	$^{14}_{7}N$	Beta decay	5730	100–50,000; up to 70,000 with special techniques	Carbon-bearing materials including wood, charcoal, bone, cloth, paper, cave deposits, underground water, and oceanic water	Extremely versatile over its age range. Particle accelerators will greatly expand uses for carbon-14 dating by extending age range covered and allowing analysis of much smaller samples.
Samarium-neodymium	$^{147}_{62}Sm$	$^{143}_{60}Nd$	Beta decay	106 billion	100 million to 4.6 billion	Whole rock (mafic igneous and metamorphic), some minerals such as garnet and zircon[2]	Especially useful for very ancient mafic rocks. Age accuracy is not easily disturbed by weathering and hydrothermal activity.

[1]Less common age.   [2]Less common material.

glacial and other events of the last 70,000 years in the disciplines of archaeology, anthropology, and geology.

The carbon-14 method is different from the others we have described. It is rarely used to date rocks because the half-life of carbon-14 is only 5730 years, making it impossible to date materials more than 70,000 years old. The great majority of rocks are much older than 70,000 years. Further, the method can be used to date only the remains of once-living organisms, not inorganic rock material.

Carbon has six protons (atomic number 6). Two of its isotopes are carbon-12, $^{12}_{6}C$, which has six neutrons, and carbon-14, $^{14}_{6}C$, which has eight neutrons. Carbon-12 is stable, whereas carbon-14 is radioactive. Carbon-14 forms continuously in the upper atmosphere when cosmic rays from nuclear reactions on the Sun bombard stable nitrogen atoms, $^{14}_{7}N$, releasing a proton and forming the radioactive isotope $^{14}_{6}C$. The atoms of carbon-14, along with the atoms of stable carbon-12, combine with oxygen to form carbon dioxide, CO_2. This carbon dioxide mixes rapidly through the atmosphere and through the oceans, lakes, underground water, and glaciers of the hydrosphere. Plants manufacture sugars and starches from this carbon dioxide, incorporating the carbon-14 into their tissues. When animals eat the plants, the carbon-14 becomes incorporated into their tissues as well.

Carbon-14 reverts very quickly to nitrogen through beta decay. However, *as long as the organism lives,* new carbon-14 continually enters the organism's tissues. The carbon-14 level in the organism thus reaches a concentration equal to the concentration of carbon-14 in the atmosphere and hydrosphere. However, as soon as the animal dies, decayed carbon-14 can no longer be replaced, and the amount of radioactive carbon in the organism declines steadily. The amount of carbon-14 is measured *indirectly* by counting the beta particles that are emitted from a given sample as it decays. For example, after about 5730 years have elapsed (one half-life), the number of beta particles being emitted per unit time by the carbon-14 remaining in the sample is only one-half that during the life of the organism.

The carbon-14 method is highly versatile. It has been used to date a wide variety of organic materials such as wood, charcoal, peat, shells, charred bones, paper, cloth, pollen, leaves, seeds, and other carbon-bearing objects in many archaeological, anthropological, and geologic studies. The age of the Dead Sea Scrolls was determined by dating the linen in which they were wrapped. The dates of charcoal from the campfires of ancient peoples has enabled anthropologists to plot their migration across the North American continent after the last Ice Age ended. The carbon-14 method has been especially helpful in determining dates for events in the last Ice Age. In some cases, we can determine with great precision when glaciers advanced into an area. For example, at Two Creeks, Wisconsin, the advancing glacier knocked over a forest and killed the trees

FIGURE 8–9
Forest buried by glacier at Two Creeks, Wisconsin. These trees were rooted in the light-colored forest soil in the bottom two-thirds of the photo. The advancing glacier felled the trees and buried them in the reddish glacial deposits at top. Date of the glacial advance (about 11,580 years ago) was obtained by dating the wood debris, using the carbon-14 method. Photo by Nicholas K. Coch.

(Figure 8–9). Carbon-14 dating of the fossil wood gave an age of 11,580 years for the glacial advance.

Carbon-14 dating currently is being revolutionized by the use of high-energy particle accelerators to measure *directly* the number of carbon-14 atoms in a sample (instead of counting beta-particle emissions). Particle-accelerator dating speeds up the analytical process greatly (hours rather than days for older techniques), and uses samples one-thousandth the size needed in present analyses. A theologically significant application of such carbon-14 dating was to analyze the famous Shroud of Turin, the cloth in which the body of Jesus was believed by many to have been wrapped. A number of laboratories analyzed minute quantities of the shroud in an attempt to reach a concordant date. The shroud dated at about A.D. 1300, indicating that it could not have been the burial cloth of Jesus.

Other Dating Methods

We have described the most widespread methods for obtaining accurate dates. However, other methods are used in more specialized situations.

Varves. In Chapter 6 we described a varve, an annual rhythmic stratification in sediments. Each varve consists of one or more coarse-grained graded laminae (summer) and a dark, fine-grained lamina (winter). Counting the number of varves in an exposure can give an idea of the length of time it took for them to accumulate. (This assumes that only one varve formed each year, and that all the varves were preserved.) Varve analysis has indicated how long glacial lakes have existed in a given area.

Tree-Ring Dating. In temperate climates, the trunk and branches of a tree usually increase in thickness by one layer during each growing season. When the tree is cut, these layers can be seen as concentric rings. Each layer usually represents a year's growth, so it is called an **annual ring**. The patterns of the rings also provide information about the region's past climate. For example, rings grown in wet years are thicker than those in dry years. By counting the rings, one also may be able to estimate how long the tree has lived.

Over a number of years, trees in any one region display a characteristic sequence of annual rings of different widths. If we examine trees of overlapping ages, a continuous record of tree-ring variation can be built up. This record is extended back in time by study of long-lived tree species such as the bristlecone pine of the American Southwest, which can live for over 4000 years. The age of a wood sample containing a number of annual rings can be dated by comparing the ring patterns with a "standard" pattern for the area. Tree-ring dating in the United States can be extended back in time to about 8254 years before the present—around 6200 B.C.

Amino Acid Dating. As fossil bones "age" there is a known rate of change in the ratio of two different forms of amino acids. After a long period of time the two amino acid types equalize in abundance. An approximate age may be determined by calculating the ratios of the two types of amino acids. This dating scheme works best for materials that have not been thermally altered and are between 200 years and 1 million years old.

Fission-Track Dating. As uranium-235 and -238 decay within a crystal, alpha particles are released. These high-velocity particles tear through the crystal, leaving **fission tracks** through the crystal structure (Figure 8–10). When the crystal surfaces are polished and etched with a strong solvent, the tracks become more visible and can be counted under a microscope. The first tracks start forming shortly after the mineral crystallizes, and the number of tracks per unit area, called the **track density,** increases with time. The track density also increases with uranium content, and so it is necessary to measure uranium content before a fission-track age can be calculated. A wide variety of crystals and glasses have been dated, and this method can be used to date rocks from those that are very young to those that are the oldest on Earth.

The Isotopic Time Scale

Geologists use a number of methods to determine the ages of rocks in an area. Igneous and metamorphic rocks can be dated directly, using the appropriate radioactive isotopes shown in Table 8–2.

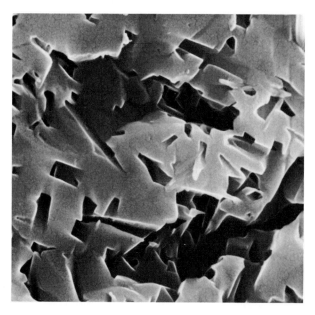

FIGURE 8-10
Fission tracks in a crystal. Photo courtesy of National Aeronautics and Space Administration.

Sedimentary rocks, the most abundant on the Earth's surface and generally the only ones that contain fossils, can be dated directly only in rare cases where they contain minerals that formed as the rock was deposited, such as glauconite. However, ages for sedimentary rocks can be obtained *indirectly* if the rocks are either cut by, or interlayered with, igneous rocks.

The techniques used in determining true ages of rocks in a section are illustrated in Figure 8-11. The rock sequence includes a number of fossiliferous and nonfossiliferous sedimentary rocks, interlayered with volcanic ash and lava flows and cut by a granite dike. The relative ages of the various layers are established by the principles of superposition, cross-cutting relationships, inclusions, and faunal succession.

Isotopic dating of volcanic ash bed F gives an age of 300 million years, and for lava flow H an age of 230 million years. By relative dating, we already have learned that fossiliferous sandstone G is younger than F and older than H, and so we can state that the sandstone is between 230 and 300 million years old. Layers A through E can be analyzed in a similar manner.

Once geologists have dated the rock layers in several areas, they often can use the information to date rocks in other areas. For example, many sedimentary rocks are not associated with datable igneous and metamorphic rocks. However, if we are trying to date a sedimentary rock that has the same fossils as those found in layer G, we know that the bed *correlates* with layer G.

Focus 8-2 asks and answers some of the basic questions we posed earlier about the age of the Earth and life upon it.

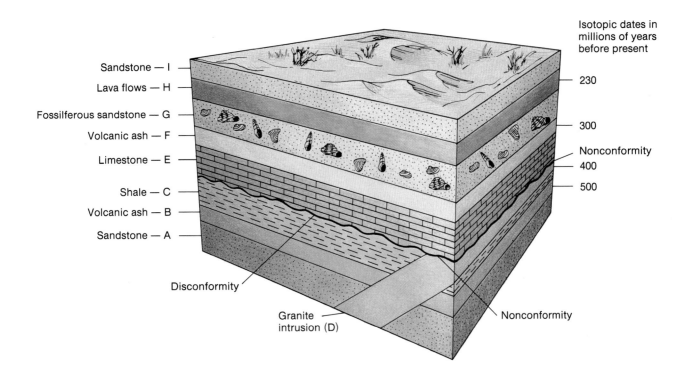

Isotopic dates in millions of years before present

Sandstone — I
Lava flows — H — 230
Fossilferous sandstone — G
Volcanic ash — F — 300
Limestone — E — Nonconformity
— 400
Shale — C — 500
Volcanic ash — B
Sandstone — A

Disconformity

Granite intrusion (D)

Nonconformity

Layer		Relative age	Actual age or age range (millions of years before present)
Youngest	I	Younger than H	Younger than 230
	H	Younger than G Older than I	230
	G	Younger than F Older than H	230–300
	F	Younger than E Older than G	300
	E	Younger than D Older than F	300–400
	D	Younger than A, B, and C Older than E	400
	C	Younger than B Older than D	400–500
	B	Younger than A Older than D and C	500
Oldest	A	Older than D Older than B	Older than 500

FIGURE 8–11
Determining the ages of a sedimentary rock series, interbedded with and cut across by datable igneous rocks (see text).

A s more field studies and isotopic dating were carried out, the absolute ages of relative time spans became subdivided more finely. Establishment of an **absolute time scale,** or **isotopic time scale** (Table 8–1, right side), now enables geologists to answer basic questions about the Earth and the evolution of life.

How Old Is the Earth?

Geologists believe that the meteorites colliding with the Earth today formed at or near the time of the formation of the Earth. Isotopic dating of meteorites gives ages around 4.6 billion years. Isotopic dating of Earth materials cannot reach back that far—see the next question.

What Is the Oldest Rock on Earth?

In October 1989, Dr. Samuel Bowring of Washington University in St. Louis reported finding rocks in the Northwest Territories of Canada that were older than any previously reported (Figure 1). The samples, collected near the Arctic Circle about 250 miles north of Yellowknife, were dated as 3.96 billion years old. However, a significant portion of the Earth's early rock record still remains unaccounted for. We may find some older rocks, but the *really* old rocks are probably long gone, recycled by Earth's internal and external processes into their component elements.

When Did Life Evolve on Earth?

Geologists have reported the existence of primitive bacterialike cells from chert deposits in northwestern Australia that are 3.5 billion years old (Figure 2). The cells contained no nuclei and were linked together like pearls in a necklace. Rocks as old as 3.8 billion years contain materials that some scientists think may be organic remains.

The chronology of evolution has been determined by radiometric dating of igneous rocks that bracket the ages of fossil-bearing sedimentary rocks. The first recognizable primitive soft-bodied animal fossils occurred in the Late Precambrian Era (Proterozoic Eon). Abundant animal fossils with hard, mineralized skeletons appear in the Cambrian Period about 570 million years ago, and humans began to appear around 1.5 to 2.0 million years ago.

FIGURE 1
Outcrop of gneiss in Northwest Territories of Canada, dated at 3.96 billion years old. This is
the oldest rock discovered on Earth thus far. Photo courtesy of Samuel Bowring.

When Did Humans Evolve?

The dating of human evolution is one of the most fascinating applications
of geochronology, because it concerns us all. Geologists, working hand in
hand with anthropologists, are providing the isotopic dates for successive
stages of human evolution. The researchers have been fortunate because
many of the sediment layers containing human and humanlike fossils also
occur interbedded with layers of volcanic ash. Radiometric dates (usually
potassium-argon) can be obtained for the ash layers, thus bracketing the age

FIGURE 2

World's oldest documented life form (bacteria) from 3.5 billion-year-old rock in western Australia. (a) Photos of two forms of bacteria. (b) Artist's rendering of what complete fossil probably looked like. Photomicrographs courtesy of J. W. Schopf.

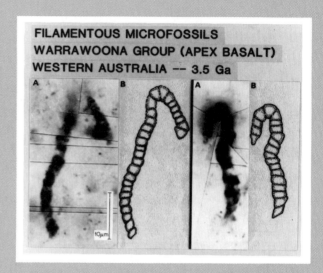

FILAMENTOUS MICROFOSSILS
WARRAWOONA GROUP (APEX BASALT)
WESTERN AUSTRALIA — 3.5 Ga

of the intervening fossil-bearing layers. We will close our discussion of geologic time by describing some research that has shed a new light on human evolution.

In 1979, anthropologist Mary D. Leakey, working in Tanzania, East Africa, discovered a set of fossil footprints (Figure 3) made by early humanlike forms called *hominids*. This find indicated that hominids, with feet similar to ours, were walking erect *half a million years earlier* than indicated by previous fossil evidence. These primitive humans had walked across a freshly deposited volcanic ash layer, and the footprints they left were preserved under overlying sediments and volcanic ash deposits.

Potassium-argon dating of biotite crystals in the overlying and underlying ash layers established the age of the footprints as *between 3.6 and 3.8 million years*. Leakey believes that when these hominids walked fully upright, they were free to use their forelimbs for other purposes. This presented new opportunities, which the brain subsequently evolved to take advantage of. As Leakey says:

> . . . those footprints out of the deep past, left by the oldest known hominids, haunt the imagination. Across the great gulf of time I can only wish them well on that historic trek. It was, I believe, part of a greater and more perilous journey, one that—through millions of years of evolutionary trial and error, fortune and misfortune—culminated in the emergence of modern man.[1]

[1]M. D. Leakey, 1979, Footprints in the Ashes of Time, *National Geographic* 155(4):457. Used with permission.

FIGURE 3
Fossil footprints in volcanic ash in
East Africa. Photo by John Reader/
Science Photo Library.

SUMMARY

1. Geologic time differs from historical time because it is measured in millions and billions of years, back to the origin of the Earth about 4.6 billion years ago.
2. Geologic events or materials (sediments, rocks) may be dated *relatively* by their relation to other events or materials and by fossil content, and *absolutely* by using radioactive isotopes.
3. Relative dating is accomplished by applying the principles of superposition, cross-cutting relationships, inclusions, and other physical criteria.
4. Fossils are used in relative dating to identify rocks of a certain age.
5. The relative geologic time scale is based on differences in evolution of fossil animals. The major subdivisions, called eras, are defined by major changes in life forms; less dramatic evolutionary changes are used to divide eras into periods and epochs.
6. Detailed studies have shown the geologic record to be incomplete at any one place, but a fairly complete record of Earth history can be obtained by piecing together the rock record from different areas. Gaps in the geologic record, called unconformities, result either from lack of deposition or from erosion of older rocks prior to the deposition of the overlying rocks.
7. Isotopic dating is based on the known, constant decay rates of certain radioactive isotopes (parents), which break down to form radiogenic products (daughters).
8. Accurate isotopic dates require that (1) the mineral must have measurable amounts of parent and daughter, (2) there must be no parent or daughter loss or gain except for the parent-to-daughter decay, and (3) there must be no daughter material present initially (or there must be some way of correcting for it).
9. When an igneous or sedimentary rock is metamorphosed, there may be a loss of daughter from the mineral, especially if the daughter is a gas like argon-40. If the daughter has been driven out of the original minerals but is still present within the rock, the age of the initial mineral formation may be obtained by the whole-rock method.
10. The absolute time scale provides absolute dates in years for the subdivisions of relative geologic time. Igneous and metamorphic rocks are dated directly by one or another isotopic method.
11. Sedimentary rocks can be dated if they are cut across by igneous rocks or are interlayered with them, or if they grade laterally into other sedimentary rocks containing fossils.
12. Isotopic dating has provided new information about the age of the Earth (4.6 billion years), the oldest rocks found so far (3.96 billion years), the oldest fossils (3.1–3.5 billion years), and evolutionary stages that culminated in modern humans.

REVIEW QUESTIONS

1. Describe the physical methods by which geologists determine if a series of rocks are in the same order as when they were deposited.
2. How can facies change be used to correlate sedimentary rocks?

3. Describe three physical methods that can be used to tell the relative ages of sedimentary rocks. Give an example for each.

4. Can an unconformity surface be a disconformity, nonconformity, or an angular unconformity at different places where it is developed? Can you think of an example?

5. What problems must we be aware of in using the potassium-argon dating method?

6. How is an isotopic age determined for a mineral?

7. A rock with a potassium-argon date of 640 million years has a date of 980 million years by the uranium-lead method. What could account for the discrepancy between the two dates?

8. How were actual dates determined for the appearance of different life forms on Earth?

9. What are the differences between the carbon-14 dating method and other isotopic methods?

10. What conditions must be met to get accurate isotopic dates?

11. If the rock record is incomplete at any one place, how was the relative geologic time scale established?

12. What principles allow fossils to be used to tell the relative ages of strata in an area?

FURTHER READINGS

Berry, W. B. N. 1968. *Growth of a prehistoric time scale*. San Francisco: W. H. Freeman & Co., 158 p.

Calvin, W. H. 1986. *The river that flows uphill*. New York: Macmillan Co. (Written as a diary while rafting down the Colorado River, the book describes the evolution of life and the changing rock record in the Grand Canyon.)

Faul, H. 1977. A history of geologic time. *American Scientist* 66:59–165.

Harland, W. B., Cox, A. V., Llewellyn, D. G., Pickton, C. A. G., Smith, A. G., and Walters, R. 1982. *A geologic time scale*. New York: Cambridge Univ. Press.

Morris, S. C. 1987. The search for the Precambrian-Cambrian boundary. *American Scientist* 75:157–67. (Describes the techniques used to define a significant era boundary.)

Nations, J. D., and Beus, S. S. 1976. Teaching historical geology in the Grand Canyon. *Journal of Geological Education* 24:77–81. (Describes a field trip down into the Grand Canyon and discusses activities and logistics for the trip. Source in part for the material in Focus 8–1.)

9

Weathering
and Soils

Volcanic neck and feeder dikes of intrusive body at Ship Rock, New Mexico. The intrusion is more resistant than the rocks that were intruded, so differential weathering has left this spectacular remnant. Photo by Danny Lehman.

The shattering of granite by frost in the winter, the blurring of the inscription on a marble tombstone, and the crumbling of the sandstone facing on a building all show that rocks exposed at the Earth's surface are subject to continual breakdown. The changes that occur in rocks and sediments as they react with the hydrosphere, atmosphere, and biosphere are called **weathering**. Weathering occurs because most of the minerals in rocks and sediments were formed under conditions that are very different from those at the surface. When these minerals are exposed at the surface, they encounter much lower temperatures and pressures, and they face a different chemical environment, namely an abundance of free oxygen, carbon dioxide, and water. Weathering breaks the rocks apart into smaller pieces and transforms many of their minerals into new ones that are more stable under surface conditions.

Weathering is a vital process on Earth. It frees from rocks elements that are essential for life. These elements are then incorporated by plants, and some are transferred to animals that eat the plants. Weathering of rocks also produces new sediments, agricultural soils, and—under special conditions—even deposits of iron and aluminum. The rate of weathering may differ for adjacent rocks; this is called **differential weathering** and is an important factor in the development of Earth's varied landscapes.

In most places on the surface, rocks are covered by a layer of loose rock and sediment called **regolith**. Regolith derived from the weathering of underlying rocks is called **residual regolith**. Regolith not derived from the rocks below but transported into the area by streams, glaciers, wind, or landslides, is called **transported regolith**. Transported regolith will be examined in more detail in Chapters 10, 11, and 13–15; here we will focus on how residual regolith develops.

Residual regolith forms by a number of different weathering processes that break down the original material. These processes can be grouped into two main categories: physical and chemical.

☐ **Physical weathering** is the *disintegration* of parent rock material into smaller and smaller pieces. However, the minerals undergo no change in structure or composition.

☐ **Chemical weathering** is the *decomposition* of the parent rock material into new minerals and soluble ions.

The relative importance of these two types of weathering depends on several factors, including the temperature range, abundance of water, and amount of plant life in a given area. In most places, physical and chemical weathering work together, each increasing the effectiveness of the other. However, for the sake of simplicity, we will discuss them separately.

PHYSICAL WEATHERING

Physical weathering processes break down rocks into smaller pieces that retain most of the characteristics of the parent materials. The size of the particles decreases, but compositions are not changed and no new minerals are formed.

Frost Wedging

Failure to drain water pipes in an unheated building before winter causes extensive damage when the water freezes within the pipes. The trapped water expands about 9% as it freezes. This can create enough pressure within the pipes to rupture them. A similar process, called **frost wedging,** occurs when water fills the crevices and pores of rocks and then is subjected to freezing temperatures. The water near the rock surface freezes first, because it is exposed to the cold air. This exerts pressure on the water remaining in the crevices, thus producing the mechanical effect of a wedge on the rocks. As ice crystals grow within the rock's fractures and pores, the pressure generated is sufficient to crack the rocks apart (Figure 9–1).

Frost wedging is the most important physical weathering process. It is most effective in temperate areas where temperatures fluctuate repeatedly above and below 0°C, forming meltwater that freezes as the temperature drops. Frost wedging is less extensive in very cold regions where freezing temperatures prevail year-round.

Animals and Plants

The expansive forces caused by growing tree roots can buckle sidewalks and break underground pipes. In a similar fashion, the growth of roots within rock

FIGURE 9–1
Frost-wedged debris on top of Mount Washington, New Hampshire. Note the angularity and poor sorting of the particles. Photo by Nicholas K. Coch.

(a)

(b)

FIGURE 9–2
Exfoliation. (a) Exfoliation dome in Yosemite National Park, California. (b) Close-up showing exfoliation sheets. Photos courtesy of King Huber/U.S. Geological Survey.

crevices pries apart the rock in a process called **root wedging**. Once the regolith is formed, burrowing animals and plant roots make more openings in it. These new openings allow more oxygen and water to penetrate, increasing the effects of chemical weathering.

Release of Confining Pressure

Many surface rocks were formed at considerable depth and later were uncovered by erosion of the overlying rock material. When these rocks were deeply buried, they were subjected to high pressures in all directions. As erosion removes more and more of the overlying rocks, the vertical pressure progressively lessens. This release of pressure allows the rocks to expand in the direction of lower pressure (upward), and sheets of rock break off at the surface. The formation of curved sheets of rock by release of pressure is called **exfoliation**. An example is the formation of exfoliation domes in Yosemite National Park, California (Figure 9–2).

Salt Crystal Growth

Salts crystallizing in rock crevices exert sufficient pressure to break off small pieces of the rock, exposing new areas to further weathering. This process occurs both in dry areas, where evaporation of surface water results in salt crystallization in surface openings, and in coastal areas, where salt spray is driven into the surfaces of rocks (Figure 9–3).

Water

The force of moving water also can be an agent of physical weathering. Raindrops falling on a partially weathered rock can exert sufficient force to dislodge loose particles. Waves drive water with great force into crevices in coastal rock exposures. The resulting pressure acts like a wedge on the walls of the crevice and can crack the rock apart.

Thermal Expansion and Contraction

When the outside of a rock is heated by the Sun, the ions in the minerals vibrate more widely around their structural positions; as a result, the outer few centimeters of the rock expand. Rocks are poor conductors of heat, so the inner part of the rock receives little. Consequently, stresses can arise between the expanding and contracting exterior and the unchang-

FIGURE 9–3
Salt crystal weathering along the California coast. Photo by Nicholas K. Coch.

FIGURE 9–4
How physical weathering increases
the surface area of a rock.

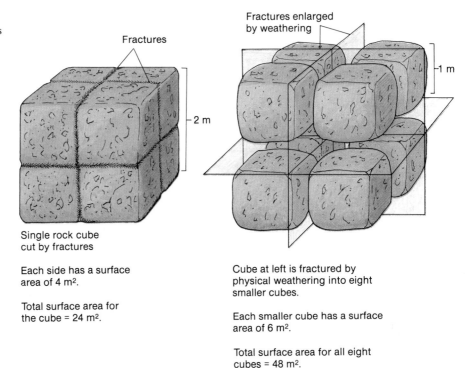

Single rock cube
cut by fractures

Each side has a surface
area of 4 m².

Total surface area for
the cube = 24 m².

Cube at left is fractured by
physical weathering into eight
smaller cubes.

Each smaller cube has a surface
area of 6 m².

Total surface area for all eight
cubes = 48 m².

ing interior. These differences in stress may result in the fracturing and "peeling off" of outer layers, particularly if the outer surface of the rock is heated to very high temperatures, as occurs in a forest fire.

In addition, each mineral in the rock expands to a different degree upon heating. For example, quartz expands three times as much as feldspar. The different expansive stresses set up among individual minerals may result in crystals breaking off at the surface of the rock.

The degree to which thermal expansion and contraction can fracture rocks is controversial. If this process is significant it would be most effective in areas of considerable daily change in temperature, such as deserts. Shattered angular rocks litter the floor of many desert areas, and in places these angular pieces can be reassembled into larger cobbles and boulders. However, in laboratory studies, experimental heating and cooling of rocks over thousands of cycles has failed to shatter them; currently it is not clear how the process works in nature. It is possible that the angular fragments on the desert floor originated from frost wedging in adjacent mountains, with the debris being carried by floods onto the desert floor.

Relationship Between Physical and Chemical Weathering

Physical weathering increases the effectiveness of chemical decomposition, because breaking a rock into smaller particles increases the total surface area exposed to chemical weathering agents. For example, consider the cube-shaped rock in Figure 9–4. Each edge of the cube is 2 m; the total surface area of the cube is 24 m². Physically weathering this cube along three mutually perpendicular fractures yields eight cubes, each with edges 1 m long. The total surface area of the rock is now 48 m², twice the surface area of the unweathered cube that we started with. In each halving of a cube edge through weathering, the surface area doubles, although the total rock volume remains the same. The smaller rock fragments weather even faster because there is now more surface area exposed to both physical and chemical weathering.

CHEMICAL WEATHERING

Chemical weathering consists of a number of reactions in which the ions that make up minerals combine with components of the atmosphere and hydrosphere. Some ions recombine to form new minerals that replace those originally present; other ions are dissolved and removed in solution from the parent rock or sediment.

The factors governing the extent and rate of chemical weathering are the same as those that play a role in increasing the rates of chemical reactions in the laboratory: (1) availability of water, (2) mixing of the reactants, (3) increase in temperature, and (4) increase in the surface area of the reactants.

Water is the most important factor in chemical weathering. In fact, in every reaction in chemical weathering, water either participates in the reaction, or carries the gases and ions that attack the rock chemically, or removes products of the reaction. Water is an especially effective weathering agent because it can exist as a fluid (rain or underground water), gas (dew or fog), or solid (ice) over the temperature ranges at the Earth's surface. In humid climates, water may partially or completely fill the fractures and pores in a rock, whereas in drier climates, it may exist only as a thin film coating individual grains.

Chemical weathering is most effective in warm climates because heat increases the rate of most chemical reactions. All other factors being equal, a 10°C increase in temperature nearly doubles the rate of chemical weathering reactions.

The most common reactions in chemical weathering involve the addition of water, carbon dioxide, and oxygen to rocks and sediments at the Earth's surface. In addition, sulfurous fumes produced by industry recently have been recognized as important factors in chemical weathering near urban centers.

The most important reaction involving water is called **hydrolysis**. Reactions involving carbon dioxide and oxygen are **carbonation** and **oxidation**, respectively.

Carbonation

Carbon dioxide in the atmosphere combines with water to form carbonic acid, H_2CO_3, by this reaction:

$$H_2O \ + \ CO_2 \ \rightarrow \ H_2CO_3$$

Water   Carbon   Carbonic
dioxide   acid

Even more carbonic acid is produced in the ground in the area around plant roots. The soil around roots is enriched in carbon dioxide by plant activities, such as respiration and the decay of organic material, to levels 10 to 1000 times the normal atmospheric concentration. As rain soaks through this regolith, large quantities of carbonic acid are formed.

The decomposition of minerals by reaction with carbonic acid is called **carbonation**. It is particularly effective in the solution of limestones because calcite ($CaCO_3$), the major mineral in limestone, is dissolved by carbonic acid in the reaction:

$$CaCO_3 \ + \ H_2CO_3 \ \rightarrow \ Ca^{2+} \ + \ 2HCO_3^-$$

Calcite   Carbonic   Calcium   Bicarbonate
acid   ion   ion
$\llcorner$ in solution $\lrcorner$

Whenever limestone occurs at the surface in areas of high rainfall and organic activity, it dissolves rapidly, and thus limestone tends to occupy topographic lows. The calcium and bicarbonate ions are then carried away in solution by water. When ions are removed from a rock or regolith in this way, they are said to have been **leached** from the material.

Hydrolysis

The polar nature of the water molecule (see Chapter 4) makes it an excellent solvent during chemical weathering, as shown in Figure 9–5. The charged ends of the molecule help break the bonds between sodium and chlorine in the halite structure (NaCl) and enable the component ions to be removed in solution.

Water's polar nature alone does not explain why it is the most important solvent during weathering. Its effect is enhanced by a process called **dissociation**, during which a molecule separates into ions:

$$H_2O \ \rightarrow \ H^+ \ + \ OH^-$$

Water   Hydrogen ion   Hydroxyl ion

The ability of the ions formed from dissociation of water to dislodge cations and anions from minerals is much greater than that of the weakly charged sides of a neutral water molecule. Not all water molecules are dissociated, but those that do help to decompose even relatively insoluble minerals. The combining of H^+ and OH^- ions with minerals is called **hydrolysis** and is the most important process in the chemical weathering of silicate minerals. For example, consider the weathering of potassium feldspar:

$$2KAlSi_3O_8 \ + \ \ 2H^+ \ \ + \ 9H_2O \ \rightarrow$$
Orthoclase   Hydrogen   Water
                              ion

$$Al_2Si_2O_5(OH)_4 \ + \ 4H_4SiO_4 \ + \ \ \ 2K^+$$
  Kaolinite   Silicic   Potassium
(new mineral)   acid   ion
└──────── in solution ────────┘

The products of this reaction are soluble potassium ions, dissolved silica in the form of silicic acid, and the residual mineral kaolinite. Similar hydrolysis of plagioclase feldspars yields kaolinite and dissolved Na^+ and Ca^{2+} ions. Kaolinite is one of a group of fine-grained sheet-silicate clay minerals produced by weathering. Clay minerals are important constituents of the regolith, and some are important in industry. For example, kaolinite is used as a filler and a coating agent in the manufacture of paper, as a thickener for foods, and as a raw material for ceramics.

Oxidation

A hammer or saw left outdoors rusts because the iron in these tools combines with the oxygen in the air. This process is called **oxidation**. It also occurs in many rock-forming minerals that contain iron—the ferromagnesian minerals such as olivine, pyroxenes and amphiboles. The weathering of such minerals is complex and may take one of several courses. Consider the chemical weathering of olivine, for exam-

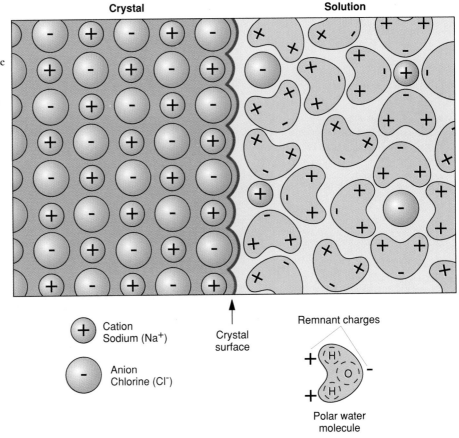

FIGURE 9–5
Role of polar water molecules in chemical weathering. Due to remnant charges on the water molecule, water is able to neutralize charges at the mineral surface and take ions into solution. Example shows water in contact with halite (sodium chloride).

Crystal   Solution

⊕ Cation
   Sodium (Na^+)

⊖ Anion
   Chlorine (Cl^-)

Crystal surface

Remnant charges

Polar water molecule

ple. The first step involves hydrolysis, which separates the iron from the magnesium and silica:

$$(Mg,Fe)SiO_4 + 3H_2O \rightarrow$$
$$\text{Olivine} \qquad \text{Water}$$

$$Mg(OH)_2 + H_4SiO_4 + FeO$$
$$\text{Magnesium} \quad \text{Silicic} \quad \text{Ferrous}$$
$$\text{hydroxide} \qquad \text{acid} \qquad \text{oxide}$$

The iron then undergoes further reactions. It may be converted into the mineral hematite by an oxidation reaction:

$$4FeO + O_2 \rightarrow 2Fe_2O_3$$
$$\text{Ferrous} \quad \text{Oxygen} \qquad \text{Hematite}$$
$$\text{oxide} \qquad\qquad\qquad \text{(ferric oxide)}$$

Hematite may then combine with water to form the mineral limonite. Such reactions, in which water is added to a mineral, are called **hydration** reactions. The hydration of hematite to produce limonite occurs by the reaction:

$$Fe_2O_3 + 3H_2O \rightarrow Fe_2O_3 \cdot 3H_2O$$
$$\text{Hematite} \quad \text{Water} \qquad \text{Limonite}$$

Reddish-brown hematite and yellow-brown limonite are found in "rust" and in weathered rock that originally contained ferromagnesian minerals.

Spheroidal Weathering

Massive, uniform rocks such as granite and basalt commonly weather into curving "peels" of crumbly altered minerals. This is sometimes referred to as "onion skin" weathering because of its appearance (Figure 9–6), but it is more properly referred to as **spheroidal weathering** because the process tends to form spheroidal boulders.

The steps that lead to spheroidal weathering are shown in Figure 9–7. Notice that chemical weathering of a block of rock is most effective at the corners and edges. At these locations, weathering proceeds from two or more directions, and the edges and corners become rounded quickly. This evolution of shape is analogous to the melting of an ice cube, in which the corners become rounded early in the melting process. As spheroidal weathering proceeds from the outside inward, new minerals form and the resulting curved weathering "rind" breaks away, exposing new rock to weathering.

FIGURE 9–6
Photograph of spheroidal weathering in an outcrop of igneous rock in Kauai, Hawaii. Photo by Nicholas K. Coch.

FIGURE 9–7
Formation of spheroidal
weathering. (a) Chemical
weathering attacks the cube all over,
but acts faster at edges because
there are two surfaces to attack, and
even faster at corners because there
are three surfaces. (b) Progressive
weathering reduces the rock cube to
a sphere, the geometric form having
the lowest surface area for a given
volume.

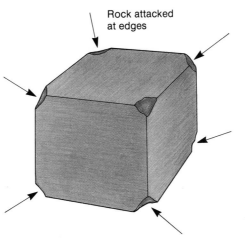
Rock attacked
at edges

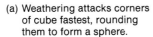

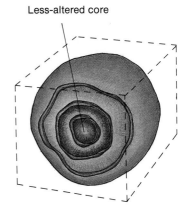
Less-altered core

(a) Weathering attacks corners
of cube fastest, rounding
them to form a sphere.

(b) Successive layers of the
sphere break off as chemical
and physical weathering continue.

Mineral Stability in Chemical Weathering

Different rocks and minerals have different susceptibilities to weathering. The relative weatherability of
common rock-forming minerals has been determined from numerous laboratory and field studies. A
comparison of this weatherability series (Figure
9–8) with the Bowen's Reaction Series (Chapter 4)
shows the relationship between conditions of mineral formation and their weathering susceptibilities.
Minerals formed at high temperatures and pressures
tend to be less stable in a weathering environment.
Thus, if an igneous rock contains both olivine and
biotite, the olivine crystals decompose more rapidly
than the biotite crystals under the same weathering
conditions. Because olivine has such high suscepti-
bility to weathering, it very rarely occurs in regolith
(see Focus 9–1). On the other hand, the greater
chemical stability of quartz, along with its abundance, makes it the commonest mineral in regolith.

Products of Chemical Weathering

Weathering processes are very effective in reducing
the large number of minerals that are found in igneous, sedimentary, and metamorphic rocks to a relatively few minerals that are stable under surface conditions. For example, most regolith is composed

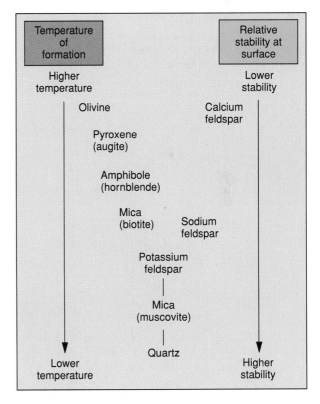

FIGURE 9–8
Relative stability of minerals in chemical weathering. Minerals
that formed under conditions most different from the Earth's
surface are the most easily weathered.

Beach colors are among the most varied features on Earth. For example, the pink beaches of Bermuda are formed from pink shell fragments, whereas many volcanic islands have black beaches composed of water-worn basalt particles and/or volcanic glass. These unusual beaches are very different from the white-to-tan quartzose and calcareous beaches found on most U.S. coasts.

The most unusual beaches in the world are the green sands found on a portion of the island of Hawaii. These sands are composed of significant amounts of olivine, one of the most easily weathered minerals. How can such an unstable mineral be so abundant in such a warm and moist climate? The answer is that Hawaii is an active volcanic site, and olivine-rich basalts are common in the area. Thus, even though the olivine weathers at a rapid rate, it is replenished by erosion of the olivine-rich basalt at a rate sufficient to ensure a continuous supply of olivine to these unusual beaches (Figure 1).

FIGURE 1
Green sand (olivine) on the island of Hawaii. Macmillan Publishing/Geoscience Resources photo.

only of quartz, iron oxides, clay minerals, organic materials, and partially weathered rock particles.

The changes in crystal structure that accompany formation of the new minerals produce changes in their physical properties. The new minerals are generally less dense than the originals, have a dull or earthy luster, and many are softer than their unweathered predecessors.

Under special conditions, significant amounts of unstable minerals can be found in sediments or sedimentary rocks. These rare occurrences result from any of the following conditions: (1) great abundance in the source rock; (2) very little transport distance between source and depositional site; (3) rapid erosion and deposition, preserving the mineral from weathering; or (4) dry or cold climate in which chemical weathering is ineffective.

What happens to all these different weathering products? Some of the minerals, both unaltered and altered, may remain in the source area to form a residual regolith. The remainder may be carried away by wind, water, or ice to form a transported regolith somewhere else (Figure 9–9).

In dry regions, the dissolved ions may never leave the area but instead may be deposited by evaporation of surface and subsurface waters. In wet regions, some of the dissolved ions may be absorbed by plants, although most are transported from the area. These ions can move with underground waters and form the cements that hold many clastic rocks to-gether. The ions also can travel in streams to the sea, where they are used in part to form the shells of marine organisms or chemical sedimentary rocks.

Differential Weathering and the Molding of Landscapes

Different rocks weather at different rates, and thus weathering shapes the landscape. Climate significantly affects this differential weathering. In humid, temperate areas such as the Middle Atlantic states, areas underlain by limestone and sandstone once had similar elevations. With time, however, the limestone areas became progressively lower as the calcite in the limestone dissolved in the humid climate. Thus relatively resistant quartzose sandstones now occur as ridges above the less-resistant limestone lowlands.

FIGURE 9–9
Products of chemical and physical weathering.

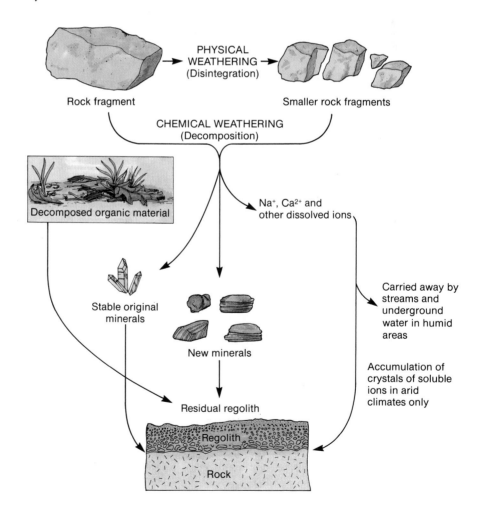

Rock fragment

PHYSICAL WEATHERING (Disintegration)

Smaller rock fragments

CHEMICAL WEATHERING (Decomposition)

Decomposed organic material

Na^+, Ca^{2+} and other dissolved ions

Stable original minerals

New minerals

Carried away by streams and underground water in humid areas

Accumulation of crystals of soluble ions in arid climates only

Residual regolith

Regolith

Rock

FIGURE 9–10
Kaibab Limestone at the top of the Grand Canyon in Grand Canyon National Park, Arizona.
In this arid climate, limestone is not dissolved, but forms cliffs. Photo by Nicholas K. Coch.

Weathering in a dry climate is quite different. In the southwestern United States, limestone deposits form cliffs and ridges as high as those underlain by sandstone, because the scarcity of water greatly inhibits limestone solution. For example, the limestones of the Kaibab, Toroweap, and Redwall Formations form the steep cliffs in the walls of the Grand Canyon (Figure 9–10). The limestones and resistant sandstones are underlain by less-resistant shales. The shales have been weathered and eroded to gently sloping terraces on the Supai Formation and the Bright Angel Shale (Focus 8–1, Figure 1). This alternation of sloping terraces and steep cliffs, produced by differential weathering, provides a spectacular view in the Grand Canyon.

SOIL

The term **soil** is used to describe loose material at the Earth's surface that has been sufficiently weathered and enriched with organic material that it can support the growth of rooted plants.

Soil Formation and Zonation

How does a soil begin to form on bare rock? Only organisms such as lichens can grow on a bare rock surface, because the lichens have limited needs for nutrients and can obtain them *directly* from the rock. Secretions released by the lichens etch the rock surface, loosening mineral particles and producing small amounts of mineral nutrients. These mineral particles, along with atmospheric dust and organic debris from dead lichens, accumulate in rock crevices as the first pockets of soil. Spores and seeds of plants then may gain a foothold in these thin soil pockets because weathering has released the mineral nutrients for their growth. The growth of the plants and trees increases the physical and chemical weathering of the rock, and the soil thickness grows steadily with time, as long as none of the soil is removed by erosion.

With the passing of time, weathering products accumulate in some parts of the regolith and are depleted in others. This leads to the formation of layers called **soil horizons,** which can be distinguished on the basis of organic content, color, mineralogy, grain size, and percentage of unweathered parent material. Soils that have been developing long enough to have distinct horizons are called **mature soils**.

Four major soil horizons are recognized. Starting with the uppermost layer, these are referred to as the O, A, B, and C horizons. These horizons are shown in Figure 9–11, which illustrates the type of soil that develops in the northeastern United States.

The uppermost layer is the **O horizon,** characterized by the accumulation of organic material and the relative paucity of mineral matter. The uppermost portion of this organic horizon is made up of fresh plant matter; the lower part contains plant matter that is more highly decomposed. The organic accumulations in the O horizon give it a dark-brown color.

The next layer is the **A horizon,** composed largely of mineral particles and organic material. In humid climates, downward percolating water charged with carbon dioxide dissolves calcium carbonate in the A layer. The removal of soluble materials such as cal-

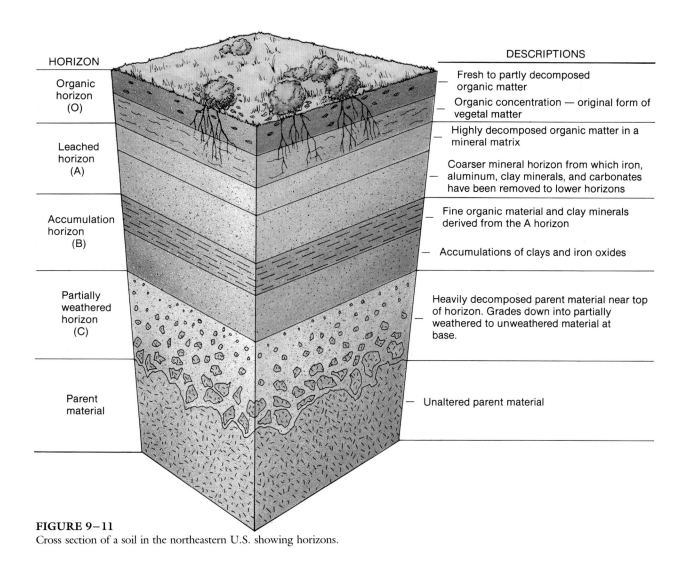

FIGURE 9–11
Cross section of a soil in the northeastern U.S. showing horizons.

cium carbonate is called **leaching,** and for this reason the A horizon sometimes is called the **leached horizon**.

The soluble carbonates, along with clay minerals and iron oxides, are transported down through the soil by percolating waters and accumulate in the underlying **B horizon,** sometimes referred to as the **accumulation horizon**. The concentrations of iron oxides in the B horizon commonly impart a yellowish-to-red color to this horizon.

The **C horizon** is composed of partially weathered parent material. At the top of the C horizon is highly weathered rock or sediment that may still preserve some of the characteristics of the parent material. The lower portion of the C horizon contains particles of the parent material.

Factors in Soil Formation

A number of factors account for the marked differences in soils from area to area. The major determiners of soil characteristics are parent materials, climate, biologic activity, topography, and elapsed time. All of these operate simultaneously, and one factor can offset or accentuate the effect of another. For purposes of discussion, we will consider each factor here as if it acted alone.

Parent Material. Parent material controls soil mineralogy to the extent that all soil minerals must either come directly from the parent materials, or have been derived from the original minerals by weathering.

However, the *relative percentages* of minerals in a soil can be very different from those in the parent material. The weathering of an impure limestone in a humid region provides an extreme example. Unlike some limestones, which are pure calcium carbonate, our example is one of many limestones containing various amounts of siliciclastic particles. The exposed limestone is weathered rapidly by water containing carbonic acid, yielding soluble ions that are promptly removed by surface and subsurface waters. Consequently, the residual regolith is greatly enriched in insoluble minerals, such as quartz and clay minerals, although they composed only a small percentage of the original rock.

Climate. Climate ranks with parent material as one of the most important factors in soil development. Climate determines the availability of water that is essential to chemical weathering, and it controls the temperature at which chemical reactions occur.

Rainfall Greater than 50–60 cm Per Year. In high-rainfall areas, calcium carbonate and other soluble compounds are dissolved by carbonic acid and removed from the soil, whereas clay minerals and iron oxides accumulate in the B horizon. Such a soil is called a **pedalfer**. (The name comes from *ped* for soil, plus *al* for aluminum clays and *fer* for ferrum, or iron.) Pedalfers are common in the eastern United States.

Rainfall Less than 50–60 cm Per Year. In low-rainfall areas, there is insufficient water to remove soluble compounds, and a soil forms that is rich in soluble compounds called a **pedocal** (*ped* for soil, *cal* from

FIGURE 9–12
Caliche soil horizon (white) with an irregular upper boundary, Texas. The caliche formed from calcium carbonate that precipitated from water percolating through the soil. Photo courtesy of Jerry L. Rives/ USDA Soil Conservation Service.

FIGURE 9–13
Caliche profile in Triassic rocks of
Connecticut. Photo courtesy of
John Hubert.

the Latin *calc,* meaning lime). A type of pedocal in which the soluble minerals actually form crusts, layers, and pore fillings is called a **caliche** (Figure 9–12).

Caliche soils are forming today in dry climates, such as the southwestern United States. When calichelike layers are found in ancient rocks, they provide a very good paleoclimatic indicator for those rocks. For example, 200-million-year-old Triassic rocks exposed in Connecticut (Figure 9–13) were found to have features with striking similarity to caliche profiles forming today in the southwestern United States. This evidence indicates that these sedimentary rocks were deposited by streams that flowed intermittently in a dry climate, and that periods of weathering and stream deposition alternated with drier periods of caliche formation.

Tropical Climate. Tropical climates with very high rainfall yield a distinctly different type of soil. Quartz may be a stable mineral in temperate and drier tropical climates, but it becomes unstable in tropical climates with high rainfall. The intense chemical weathering that characterizes such climates can result in

solution of quartz and removal of silica from clay minerals. Soluble silica is washed from the soil by heavy rainfall, and the residual regolith becomes enriched in the oxides and hydroxides of iron and aluminum. Such a soil is called a **laterite** (Figure 9–14).

FIGURE 9–14 (right)
Laterite soil in western Australia. Photo courtesy of Gene Alexandrov.

The type of metal concentrated in the laterite is determined by the parent rock. For example, deep weathering of a granite would enrich the regolith in aluminum; deep weathering of a basalt would yield an iron-rich regolith. High-grade aluminum and iron ores, formed during the process of laterization, are mined in Jamaica, Australia, Brazil, and other nations. The major source of the world's aluminum is an aluminum-rich laterite called **bauxite**.

Weathering of granite and basalt provides a good example of how rocks of very different mineralogical compositions can produce very similar regoliths in the same climate. In Figure 9–15, the original mineralogy of these rocks is shown on the right and left sides. In the center of the figure are the weathering products, which are similar. The end product of weathering in both cases is a reddish, clay-rich regolith.

Biologic Activity. Plants and animals play a significant role, both in initiating soil formation and in aiding its development. The root systems of higher plants such as grasses and trees, along with the burrows of animals, provide passages through which oxygen and solutions can move. Vegetation also acts to hold the soil together through its root system, thus preventing soil from being eroded and retaining moisture around the roots between rainfalls.

The type of vegetation, such as grasses or trees, also influences the type of soil that will develop. Dead grass material is incorporated more easily into the soil than woody tree remains, resulting in a higher organic content in soils of a grassland region than in soils of a forested region. This is the mechanism by which the fertile soils of the American Midwest were produced.

Topography. Soils that develop on slopes can be quite different from soils that form on flat sections of the same area. Soils that develop on slopes generally have a lower organic content, are thinner, and are coarser grained than soils that form on the adjacent flat areas. This is because the rain falling on steeper slopes has a greater tendency to run off along the

FIGURE 9–15

Differences in weathering products of granite and basalt in a warm, humid climate. Although the rocks differ greatly in original appearance and mineralogical composition, their weathering products are quite similar. The reddish, clayey regolith produced in both cases is very similar in appearance, although the basalt regolith contains no quartz because that mineral is absent in the original rock. However, there is a significant difference in the types of soluble ions produced in each case.

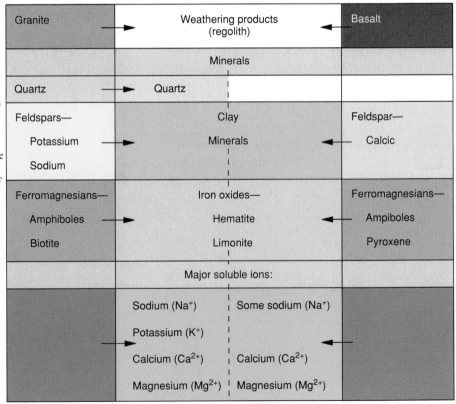

WEATHERING ON THE MOON AND MARS: A PLANETARY PERSPECTIVE

How would rocks weather in an environment different from any on Earth? The *Apollo* lunar missions (1969–1972) and the *Viking* mission (1976) to Mars have lent new perspective to weathering on other planetary bodies.

Weathering on the Moon is purely physical disintegration, because there is no water, oxygen, or biologic activity to aid chemical decomposition. The absence of chemical weathering is reflected in the color of lunar soils. They are typically gray and purplish-gray, with no traces of the orange and red iron oxides so common in soils on Earth (Figure 1).

Rocks (and regolith) exposed on the lunar surface are constantly being reworked by the impacts of meteorites ranging in size from many kilometers to fractions of a millimeter (Figure 2). Larger meteorite impacts excavate deeper and wider craters, throwing fresh rock onto the lunar surface to be reworked by subsequent meteorite impacts. The smaller impacts rework the regolith continually, mixing it and progressively breaking down the rock and mineral fragments.

The great heat generated by meteorite impacts is sufficient to fuse the local rocks or regolith into dark-colored molten glass. Sometimes this glass "splashes" onto nearby rocks, forming glass-coated boulders. Smaller glass "blobs" cool into distinct glass droplets which may cement existing soil particles together to form uniquely lunar regolith particles called **agglutinates** (Figure 3).

FIGURE 1
Poorly sorted lunar regolith at the *Apollo 14* landing site. Photo courtesy of National Aeronautics and Space Administration.

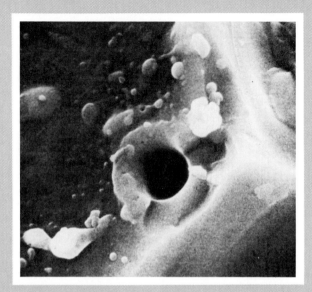

FIGURE 2
Micrometeorite crater drilled through an individual crystal.
Photomicrograph courtesy of National Aeronautics and Space
Administration.

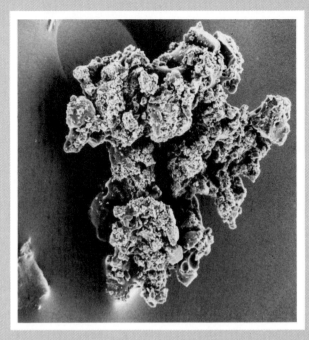

FIGURE 3
Agglutinate, composed of lunar sediment particles which were
cemented together by glass droplets formed as a result of
micrometeorite impacts. Photomicrograph courtesy of National
Aeronautics and Space Administration.

The first television pictures transmitted from Mars by *Viking Lander* cameras on July 20, 1976, showed that weathering on Mars is quite different from that on the Moon. Evidence that we will present later in this book indicates that water was available on the Martian surface in the past. The presence of water aided both physical and chemical weathering. The angular particles seen in *Viking Lander* photographs of the Martian surface (Figure 4) suggest that physical weathering is (or was) an important weathering process. However, the distinctly red color of the soil in several localities indicates that chemical weathering definitely occurred at some time on the Martian surface (Figure 5).

FIGURE 4
Angular particles in the regolith on the Martian surface. Photo courtesy of National Aeronautics and Space Administration.

FIGURE 5
Red regolith on the Martian surface. Photo courtesy of Raymond E. Arvidson. (Washington University, St. Louis, Missouri).

slope than to sink in. Such increased surface drainage removes both organic material and finer particles, leaving the coarser particles behind.

Elapsed Time. Soil-particle decomposition, soil-horizon development, and soil thickness increase with the passage of time, if all other factors are equal. This principle has considerable application in determining the relative ages of surficial deposits in geologic field studies. For example, the degree of decomposition of boulders in several glacial deposits in the northern Sierra Nevada mountains has been used to determine the relative age of each deposit.

For an interesting view of soil development under unusual atmospheric conditions, please see Focus 9–2.

Soils and Humans

Soil and water are the two basic resources for civilization. Although great areas of the Earth's surface are covered by regolith, only a small part of this area has productive soils. The tremendous U.S. agricultural production is a direct result, in part, of the geologic processes that formed its soils. Glaciation during the Pleistocene Epoch left finely ground mineral deposits over most of the U.S. Midwest. This material weathered readily after the glaciers retreated, yielding a thick, rich soil in a relatively short time, geologically speaking (10,000 years or so). The soil's fertility was increased by the drier climate, which favored grassland development over forests. Grasses have small roots that penetrate the ground

FIGURE 9–16
Gullying near a housing development in Franklin County, Ohio. Removal of vegetation increases soil erosion, lowers the land surface and causes gullies. Photo courtesy of U.S. Department of Agriculture.

FIGURE 9–17
Conditions in the Dust Bowl, Oklahoma. Photo courtesy of U.S. Department of Agriculture.

more thoroughly, and they decompose faster; both factors aid in soil formation.

Soil is usually a thin layer (generally 3 m or less), and so soil erosion can be a serious problem. Erosion of the overlying soil may expose many feet of underlying regolith, but that material is useless for agriculture because the minerals in it have not been sufficiently weathered to release the elements required by plants.

Soil Erosion. Soil erosion has become a serious problem in many places in the world as natural and human-caused events have resulted in drastic soil loss (Figure 9–16). Areas have been cleared of trees for construction of roads, homes, or factories; to obtain wood; or to plant crops. This destruction of tree cover is unfortunate, because the vegetation holds the soil through its root system. When vegetation is removed, for whatever the reason, there is a potential for soil erosion. Trees also act as windbreaks to inhibit wind erosion. Eroded soil washes from the fields and causes additional problems as the sediment covers adjacent areas, clouds lakes and streams, clogs culverts and flood-prevention canals, and causes numerous other problems.

The Dust Bowl. One of the greatest American population migrations in this century was the result of severe soil erosion in Texas and Oklahoma in the 1930s. A combination of drought, clearing of trees that had served as windbreaks, planting crops that were ill-suited to the soil, and improper cultivation created the infamous "Dust Bowl." Conditions in the Dust Bowl were harsh indeed (Figure 9–17).

Frequent windstorms kept people indoors, and even then the fine dust would enter homes through tiny crevices and cloud the indoor air, year after year. As winds eroded the dry soil, many families lost their livelihood and left the area. This migration was chronicled by John Steinbeck in his novel, *The Grapes of Wrath.*

The Dust Bowl generated remedial actions that returned the area to productivity once again. The region is now irrigated, keeping the soil moist and drought resistant. Farmers have planted lines of trees as windbreaks and have adopted cultivation improvements such as contour plowing.

SUMMARY

1. Weathering is a combination of physical and chemical processes that break down rocks and sediments to produce materials stable at the Earth's surface.
2. Physical weathering disintegrates rocks into smaller particles with no change in original mineralogy. Major

physical weathering changes include the growth of ice and salt crystals in rock crevices, root wedging, release of confining pressure, and thermal expansion and contraction.

3. Physical weathering increases the surface area of weathered debris, which increases the efficiency of chemical weathering.

4. Chemical weathering decomposes rocks and causes significant changes in the original mineralogy. Some minerals, such as quartz, remain unchanged in weathering in temperate areas; most others are changed into new minerals.

5. Chemical weathering reduces the large number of minerals in igneous, metamorphic, and sedimentary rocks into a residual regolith composed largely of quartz (in temperate climates and only if present in the original material), clay minerals, and iron oxides.

6. Soluble ions are an important product of chemical weathering. In humid areas, soluble ions are removed by surface and subsurface water and accumulate elsewhere, generally in the ocean. In dry areas, soluble salts are not removed from the area and can be deposited as layers and crusts within the regolith.

7. Minerals have different susceptibilities to weathering. In general, those minerals that formed at higher temperatures are less stable at the Earth's surface.

8. The differential weathering of minerals is a major agent in the formation of landscapes.

9. In humid climates, limestones and other soluble rocks are dissolved and underlie low-lying areas. In arid climates, these same rocks are resistant and stand above the surface as ridges and cliffs.

10. One of the major effects of weathering is the production of soils. Soil is a residual regolith that has been weathered sufficiently to support the growth of plants.

11. Weathering processes form a series of soil horizons. From the surface downward, they are O (organically rich), A (mineral material leached of soluble components), B (accumulations of material removed from the A horizon), and C (partially weathered parent material).

12. The major factors in determining the type of the soil that will develop are mineralogy of the parent material, climate, biologic activity, topography, and elapsed time.

13. In humid tropical areas with high rainfall, quartz is weathered and the silica removed. Under such conditions, iron and/or aluminum can be concentrated to form a lateritic soil. A tropical soil rich in aluminum is called bauxite.

14. In arid areas, deposits of soluble minerals precipitated in place form a caliche soil.

15. Weathering is important in recycling the elements, preparation of rock material for erosion and transport, formation of agricultural soils, development of topography, and concentration of iron and aluminum in economically significant amounts.

REVIEW QUESTIONS

1. How does physical weathering facilitate chemical weathering?

2. How do the products of chemical and physical weathering differ in appearance, composition, and size?

3. Why is water so important in both chemical and physical weathering?

4. Describe differences in the character of regolith derived from weathering a granite in the following climates:
   a. Subarctic.
   b. Warm temperate.
   c. Hot, humid tropical.

5. Describe differences in weathering of an impure limestone in a warm, humid climate and a warm, dry climate.

6. What properties of water make it so effective in chemical weathering?

7. Explain the differences in the origin of residual and transported regolith?

8. What conditions would supply sufficient carbonic acid for weathering of the underlying rocks?

9. Describe how significant amounts of an unstable mineral, such as olivine or pyroxene, could be preserved in sediments and sedimentary rocks.

10. Describe how each of these factors contributes to the type of soil that will develop:
    a. Parent material.
    b. Climate.
    c. Topography.
11. How might you use weathering to determine the relative ages of a series of similar deposits in the same area?
12. Lunar and Martian regoliths both contain large and small particles. Did they both form in the same way? Explain.

FURTHER READINGS

Birkeland, P. W. 1976. *Pedology, weathering and geomorphological research*. New York: Oxford Univ. Press, 285 p. (A detailed treatment of weathering and soils.)

Hunt, C. B. 1972. *Geology of soils: Their evolution, classification and uses*. San Francisco: W. H. Freeman & Co. (A good reference for starting a study of soils.)

Likens, G. E., and Bormann, F. H. 1974. Acid rain: A serious regional environmental problem. *Science* 184:1176–79. (A discussion of the effects of acid rain on the environment.)

Lockeretz, W. 1978. The lessons of the Dust Bowl. *American Scientist* 66:560–69. (A detailed description of the conditions that led to the Dust Bowl.)

McKay, D. S., Fruland, R. M., and Heiken, G. H. 1974. Grain size and the evolution of lunar soils. In *Proceedings of the 5th Lunar Science Conference*. Vol. 1, 887–906. (Describes the formation of soils on the Moon and how they change with time.)

McNeil, M. 1964. Lateritic soils. *Scientific American* 211:96–102. (Describes the origin and geographic distribution of lateritic soils.)

Soil Survey Staff. 1975. *Soil taxonomy: A basic system of soil classification for making and interpreting soil surveys*. U.S. Department of Agriculture handbook 436. (A detailed description of soil classification.)

U.S. Department of Agriculture. 1976. *Know the soil you build on*. Soil Conservation Service, Agricultural Information Bulletin 320, 13 p. (A very useful booklet that describes how soil type is related to our everyday lives. Well illustrated and includes a discussion of soil maps and surveys.)

10
Mass Movements

Landslide on steep slope in Hong Kong, June 18, 1972. The landslide was preceded by two days of heavy rain; 67 people were killed and a number of structures were destroyed. Photo courtesy of Government of Hong Kong, Geotechnical Control Office.

Cliffs overlooking the Great Lakes collapse after their bases have been eroded by high lake levels. In Los Angeles, heavy winter rains are followed by mud flowing down canyons. In the Pittsburgh area, home foundations collapse into old mine workings below. In New York, blocks of rock loosened by frost action fall from the walls of a roadcut onto an interstate highway. Such events, commonly called "landslides," are among the most spectacular of all geologic processes. Geologists use the term **mass movements** for processes in which masses of material move down slopes under the force of gravity.

Mass movements are one type of **erosion,** the removal of surface materials by geologic agents such as water, wind, or ice. However, mass movements are different from other erosion agents in several ways. The most important difference is that entire masses of material rather than discrete particles are eroded. Mass movements of one kind or another occur in many land environments, in parts of the ocean basins, and have been identified on the Moon and Mars. They are vital in moving rock and regolith down slopes and into transportation systems such as glaciers, streams, rivers, and coastal currents, which then can move these sedimentary particles great distances. Mass-movement processes differ from other erosional agents in additional ways; the transport

distance is generally small, and the amount of change in the sediment is minor.

Some mass movements occur so fast that the initial roar of the rock masses ripping loose from a slope is followed in a minute or less by the particles cascading onto the land at the base of the slope. Other types of mass movements are slower, with the regolith flowing over days, weeks, or even months. Still other types of mass movements are so slow that the only way to sense their occurrence is by noting that originally vertical features such as trees, utility poles, and fences have been tilted or moved downslope over a period of years.

Why do materials on some slopes move downslope so rapidly, whereas materials on other slopes move so slowly that movement can be documented only over a long period of time? This question can be approached by comparing those factors that tend to aid downslope movement, the **driving forces,** with those that prevent downslope movements, the **resisting forces.**

FACTORS IN MASS MOVEMENTS

Gravity is the greatest driving force in mass movements. Others are more passive, weakening the material so that gravity can move it downslope more

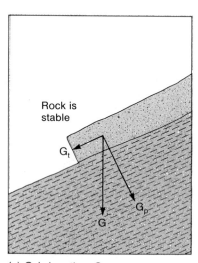

(a) G_t is less than G_p

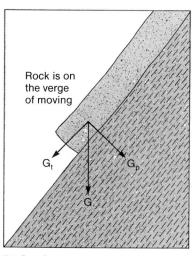

(b) $G_t = G_p$

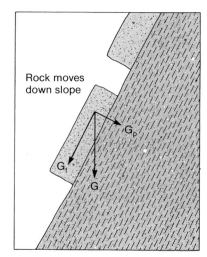

(c) G_t is greater than G_p

FIGURE 10–1
Effects of gravitational forces on a block of rock resting on slopes of different inclinations.

easily. Water content can be a driving force under one set of conditions and a resisting force under a different set. We will start our discussion with gravity, because it is the major driving force in all mass movements.

Gravity and Its Effects on Slope Materials

The role of gravity in triggering mass movements can be seen by considering the forces acting on a block of rock exposed on slopes of different inclinations. The steeper the slope, the greater the tendency for particles to move to lower positions. This principle is illustrated in Figure 10–1.

The force of gravity, G, is directed toward the center of the Earth. This gravitational force may be considered as having two components: one acts parallel to the slope, G_t, and the other acts perpendicular to the slope, G_p. On gentle slopes, such as in Figure 10–1(a), G_p is greater than G_t and the block does not move.

At some intermediate slope, G_t equals G_p, and the particle is on the verge of moving—Figure 10–1(b). Such a slope with its potential for mass movement is called an **unstable slope**.

At any steeper slope angle, G_t is greater than G_p, and there is a tendency for the block to move

downslope—Figure 10–1(c). This tendency is resisted by the strength or cohesiveness of the material and by the force of friction. When these resisting forces are exceeded by the gravitational component acting down the slope G_t, the block can break loose and move. The actual movement of slope materials is called **slope failure**. As the particles start to move downslope, the gravitational potential energy is converted into kinetic energy, the energy of movement.

Orientation of Rock Layers

We mentioned that steeper slopes tend to be more unstable than gentler ones. However, orientation of rock layers influences slope stability. In some situations, even steep slopes can be quite stable so long as the rock layers are inclined *away* from the slope (Figure 10–2). In other situations, though, if rock layers are inclined *toward* the slope (i.e., in the downslope direction), slope stability decreases. Thus, slope stabilities on opposite sides of a highway cut may be quite different—Figure 10–2(a).

Sometimes the rock layers are inclined away from the slopes, tending to add stability, but *fractures* within the rock are inclined *downslope*, making it quite unstable—Figure 10–2(b). You can see why detailed geologic studies are necessary before exca-

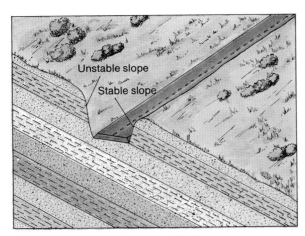

(a) Rock layers are inclined toward highway (unstable slope) and away from highway (stable slope) on opposite sides of the roadcut.

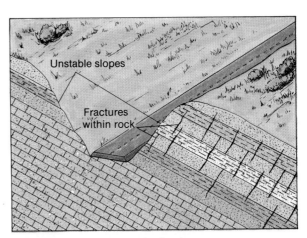

(b) Rock layers on right are inclined away from the roadcut, but fractures within the rocks slope toward the roadcut.

FIGURE 10–2
Slope stability depends on the orientation of rock and sediment layers. When either layers or fractures are inclined *toward* a slope, that slope is unstable and subject to mass movements.

FIGURE 10–3
Cracks in land surface underlain by shrink-swell clay, Texas. Photo courtesy of Jerry L. Rives/USDA Soil Conservation Service.

vations are made in a slope; a geologist must determine the potential for slope failure.

Water

Water is extremely important in many mass movements, but its role is complex. It can promote movement in two ways: as an *active* driving force, by increasing the weight of the sediment or rock (this is called loading), or *passively,* by decreasing the strength of the rock or sediment through reducing friction and cohesion between particles. (Conversely, small quantities of water in a sediment may enable it to resist mass movements.)

Loading. Water acts as a driving force by increasing the weight of slope material. A stable mass of dry sand exposed on a slope can have up to 35% of its volume composed of dry pores. After a prolonged period of rain, these pores may be completely filled. This increases the weight of the sediment significantly and thus increases the gravitational force driving the sediment down the slope. Many mass movements occur during or shortly after prolonged rainfall because of the filling of pore spaces by water.

Reducing Rock Strength. Water can reduce the strength of rock in several ways. As it circulates through the pores of some rocks, water may dissolve soluble cementing materials such as calcium carbonate, reducing cohesion and allowing the grains to more easily move past one another downslope. Water can react with clay minerals to soften them, reducing frictional forces acting between layers. The force exerted by ice-crystal growth in rock crevices can loosen the rock by frost wedging (see Chapter 9). This can result in rockfalls and rock slides during thaws.

Expanding Clays. Some sediments, called **bentonites,** have the ability to absorb large quantities of water, swelling up to as much as *eight times* their original volume. Bentonite is composed of clay minerals formed by the chemical alteration of glassy igneous rocks such as volcanic ash and tuff, either in weathering or in the early stages of sediment burial.

When water from rainfall, lawn-watering, leaking pipes, or whatever enters ground containing bentonite, the bentonite layers swell and exert great pres-

FIGURE 10–4
Formation of quick clay as freshwater flushes saltwater from pore spaces between clay particles. (a) Ions in salty pore waters hold clay together in an open structure. (b) Collapse of clay structure results in an excess of water and liquefied clay starts to flow.

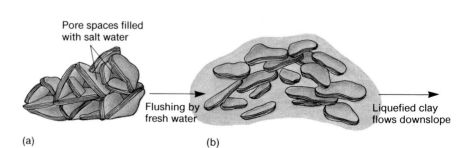

Pore spaces filled with salt water

Flushing by fresh water

Liquefied clay flows downslope

(a)   (b)

FIGURE 10–5

Landslide at Saint-Jean-Vianney, Quebec, May 1971. This disaster, which took about 30 lives and caused the loss of many homes, resulted from failure and seepage in Pleistocene clays. Photo courtesy of Geological Survey of Canada.

sure on the overlying material. Between rains, the layers dry out and contract, forming large cracks in the surface that can damage overlying structures (Figure 10–3). Sediments of this type are commonly referred to as "shrink-swell" clays; they are an enormous problem in building foundations in many parts of the United States.

If water-saturated bentonite is on a slope, the material provides a slick surface which reduces friction and facilitates the downslope movement of any overlying layers.

Liquefaction of Clays. Some clays can be transformed quickly from solids to liquids under certain conditions. Rapidly liquefiable clays are called **quick clays**. One type of quick clay is formed originally from clay minerals that accumulated in saltwater (Figure 10–4).

Ions in the salty pore waters hold the clay minerals together as aggregates, forming an open "house of cards" structure—Figure 10–4(a). When such clays are exposed by erosion, they are subject to the subsurface flow of freshwater through their pores. This freshwater flushes the saltwater and ions that held the open clay structure in place, allowing the structure to collapse and making the formerly solid clay unstable—Figure 10–4(b). If the slope is subject to any vibrations, the clay structure collapses and the

clay is transformed quickly into a viscous fluid, which flows downslope. Quick clay of this type is a serious environmental problem in many areas, including the St. Lawrence River valley of Canada (Figure 10–5).

Cohesive Forces

In some cases, water acts as a resisting force. As long as the pores in a sediment are not filled *completely*, water films make the particles cohesive. **Cohesion** is the ability of particles to attract and hold each other. The thin films of water lining the pores develop surface tension due to the attraction that water molecules have for each other. **Surface tension** is a force that acts parallel to the water surface and pulls on it (Figure 10–6). Because the water films also are attracted to the surfaces of the particles, the effect of surface tension is to pull the particles together.

Surface tension enables you to build a castle from wet sand at the beach. The castle can remain standing even though it has vertical walls, because the water's surface tension in the partially filled pores holds the sand together. As the sand dries, it loses cohesion, and the castle begins to crumble. When the tide rises and submerges the base of the castle, the pores become fully saturated and the castle crumbles into the water.

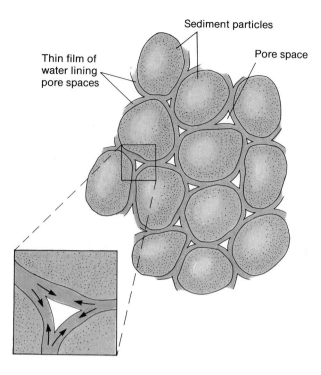

FIGURE 10–6
Surface tension—in the thin films of water that line the pores—holds partially saturated sediments together. Inset detail shows how surface tension pulls on both the water surface and the sediment particles, pulling the particles together.

Angle of Repose

Have you ever added dry sand to a pile and watched the sides collapse periodically as it builds up? The resulting pile is higher, but its sides have a slope with the same inclination as the original mound (Figure 10–7). The maximum angle at which granular materials can be piled is called the **critical angle of repose**. For most sediments, the critical angle is between 25 and 45 degrees.

Particle size is the dominant factor in determining the critical angle of repose, but other factors are also important. Angular particles can interlock along their rougher edges and assume higher angles than more rounded particles of the same size. Poorly sorted sediments have a steeper angle of repose because the smaller particles fit between the larger ones, permitting all of the particles to be stable at a higher angle. Partially saturated sediments have a steeper angle of repose because of the surface tension of the water films in the pores (Figure 10–5).

Removal of Slope Support

Material on a slope is supported by the material at its base. Whenever natural or human activities remove a portion of the lower slope (base), this not only changes the angle of repose but reduces the support for the upper part of the slope. Many landslides have occurred because of these activities. For example, river erosion may undermine the sides of a stream valley and cause a cave-in of the stream bank. In a similar fashion, the excavation of a building lot on a slope can cause a landslide into the development.

Particle Packing

The way that particles are arranged in a deposit can affect slope stability. Arrangement of particles in a deposit is called the **packing** (Figure 10–8). There are two extremes of packing:

- □ **Cubic packing** occurs when the grains are positioned so that their centers are directly above those of the grains below—Figure 10–8a). Cubic packing is the loosest type and has the most pore space. It occurs wherever material has been dropped or bulldozed into space with little reworking.
- □ **Rhombohedral packing** occurs when the centers of grains are located over the spaces between the grains below—Figure 10–8(b). It is the tightest form of packing and has the least pore space. Rhombohedral packing occurs in nature where material has been vibrated into place or deposited by a force acting parallel to the surface of deposition, such as a water or wind current.

Slope stability can be affected by a change in packing, because a change from loose to tight packing results both in a decrease in volume and lowering of the surface. You can test this yourself by partially filling a container with any granular material (rice, sand, marbles, and so forth). These particles have cubic packing. Mark the level of the upper surface and shake the container sideways. The particle sur-

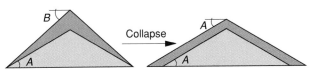

(a) Stable pile of sand with side slope angle (*A*) equal to the critical angle of repose for that size of sand grain.

(b) More sand is added, forming a steeper side slope angle (*B*). This is greater than the critical angle of repose for that size of sand grain. The pile is now unstable and the sides collapse.

(c) A higher stable pile forms with side slope angles equal to the critical angle of repose (*A*) for that size of sand grain.

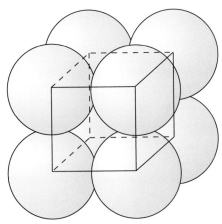

(a) Cubic packing

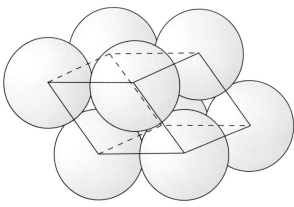

(b) Rhombohedral packing

FIGURE 10–8
Types of packing in granular materials: (a) cubic packing, (b) rhombohedral packing.

face will drop as the grains assume a more nearly rhombohedral packing.

In nature, the surface of loosely packed sediments also will drop if they are shaken by ground movement, such as an earthquake; construction activities, such as blasting; or highway traffic. A structure built on such a surface may be damaged as its foundation settles. Another effect of a change from cubic to rhombohedral packing is a reduction in pore space. This quickly expels some of the pore fluids and can cause a material to liquefy into a slurry. Structures built upon such material can be damaged through losing support for their foundations.

TYPES OF MASS MOVEMENTS

The best classification of mass movements is that proposed by D. J. Varnes in 1978. We will use a simplified version of this scheme.

Three major types of mass movements can be distinguished, based on the type of movement involved:

1. **Falls** involve sediment and rock that move through the air and land at the base of a slope.
2. **Slides** are movements of rock or sediment as a unit, principally along one planar surface.
3. **Flows** are plastic or semiliquid movements of rock or sediment, either in air or water.

Further subdivisions are based on the material involved and the speed of movement. The major mass movement processes are described and illustrated in Table 10–1.

TABLE 10-1

Characteristics of major mass-movement processes. After "Landslides," 1978, with permission from the National Academy of Sciences, National Academy Press, Washington, DC.

Mass Wasting Type	Character of Movement	Subdivision	Speed and Type of Material
Falls	Particles fall from cliff and accumulate at base	Rockfall —see (a) below	Extremely rapid; develops in rocks
		Soilfall	Extremely rapid; develops in sediments
Slides	Masses of rock or sediment slide downslope along planar surface	Rockslide —see (b) below	Rapid-to-very rapid sliding of rock mass along a flat inclined surface
		Slump —see (c) below	Extremely slow-to-moderate sliding of sediment or rock mass along a curved surface

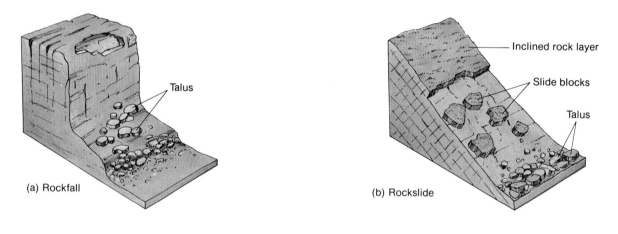

(a) Rockfall

Talus

Inclined rock layer

Slide blocks

Talus

(b) Rockslide

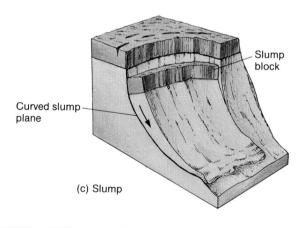

Slump block

Curved slump plane

(c) Slump

Mass Wasting Type	Character of Movement	Subdivision	Speed and Type of Material
Flows	Displaced mass flows as a plastic or viscous fluid	Creep —see (d) below	Extra-slow movement of surface regolith and rock
		Solifluction —see (e) below	Very slow-to-slow movement of water-saturated regolith as lobate flows
		Mudflow —see (f) below	Very slow-to-rapid movement of fine-grained sediment and rock particles with up to 30% water
		Debris flow —see (g) below	Very rapid flow of debris; commonly starts as a slump in the upslope area
		Debris avalanche —see (g) below	Extremely rapid flow; fall and sliding of rock debris

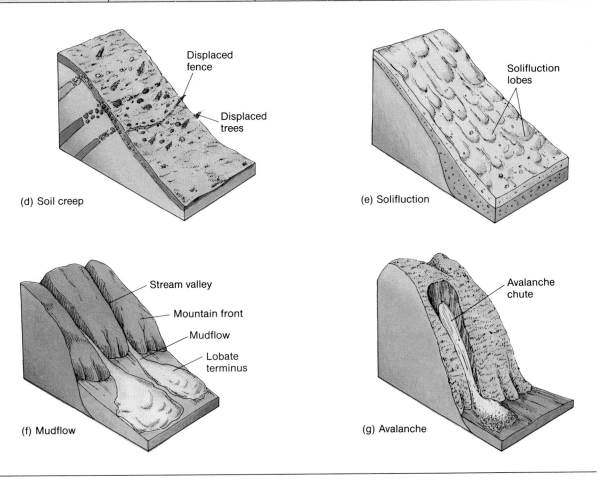

(d) Soil creep

Displaced fence

Displaced trees

(e) Solifluction

Solifluction lobes

(f) Mudflow

Stream valley

Mountain front

Mudflow

Lobate terminus

(g) Avalanche

Avalanche chute

Falls

The fall of rock particles through the air from a cliff is called **rockfall** (Figure 10–9). Rockfall can be a dry process, triggered by root wedging, or a wet process, triggered by frost wedging in which ice formation in crevices loosens the rock particles so that they fall from the cliff face. Rockfall is an extremely rapid process and thus is a serious geologic hazard. It can demolish structures near the bases of cliffs and is a danger to motorists where a highway cuts through rocks. The pile of rock fragments deposited by rockfall at the base of a slope is called a **talus pile** (Figure 10–10).

Slides

Movement of rock or sediment along a planar surface is called a **slide**. Sliding movements are differentiated by the character of the planar surface along which slope failure occurs.

Rockslide. The downslope movement of rock masses along a planar surface is called a **rockslide**. The sliding surface is commonly a bedding plane, but rockslides have developed on a wide variety of other planar surfaces, such as exfoliation sheets or fractures cutting across layered rocks—Figure 10–2(b). Any area where such planar rock surfaces are inclined toward an open space, such a road cut or valley, has the potential for rockslide development—see the unstable slopes in Figure 10–2(a) and (b), and see Focus 10–1.

Slump. Sliding of a mass of material along a *curved* surface is a **slump**. Although slumping is most common in unconsolidated sediments, it also occurs in some poorly consolidated rock sequences. One of the commonest causes of slumping is erosion at the base of the slope, which removes support for the material above. This erosion may be due to natural processes, such as the undermining of a river bank by streamflow or the cutting away of a coastal cliff base by

FIGURE 10–9
Rockfall of jointed, resistant cap rock at edge of plateau, Grand Canyon National Park, Arizona. Large, angular boulders have fallen to the slope below. Photo by Nicholas K. Coch.

FIGURE 10-10
Talus slope at base of cliff, Grand Canyon National Park, Arizona. Rock fall in the upper part
of the cliff has exposed unweathered rock. Photo by Nicholas K. Coch.

storm waves (Figure 10–11). When the slope fails, the slump block rotates downward and a scarp (cliff) is formed at the top of the slope.

Slumping is an especially serious problem where structures are built on bluffs above the shoreline. Wave energy erodes the bases of the cliffs, causing them to fail (Figure 10–11). In many cases, homes built along the bluffs are lost and the scarp recedes landward, to pose a new problem for the next line of houses behind the cliff (Figure 10–12).

Flows

Flows are mass movements in which the material has a plastic or semiliquid behavior resembling that of a viscous fluid. In many cases, mass movements that start as falls, slides, or slumps are transformed into flows farther downslope. Flows exhibit a wide range of characteristics; among the mass movements, flows include the wettest, driest, slowest, and fastest. We will discuss flows in order of increasing speed of movement.

Creep. You may have noticed a downslope displacement of trees, fences, or utility poles over a long period of time. This downslope displacement is the result of **creep,** an extremely slow downslope movement of regolith, soil, and rock under the influence of gravity (Figure 10–13). Creep may be the least spectacular of all mass movements, but its continuity of operation and its action over such an extensive part of the Earth make it the most important mass-

FIGURE 10–11
Slumping in coastal cliffs. (a) Coastal cliff before a storm. (b) Storm waves erode the cliff base, undermining it and causing a slump block to rotate downward as the slope above fails. (c) Slumped material is reworked by the waves and the cliff recedes landward in the process.

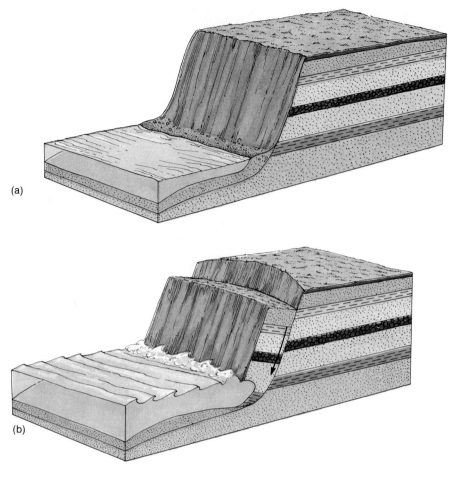

(a)

(b)

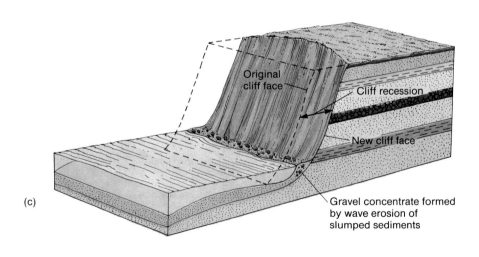

Original cliff face

Cliff recession

New cliff face

Gravel concentrate formed by wave erosion of slumped sediments

(c)

FIGURE 10–12
Damage caused by slumping of a coastal cliff on Cape Cod, Massachusetts. Home at top is endangered. Photo courtesy of S. P. Leatherman.

FIGURE 10–13
Soil creep in Santa Monica Mountains, California. The creep has displaced layers downslope at the top of the outcrop. Photo by John Shelton.

THE VAIONT DAM DISASTER

O ne of the most tragic geologic catastrophes in recent years was the 1963 rockslide associated with the building of a massive dam across the Vaiont River in northern Italy. This disaster demonstrates the sequence of development of a large mass movement and the tragic human consequences. Our discussion is based on the detailed synthesis of English and foreign reports of the event written by Frank Fletcher (see Further Readings at end of this chapter). Could this disaster have been avoided?

The Vaiont River flows along the base of a deep, glacially eroded, U-shaped valley which opens into the much broader Piave River valley opposite the town of Longarone (Figure 1). A large landslide scar on the northeastern side of the Vaiont River valley indicates that large-scale mass movements had occurred on the valley walls in the recent past.

FIGURE 1

The Vaiont Dam area and Piave River valley, showing surface features mentioned in text. Section line *A–A'* is detailed in Figure 2. After Kiersch, 1964, "The Vaiont Reservoir Disaster," *Civil Engineering* 34(3). Reprinted with permission of American Society of Civil Engineers.

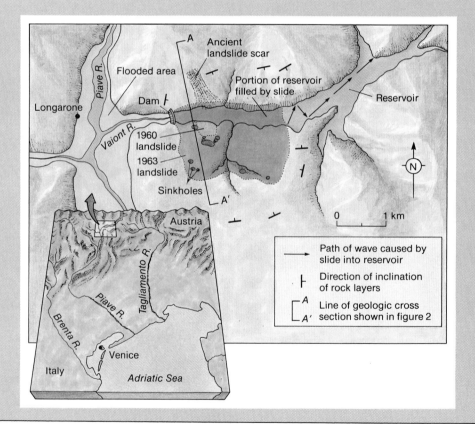

The valley walls are underlain by a number of limestone formations ranging from thick to very thin layers and interbedded with shaly limestone (Figure 2). The rock layers on both sides of the valley are inclined *toward* the center of the valley.

Construction began in 1956 on a concrete arch dam across the valley, 261 m high, to create a large lake whose waters would be used for hydroelectric power generation. In the summer of 1957, it was noted that concrete being poured for the dam foundations was disappearing into the rocks below, suggesting that there were *fractures* or *solution cavities*, or both, in the limestone. By 1959, the dam was nearing completion and the reservoir was allowed to fill in stages.

Raising the water level in the reservoir was to have profound effects on the stability of the adjacent slopes. On November 4, 1960, a mass of rock

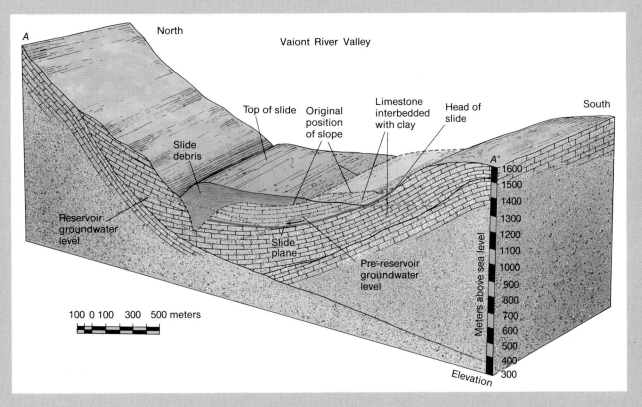

FIGURE 2
Section of Vaiont valley along line *A–A'* in Figure 1. All layers except the base are limestones. The mass that slid (shaded) was limestone interbedded with clay. After Kiersch, 1964, "The Vaiont Reservoir Disaster," *Civil Engineering* 34(3). Reprinted with permission of American Society of Civil Engineers.

and soil with a volume of 700,000 m³ slid off the south slope into the reservoir, generating a wave 2 m high that spread across the lake surface (Figure 1). Engineers did some studies and model analyses and concluded that the reservoir could be safely filled to the planned level. As the reservoir level rose, the rate of surface creep, a normally very slow process, increased dramatically. On September 4, 1963, the creep rate was 3–6.5 mm/day; by September 15th it was 12 mm/day; and it rose to 40 mm/day by October 6th. New cracks began to form on the slope. If that were not enough, the slope material was becoming saturated because rainfall in August and September had been *three times* greater than that recorded for the same time period during the previous 20 years.

The engineers now realized that mass movement was inevitable, but they felt confident that slope materials would come down as a series of pieces and blocks, because that had been the previous behavior of large landslides in the region. In addition, scale-model experiments indicated no danger from a wave formed in the lake by displacement of the lake water by a landslide.

We can now see that a number of natural and human-induced changes were combining to create a mass movement of gigantic proportions. Postglacial stream erosion by the Vaiont River had eroded a steep-walled inner valley within the broader glacial valley (Figure 2). Stream erosion of the slope's base removed lateral support from the fractured, steeply inclined rocks within the valley walls (Figure 2).

Water played a considerable role in increasing the driving forces on the slope. The exceptional rainfall had increased the weight of the rocks as it filled the fractures and solution cavities in the limestone. Cohesion within the clayey limestone beds was being reduced as water saturated the clay minerals. Water within the rock also was being subjected to great pressure from the rainfall entering the rocks above, plus the sideward pressure into the rock by the rising reservoir level. This excess pressure on the water in the rock pores decreased the frictional resistance and in effect made the rock "buoyant," somewhat like a beached boat later picked up by a rising tide.

It was now clear to the authorities that mass movement on the south slope was inevitable and that people should be evacuated from the surrounding areas. The evacuation order was signed shortly before 10:00 P.M. on October 9, 1963—but too late! Thirty-nine minutes later, at 10:39, 300 million m³ of fractured rock and water-laden limestone tore loose from the top of the south slope and roared downward toward the reservoir.

The mass slid down the slope *as one intact block* along a plane underlain by slick, clayey limestone beds. The mass slid quickly into the reservoir, displacing about 50 million m³ of water and forming a colossal wave 200 m high. The wave spread eastward along the lake and quickly wiped out lakeshore villages (Figure 1). It also spread westward toward the dam, but the well-designed structure took the full force without failing. The wave rose

FIGURE 3

The Vaiont Dam as it looks today, viewed looking east into the gorge. Photo courtesy of F. W. Fletcher.

over the dam and plunged into the narrow river gorge below (Figure 3). A violent and destructive wind swept the valley in advance of the floodwaters, stripping leaves from the trees.

The confined floodwaters surged out of the narrow Vaiont River gorge in a tremendous torrent, crossing the wide Piave River valley (Figure 1) and destroying the village of Longarone, with a loss of 1450 lives. The total loss was to be much higher as the wave also surged upstream along the Piave River valley to damage villages to the north (Figure 4) and spread as a massive flood across villages downstream of Longarone.

The Vaiont Dam disaster is an excellent example of the complex geologic problems that need to be considered in the design and emplacement of large structures. It demonstrates that even a well-designed engineering structure can be quite unsafe *if it is located in an area of geologic instability*. The tragic loss of life and property could have been prevented if the significance of the area's past geologic history and the potential changes that would be brought about by dam construction had been realized and acted upon earlier.

FIGURE 4

Upper half of Vaiont Dam, showing a portion of the landslide mass which filled the reservoir. Photo courtesy of F. W. Fletcher.

The conclusion of Frank Fletcher's description of the Vaiont Dam disaster puts the event in a human perspective. It is based on his field studies many years after the disaster, and the trial of the officials involved.

> Along the sides of the new road one comes across a crumbling stone wall, all that remains of someone's home, or some simple memorial to the victims—a crude wooden cross in a meadow or, in a shallow rock hollow, a small, carefully placed cluster of fresh, native alpine flowers or, sometimes, just a short list of names, often with the same last names, on a plaque. At the west end of the valley stands the dam, still the highest double-arch dam in the world; but no electricity is produced and exported to Italy's prosperous industries. Three hundred million cubic meters of rock and soil fill the reservoir and 1899 people are dead. On November 24, 1968, as the trial began far away in the Abruzzi [region of southern Italy], the total came to 1900. [Chief] Engineer Mario Pancini, his bags packed for the trip to [court in the city of] L'Aquila, taped the cracks around the doors of his Venetian room and turned on the jets of his gas range.[1]

---

[1] F. Fletcher, 1970, A terrifying equality: The story of the Vaiont Dam disaster, *Susquehanna University Studies* 8(4):300.

movement process in terms of *total volume* of material moved downslope each year.

Creep is aided by expansion and contraction of soil due to heating and cooling, freezing and thawing, or wetting and drying. For example, repeated cycles of freezing and thawing can move particles downslope in a process called **frost creep**. Each day, water from melting soil ice forms a thin film under the rock particles (Figure 10–14). This thin water film freezes at night, expanding in the process, pushing out the particles at right angles to the slope. When the ice melts the next day, the particles settle to the slope *parallel to the gravitational force* (G in Figure 10–1). Each of these cycles displaces the particles farther downslope. This process is particularly effective where the freeze-thaw cycle occurs frequently, as in high latitudes or in mountains at lower latitudes.

Solifluction. The downslope movement of water-saturated regolith is called **solifluction**. Movement rates are faster than in creep and may reach up to a few centimeters per year. Solifluction may occur in any climate in which regolith becomes saturated with water. However, it is most common in cold climates where the upper part of the regolith freezes and thaws periodically. Many cold areas are underlain by permanently frozen ground called **permafrost**. Dur-

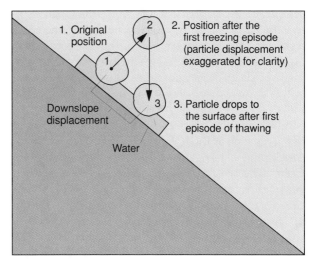

FIGURE 10–14
Mechanism of frost creep on slopes that are subject to freezing and thawing. Freezing water literally lifts particles, then lowers them during thaw.

ing the warmer parts of the year, the uppermost part of the ground thaws and releases the water within the sediment. This soggy mass of soil then can flow downslope over the permafrost below. Solifluction creates a topography characterized by curved, lobate flows on the surface (Figure 10–15).

FIGURE 10–15
Solifluction features, Nivot Ridge, Front Range, Colorado. Photo courtesy of Mitchell Algus.

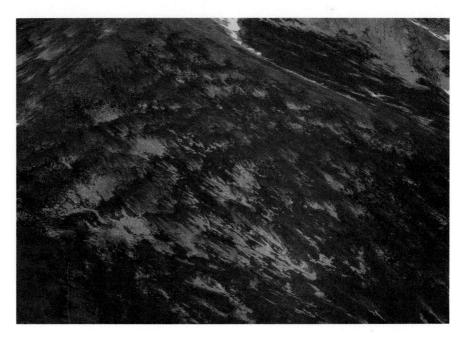

Mudflows. Flows that contain significant water (up to 30%) and a large proportion of fine-grained material are called **mudflows**. Mudflows are common on slopes in semiarid areas where infrequent but very intense, short-lived rainstorms quickly convert regolith into a mass of viscous mud and rock that moves downslope at high velocities (Figure 10–16). Mudflows can be a serious environmental problem in dry areas such as the Los Angeles basin, where urban development has spread toward steep mountain fronts. The problem becomes especially serious when heavy rains follow a period of drought or forest fires that remove slope vegetation. Severe rains during 1978 in southern California triggered mudflows in several areas. Similar storms had occurred in 1952, 1958, 1962, and 1969. Each successive storm did more damage because of the increase in population, buildings, and dwellings that had occurred since the previous mudflow activity.

Debris Flows. Mass movements in which rock debris and regolith flow very rapidly downslope are called **debris flows**. Many debris flows start as slumps or slides, but are transformed into flows downslope as the mass breaks up and mixes with air and water.

The Alaska earthquake of March 1964 triggered a massive debris flow from a mountain bordering the Sherman Glacier. The debris flow traveled 5 km from its source and deposited a layer of debris 1.5 m thick as it swept across the glacier without disturbing the fresh snow on the glacier's surface (Figure 10–17). The debris that covered the glacier showed little sort-

FIGURE 10–16
Mudflow northeast of Doublespring Pass, Idaho. Photo by John Shelton.

FIGURE 10–17
Aerial view of the 1964 earthquake-induced debris flow across the Sherman Glacier in Alaska.
Photo courtesy of George Plafker/U.S. Geological Survey.

ing with distance, and the slope across which it moved was inclined by only a few degrees.

This raised an interesting question. How do unsorted masses of debris travel so far over gentle slopes? It is now believed that at least a few debris flows are able to travel great distances over gentle slopes because they ride *on a cushion of air*. The air cushion is formed when tumbling rock debris traps air beneath itself. The presence of an air cushion enables the flow to move gently over land surfaces in much the same way that a "hovercraft" vessel skims at high speeds over the waves.

Debris Avalanches

The general term **avalanche** is used for the most rapidly flowing, sliding, and falling mass-movement processes. Anyone who has seen pictures of a snow avalanche realizes that it takes only a few minutes for

the material to travel from high on the mountain down the slope and across the valley below.

Very rapid to extremely rapid movements of rock and sediment are referred to as **debris avalanches**. Many debris avalanches are characterized by semicircular heads with elongate tongues of debris extending downslope. (Figure 10–18).

SEDIMENT DEPOSITED BY MASS MOVEMENTS

The collective term for all sediments deposited by mass movements is **colluvium**. The short distance of colluvium transport results in minimal rounding of particles and a mineralogical composition very similar to that of the source materials. The dense and viscous nature of most mass movements results in little sorting of the sediment particles and an absence of layering (in most cases).

FIGURE 10–18
Debris avalanche chutes and talus pile in Lee Vining Canyon, California. Photo courtesy of King Huber/U.S. Geological Survey.

SUMMARY

1. Mass movements involve transporting masses of material down slopes under the influence of gravity.

2. *Driving forces* cause mass movements; *resisting forces* retard them.

3. The steeper a slope, the greater is the tendency for mass movements down that slope. The actual movement of slope materials is called slope failure.

4. A slope having the potential for mass movements is called an unstable slope.

5. Wherever rock layers are inclined toward a slope, that slope is potentially unstable.

6. Water can act as both a driving and a resisting force in mass movements.

7. Water acts as a driving force by increasing the loading, reducing the rock strength, increasing the volume of expanding clays, and liquefying quick clays.

8. Water acts as a resisting force when the pore spaces of a sediment are filled partially. The surface tension between the water and the particle surfaces holds the grains together.

9. When particles that have a cubic packing are shaken, they assume a rhombohedral packing with a decrease in volume. This also results in a decrease of pore space and the expulsion of part of the pore fluids, as well as subsidence of the surface above.

10. When granular particles are piled up at an angle higher than their critical angle of repose, the slope fails.

11. Mass movements are of three types—falls, slides, and flows. Falls involve sediment and rock that move through the air and accumulate at the base of the slope. Slides are movements of rock and sediment along planar surfaces. Flows are plastic or semiliquid movements of rock and sediment in either air or water.

12. The apron of rockslide debris at the base of a slope is called talus.

13. Rockslides occur along a flat plane; slumps occur along a curved plane.

14. Although creep is the slowest of mass movements, it delivers more sediment to transporting systems than any other mass-movement process.

15. Solifluction and frost creep are mass movements common in areas subject to freezing and thawing. Solifluction requires an underlying base of permafrost; frost heaving does not.

16. Mudflows contain significant water (9–30%) and a large proportion of fine-grained material. They are especially common in drier areas.

17. Very rapid movements of rock debris and regolith are called debris flows. Some debris flows may move over a cushion of air, which greatly facilitates their spread across gentle slopes.

18. The most rapid mass movements are debris avalanches. Debris avalanches are characterized by semicircular heads with elongate tongues of debris extending downslope.
19. Material transported by mass movements is called colluvium.
20. The short transport distance of mass movements results in distinct properties for the resulting colluvium. Colluvium has minimal particle rounding and sorting and a mineralogical composition very similar to that of the original material.
21. Mass movements are important because they move material off slopes and into transportation systems such as rivers, glaciers, and coastal currents.
22. The catastrophic nature of many mass movements makes it vital to evaluate potential mass-movement problems in detail prior to the development of an area.

REVIEW QUESTIONS

1. How do mass movements differ from other types of transportation processes?
2. How can water act as a resisting force in mass movements?
3. What is always a driving force in mass movements?
4. Describe three different ways in which water triggers mass movements.
5. How could you recognize an area that might be underlain by "shrink-swell" clays?
6. What geologic conditions form quick clays?
7. Why does removal of material from the base of a slope trigger a mass movement?
8. What factors affect the critical angle of repose of a sediment?
9. What are the differences among the mass movements characterized as falls, slides, and flows?
10. What is a talus slope, and what type of packing would the underlying particles have?
11. How do rockslides and slumps differ?
12. Describe how a slump can form in a coastal cliff or along a river bank.
13. Creep is a slow process, and yet it is the most important of all mass-movement processes. Why?
14. Frost creep and solifluction both involve freezing. How do these two processes differ?
15. Why would mudflow be an especially dangerous problem in dry areas that are experiencing urbanization?
16. Some debris flows can move considerable distances over low slopes and with minimal disturbance of the underlying material. How is this possible?
17. What are the characteristics of colluvium, and why does it have these properties?
18. What factors could have been used to predict the Vaiont River valley rockslide before the dam construction started?
19. Describe the roles that water played in triggering the Vaiont landslide.

FURTHER READINGS

Briggs, R. P., Pomeroy, J. S., and Davies, W. E. 1975. *Landsliding in Allegheny County, Pa.* U.S. Geological Survey Circular 728, 18 p. (Describes different types of mass movements based on what causes them and how they can be recognized.)

California Division of Mines and Geology. 1979. Southern California landslides—1978. *California Geology* 32(1). (A series of articles describing a variety of mass movements.)

Fletcher, F. 1970. A terrifying equality: The story of the Vaiont Dam disaster. *Susquehanna University Studies* 8(4):271–300. (A beautifully written description of the events that led up to this disastrous rockslide.)

Kerr, P. F. 1963. Quick clay. *Scientific American* 209(5):132–42. (Describes the origin of quick clays.)

Luchitta, B. K. 1978. A large landslide on Mars. *Geological Society of America Bulletin* 89:1601–09. (Describes a large Martian landslide and compares it with similar features on Earth.)

Pestrong, R. 1974. *Slope stability.* Council on Education in the Geological Sciences, Publication No. 15 (McGraw-Hill Concepts in Introductory Geology Series), 65 p.

Shreve, R. L. 1968. *The Blackhawk landslide.* Geological Society of America Special Paper 108. (Describes an ancient landslide and the evidence suggesting that it traveled on a cushion of trapped air.)

Varnes, D. J. 1978. Slope movement types and processes. In *Landslides: Analysis and control.* Transportation Research Board, National Academy of Sciences, National Research Council Special Report 176, p. 11–33.

11

Streams and Stream Sculpture

Sunrise over the Colorado River at Dead Horse State Park,
Utah. Photo by Kirkendall/Spring.

ater in its many forms is the most important agent of erosion on the Earth's surface, as well as being essential to life itself. The turquoise waters of the Florida Keys, the steam jetting from Old Faithful geyser in Yellowstone National Park in Wyoming, the Malaspina Glacier in Alaska—all are forms of water on the Earth's surface.

THE HYDROLOGIC CYCLE

Figure 11–1 shows the relative amounts of water existing in various forms and places on the Earth's surface at present. Most is found in the oceans (97.3%,) followed by glaciers (2.14%). All other forms of water make up 0.56% of the total water distribution. The total amount of water in, on, and above the Earth remains almost constant.

In this chapter we will focus on *water that moves in channels on the surface,* features that we call **streams**. As Figure 11–1 shows, only 1/10,000 of the world's water is in streams, but streams are extremely important. They are the major agents of erosion in both wet and dry regions; they are the major transporter of sediment worldwide; and they are the source of drinking water for many communities. This chapter deals principally with streams in humid areas; stream activity in dry areas will be covered in Chapter 14.

Water is constantly changing from one form to another. For example, right now, glaciers are melting to form liquid water, ocean water is evaporating to form clouds of water vapor, and water vapor is freezing to form ice crystals and snow. These changes of

water from one state to another are called the **hydrologic cycle** (Figure 11–2).

The hydrologic cycle is driven by solar energy, which evaporates water from the surface of the oceans, lakes, and rivers. This water vapor rises into the atmosphere until it cools and condenses to form either rain or snow, depending on the temperature. When the rain falls, some of it sinks into the ground and some flows along the surface. When snow falls, it may melt within a season or form glaciers if it accumulates from year to year.

The time necessary for the return of a water molecule to the atmosphere is highly variable and depends on the path taken. For example, some raindrops never reach the ground but are evaporated back into the air as they fall. Some of the water that soaks into the ground may not appear on the surface again for months, years, or even centuries. Other water molecules may be absorbed by the roots of plants and trees and returned to the atmosphere by the plants in a matter of days, through a process that biologists call *evapotranspiration*.

SURFACE VERSUS SUBSURFACE WATER

What happens to raindrops falling onto a hill or snow melting on the ground? Some of the water will sink into the ground; the rest will flow along the land surface.

The sinking of water into the ground is referred to as **infiltration**. Each surface material has a given **infiltration capacity,** which is a measure of the max-

FIGURE 11–1
Distribution of the Earth's water. After Nace, U.S. Geological Survey Circular 536.

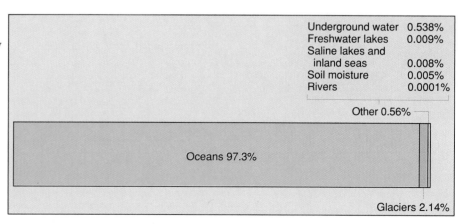

Underground water	0.538%
Freshwater lakes	0.009%
Saline lakes and inland seas	0.008%
Soil moisture	0.005%
Rivers	0.0001%

Other 0.56%

Oceans 97.3%

Glaciers 2.14%

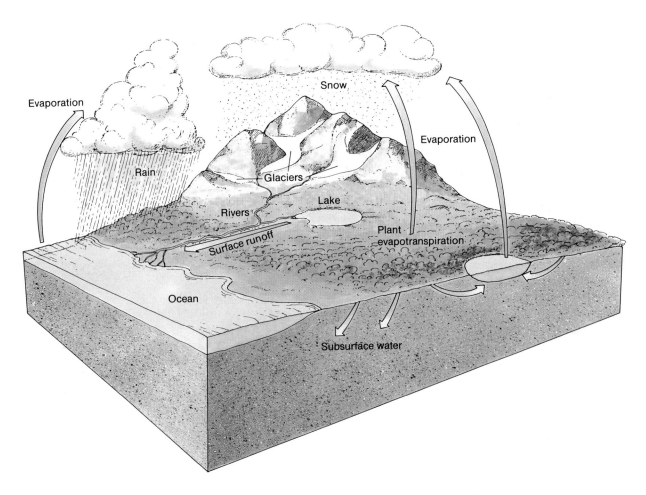

FIGURE 11–2
The hydrologic cycle.

imum flow rate of water through that material over time. Infiltration capacity is determined by the *porosity* of the material, the *slope*, and the *vegetation*. Surfaces with more openings, such as rock fractures or pore spaces between rock grains, will have a higher infiltration capacity. Steeper slopes limit infiltration because the water flows faster than it can sink into the ground. Greater amounts of vegetation slow the flow and allow more of it to infiltrate. The water which infiltrates the ground forms the underground water system, which we will describe in the next chapter.

If the rate of rainfall exceeds the infiltration capacity in an area, the excess water moves across the surface in streams. If the stream channels cannot transport that much water, the excess moves over the land surface as a layer referred to as a **sheet flow.**

STREAM ENERGY AND HOW IT IS USED

Water that has collected at a high elevation has potential energy (see Chapter 2). As it begins to flow downslope under the influence of gravity, the potential energy is transformed into kinetic energy, providing the energy needed for stream erosion and transport.

Some streams flow slowly across gentle slopes; others rush in torrents down steeper ones. The steepness of a stream slope is measured by its **gradient,** which is the drop in elevation of the stream bed over a given horizontal distance (Figure 11–3). Most streams have steep gradients near their headwaters and gentler gradients downstream. This gradual decrease in stream gradient from upstream to down-

FIGURE 11-3

Cross section along the length of a stream from its headwaters to the ocean, showing gradient and longitudinal profile. The longitudinal profile is concave upward, with a steeper upstream gradient and a gentler downstream gradient.

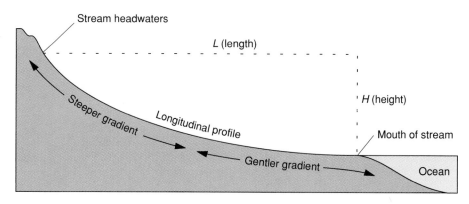

stream results in a "concave-up" profile (Figure 11-3). A cross section down the middle of a stream from its headwaters to some point downstream is called a **longitudinal profile** (Figure 11-3).

Streams continue to flow downslope—eroding, transporting, and depositing—until they reach a level at which they can no longer erode downward into the land surface. This level is called the **base level**. The *ultimate* base level is the ocean (sea level); however, many streams encounter *local* base levels in their paths downslope. Often, local base levels include the surfaces of lakes or of reservoirs behind dams. Some streams never reach the ocean, and they deposit their loads in continental basins from which there is no outlet. These basins may have floors that are well above sea level, or in the case of California's Death Valley, below sea level. The relationship among a stream's energy, local base levels, and stream behavior can be seen by looking at the effects of constructing a dam across a stream (Figure 11-4). The dam impounds a lake upstream and the reservoir surface serves as a local base level for streams flowing into the reservoir.

Streamflow decreases abruptly as it enters the reservoir. It loses so much kinetic energy that it can no longer carry sediment. As a result, some sediment is deposited in the channel immediately upstream from the lake, creating a new gradient that is more gently sloping than the original. In addition, sediment is deposited in the upstream part of the reservoir.

Although the reservoir water has no kinetic energy, it does have potential energy because the reservoir surface is well above sea level. When the relatively sediment-free water in the downstream part of the reservoir is released from the dam to create

hydroelectric energy or for flood control, its potential energy is transformed into kinetic energy once again. The transportation of sediment requires energy, and because this water now carries very little sediment, more of its energy can be used to erode the channel downward. This accelerated erosion steepens the stream channel locally downstream from the dam. The stream then continues downslope, locally eroding, transporting, and depositing sediment until it reaches sea level. The more kinetic energy a stream has, the more material it can erode and transport. When kinetic energy of a stream decreases, it deposits this sediment.

STREAMFLOW

Characteristics of the flowing water, and even the shape of the stream channel itself, can change considerably with time. Several variables are used to describe streamflow; two of them are:

☐ **Discharge** is the volume of water passing a given point in a given time.
☐ **Velocity** is the rate of water flow.

Another set of variables is used to describe changes in the cross-sectional area of flow:

☐ **Stream width** is the horizontal distance measured across the top of the water surface from bank to bank.
☐ **Stream depth** is the vertical distance measured between the water surface and the stream bed.

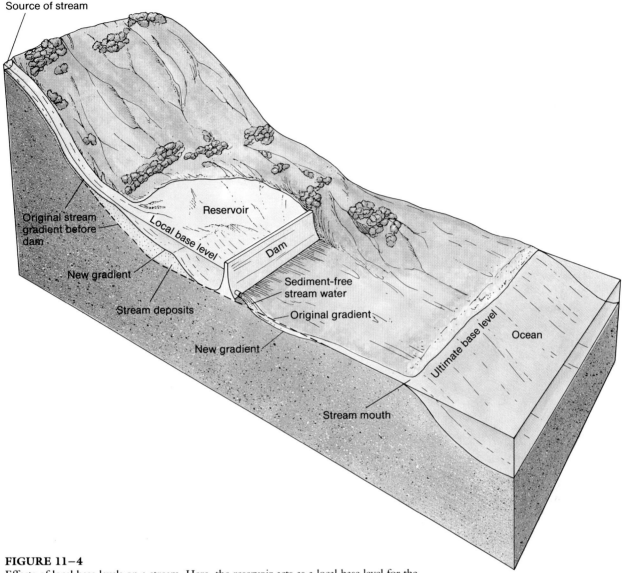

FIGURE 11–4
Effects of local base levels on a stream. Here, the reservoir acts as a local base level for the stream. Streamflow decreases in velocity as it enters the reservoir, resulting in deposition of its sediment load. The deposition of sediment load immediately upstream from the reservoir decreases the stream gradient locally. Sediment-free water exiting the base of the dam has sea level as a base level; it erodes its bed, steepening its stream gradient locally.

Discharge

The discharge of most streams changes appreciably from drought to flood periods. Thus, in times of heavier rainfall or snowmelt, we would expect an increase in stream discharge. Increasing stream discharge also increases the ability of the stream to erode its bed and to transport material from that area. The discharge of larger streams increases further in a downstream direction, independently of the weather, because more and more tributaries add their drainage to the major stream.

Velocity

Water velocity varies widely at different places in the flow. Maximum velocities occur in the center of the channel and at a depth that ranges from just below the stream's surface to one-fourth of total depth—Figure 11–5(a) and (b). The lowest velocities occur near the bottom and on the channel sides—places where frictional effects are greatest—Figure 11–5(b). The drop in velocity near the surface reflects the frictional drag of air on the stream surface—Figure 11–5(a).

All other things being equal, velocity is higher in streams with steeper gradients, with smooth and straight channels, and with high discharges. The last point may not be as obvious as the other two. As stream discharge rises after a rainfall, the stream must flow faster to move this greater discharge through the same channel.

Relationship Between Streamflow and Channel Variables

U.S. Geological Survey geologists Luna B. Leopold and Thomas Maddock used a great number of stream measurements to learn the basic relationships between streamflow variables (discharge and velocity) and channel cross-section variables (width and depth). They showed that, as the discharge increases, there are predictable increases in the width, depth, and velocity of streams. These variables are related by the formula:

$$\underset{(m^3/sec)}{\text{Stream discharge}} = \underset{(m)}{\text{width}} \times \underset{(m)}{\text{depth}} \times \underset{(m/sec)}{\text{velocity}}$$

Such a system, in which a change in one part is balanced by a change in another part, is said to be in **dynamic equilibrium**.

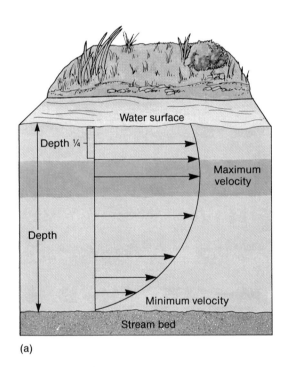

(a)

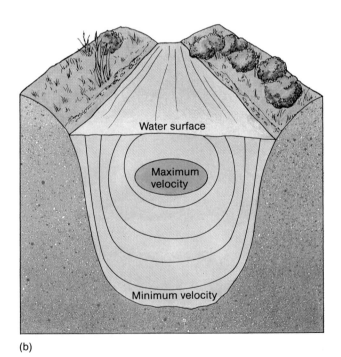

(b)

FIGURE 11–5
Variations in stream velocity with depth. (a) Section parallel to streamflow. The maximum stream velocity is located about one-fourth of the depth below the stream surface. The length of each arrow is directly proportional to the velocity of water flow at that depth. (b) Section perpendicular to streamflow. As shown in (a), the maximum stream velocity is within the center of the stream, below the surface to about one-fourth of the total depth.

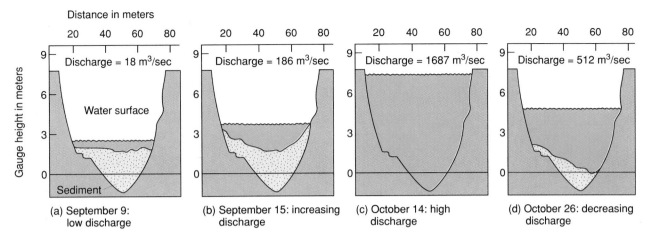

FIGURE 11–6

Changes in the water surface and stream-bed elevation with increasing discharge in the San Juan River, near Bluff, Utah, September–October 1941. (a) During low discharge the stream flows slowly over thick channel deposits. (b) A rise in discharge is accompanied by increasing width, depth, and velocity. (c) High discharge results in erosion of the bed material so that the stream reaches its maximum cross-sectional area. (d) Decreasing discharge results in a deposition of stream deposits and lowering of water-surface elevation. After *Fluvial Processes in Geomorphology*. By L. Leopold et al. Copyright © 1964 by W. H. Freeman and Co. Used with permission.

The dynamic equilibrium that operates in streams can be illustrated by changes that occurred in the channel of the San Juan River near Bluff, Utah, as discharge increased markedly between September and October 1941 (Figure 11–6). The rising discharge was accompanied by higher velocities, a rise in the elevation of the water surface (increasing stream width and depth), and a scouring of material in the stream bed (increasing the depth) as shown in Figure 11–6(a)–(c). As the stream discharge began to decrease—Figure 11–6(d)—the velocity decreased and the water level dropped while the channel bed was built up once again. These adjustments brought the stream channel into equilibrium with the lower discharge.

Flooding

A discharge that fills the stream channel is called a **bankfull discharge** (Figure 11–7). When the discharge exceeds the bankfull level the stream spills out of its channel and flooding begins. The area adjacent to a stream that is periodically flooded is the **floodplain** (Figure 11–7). Floods are natural events

that are to be expected along most streams every few years.

The low infiltration capacity of rock exposed on steep mountain slopes results in much greater runoff than infiltration. Heavy, sustained rainfalls in such areas can result in severe flooding. An example of this type of flood occurred in Rapid City, South Dakota on June 9–10, 1972. The city is situated on the eastern edge of the Black Hills. On June 9, easterly winds forced local air to higher altitudes, where it cooled and condensed to rain. The weak winds at higher altitudes prevented the rainstorm from moving out of the area, and by midnight 10–30 cm of rain had fallen over the area.

Rapid Creek, which rises in the Black Hills flows through the middle of Rapid City, increased to a peak discharge of 883 m³/sec by 11:15 P.M. The torrent rushed downstream, further increasing in discharge as more and more tributaries joined the flow of Rapid Creek. The floodwaters overtopped a small recreational dam at Canyon Lake near the edge of Rapid City and rushed into the town. The discharge greatly exceeded Rapid Creek's channel capacity, and the floodwaters spread quickly over the city, killing

FIGURE 11–7
Variables used to describe streams.
Note changes in the width *w* and
depth *d* as discharge increases.

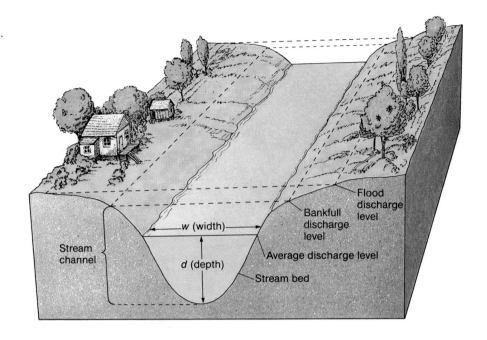

237 and causing $79 million damage. Some 1335 houses and trailers and 5000 automobiles were destroyed in the 10 hours during which the flood surged through the area.

Heavy rains are not the only cause of flooding. Floods can occur when snow cover melts more rapidly than usual. Others happen when a hurricane hits a shoreline and drives coastal waters inland, backing up streams and causing them to overflow their banks.

Flooding and Urban Development. When rain soaks a vegetated area, a stream's discharge increases slowly and the peak discharge occurs well after the rain has started to fall. The time between the greatest rainfall and the maximum stream discharge is called the **lag time**—Figure 11–8(a). Lag time exists because of the time needed for rain to exceed the infiltration capacity of the soil, so that significant runoff can begin. It also takes time for the water that sinks into the ground to flow underground until it reaches a stream.

Increased flooding occurs wherever natural vegetation has been removed for development. During urbanization, extensive surfaces are paved (roads, parking lots), vegetation is removed, and artificial modifications to stream channels are made. These changes affect the lag time of streams in the area. Figure 11–8 shows how development of an area affects the lag time. The bar graph shows distribution of rainfall with time; the bell-shaped curves show the rise and fall of stream discharge over the same length of time. The gentler curve in (a) represents a natural stream in the area, whereas the steeper curve in (b) represents a stream in an urbanized area.

Compare the two curves and note that urbanized streams not only reach their peak discharges sooner (a decrease in lag time), but they also have a greater peak discharge than similar streams in natural areas. Paving of surfaces in urbanized areas prevents water from sinking into the ground, forcing it to flow quickly over the surface into the stream. This both decreases lag time and increases peak discharge of the stream.

TRANSPORT OF SEDIMENTS BY STREAMS

Some aspects of sediment transport were described in Chapter 6. Here we will provide a more detailed discussion.

The total amount of material carried by a stream is called its **load**. Figure 11–9 shows the different

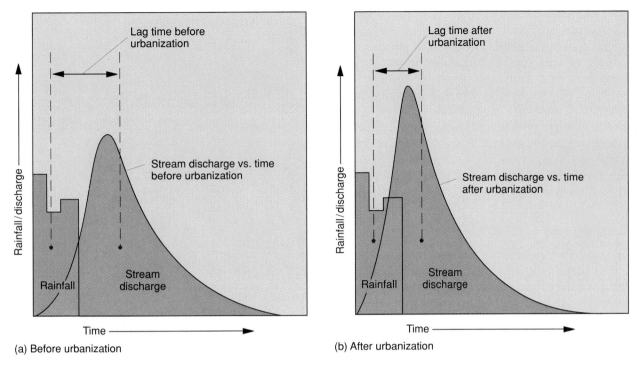

(a) Before urbanization

(b) After urbanization

FIGURE 11–8
Stream hydrographs of an area (a) before urbanization and (b) after urbanization.
Urbanization results in a decrease in lag time and an increase in peak discharge.

FIGURE 11–9
Types of stream load.

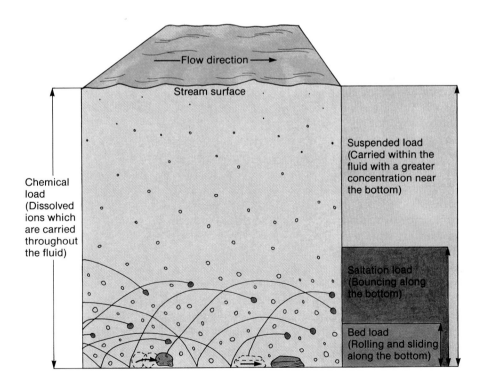

ways that streams carry sediment. Part of the load is transported as dissolved ions, whereas the rest is transported physically as particles. The physical load consists of finer particles suspended in the stream-flow (*suspended load*) and larger particles which bounce (*saltation load*) or roll and slide (*bed load*) across the bottom of the stream.

The amount of dissolved chemical load is determined by the climate and the solubility of rocks exposed in the area. The amount of physical load is determined by a number of factors that are described in the next two sections.

Capacity and Competence

The **capacity** of a stream is the *amount* of sediment it carries past a point in a given amount of time. The **competence** of a stream is a measure of the *largest particle* size that the stream can transport. An increase in stream velocity increases both the competence and capacity. For example, a slow-moving stream may be transporting moderate amounts of clay, silt, and fine sand; it is said to have low capacity and competence. As the velocity increases in a flood, the stream can transport greater amounts of sediment (increased capacity) and larger particle sizes (increased competence).

Laminar and Turbulent Flow

The paths that individual water particles take in a stream are called **flow lines**. The changes in flow lines as velocity changes are shown in the longitudinal cross sections of Figure 11–10. At very low velocities, the flow lines are almost parallel and there is little vertical movement in the stream—Figure 11–10(a). Such a water flow is called **laminar flow**.

Stream behavior changes markedly as the velocity increases—Figure 11–10(b). The flow lines become irregular and vertical mixing becomes common as swirling, whirlpool-like masses of water called **eddies** extend up and down through the streamflow. Eddies move with the current, dying out and forming again in another place. Streamflow characterized by extensive vertical mixing is called **turbulent flow**. In turbulent flow, local movement of the water can be in any direction, but the *net* movement of the water is downstream.

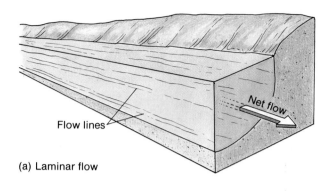

(a) Laminar flow

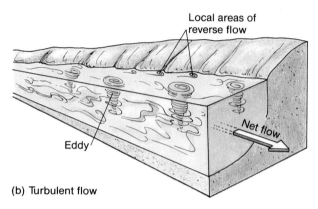

(b) Turbulent flow

FIGURE 11–10

Types of streamflow. (a) *Laminar* flow lines at low velocities are relatively straight and parallel, with no vertical mixing. (b) *Turbulent* flow lines at high velocities are highly irregular, with extensive vertical mixing. Local flows are in all directions, but net flow is downstream.

The presence of either laminar or turbulent flow is not dependent on velocity alone but also on the roughness of the stream channel and the depth. A rough stream channel interferes with laminar flow lines and can initiate turbulent flow at a lower velocity than would be possible in a smooth channel. In a similar fashion, shallower depths result in more turbulent flow for a given velocity and channel roughness. Velocities of most streams in nature are fast enough to be turbulent. The multidirectional water movements that occur in turbulent flows aid both in eroding particles from the stream bed and in keeping them suspended within the streamflow so that they can be transported downstream.

Erosion and Deposition of Stream Sediments

Particles are eroded from the stream bed when they are hit by bouncing particles or when turbulent eddies scour the bottom and carry them upward into the higher velocities of the central part of the stream. The eroded grains are held in the main flow by the turbulence and the upward flow in the eddies which act against the gravitational force that otherwise would make them settle out of suspension. The net downstream flow then transports them from the area.

As the stream velocity and turbulence decrease, particles are deposited. Larger particles are deposited first, followed by progressively smaller and smaller particles. Stream velocity changes usually are gradual, and successive layers of different grain sizes are deposited within the same bed, as illustrated in Figure 11–11.

STREAM CHANNELS

Most stream channels fall into one of three categories (Figure 11–12):

1. A **straight channel** is the most direct course between two points on the stream—Figure 11–12(a).
2. A **meandering channel** includes a series of sinuous curves—Figure 11–12(b).
3. A **braided channel** breaks up into numerous smaller channels separated by islands or sand-

bars. It is named for its similar pattern to braids of a rope—Figure 11–12(c).

Different parts of the same stream may have different channel patterns due to changes in local conditions. In addition, a given stream segment may have different channel patterns at different times of the year, because discharge or sediment supply vary seasonally. The ability of a stream to change its channel pattern is another example of the dynamic equilibrium mentioned earlier in this chapter.

Straight Channels

Straight channels consist of alternating deep areas called **pools,** and shallow areas called **riffles**—Figure 11–13(a). These pools and riffles do not have a random location along the stream but occur at regular intervals; the space between riffles and pools is five to seven times the stream width—Figure 11–13(b). As the arrows show, the flow in straight stream segments is not parallel to the channel walls but moves from side to side. Thus, the deepest part of a straight stream channel is not necessarily at the center but varies slightly back and forth between the straight channel walls.

We have seen that flow lines tend to meander, even in a straight stream. Can a straight stream evolve into a meandering one with time? We mentioned in Chapter 1 that geologists commonly build models of geologic systems to study what might happen under natural conditions. The U.S. Army Corps of Engineers Waterways Experiment Station at Vicksburg, Mississippi is the foremost facility world-

FIGURE 11–11
As stream velocity decreases, sediment layers of different particle size are formed. (a) Stream energy is sufficient to carry all sediment particles in suspension. (b) Stream energy decreases and particles having the highest settling velocity (coarse sand) are deposited. (c) Stream energy decreases still further and medium-sized sand particles are deposited as a second bed.

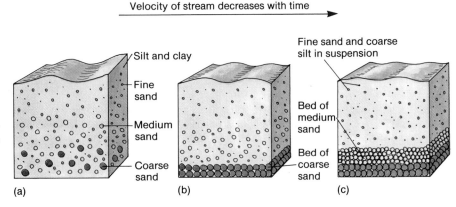

Velocity of stream decreases with time →

Silt and clay
Fine sand
Medium sand
Coarse sand
(a)

(b)

Fine sand and coarse silt in suspension
Bed of medium sand
Bed of coarse sand
(c)

FIGURE 11–12
Types of stream channel patterns.
(a) Straight; (b) meandering; (c) braided.

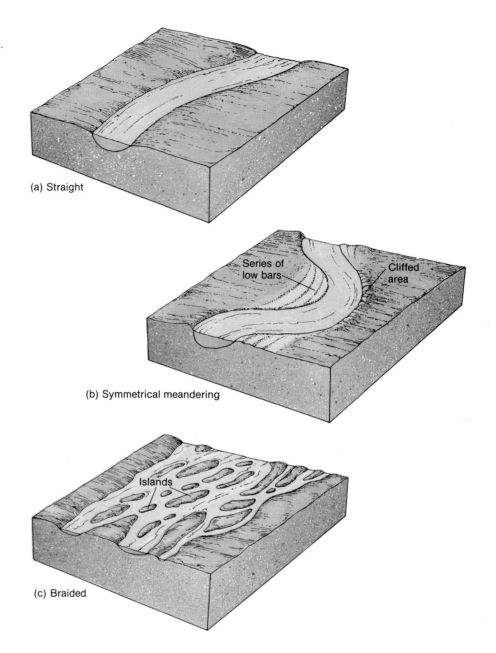

(a) Straight

(b) Symmetrical meandering

Series of low bars

Cliffed area

(c) Braided

Islands

wide in the use of models to study the behavior of water bodies (harbors, lakes, coastlines, and rivers). These models cover acres of land and have proven most useful in predicting changes that occur in streams with time. Figure 11–14 shows an experiment conducted there to observe changes over time in a straight channel excavated in a bed of sand. All conditions were kept constant, but the changes shown in this figure show how the stream slowly evolved into a meandering one. Where the straight-channel pattern persists in nature, it must be due to local geologic and topographic conditions that strongly resist the natural meandering tendency.

Factors that favor maintenance of straight stream segments include the presence of resistant rocks in the channel, well-developed linear (straight) frac-

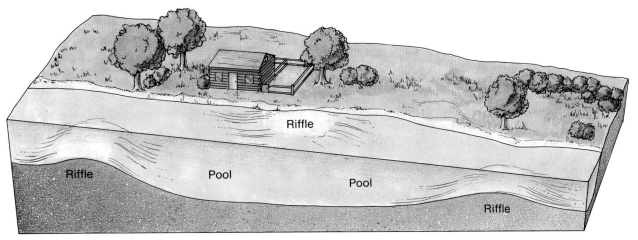

(a) Side view (cross section)

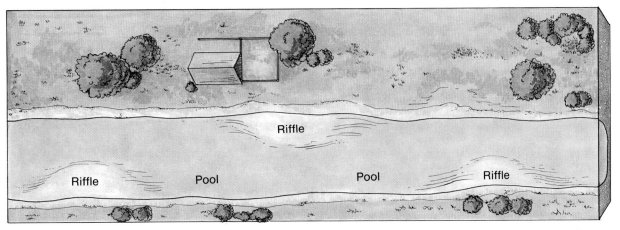

(b) Top view (map)

FIGURE 11–13
Alternating shallows (riffles) and pools (deeps) occur along straight stream channels. (a) Side view showing variation in bed topography along the length of the stream. (b) Aerial view showing the position of riffles and pools along the length of the stream. Note in (b) that the streamflow moves in a curving path from side to side, although the sides of the channel are straight. (The spacing between pools and riffles is five to seven times the stream width, but the spacing has been shortened here for illustrative purposes.) After "River Meanders," Leopold and Langbein, Copyright © 1966 by *Scientific American*, Inc. All rights reserved.

tures in the rocks, steep gradients, and areas of active uplift of the land. Hard rocks inhibit the sideward erosion of channel walls into meanders. Well-developed linear fractures in rocks provide a path along which erosion is likely to occur; thus the stream aligns itself with the fractures (Figure 11–15). High gradients result in a greater tendency for the water to move downslope by the most direct path—a straight line.

When an area is being uplifted, the gradient increases. This change in base level increases the potential energy of the stream. As a result, the stream's greater energy erodes the channel in the most direct path downslope.

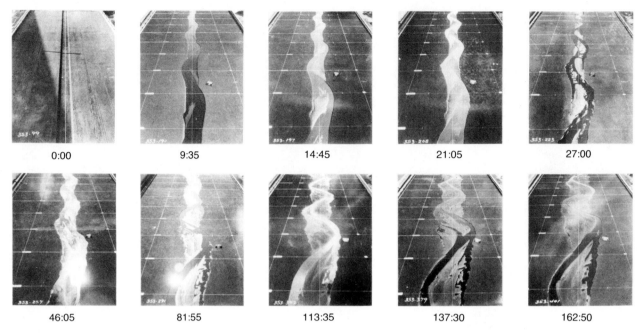

0:00	9:35	14:45	21:05	27:00
46:05	81:55	113:35	137:30	162:50

FIGURE 11-14

Sequential evolution of a straight channel into a meandering one on an experimental stream table at the U.S. Army Corps of Engineers Waterways Experiment Station at Vicksburg, Mississippi. The initial straight channel slowly changes into a meandering one with the development of point bars and undercut curves. The width of the stream valley increases as the meandering channel migrates across the floodplain, eroding into the valley walls and leaving behind point-bar deposits on the floodplain. Photo courtesy of U.S. Army Corps of Engineers.

Meandering Channels

Most streams have meandering patterns. We know that straight stream segments can evolve into meandering ones, but we must look in more detail at the water flow in channels to see *how* this happens.

The idealized flow-line pattern of a meandering stream is shown in Figure 11-16(a). All flow lines are directed downstream in straight sections A, C, and E. As the stream flows around curved parts B and D, the flow is pushed toward the outer bank, elevating the water surface there—Figure 11-16(b). (This is similar to what happens when you drive a car around a sharp curve; there is a tendency for the car to skid toward the outside of the

curve.) In the stream, there is a compensating return flow of water along the bottom toward the inside of the meander, resulting in flow lines directed to the side of the stream—Figure 11-16(b).

Once we understand the flow patterns, we can explain how meanders develop from curved sections of streams. The concentration of flow and the higher velocities developed on the outside of the curves result in further erosion of the outside stream bank, thereby accentuating the curve. Such erosion at the sides of a stream is called **lateral erosion**. A portion of the sediment eroded from the outside of the meander bend is moved along the bottom toward the inside of the meander—Figure 11-16(b). This sediment is deposited as bars, which grow on the inside

FIGURE 11-15 (opposite)

Silver Cascade at Crawford Notch, New Hampshire. The stream segment here is quite straight because it flows parallel to fractures in granite and has a steep gradient. Photo by Nicholas K. Coch.

FIGURE 11–16
Formation and movement of
meanders. (a) A meandering stream,
showing changes in flow directions
in straight portions A, C, and E and
in curved portions B and D. (b) A
vertical section along section line
X–X', perpendicular to flow across
a curved section of the stream.
Surface flow is directed to the
outside of the meander and bottom
flow is directed to the inside of the
meander. These flow conditions
result in erosion on the outside of a
meander bend and deposition on
the inside of the curve.

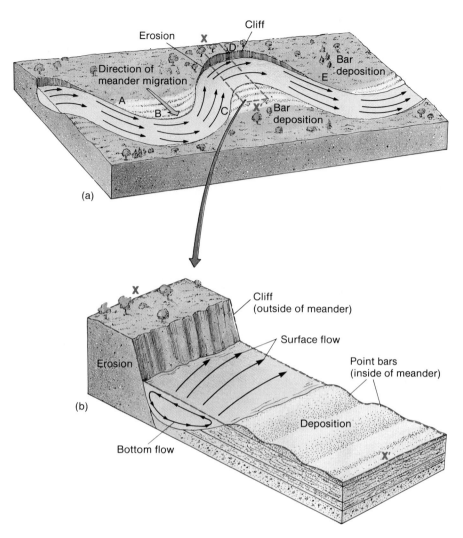

of the meander. Such a bar grows outward into the stream channel as the opposite bank is being cut back by lateral erosion. The result of this combination of erosion and deposition is that the channel migrates back and forth across the floodplain.

Conditions favoring the development of a meandering channel pattern include easily erodible bank materials and a gentle gradient. Streams with gentle gradients are close to base level; they have less of a tendency for vertical downcutting and more of a tendency for lateral erosion. Lateral erosion is most marked when the bank materials are unconsolidated sediments.

Meander Cutoffs and Oxbow Lakes. Many aerial photographs of meandering streams show complex

arc-shaped deposits laid down by the stream as it migrates across its floodplain (Figure 11–17). How these features develop is illustrated in Figure 11–18: as the stream migrates laterally across the floodplain (a), a meander may become so sinuous (b) that the stream eventually cuts across the narrow neck, separating two meander curves (c). Such an abandonment of a former meander curve leaves a crescent-shaped lake on the floodplain called an **oxbow lake.** Many of these subsequently fill with vegetation and fine suspended sediments from later flooding.

Meander cutoffs can result in significant geographic dislocation, because many political boundaries are located within river channels. A striking example is the case of Carter Lake, Iowa. The Missouri River serves as the boundary between the states

FIGURE 11–17
Aerial view of meandering Red River floodplain, Louisiana, showing point bars, oxbow lakes, and meandering channels. Photo courtesy of U.S. Army Corps of Engineers.

of Nebraska and Iowa (Figure 11–19). Up until 1877, the town of Carter Lake was located on the inside of a meander bend on the Iowa side of the river, just across from Omaha, Nebraska. In 1877, the Missouri River cut across its own meander loop to isolate the town of Carter Lake *within the state of Nebraska*. The abandoned channel of the river was transformed into an oxbow lake (Carter Lake), and the town and surrounding region were no longer physically continuous with the state of Iowa.

Braided Channels

Braided channels tend to develop wherever streams are unable to carry the coarser part of their sediment load and thus deposit it temporarily within the channel as islands or bars (Figure 11–20). Braided streams are one more example of dynamic equilibrium in streams. The excess sediment is deposited temporarily within the channel until the stream has

sufficient discharge to move it once again. As the discharge increases, the braided stream channel may become a meandering channel, or even a straight channel.

Braided stream channels can develop where the stream gradient decreases abruptly, and the velocity of the stream drops. They also can form where the discharge decreases, through evaporation or infiltration into the ground. Braiding also occurs when more sediment is added to a stream than its discharge can move; this may happen when the bank material is easily eroded or when a high-gradient tributary or a landslide add a great quantity of sediment to the stream.

Braided channels develop under a wide variety of topographic and climatic conditions. Fast-flowing mountain streams may develop braided patterns at the abrupt decrease in gradient at the base of the mountain. This is especially marked in drier climates where there is not only a decrease in gradient at the

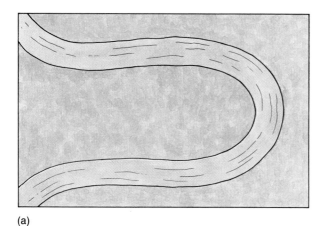

(a)

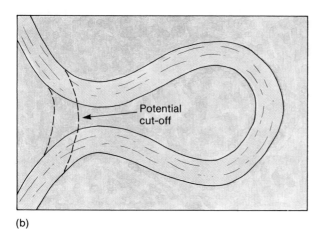

Potential cut-off

(b)

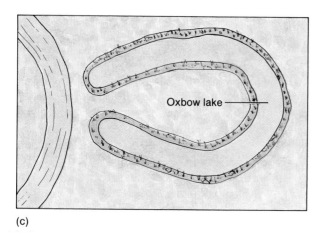

Oxbow lake

(c)

FIGURE 11–18
Formation of oxbow lakes by meander cutoff. (a) Stream meandering over its floodplain. (b) Stream meander becomes so sinuous that only a narrow neck of land separates the two channels. (c) Stream erodes through the neck by lateral erosion and forms a new, shorter course. The abandoned part of the channel becomes isolated as an oxbow lake.

base of the mountain but also a greater evaporation rate in the hotter, drier air of the desert valleys.

Braided streams also are common near glaciers. The large volume of sediment released at the front of a melting glacier usually is more than the available water can carry, and the stream develops a braided pattern.

STREAM EROSION

Streams erode the surface in several different ways. In some cases, the water force alone is sufficient to dislodge pieces of rock or sediment. In other cases, particles carried within the flow can abrade rock on the channel floor in the same way that sandpaper smooths a piece of wood. Turbulent eddies aid erosion in several different ways. Rock fragments that are carried in turbulent eddies can drill out circular cavities in the stream bed called **potholes** (Figure 11–21). Particles of sediment or rock may be drawn from the bottom up into the flow as the eddies scour the bottom of the stream.

Erosion of a stream bottom is **vertical erosion**. The rate of downcutting increases with the velocity of the stream and the erodibility of the channel material. A stream flowing through granite (Figure 11–15) may take thousands of years to modify its channel, whereas a stream flowing through unconsolidated sands and gravels may undergo significant enlargement in one flood. Stream erosion is most marked in areas where there is little vegetation, such as deserts, and in areas where vegetation has been removed by construction or killed by pollution. This is because vegetation resists the flow of water, inhibiting erosion of the surface.

River erosion is greatly aided by the mass movement processes we described in Chapter 10. We like to think that the Colorado River directly eroded most of the Grand Canyon by downcutting, but the true story is quite different. The Colorado River does cut down into its bed by abrasion, and it does remove sediment and rock by water flow. However, most of the erosion of its banks occurs as the stream migrates to the side, undermines the bank, and triggers a slump or rockslide in the weathered bank material. This erosion of the banks of a stream is **lateral erosion**. Large volumes of slope materials are thereby eroded by *mass movement processes* which

280 | STREAMS AND STREAM SCULPTURE

FIGURE 11–19

Map showing meander cutoff at Carter Lake, Iowa. A change in the Missouri River channel resulted in the town of Carter Lake, founded on the Iowa side of the Missouri, being located on the Nebraska side of the river! Note that the Nebraska/Iowa state boundary follows the old meander, so the town of Carter Lake remains in Iowa, politically if not geographically. After map in *New York Times* April 21, 1988, copyright © 1988 by the New York Times Company. Reprinted by permission.

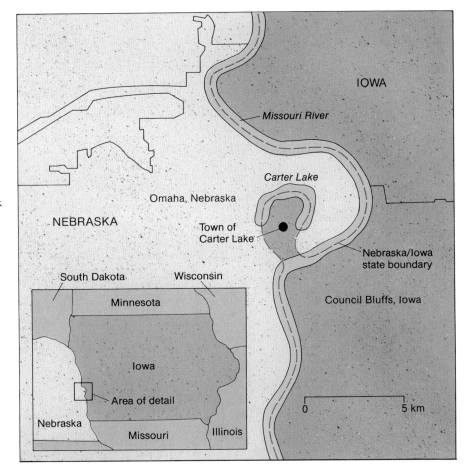

FIGURE 11–20

Aerial view of braided stream system in Colombia, showing multiple channels, bars, and islands. Photo courtesy of Deborah Harden.

FIGURE 11–21
Potholes in the bed of the James River at Richmond, Virginia, exposed during low river stage during autumn 1970. Photo by Nicholas K. Coch.

widen the canyon. Thus, the Colorado River serves mainly to *transport* the material from the area. This example illustrates how weathering, mass movements, and streams work together to form landscapes.

Streams usually cut through sequences of rocks that have different resistances to stream erosion. When a hard rock overlies a soft one, the softer rock is eroded more rapidly, resulting in a waterfall (Figure 11–22). The tremendous turbulence generated at the base of the falls results in accelerated erosion of the weaker rock and removal of the support for the rock above. The unsupported cap rock breaks off and accumulates as talus at the base of the falls, and the falls recedes upstream a short distance with each such event.

Niagara Falls, on the Niagara River between Lake Erie and Lake Ontario, is capped by a resistant dolomitic limestone underlain by more erodible shale.

Undermining of the dolomite by stream erosion of the shale has resulted in rockfalls of the cap rock and a steady upstream recession of Niagara Falls. The average rate of retreat of the falls between 1850 and 1950 was a little over 1 m per year. As the falls move upstream, they leave behind a steadily lengthening gorge (Figure 11–22). Eventually the falls will migrate upstream until Lake Erie and Lake Ontario are connected directly.

Wherever a stream cuts down into resistant rock, it forms shallows and rapids where the stream velocity increases. Rapids also form where debris flows from tributaries narrow the main channel flow and cause the stream to flow much faster so that the same discharge can move through the area. You can see the same effect as you squeeze the end of a hose. As the opening decreases, the water flows faster. These conditions form the "whitewater" rapids that make stream rafting so exciting (see Focus 11–1).

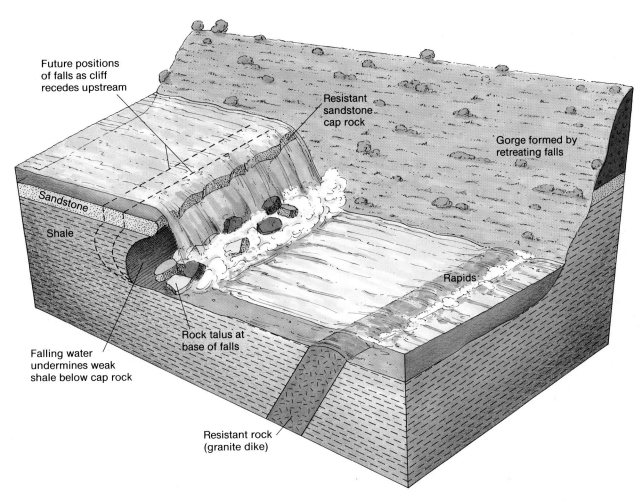

Future positions
of falls as cliff
recedes upstream

Resistant
sandstone
cap rock

Gorge formed by
retreating falls

Sandstone

Shale

Rapids

Rock talus at
base of falls

Falling water
undermines weak
shale below cap rock

Resistant rock
(granite dike)

FIGURE 11–22
Formation of waterfalls and rapids in a river.

STREAM DEPOSITION

Streams transport and deposit a wide range of sedimentary particles through the processes described in Chapter 6. In general, sediment is deposited by a stream whenever its velocity decreases so that it is unable to transport particles of a given size. Material may be eroded again when the velocity increases, as in a flood. The collective term for all the deposits made by streams is **alluvium**. Alluvium can be deposited within the stream channel, on the floodplain, or at the point where the river enters a lake or the ocean.

Floodplain Deposits

The alluvium under the floodplain of a meandering river is composed of bar sediments deposited as the channel migrates laterally and finer-grained sediments deposited from suspension when floodwaters cover the floodplain.

Arc-shaped bars that are deposited on the insides of meander bends are called **point bars**. We previously described the lateral erosion and deposition that occurs at meander bends; the process of point-bar formation is indicated in Figure 11–16(b). Point-bar deposits include gravels and coarse sands

TAMING THE COLORADO RIVER IN THE GRAND CANYON

Anyone watching boaters shooting the impressive rapids of the Colorado River in the Grand Canyon region would think that the Colorado was a wild river, but nothing could be further from the truth. The discharge of the Colorado is completely controlled by a system of dams, and river level and velocity are largely a function of the amount of water released from the dams.

However, prior to the 1930s, the Colorado was indeed a wild river. It rose in the Rocky Mountains and flowed southwestward toward the Gulf of

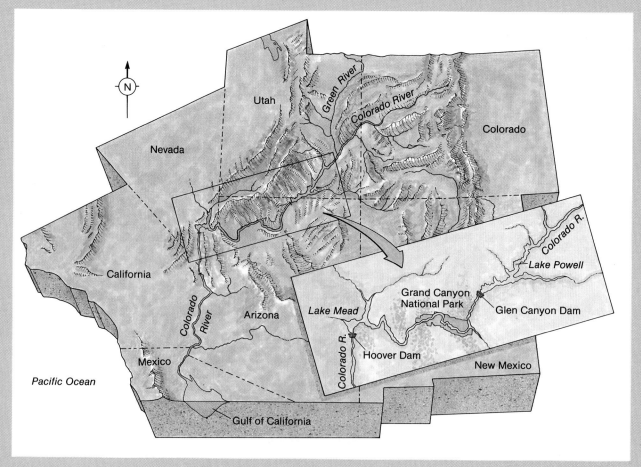

FIGURE 1
Southwestern United States, showing the course of the Colorado River and features mentioned in text.

California (Figure 1). By gathering the drainage of other large rivers along its path, it had sufficient discharge to cross a desert area (with its very high evaporation rates) and still have significant water in its channel.

The river was noted for wide ranges in discharge. During flood stages, it roared through the canyon, removing any material that had been supplied by tributary streams since the last flood, carrying the sediment downstream to its delta in Mexico. Evidence of these great floods is still visible along the sides of the Grand Canyon (Figure 2). When the river was wild, there were fewer sandbars along its shores, and little vegetation could take root near the river between floods.

By the 1930s the natural state of the Colorado was on a collision course with the growing needs of the region for water and hydroelectric power. Hoover Dam was completed across the Colorado south of the Grand Can-

FIGURE 2
Evidence of past floods in the Grand Canyon above the prese. level of the Colorado River. The driftwood on the large boulder in the stream was deposited by a receding flood prior to the building of the Glen Canyon Dam in 1956. Photo courtesy of Fred Wolff.

FIGURE 3
Cloudy waters of the Colorado River entering the northern end of Lake Mead. Photo by Nicholas K. Coch.

yon in 1936, and it shortly began to change the regimen of the river. Sediment that previously had been transported southward into Mexico began building a delta into Lake Mead, behind Hoover Dam. Then and now, reddish-brown Colorado River water enters the northern part of Lake Mead and deposits its sediment (Figure 3). As the sand, silt, and clay are deposited, the lake water clears.

Colorado River control was completed with the building of the Glen Canyon Dam north of the Grand Canyon in 1956, and impoundment of the Colorado's waters behind the dam in Lake Powell by 1964. The discharge, velocity, and even the temperature of the river water in the Grand Canyon are now almost completely controlled by the dam operators. When the dam gates are opened, a flow of icy cold, clear water starts down Marble Canyon and into the Grand Canyon. The water is cold because it comes from the bottom of the Lake Powell reservoir behind the dam. Although air temperatures along the river in the canyon reach well over 100°F in summertime, few can swim for more than a minute in that cold water.

The released water is initially clear, but clouds as the first tributary supplies sediment to the river (Figure 4). Frequent water releases from the Glen Canyon Dam raise the surface of the Colorado a few feet, but nowhere near the flood levels of the past. Consequently, much less sediment is removed now, and sandbars, sand plains, and vegetation along the banks have become much more common.

However, torrential flooding still occurs down the steep tributary canyons. These floods contribute massive supplies of sediment to the river below, but it no longer has the power to remove them. This debris (Figure 5)

FIGURE 4
Clear waters released from Lake Powell through the Glen Canyon Dam at Lees Ferry, Arizona. First turbid water from local streams enters the Colorado River from the right, near top of photo. Photo by Nicholas K. Coch.

FIGURE 5
Debris flow from side canyon narrows the Colorado River in Marble Canyon, forming rapids.
Sand is deposited in bars along the river as the rapids die out below the flow restriction. Photo
by Nicholas K. Coch.

forms fans which narrow the channel and constrict the river flow, locally increasing its velocity and forming rapids.

Other rapids exist where resistant rocks cross the channel or where an abrupt drop of the channel floor occurs. These types of rapids existed before the river was dammed. The most exciting rapid on the Colorado River is Lava Falls (Figure 6) where the riverbed abruptly drops over 11 m. This rapid is rated a "10" on the rafting difficulty scale and is a thrilling finale to most Colorado River raft trips. However, rafters must keep close tabs on the water releases from the Glen Canyon Dam to safely navigate the submerged hazards of the rapids.

FIGURE 6
Rafters in Lava Falls rapids, Grand
Canyon National Park, Arizona.
Photo courtesy of Deborah Berg.

at the base, grading upward into cross-bedded sands and finally into fine sand and silt near the top of the section. Arcuate remnants of old point bars are common features of floodplains (Figure 11–17).

Sediment is deposited on the floodplain when the stream discharge exceeds the bankfull discharge. As soon as the river spills from its channel, it slows and the coarser suspended load (sand and coarse silt) is deposited near the top of the channel banks. Successive floods build a low ridge (4–5 m) on either side of the channel bank. These low ridges atop the channel banks are called **natural levees** (Figure 11–23). The floodwaters carry the finer sediment particles (silt and clay) in suspension over the levee and across the floodplain. As the waters spread across the floodplain, the velocity decreases further, and the remaining sediment particles are deposited. The greater thickness of sediment deposited near the stream results in a floodplain which slopes away from both sides of the stream channel (Figure 11–23).

If the force of the floodwaters exceeds the strength of the levees, they are breached, and the coarse sediment, normally deposited within the

channel or on the levees, is deposited instead on the floodplain. In many areas, levees have been reinforced and built higher to protect population centers and industrial areas. Nowhere is this more important than in New Orleans, where the Mississippi River is confined between levees 7.5 m high. The French Quarter bordering the Mississippi River is among the highest areas in New Orleans, but it is only 4.3 m above sea level. The land slopes downward from there, and much of New Orleans is about 1.8 m *below* sea level. Visitors to the French Quarter are frequently startled upon hearing boat horns and then looking *up* to see the large ships passing them along the Mississippi River.

Deltas

Where a stream enters a standing body of water, such as a lake, the ocean, or a reservoir, its velocity drops, and most of the stream sediment is deposited within a short distance. This deposit is called a **delta**. On larger deltas, such as the Mississippi Delta now being built into the Gulf of Mexico, the main channel

FIGURE 11–23
Mississippi River levee and floodplain sloping away from it, 24 river miles above Greenville, Mississippi. The natural levee has been built up to protect inhabited areas on the floodplain. Photo courtesy of U.S. Army Corps of Engineers.

branches into a number of smaller channels, called **distributaries,** that flow across the delta surface. The distributaries commonly are separated by lakes, bays, marshes, or swamps (Figure 11–24). As the distrib-

utaries cross the delta, their bed load and suspended load are deposited on the delta surface. The coarser sediment is deposited nearer the land, whereas the finer sediment is carried in suspension far out into

FIGURE 11–24
Aerial view of Mississippi River distributary channels at the edge of the Mississippi River delta. Photo courtesy of U.S. Army Corps of Engineers.

the body of water into which the delta is advancing. This pattern of sedimentation results in a distinctive sequence of layers that become progressively finer from the delta out into the basin (Figure 11–25).

The upper surface of the delta is covered with thin beds of distributary sands and the finer organic-rich sediments of the interdistributary areas. These thin, horizontally layered sediments covering the delta top are called **topset beds**. They grade into finer-grained, thicker, inclined **foreset beds** that cover the submerged front of the delta surface. Each foreset bed grades, in turn, into a thin, fine-grained, horizontally bedded layer, the **bottomset bed**.

Bottomset beds extend from the delta front across the bottom of the water body into which the delta is advancing. Bottomset beds are made up of material derived in part from the river and in part from sedimentation within that body of water. For example, bottomset beds of the Mississippi Delta are composed in part of Gulf of Mexico sediments and shells of marine organisms and in part from suspended material carried into the Gulf by the force of the Mississippi River.

The edges of some deltas, such as the Mississippi, protrude like the feet of a bird far into the body of standing water (Figure 11–24). Other deltas, such as the Nile River Delta in Egypt, protrude very little and are triangular, with the base of the triangle along the delta front and the apex pointed upstream (Fig-

ure 11–26). Why is there such a difference in the shapes of delta fronts?

The shape depends on the balance between the rate of supply of river sediment and the removal of the sediment at the outer part of the delta by waves. If more sediment is supplied to the front edge of the delta by the river than can be eroded by waves along the shoreline, the delta will extend well out past the shoreline. This is the situation in the Mississippi Delta (Figure 11–24). On the other hand, any changes in the stream which reduce the supply of sediment have the opposite effect. For example, the Nile Delta (Figure 11–26) has always protruded very little into the Mediterranean Sea because the sediment supply in that dry climate always has been seasonal and allowed time for the coastal waves and currents to erode previous deposits. Building of the Aswan Dam across the Nile has trapped much of the river's sediment load in Lake Nasser and reduced the sediment supply to the Nile Delta. This has resulted in increased erosion of the front of the delta.

STREAM SYSTEMS

Mighty rivers start with raindrops on a slope. The drops soon begin to erode shallow channels where small depressions already exist in the surface and where the turbulence in the water is stronger. These

FIGURE 11–25
Cross section of a delta showing deposits.

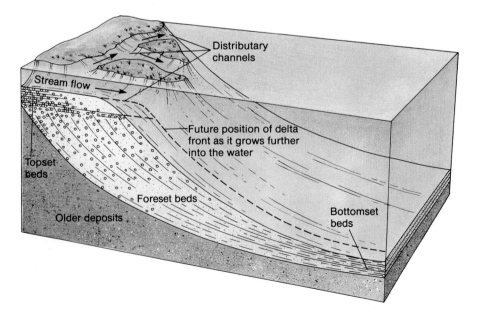

FIGURE 11–26
Satellite image of Nile River delta. Photo courtesy of National Aeronautics and Space Administration.

FIGURE 11–27
Rill system forming on new slope.
Photo courtesy of P. Carrara/U.S.
Geological survey.

shallow, temporary channels are called **rills** (Figure 11–27). In subsequent rainfalls, some of the rills are abandoned, whereas others are deepened and joined to form a branching system of stream channels. The area drained by any one of these branches is called a **drainage basin** (Figure 11–28).

Drainage basins vary greatly in size, depending on the number of streams involved. For example, a trib-utary creek to the Missouri River can have a drainage basin area of only a few km^2, whereas the area of the entire Missouri-Mississippi River basin is 3,221,183 km^2. The drainage basin of one stream is separated from that of another by a higher intervening area called a **drainage divide** (Figure 11–28).

As the stream grows, it slowly erodes back into the areas within its drainage basin, dissecting them.

FIGURE 11–28
Aerial view of drainage basins (B) and divides (D). Photo courtesy of J. R. Balsley/U.S. Geological Survey.

This carving of undissected areas at the upstream ends of stream segments is called **headward erosion**. Sometimes streams with steeper gradients erode headward through drainage divides and intercept the headwaters of other streams having gentler gradients in the adjacent basin, thus incorporating them into their own drainage. Such an interception and incorporation of another stream's drainage is **stream capture**.

Streams from the smaller drainage basins join with others in an ever-widening branching pattern to form stream systems with higher and higher discharges, until master streams are formed, such as the Amazon, Congo, Nile, and Mississippi.

Stream Networks

A large number of interconnected streams is called a **stream network**. If you have carefully observed the Earth's surface during a long plane flight, you may have noticed the stream patterns and that they differ from place to place. The stream networks shown on an aerial photograph or map provide the geologist with valuable clues about the characteristics of the surface over which the streams are flowing (Figure 11–29). There are several distinctive stream-system patterns:

☐ **Dendritic pattern**—in areas where the surface materials are uniform in composition and are not cut by any planar structures such as joints and faults, streams can erode headward equally in all directions. This results in a branching dendritic pattern—Figure 11–29(a). Dendritic patterns are the most common type of stream network.

☐ **Rectangular pattern**—in areas where relatively uniform rocks are cut by sets of parallel joints, streams follow these zones of weakness to develop a system characterized by numerous angular turns—Figure 11–29(b).

☐ **Trellis pattern**—where a series of parallel hills and valleys exist, a trellis network forms, consisting of long stream segments along the valleys fed by short tributary stream segments that cut through drainage divides at right angles—Figure 11–29(c). The long segments result as the streams erode headward along belts of less-resistant rock, producing the valleys. The short segments result as streams flow down the ridges separating the two drainage basins.

☐ **Radial pattern**—streams draining away from a high area, such as a volcano, form a radial pattern in which they radiate downward from all sides of the central peak—Figure 11–29(d).

☐ **Distributary network**—this stream pattern, in which the stream channels branch out in the *downstream* direction, is common on the surfaces of deltas—Figure 11–29(e).

STREAM SCULPTURE

Landscapes produced by streams consist of both erosional features (valleys) and depositional features (bars, floodplains, and deltas). However, the relative percentages of erosional and depositional features vary from region to region. We will now look at how streams create landscapes, and how these landscapes evolve with time.

Evolution of Stream Landscapes with Time

Streams evolve through a series of stages as they erode down to base level and the energy available for stream erosion decreases. Each stage is marked by differences in altitude between the stream channel and the top of the adjacent drainage divides. A measure of relative altitude difference is called the **relief**.

W. M. Davis (1850–1934), a noted American geologist, referred to three successive stages of stream landscapes as *youth, maturity,* and *old age*. We will not use these terms here because they imply that time is the *only* factor in landscape evolution. Many other factors, such as the type of rock, its structure, and stability of the land surface also are important in changing the relief of an area. For example, it may take several million years to develop a stage of "youth" in a granite region, whereas in the same million years, an area of more easily erodible shale and limestone may evolve into a "mature" landscape.

We will describe landscape evolution more objectively, by referring to the *relative stages of relief* (high, moderate, and low) developed in an area as landscape evolution proceeds. To simplify our discussion of stream landscape evolution, we will start with a

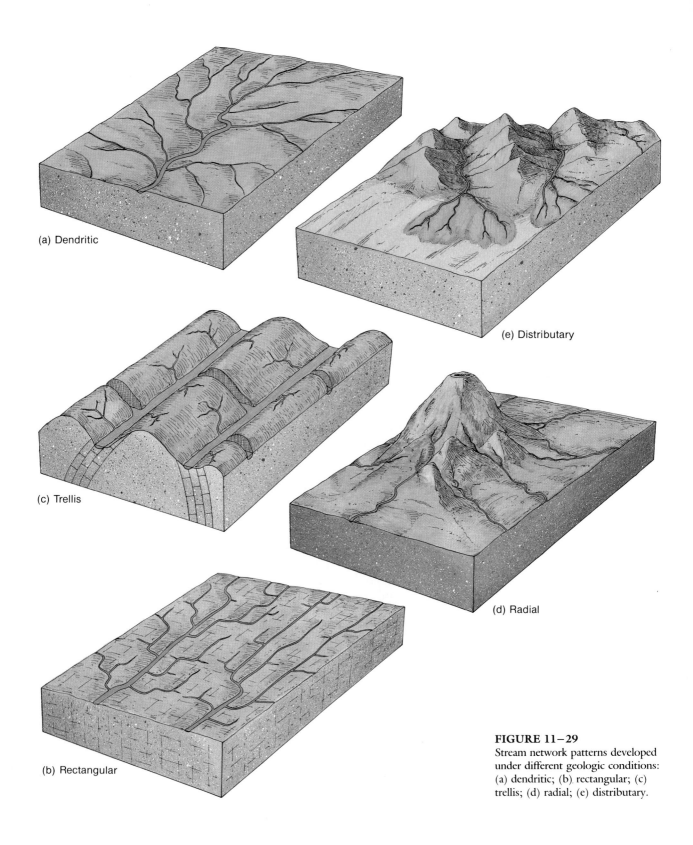

(a) Dendritic

(e) Distributary

(c) Trellis

(d) Radial

(b) Rectangular

FIGURE 11–29
Stream network patterns developed under different geologic conditions: (a) dendritic; (b) rectangular; (c) trellis; (d) radial; (e) distributary.

region of low relief that has been uplifted above base level (Figure 11–30). The rocks in this area are uniform in composition, and the humid climate and existing base level do not change with time.

As the streams cut down through the land surface, the relief grows greater and greater. In this *high relief stage,* the vertical distance (the relief) between the stream channel and the tops of adjacent stream divides reaches a maximum—Figure 11–30(a). Stream downcutting is more active than lateral ero-

sion. Therefore, the width of the stream channel is very close to the width of the stream valley, and a significant portion of the upland surface is still at the original altitude.

As the downcutting stream gets closer and closer to base level, lateral erosion increases; as a result, the width of the valleys far exceeds the width of the stream channels—Figure 11–30(b). A combination of weathering and mass movements triggered by lateral erosion of the streams reduces the elevation of

FIGURE 11–30
Stages of stream erosion in an area where all factors have remained constant over time: uniform rocks, unchanging climate, and unchanging base level. (a) *High relief stage:* vertical downcutting is dominant, the relief is maximum, and the valley width is close to the channel width. (b) *Moderate relief stage:* lateral erosion dominates over vertical downcutting, and valley width is much greater than channel width. (c) *Low relief stage:* streams meander over a very wide floodplain, and drainage divides have very low relief.

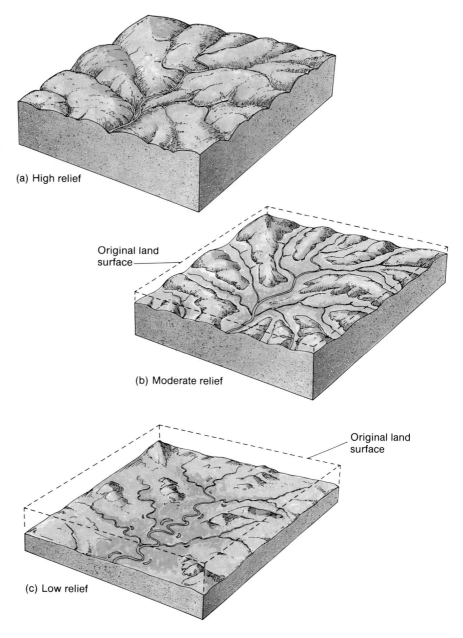

(a) High relief

Original land surface

(b) Moderate relief

Original land surface

(c) Low relief

FIGURE 11–31
Cross section view of peneplain, Grand Canyon National Park, Arizona. The peneplain developed on the eroded surface of the Vishnu Schist prior to deposition of the overlying Tapeats Sandstone. Photo by Nicholas K. Coch.

the drainage divides, and the region attains a *moderate relief stage*. Only a few areas on the drainage divides preserve the original upland surface elevations.

Continued erosion reduces the area to a *low relief stage,* in which streams meander over exceptionally wide floodplains. The altitude of the intervening drainage divides has been greatly reduced to where they are now only rolling lowlands—Figure 11–30(c). Some of the streams cut through the low drainage divides and capture the drainage of others so that fewer, but larger, streams dominate the landscape. Eventually, the intervening drainage divides will be almost completely eroded away, and the landscape will be reduced to a featureless plain of very low relief. Such a stream-eroded surface of low relief is a **peneplain** (from Latin *pene,* "almost" plain).

Development of a peneplain requires that the base level remain constant for an exceptionally long period, so peneplains are relatively rare. The geologic record in the Grand Canyon is remarkably exceptional in that two peneplains of very different ages are preserved in the walls of the Grand Canyon. Figure 11–31 shows one of these peneplains.

Factors in the Evolution of Stream Landscapes

The *actual* evolution of stream landscapes is much more complicated than in the general example given in the previous section. This complexity arises because an area usually contains different types of rocks and its climate and base level usually have changed one or more times as the landscape evolved. For example, a change from a dry to a humid climate would have been accompanied by a great increase in stream discharge, as well as an increase in weathering and mass-movement activity. This would result in a widening of the valleys and greater downcutting into the valley floor.

Today, parts of the United States are sinking (the Gulf Coast), while other parts are rising (the Colorado Plateau). Further, during the history of the Earth, sea levels have changed. When a stream's gradient is increased either by elevation of the land or by a drop in sea level, its energy increases and the stream begins to erode its channel actively once again. A stream which begins vertical downcutting once again because of a change in base level is called a **rejuvenated stream**. For example, streams on the Colorado Plateau once meandered across an area of low relief. Then, as the Colorado Plateau was uplifted in the last million years, they began to downcut actively. This downcutting has formed **entrenched meanders,** where the channel is bordered by steep valley walls and the valley width is close to that of the stream channel (Figure 11–32).

The opposite situation occurs when a decrease in stream energy is caused either by a drop in the land surface or by a rise in sea level. The result is accelerated deposition of alluvium. Streams that deposit alluvium actively within their valleys are called **aggradational streams**.

An aggradation-rejuvenation sequence is shown in Figure 11–33. Initially, sea level is low, the stream is downcutting actively, and it is depositing only a thin layer of alluvium across its floodplain—Figure 11–33(a). A subsequent rise in sea level results in aggradation and filling of the valley with a thick deposit of alluvium—Figure 11–33(b). A later drop in sea level results in rejuvenation—Figure 11–33(c), with the rejuvenated stream actively cutting downward into the thick valley fill that it deposited in the previous aggradational cycle. In many cases, excavation of the valley is not complete, and portions of the fill are preserved along the valley walls. These flat-topped erosional remnants preserved above the level of the stream are called **stream terraces**—Figures 11–33(c) and 11–34.

When we consider the combined effects of different durations of time in which the land was exposed to erosion, surface composition, climate, and base level changes, it is easy to understand why there are such great differences in the relief and the type of landscape development we see in different places on the surface of the Earth.

FIGURE 11–32
Aerial view of entrenched meanders on the lower Green River, Utah. Photo courtesy of Deborah Harden.

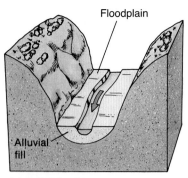

Floodplain

Alluvial fill

(a) Initial (low) base level

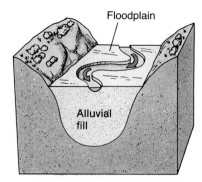

Floodplain

Alluvial fill

(b) Rise in base level

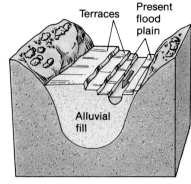

Terraces   Present flood plain

Alluvial fill

(c) Drop in base level

FIGURE 11–33
Stream aggradation and rejuvenation, resulting from a change in base level. (a) Initial (low) base level (often sea level). Stream is actively downcutting and depositing a thin alluvial fill as it excavates a valley in older materials. (b) Rise in base level. The stream deposits a thick section of alluvium as it aggrades because its gradient has been reduced. (c) Drop in base level. The stream now erodes into its alluvial fill as it adjusts to a lower sea level. Flat remnants of the fill are left as terraces above the level of the present floodplain.

FIGURE 11–34
Stream terraces of Charley River, Eagle District, Yukon Region, Alaska. Photo courtesy of E. E. Brabb/U.S. Geological Survey.

SUMMARY

1. Water exists in solid, liquid, and gaseous forms on the Earth.
2. Each form of water is transformed into another as the water moves from the ocean, to the atmosphere, to the land and, back to the ocean. This is the hydrologic cycle.

3. Virtually all of the Earth's water is stored in the ocean (97.3%), glaciers (2.14%), and underground water (0.56%). Water stored in rivers, lakes, and soils makes up the remaining 0.02%.
4. Streams represent that portion of the hydrologic cycle where surface water moves in channels under the influence of gravity.
5. The potential energy of a stream is related to the height of its source above sea level. The stream's po-

tential energy is converted into kinetic energy as it flows downslope.

6. Streams have a longitudinal profile that is concave upward, with a steeper upstream slope and a gentler downstream slope.

7. Streams continue to erode the land until they reach base level. The base level may be a local one, such as a lake or reservoir, or the ultimate base level—sea level.

8. A stream's discharge is in a dynamic equilibrium with its width, depth, and velocity. Any change in discharge must be balanced by a compensating change in one or more of the other variables.

9. Flooding occurs when discharge exceeds bankfull discharge, causing the stream to overflow its banks and deposit sediment across its floodplain.

10. Urbanization has decreased the lag time (time between peak rainfall and peak discharge) in many areas, resulting in increased peak discharge and flood frequency.

11. Streams may have straight, meandering, or braided channel patterns, depending on the nature of the surface materials, the slope, and the stream's available energy to transport materials.

12. Stream erosion is carried out by the force of the flowing water, by sedimentary particles carried in the water, and by mass movements such as slumping as the stream undermines its banks.

13. Falls occur where hard, resistant cap rock overlies weaker rocks. The falls retreat upstream as erosion of the weak rock undermines the resistant cap rock above. Rapids occur where obstructions decrease the depth locally or constrict the channel width.

14. Streams deposit sediment within their channels, on their floodplains, and in deltas built into standing bodies of water. Alluvium is deposited within the channel as sandbars and on the insides of meander bends as point bars.

15. Levees form atop stream channel banks when floodwaters spill out of the channel and rapidly drop sediment from suspension.

16. Floodplain deposits are formed by lateral deposition of sandy point-bar sediments and by deposition of suspended silt and clay particles from floodwaters that periodically cover the floodplain.

17. Deltaic deposits built into standing water are made up of (1) coarser-grained topset beds and (2) finer, inclined foreset beds, which overly (3) darker, finer-grained bottomset beds. The shape of the delta front is a function of the balance between the rate of sediment supply by the river and the rate of sediment removal by waves and currents along the delta shoreline.

18. Stream landscapes undergo a sequential development over time. This sequence involves a decrease in regional relief and the ratio of stream channel width to valley width. If the land surface remains stable, stream erosion eventually can reduce it to a peneplain.

19. Changes in base level, resulting from uplift or subsidence of the land, or rising or falling of sea level, alter the behavior of streams. As base level changes, streams either may aggrade and fill their valleys with alluvium, or may downcut actively into the valley floor.

20. The great differences in stream landscapes from place to place are due to differences in surface materials, changes in climate and base levels, and the amount of time that surfaces have been subjected to erosion.

REVIEW QUESTIONS

1. Describe the hydrologic cycle and indicate its geologic significance.

2. What factors determine the infiltration capacity of a surface material?

3. Describe how the relationship between rainfall rate and infiltration capacity affects the relative amounts of water supplied to surface and underground water systems.

4. What conditions form a sheetflood?

5. What differences in stream behavior would result from uplift of an area?

6. Describe changes that occur in stream behavior after a dam is built across it.

7. What changes occur in stream behavior as discharge increases?

8. What effect would each of the following have on stream lag time?
   a. Planting grass and trees on a bare slope.
   b. Constructing homes, factories, and roads in a previously naturally vegetated area.

9. Describe several different sets of conditions that would result in stream flooding.

10. What factors determine whether a straight, meandering, or braided stream will develop on a given surface?

11. Describe a condition under which each of these stream channel transformations could occur:
    a. Straight channel to meandering channel.
    b. Meandering channel to braided channel.

12. Describe the formation process for each of these stream features:
    a. Levees.
    b. Point bars.
    c. Oxbow lakes.

13. What does the presence of potholes in rocky stream beds tell us about the way that water flows in stream channels?

14. What geologic conditions result in the formation of falls and rapids?

15. Contrast the characteristics of the following stream deposits, and account for their differences:
    a. Channel.
    b. Floodplain.
    c. Oxbow lake.
16. What does each of these stream-network patterns tell us about the geology of an area?
    a. Dendritic.
    b. Trellis.
    c. Radial.
17. How do you account for the differences in the characteristics of topset, foreset, and bottomset beds in a delta?
18. Describe the changes that occur in the relief of an area as streams continue to cut into the land surface.
19. What are stream terraces? What do they tell us about the geologic history of an area?
20. What happens to the land surface as sea level rises? As it falls?

FURTHER READINGS

Belt, C. B., Jr. 1975. The 1973 flood and Man's constriction of the Mississippi River. *Science* 189: 681–84. (Describes how human modifications of the Mississippi River have changed its behavior.)

Hack, J. T. 1960. Interpretation of erosional topography in humid temperate regions. *American Journal of Science* 258A: 80–97. (Describes the factors working to produce changes in landscapes in humid areas over time.)

Mississippi River Commission and U.S. Army Corps of Engineers Division for the Lower Mississippi River Valley. 1976. *Flood control in the lower Mississippi River valley*. 43 p. (A detailed and well-illustrated treatment of the methods used to control the Mississippi River.)

Rabbitt, M. C. 1978. *John Wesley Powell's exploration of the Colorado River*. Washington: U.S. Geological Survey, U.S. Govt. Printing Office O–261–226, 28 p. (A fascinating description of one of the greatest river adventures of all time. Many excellent drawings illustrate the rigors of the trip and the canyon landscapes.)

Ritter, D. F. 1986. "Fluvial processes"; "Fluvial landforms." Chapters 6 and 7 in *Process geomorphology*. 2d ed. Dubuque, IA: William C. Brown Co., 579 p. (A modern, detailed treatment of all aspects of streams.)

U.S. Geological Survey. 1981. "Hazards from flooding." Chapter 3 in *Facing geologic and hydrologic hazards—Earth science considerations*. U.S. Geological Survey Professional Paper 1240—B. 108 p. (Describes all the varieties of floods.)

12
Groundwater

Carlsbad Caverns National Park, New Mexico. Photo ©
Norman R. Thompson, 1990.

prings flow from the face of a rock cliff; streams continue to flow for weeks after the last rain; caverns form underground in limestone areas; and water bubbles to the surface in a palm-fringed oasis in the middle of a desert. These phenomena all are related to **groundwater,** water that accumulates and migrates below the Earth's surface. This type of water is less apparent than the surface water we discussed in Chapter 11, but it is estimated that 68 times more groundwater exists than all the surface water in streams and lakes.

Groundwater is one of our most important resources. It provides drinking water for 48% of the U.S. population and irrigation water for agriculture in dry areas such as the American West and Southwest. Dependence on underground water varies from state to state. For example, subsurface water makes up 98% of the water used in New Mexico but only 30% of that used in Maryland and Pennsylvania. To understand how water accumulates and migrates underground, we must first look at the characteristics of rocks and sediments that enable water to pass through the Earth's surface and move below it.

GROUNDWATER ACCUMULATION

Not all rocks and sediments have the same potential for yielding groundwater. For example, if you dig a well into sand, gravel, sandstone, or limestone, you will probably obtain water. On the other hand, wells drilled into shale, or into unfractured igneous or metamorphic rock, usually yield only small quantities of water.

Groundwater potential is greatest in rocks or sediments with high **porosity**. Recall from Chapter 6 that porosity is the volume of the pores compared to

FIGURE 12–1

Types of porosity developed in sediments and rocks. (a) Sand, (b) sandstone, (c) granite, (d) limestone.

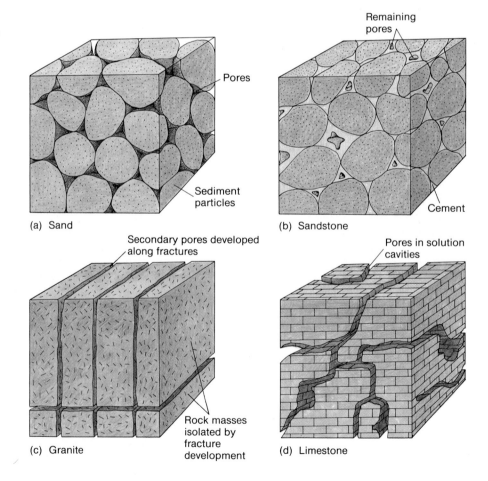

(a) Sand

(b) Sandstone

(c) Granite

(d) Limestone

the total volume of the sediment or rock. Some different types of porosity are shown in Figure 12–1. Coarse-grained sediments, such as sand and gravel, have pores among the sedimentary particles— Figure 12–1(a). Such sediments can have porosities as high as 35% and thus have great potential for storing and transmitting groundwater. However, when sediments are lithified to form sedimentary rocks, the pore spaces become filled with cement or matrix that decreases the porosity and groundwater potential— Figure 12–1(b).

In contrast, many chemical sedimentary rocks, igneous rocks, and metamorphic rocks have no initial porosity because they are composed of tightly interlocking crystals. However, these crystalline rocks may develop porosity if they are subsequently fractured— Figure 12–1(c). In this case, pores in fractured rocks are not among sedimentary particles or crystals but are developed among large *masses* of rock. Water may be obtained from fractured crystalline rocks if a well can be drilled to intersect these rock fractures.

Limestones can develop a special kind of porosity because they are composed of the mineral calcite, $CaCO_3$, which is soluble in acidic groundwater— Figure 12–1(d). As the limestone dissolves, cavities form, providing places for water to accumulate where none had previously existed. We will return to this topic later in the chapter when we discuss the origin of caverns.

GROUNDWATER MOVEMENT

Although surface water, such as a stream, can flow quite rapidly along the surface, groundwater in the same area flows much more slowly because it lacks a wide channel to flow through. Groundwater must pass through tiny pores to get where it is going.

Permeability

For groundwater to migrate, the sediment or rock not only must be porous, but the pore spaces must be *interconnected*. The capacity of a rock or sediment to transmit a fluid through interconnected pore spaces is called **permeability**. For example, well-sorted gravel is quite permeable, whereas unfractured granite is not. However, porous materials can

be impermeable if the pores are not connected. Clays are an example of a highly porous material that is almost impermeable. They may have porosities of over 50% and hold large volumes of water within their pores when saturated, but the pore spaces in clays are very small and few are connected, resulting in a low permeability for the sediment.

A permeable rock or sediment capable of transmitting groundwater is called an **aquifer**. An impermeable sediment or rock is called an **aquiclude**.

Flow Rates

Groundwater moves under the influence of gravity from high to low areas in a number of ways. In sediments and sedimentary rocks, countless minute "threads" of water wind their way *slowly* through the tiny pores. In contrast, in fractured or dissolved rock where the pores are large cavities, groundwater moves much faster.

The rate of groundwater movement was computed first by the French engineer Henry Darcy in 1856. Darcy showed that the discharge of water through an aquifer increases with the aquifer's permeability, its cross-sectional area, and the difference in elevation between where water enters and leaves the aquifer. He expressed these relationships in an equation:

$$\text{Discharge} = \frac{\substack{\text{Cross-}\\ \text{sectional}} \times \text{Permeability} \times \substack{\text{Vertical} \\ \text{distance}}}{\text{Distance of transport}}$$

This relationship is called **Darcy's law** and is illustrated in Figure 12–2. Thus, for a given aquifer size (cross-sectional area and length of aquifer), the groundwater flow rate increases both with permeability and with the inclination of the aquifer.

The well water that you drink today may have entered the ground long before you were born. Most groundwater flow is extremely slow because the portion of the water closest to the pore walls is slowed by the molecular attraction—**surface tension**— between the water and the walls (see Figure 10–6). Water in the center of the pores is affected less and moves faster. In general, the rate of groundwater flow increases as the pore size increases, because a greater proportion of the flow is unaffected by the pore walls. When the pores are very large, such as in

FIGURE 12–2
Groundwater rate of flow can be predicted by Darcy's law.*

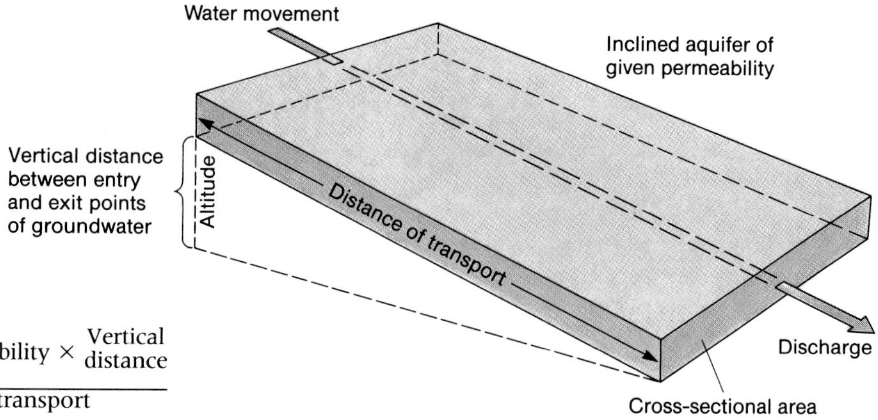

$$*\text{Discharge} = \frac{\genfrac{}{}{0pt}{}{\text{Cross-}}{\text{sectional}} \times \text{Permeability} \times \genfrac{}{}{0pt}{}{\text{Vertical}}{\text{distance}}}{\text{Distance of transport}}$$

a cavernous limestone, the water even may move in turbulent flow, as in many streams on the surface.

Groundwater flow rates generally are only a few centimeters per day, although higher and lower rates may occur under special conditions. For perspective, consider that with a flow rate of 2 cm/day, groundwater would move only 7 m/yr, whereas with a flow rate of 1 m a day, movement would be 365 m/yr. These slow rates of groundwater flow have considerable environmental significance. It is this slow rate of flow through interconnected pore spaces that allows oxygen, soil microorganisms, and bacteria sufficient time to break down any harmful substances, such as sewage, which may have been introduced into the groundwater. The slow rate of groundwater flow also is important to consider when we think of the very large volumes of water being pumped from aquifers each day. When groundwater is pumped from the ground faster than it can be replenished, wells will go dry.

UNCONFINED AQUIFERS

Rainfall enters the ground and moves downward under the influence of gravity until it is stopped by an impermeable layer. The groundwater then begins to accumulate above this barrier, gradually filling the pores upward toward the surface. If the rainfall is sufficient and spread over a long period of time, all the pore spaces in the ground may become saturated. However, permeable rocks and sediments usually are not saturated all the way to the surface, and the groundwater is distributed in several distinct *zones,* as shown in Figure 12–3.

A small portion of the rainfall infiltrating through the ground is retained within the root system of plants at the surface. This thin belt of partially saturated pores within the root system is called the **soil water zone.** Below this zone is the thicker **aeration zone,** in which pores are partially filled with water in transit between the surface and the material below. The lowest zone is the **saturation zone,** where all the pores are filled with water. The saturation zone extends downward to the impermeable materials, below which water movement stops.

The contact between the aeration zone and the saturation zone is called the **water table**. The water table in any area fluctuates up and down seasonally and as a result of the balance between groundwater pumping and replenishment of groundwater by rainfall. Replenishment of water to the groundwater is called **recharge**. Recharge is greater in the heavier rainfalls of spring and fall, and the water table rises in elevation. In summer and winter recharge is generally lower, and the water table drops. In addition, if groundwater is pumped out of the ground in greater volumes than can be replaced by recharge, the elevation of the water table drops.

Water table elevations also reflect longer-term climatic changes. During a time of extreme drought, the water table can drop below the bottoms of local wells, causing them to "dry up" (Figure 12–3). A prolonged drought in the northeastern United States during 1963–1966 limited groundwater recharge at a time when many new homes were being built in

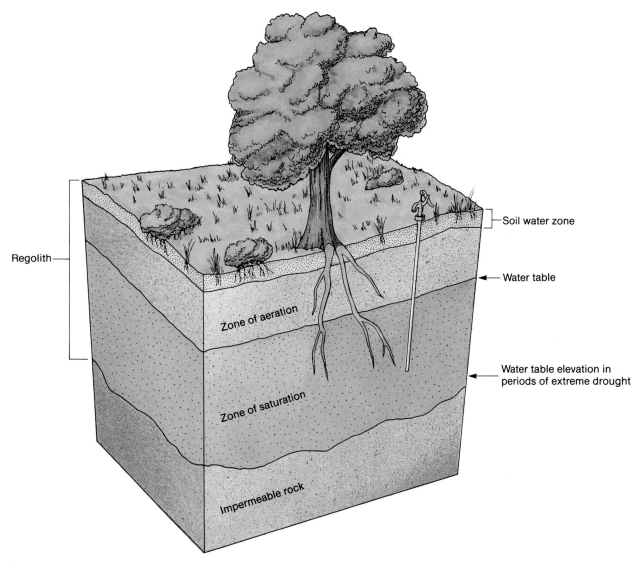

FIGURE 12-3
Vertical distribution of groundwater. The contact between the zone of aeration and zone of saturation is the water table. The vertical position of the water table fluctuates up and down seasonally and as a result of the balance between groundwater pumping and replenishment of groundwater by rainfall.

suburban Nassau County and Suffolk County on Long Island, New York. Average water table elevations dropped 2 m during the drought. Builders included basements in many of these new homes because they found no water during their foundation excavations. When recharge increased in the more humid—and normal—climate of the 1970s, the wa-

ter table rose and basement flooding became a serious problem for many.

There is more than one type of aquifer. We have been describing **unconfined aquifers,** which have no aquicludes above them and have a water table that is free to rise and fall and whose surface is at atmospheric pressure. Most shallow wells are drilled into unconfined aquifers. Another type, the confined aquifer, will be discussed later in this chapter.

Shape of the Water Table

The water table is a three-dimensional surface that is highest in elevation beneath hills and lower in elevation under valleys (Figure 12–4). So far we have been describing groundwater movement through uniformly permeable material, but what happens when downward-percolating water encounters an aquiclude? The water accumulates above the aquiclude to form a local water table that is perched above the regional (lower) water table (Figure 12–4). Such a local water table is called a **perched water table** (Figure 12–4). The more subdued water table that underlies the perched water table is fed by water that sinks through the sides of the hill below the aquiclude.

If the perched water table is on a hill, groundwater may flow along the top of the aquiclude and emerge on the side of the hill as a **spring**. A spring is any natural discharge of groundwater that appears on the ground surface or that flows up through the bottoms of water bodies such as lakes, rivers or even the ocean. Groundwater is insulated from surface variations in temperature and therefore maintains a generally constant temperature year-round. Conse-quently, spring water may be cooler than local surface waters in summer, and warmer in winter.

Springs usually are small features but may be so large that they supply water for homes, industry, or entire towns. For example, Big Spring in Cumberland County, Pennsylvania is the third largest spring in that state. The spring, which issues from a limestone aquifer, has a flow rate that shows little variability and maintains a nearly constant temperature year-round. The flow varies between 11,000 and 16,000 gallons per *minute,* and the temperature varies between 10.5° and 11.5°C. Some perspective on the flow of this spring is given by the fact that one of its users, the Pennsylvania Fish Commission, pumps 6 to 10 *million* gallons of water from the spring *each day* to support a trout hatchery!

Effects of Excessive Discharge on the Water Table

The vertical elevation of the water table is a function of the balance between recharge (precipitation, streams) and discharge (pumping of wells). It is important to remember that pumping usually depletes water from rock or sediment pores at a far greater rate than they can be refilled by groundwater flow. This excessive withdrawal results in a conically shaped depression in the local water table called a

FIGURE 12–4
Elevation of the water table varies with changes in surface topography. An impermeable bed (aquiclude) can cause a perched water table (center); groundwater may migrate across the top of the aquiclude and exit the hillsides as springs.

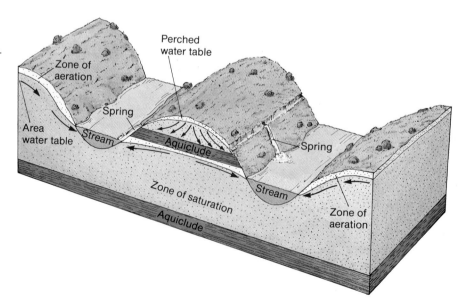

FIGURE 12–5

A cone of depression forms in the water table when local discharge exceeds the recharge. With continual pumping, the expanding conical depression may draw the water table below the level of adjacent wells, causing them to dry up temporarily.

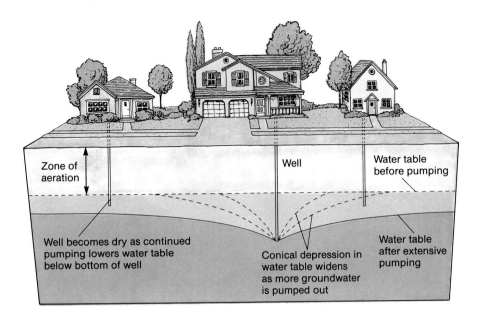

Zone of aeration

Well

Water table before pumping

Well becomes dry as continued pumping lowers water table below bottom of well

Conical depression in water table widens as more groundwater is pumped out

Water table after extensive pumping

cone of depression (Figure 12–5). The cone of depression shown in the figure is a local one, only a few tens of meters across at the top. On a larger scale, pumping for public water supply as well as for agricultural and industrial use can produce *regional* cones of depression in the water table that may be several kilometers across.

CONFINED AQUIFERS

Aquifers that are confined above and below by impermeable materials, and that contain groundwater under pressure significantly greater than atmospheric pressure, are called **confined aquifers**. Conditions necessary for the formation of a confined aquifer are shown in Figure 12–6. Water enters in a humid recharge area at the edge of the aquifer (a mountain range in this case) and flows downward by the force of gravity. Its flow is confined by aquicludes above and below. In some cases the confining aquicludes are fractured, and in these places, the aquifer receives additional recharge from above. The water in the deeper (downslope) parts of the aquifer is under great pressure from the weight of the water upslope from it.

When a well is drilled into the confined aquifer, the pressure is relieved and the water rises upward toward the ground surface, and sometimes even above it. Wells in which the water rises above the ground surface without pumping are referred to as **artesian wells**.

Theoretically, the water in an artesian well should rise to the same elevation as where it entered the aquifer in the recharge area, because "water seeks its own level." This theoretical "artesian surface" is shown in Figure 12–6. However, the water actually rises to a level *below* the theoretical level because energy was lost to friction as the water moved through the pores of the aquifer (see actual artesian surface in Figure 12–6). In the early stages of water withdrawal from artesian aquifers, little or no pumping may be required; but as more and more wells tap the aquifer, the pressure drops and the groundwater eventually must be pumped to the surface.

RELATIONSHIP BETWEEN SURFACE WATER AND GROUNDWATER

Have you ever wondered how a stream can continue to flow month after month during a drought? The answer is that, in humid areas, the groundwater table recharges the stream—Figure 12–7(a). In Chapter 11, we mentioned that groundwater provides water to streams. The amount of groundwater that a

FIGURE 12-6
Geologic section showing
unconfined aquifer, confined aquifer
with flowing artesian well, and
theoretical and actual surfaces to
which the confined groundwater
rises.

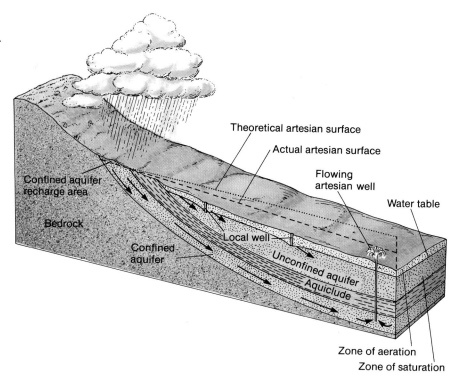

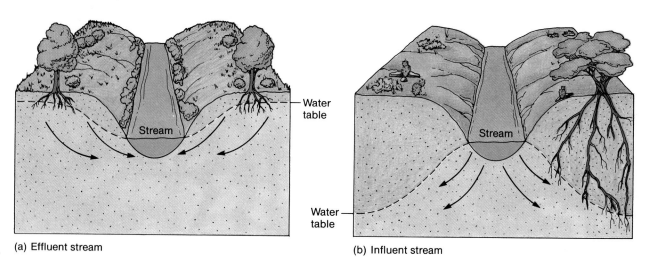

(a) Effluent stream

(b) Influent stream

FIGURE 12-7
Relationship between surface water and groundwater in wet and dry climates. (a) *Effluent*
stream is supplied by continual base flow from a water-table aquifer is characteristic of
continually flowing streams in humid climates. (b) *Influent* stream supplies water to adjacent
areas and is characteristic of through-flowing streams in dry areas. Influent flow supports belts
of vegetation parallel to the stream banks.

stream receives from the water table is called the **base flow** of the stream. The local groundwater table slopes toward the stream, and the stream receives its base flow from springs on the sides and bottom of the stream bed. Streams that receive a portion of their discharge from groundwater are called **effluent streams** because the water flows *out* of the groundwater system into the stream.

The relationship between surface water and groundwater in dry areas is reversed. In dry climates, the local water table slopes away from the stream and the stream recharges the groundwater supply—Figure 12–7(b). Streams that supply water to the groundwater table are referred to as **influent streams** because the water flows *into* the ground from the stream.

The Colorado River is an influent stream which originates in the Rocky Mountains and flows southward through the deserts of the American Southwest until it empties into the Gulf of California. How can the Colorado River continue to maintain its flow as it crosses one of the driest areas in America? As it flows through this area it loses large amounts of water through evaporation into the air and through leakage into the ground. It continues to flow because the discharge that it brings out of the mountains, plus additions of water from springs and tributaries, exceeds its water losses. In many dry areas, the influent flow from a stream into the ground supports a narrow belt of vegetation that borders each side of the channel.

Many streams, especially in dry areas of the western United States, do not flow all of the time. Thus, streams can be classified on the basis of their continuity of flow:

☐ **Perennial streams** flow year-round, supplied largely by base flow—Figure 12–7(a). They are more common in wet regions, such as the eastern United States. However, in the drier parts of the western United States, there are perennial streams wherever the amount of water in the stream exceeds the losses from infiltration or evaporation.

☐ **Intermittent streams** flow part of the year, receiving a portion of their water from the local water table when it is high enough.

☐ **Ephemeral streams** are dry most of the year because the water table is so far below their channels. Ephemeral streams transport water through an area very quickly, usually as a flash flood. The only vegetation that borders such ephemeral or intermittent streams consists of plants with deep roots that can reach the water table below.

GROUNDWATER SYSTEMS

Some aquifers are local because they are limited in area or porosity. These local aquifers supply water sufficient only for small populations or light industries. Other aquifers are much greater in areal extent and can underlie entire states or even regions. These regional aquifers can transmit great volumes of groundwater to a number of population centers and agricultural areas.

Local Aquifer Systems

Most local groundwater resources are derived from unconfined aquifers in permeable sediments or rocks (Figure 12–6). The most commonly tapped aquifers are those that developed in ancient or modern stream deposits, because they are so abundant and usually are quite permeable. For example, in the Midwest and New England, large volumes of groundwater are obtained from sand and gravel originally deposited by streams from melting glaciers. Other communities obtain groundwater from wells drilled in the floodplains or islands of modern rivers.

Local groundwater supplies also can be obtained from impermeable rocks that are cut by interconnected fractures (Figure 12–8). The groundwater trapped in fractured rock differs from that trapped among the sedimentary particles in sands and gravels in one important way: groundwater moving through the large pore spaces in fractured rocks can move much faster than groundwater moving among sedimentary particles. This means that pollutants introduced into an aquifer in fractured rock can travel great distances in less time, polluting other groundwater supplies rapidly.

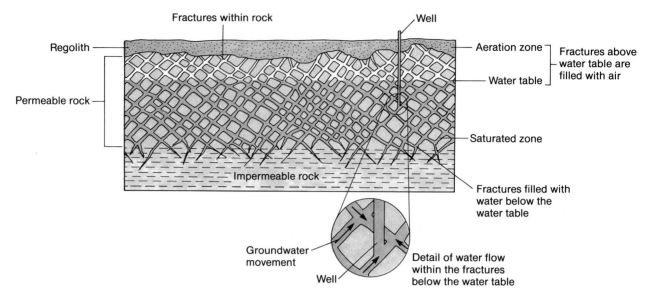

FIGURE 12–8
Groundwater distribution in fractured rock.

Regional Aquifer Systems

Water supplies for many large cities in the United States are obtained from regional-scale aquifers, either unconfined or confined. Different types of regional aquifers underlie coastal plains, islands, volcanic areas, basins between mountain ranges, deserts, and areas underlain by limestones. Each of these regional aquifers has a wide areal extent, but the mode of groundwater occurrence in each is quite different. We will discuss each of these aquifer types briefly, because each tells us something different about groundwater occurrence, quantity, and quality in different areas.

Coastal Aquifers. Many cities on the Atlantic coastal plain and Gulf Coastal plain of the United States obtain their drinking water from aquifers that are recharged inland and extend under the coastal plains into the ocean. The structure of these aquifers is similar to that of the generalized unconfined aquifer in Figure 12–6. However, coastal aquifers differ from the generalized model in one respect: the pores in their oceanward ends are filled with saltwater. Aquifers that are partially intruded by saltwater and that extend under coastal areas are called **coastal aquifers**. The freshwater is lighter and less dense and therefore tends to overlie the saltwater in areas where the two are in contact (Figure 12–9).

The penetration of saltwater into aquifers is called **saltwater encroachment**. Saltwater encroachment in a coastal aquifer increases when recharge decreases, or when freshwater withdrawal increases. The saltwater contact shown in Figure 12–9 represents the position of the freshwater-saltwater contact at a time of high rainfall. The pressure of the freshwater keeps saltwater from rising up the aquifer. During this period, coastal wells draw freshwater from both the upper and lower artesian aquifers, and freshwater under pressure migrates downward and emerges as artesian springs, which flow into the saltwater bay. Freshwater artesian springs are common anywhere that freshwater under pressure is driven upward into an open body of water.

The balance between freshwater and saltwater in coastal aquifers changes markedly in periods of drought. The reduced freshwater recharge allows saltwater encroachment to the position indicated by the dashed lines in Figure 12–9. During a drought, the coastal wells would be pumping saltwater from the aquifers, and there would be no freshwater springs into the bay. (The island shown in Figure 12–9 preserves its own freshwater supply.)

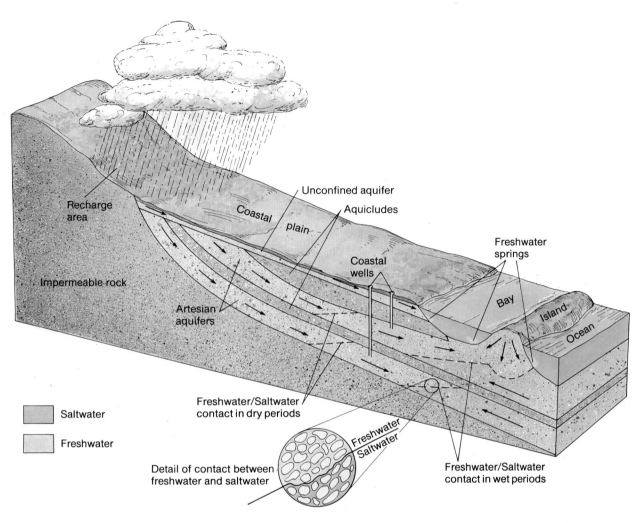

FIGURE 12–9
Freshwater-saltwater relationships in coastal aquifers. During periods of high rainfall,
freshwater displaces saltwater in the aquifers, and freshwater springs occur in the bays. During
periods of drought, saltwater moves up the aquifers (dashed lines show freshwater/saltwater
boundaries) and contaminates coastal wells.

Island Aquifers. Have you ever wondered how
some islands can have abundant natural supplies of
freshwater, even though they are surrounded by saltwater? The answer lies in the difference in *density*
between freshwater and saltwater. So long as there is
rainfall and the island surface is permeable, freshwater will infiltrate into the ground. The weight of this
freshwater acts like a piston, displacing saltwater
from the pores under the island. After a while this
mechanism builds up a lens-shaped body of freshwa-

ter that overlies the denser saltwater in the pores
beneath the island (Figure 12–10).

The depth of freshwater under the island can be
calculated if the height of the water table above sea
level is known. Due to the difference in freshwater
and saltwater densities, a column of freshwater 41 m
high has the same mass as a similar column of saltwater only 40 m high. This relationship allows us to
determine the thickness of the freshwater at any
point on the island. For every meter the freshwater

FIGURE 12–10
Geologic section through a permeable island showing the vertical distribution of freshwater and saltwater. The lens-shaped freshwater body sits atop the saltwater below. For every meter the freshwater table is above sea level (h), there is 40 m of freshwater below that point.

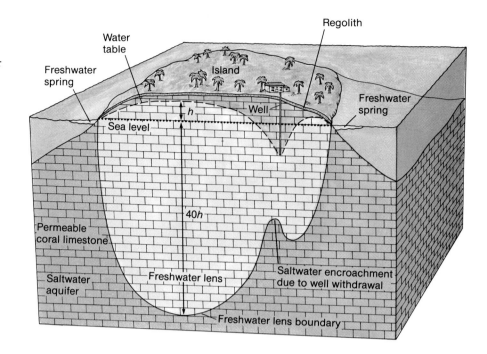

FIGURE 12–11
Geologic section showing groundwater distribution beneath a desert. The oasis forms above a break in the confined aquifer below, which allows groundwater to reach the surface as artesian springs.

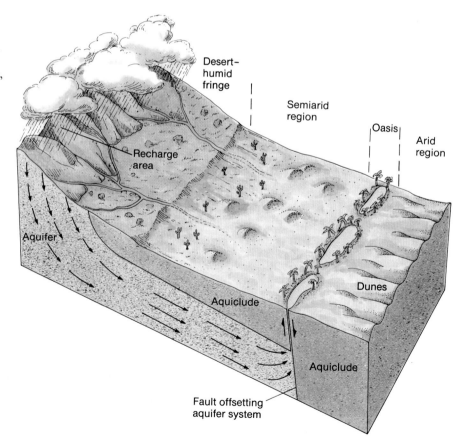

table is elevated above sea level, there is 40 m of freshwater below. For example, if the water table in the center of the island is 3 m above sea level, the total thickness of freshwater below that point is 3 m × 40 = 120 m.

The water table follows the contours of the land, so it has a lower altitude near the ocean, and the thickness of freshwater there is correspondingly less. This relationship gives the freshwater reservoir beneath a permeable island the lens shape seen in Figure 12–10. In dry periods, or periods of high groundwater withdrawal, the water-table elevation drops, the size of the freshwater lens decreases, and saltwater encroachment occurs in wells on the island. Island aquifers are sources of water for people living on islands along the Atlantic and Gulf Coasts and the many coral islands of the South Pacific. On low-lying islands, such as the Bahamas, the low elevations result in a thinner freshwater lens, which is much more susceptible to saltwater encroachment resulting from low recharge. Groundwater supplies on such islands must be supplemented by processing well water to remove salt or by catching rainwater on roofs and channeling it downward into underground storage tanks called cisterns.

Volcanic Aquifers. Most igneous rocks are relatively impermeable; however, some extrusive igneous rocks such as basalt can be quite permeable. Spaces in basalt result from shrinkage cracks, lava tunnels, gas vesicles, and molds of trees engulfed by the lava flows. Layers of altered volcanic ash and soils that developed between successive basaltic flows now act as aquicludes that confine water within the layers of basaltic flows.

Large volumes of groundwater are obtained from volcanic aquifers under extensive basaltic plateaus in Idaho, other areas of the northwestern United States, and the Hawaiian Islands.

Alluvial Basin Aquifers. Basins that are bordered by mountains and filled with stream deposits store large volumes of groundwater in thick aquifers underlying the basin. These are recharged by rainfall entering the valley at the edges of the basin. In some basins, additional recharge is provided by influent streams. Groundwater is obtained from basin aquifers in many areas of Nevada, western Utah, western

Colorado, southern Arizona, and southwestern New Mexico.

The most intensive use of alluvial basin aquifers is in California. About 40% of the state is underlain by basins that contain groundwater. These underground water resources are especially important in supplying drinking and agricultural water during drought periods when surface water supplies are minimal, such as the drought of 1976–1978. The combined storage capacity of all of California's groundwater basins has been estimated to be nearly 30 times the total surface-water storage capacity of all the reservoirs in the state. However, only about 11% of this aquifer water is both of acceptable quality and economical to withdraw. Excessive groundwater withdrawal from basin aquifers in California has created serious environmental problems, several of which will be discussed later in this chapter.

Desert Aquifers. Many movies about deserts show oases with patches of vegetation, palm trees, and pools of water. How can oases exist in such dry areas? The answer is that many oases occur where there is a break in the rock that confines the aquifers below, allowing artesian springs to rise to the surface (Figure 12–11).

A necessary condition for formation of a desert spring is the presence of a confined aquifer below, with a recharge area at a higher elevation on the edges of the desert. The sandstone aquifer shown in Figure 12–11 is inclined under a desert. It was broken by earth movements, and parts of it were displaced along a fault. This displacement moved the aquifer against an aquiclude, stopping the lateral movement of the groundwater. The trapped water was then forced upward by artesian pressure, moving upward along the fault and exiting as an artesian spring at the surface. This created a local water table in the oasis, the presence of which is indicated by permanent vegetation and the year-round bodies of standing water on the surface.

Many deserts do not contain springs or oases, either because there are no aquifers below or because the aquifers are not fractured, preventing the groundwater from reaching the surface. Not all desert springs contain drinkable water, because the groundwater in some areas dissolves harmful minerals on its way to the surface.

Limestone Aquifers. Limestone aquifers differ from the other types because groundwater dissolves the limestone in some places and deposits some of the dissolved calcium carbonate in others. Groundwater migrating through fractures in limestone gradually widens the fractures by dissolving the rock (Figure 12–12), making it possible for even more groundwater to move through the rock.

Local supplies of groundwater are present in many areas of the United States that are underlain by limestone. The major regional limestone aquifers are the Roswell Artesian basin, developed in Permian limestones in eastern New Mexico, and the Florida aquifers developed in Tertiary and Quaternary limestones beneath the Florida peninsula and in adjacent states.

EROSION AND DEPOSITION OF LIMESTONE BY GROUNDWATER

The ability of groundwater to dissolve limestone and deposit calcium carbonate elsewhere creates a wide range of underground and surface features.

Solution of Calcium Carbonate

Limestone is insoluble in pure water, but dissolves readily in water that has been acidified by the addition of carbon dioxide. Water infiltrating the ground surface picks up carbon dioxide from decaying vegetation in the uppermost part of the soil. The solution of limestone may be represented by the reaction:

$$CaCO_3 \ + \ H_2O \ + \ CO_2 \ \rightarrow \ Ca^{2+} \ + \ 2HCO_3^-$$

| Limestone | Water | Carbon dioxide | Calcium ions | Bicarbonate ions |

This reaction is reversible. If the carbon dioxide in the solution is increased, the reaction goes toward the right (more limestone can be dissolved). Conversely, if carbon dioxide is lost, the reaction moves toward the left and calcium carbonate can be deposited from the solution.

The rate of limestone solution increases where rocks are fractured because a greater surface area is exposed to solution. Not all limestones are pure calcium carbonate; most contain small amounts of insoluble minerals, such as quartz grains and clay min-erals. Solution of such impure limestones results in the accumulation of these insoluble minerals. At least part of the "cave earth," a sediment found in many caves, results from the concentration of insoluble materials during the solution of impure limestones.

Solution of limestone is believed to occur most rapidly at the water table and just below it. Rock solution downward is accelerated when the water table is lowered, because the groundwater flowing through many openings in limestone is connected into the regional drainage system; therefore, any lowering of base level for the regional streams would drain the caves to a lower level. This lowering of the water table results in renewed downward solution of the limestone.

Limestone Solution Features

Underground solution of limestone can result in large cavities, which usually are interconnected. These features can range from openings just wide enough to crawl through, to wider features called **caves,** to — under unusual conditions — massive **caverns** such as those at Carlsbad Caverns in New Mexico, Mammoth Cave in Kentucky, and Luray Caverns in Virginia. Some caves are systems of long, intersecting tunnels; others consist of much larger passageways opening into high-roofed chambers. In areas of flat-lying limestones, the horizontal part of many caves develops through solution along the bedding planes, whereas the high chambers are developed by solution along vertical fractures.

The diagram of Mammoth Cave in Kentucky (Figure 12–13) shows how structural features in the limestone determine the shape of the cavern system. The elongate, gently sloping "passageways" are developed by solution along bedding planes. The high chambers called "domes" result from solution of the limestone along vertical fractures.

Underground solution of limestone may weaken the support for overlying beds sufficiently that the surface may collapse into the cavity below. A circular depression in the surface resulting from underground solution of limestone is called a **sinkhole** (Figure 12–14). Although sinkhole formation is a natural process that has always occurred, the *rate* of sinkhole formation has increased markedly in rapidly urbanizing areas underlain by limestone, such as Florida. The reason is that increased groundwater

FIGURE 12–12
Solution cavities in limestone, England. Solution along vertical fractures has formed deep linear cavities between blocks of limestone. Photo courtesy of Florida Sinkhole Research Institute/ University of Central Florida.

withdrawal has resulted in lowering of the water table. As the water table drops, solution continues deeper and deeper into the limestone, and more and more cavities are formed. As more structures are built on this weakened rock, the frequency of collapse increases.

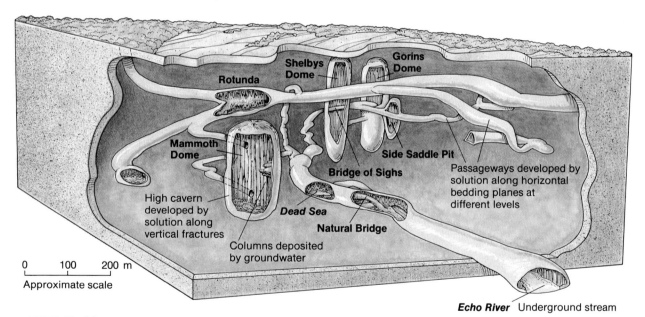

FIGURE 12–13
Cross section of Mammoth Cave, Kentucky. This cavern system contains a wide variety of limestone erosional and depositional features. The high-roofed "domes" follow vertical fractures, and the passageways at different levels follow the horizontal bedding planes of the limestone. Certain passageways have been cut away in the diagram to show their internal features. After *Geomorphology* by Lobeck, A. K., © 1939. Used with permission of McGraw-Hill Book Company.

Karst Topography. Areas underlain by limestone contain a distinctive set of surface features that result from subsurface solution. These features collectively are called **karst topography,** named from the area in Yugoslavia where they are particularly well developed. The formation of these features is shown in the block diagrams of Figure 12–15.

The area shown is capped by a solution-resistant sandstone layer into which perennial streams have cut channels—Figure 12–15(a). Subsurface solution of the limestone has removed surface support, and sinkholes have begun to form across the surface. More and more of the surface drainage is being diverted underground via the sinkholes, sometimes creating "disappearing streams," and underground solution is beginning to form a system of caves and passageways below the surface.

With the passage of time, the expanding sinkholes begin to join, forming a system of **dry valleys**—Figure 12–15(b). Rainfall flows briefly along the surface into sinkholes and fractures and then flows underground, perhaps emerging on the surface at a lower elevation. The presence of sinkholes, the absence of surface drainage, and the presence of dry valleys are characteristic features of karst topography.

As the surface is continually lowered by solution and collapse, part of the undissected surface may be preserved as a **natural bridge**—Figure 12–15(c).

Very little of the original land surface is present and most of the present surface is covered by sinkholes and dry valleys.

In the last stage of karst topography development, vertical downcutting and solution have slowed considerably because an underlying insoluble layer has been reached—Figure 12–15(d). Perennial drainage channels may now develop on the surface once again. These surface streams flow between residual masses of limestone called **karst towers,** which rise majestically above the lowlands. Karst towers form

FIGURE 12–15 (opposite)
Development of karst topography in a humid region. It is important to note that the sequence of features developed here assumes no change in the regional base level over time.
(a) Surface drainage is established on solution-resistant rocks. Solution of the underlying limestone by groundwater has started to form sinkholes into which some of the surface drainage has been diverted. (b) Very little surface drainage remains as more and more of the limestone is exposed by collapse of the overlying sandstone. Interconnecting of subsurface solution cavities begins to form a cave system. (c) Sinkhole expansion has formed solution valleys, and the cave system has been enlarged into a cavern system. Some areas are still capped by sandstone. (d) The area has been dissolved down to one of low relief. Surface drainage has become established once again on underlying solution-resistant rock. These streams flow between isolated karst towers and hills of less-soluble limestone. After *Geomorphology* by Lobeck, A. K., © 1939. Used with permission of McGraw-Hill Book Company.

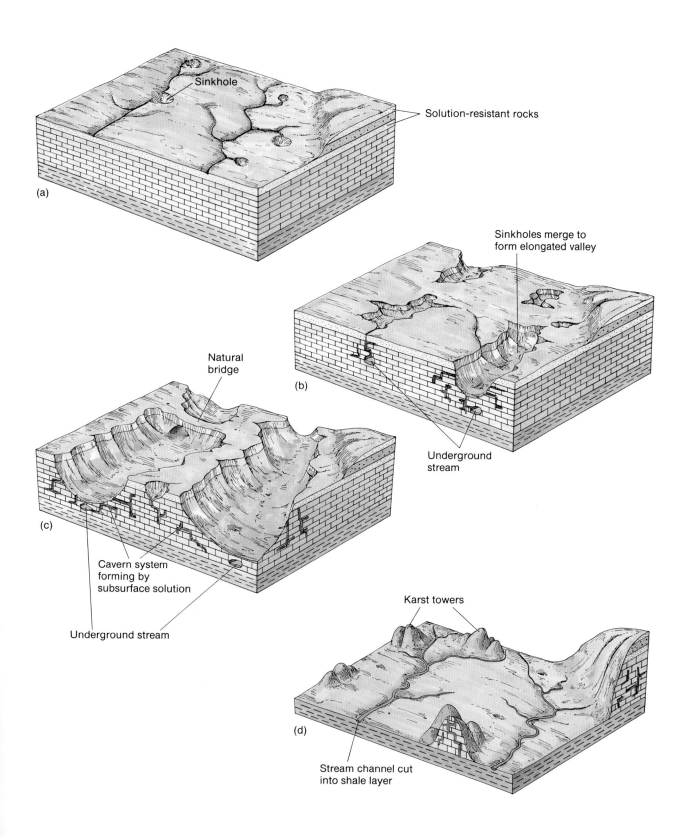

(a)

Sinkhole

Solution-resistant rocks

(b)

Sinkholes merge to
form elongated valley

Underground
stream

Natural
bridge

(c)

Cavern system
forming by
subsurface solution

Underground stream

(d)

Karst towers

Stream channel cut
into shale layer

EROSION AND DEPOSITION OF LIMESTONE BY GROUNDWATER | 319

dramatic landscapes in Puerto Rico, Jamaica, Vietnam, and especially in China, where they are traditional features in classic Chinese art.

Limestone Depositional Features

Under the right conditions, percolating groundwater also may *deposit* calcium carbonate. The migrating groundwater contains dissolved carbon dioxide as well as calcium and bicarbonate ions from the solution of limestone. As this groundwater migrates through the limestone and drips slowly into larger cavities, the carbon dioxide is released into the cave air and the ions are deposited as calcium carbonate. The reaction is:

$$
\underset{\substack{\text{Calcium}\\\text{ions}}}{Ca^{2+}} + \underset{\substack{\text{Bicarbonate}\\\text{ions}}}{2HCO_3^-} \rightarrow \underset{\substack{\text{Calcium}\\\text{carbonate}}}{CaCO_3} + \underset{\text{Water}}{H_2O} + \underset{\substack{\text{Carbon}\\\text{dioxide}\\\text{(escapes)}}}{CO_2}
$$

SOLUTION   DEPOSITION

This reaction is merely the reverse of the one shown earlier in this chapter for the solution of limestone. Recall our statement that, if carbon dioxide is lost, the reaction moves to the side favoring the deposition of calcium carbonate.

Calcium carbonate is deposited by groundwater in widely varied forms. Water dripping from cave ceilings deposits needlelike **stalactites,** which extend downward from the cave ceiling (Figure 12–16). Water dripping down stalactites and onto the cave floor below deposits calcium carbonate as broad, pinnaclelike **stalagmites.** Where stalactites and stalagmites grow together, broad **pillars** are formed. Water dripping and running across the cave walls and floors deposits layers of a banded form of calcium carbonate called **travertine.**

THERMAL SPRINGS AND GEYSERS

So far we have discussed groundwater springs in which the water generally is cooler than surface wa-

FIGURE 12–16
Cavern interior, Mammoth Cave National Park, Kentucky, showing stalactites, stalagmites, and travertine on cavern walls. Photo courtesy of National Park Service.

FIGURE 12–17
Thermal spring and mineral deposits in Yellowstone National Park, Wyoming. Photo courtesy of National Park Service.

ters. However, some **thermal springs** are much warmer than the surface waters in their area. Thermal springs are formed where a magma chamber lies at shallow depths and heats the groundwater as it travels upward toward the surface. The hot water can contain high concentrations of dissolved ions derived from solution of the underlying rocks. As the water reaches the surface, it cools, and many of the ions are precipitated as layered mineral deposits around the thermal spring (Figure 12–17).

Groundwater that emerges as a column of steam and hot water is called a **geyser**. The best-known geyser in the United States is Old Faithful in Yellowstone National Park. In many cases, the plumbing of geysers is complex; what follows here is a simplified explanation.

The process begins as a narrow, vertical opening in the rock fills with groundwater from the surrounding area. The groundwater is heated by the magma chamber. The increasingly higher column of water exerts more and more pressure on the water at the base of the column. This increased pressure raises the boiling point of the water at the base, just as the lid on a pressure cooker does. Boiling of the basal water starts either when the pressure is reduced or when the temperature is increased. As the basal water begins to boil, steam forms. The steam pushes the water above it upward, which reduces the pressure on the basal water even more, and the overlying water begins to boil in turn. Soon, a jet of steam and hot water shoots above the surface as a geyser. After the geyser erupts, the rock fractures fill with more groundwater and the process starts again, commonly in a regular period such as Old Faithful's 67-minute cycle.

GROUNDWATER RESOURCES

Groundwater is one of our most important resources. In fact, in areas with little rainfall, it may be the only source of water. Increasing development in formerly rural areas has placed great demands on groundwater resources by depleting aquifers at a much greater rate than they can be recharged by groundwater flow. Another problem is that groundwater is being polluted with increasing frequency (Focus 12–1). In this section we will consider how groundwater resources are developed, how they are being recharged, and some types of pollution that are reducing the quality of the groundwater in many regions.

Development of Groundwater Resources

The water needs of a region change markedly as it undergoes a transition in population density from rural, to suburban, to urban. Rural areas generally obtain groundwater from shallow wells that tap unconfined aquifers. This type of groundwater supply is adequate, so long as withdrawal is modest and wells are far enough from waste-disposal sites (cesspools and septic tanks) so that the shallow groundwater supplies are not contaminated.

As an area becomes more urbanized, the unconfined surface aquifers become contaminated by runoff from streets and pavements, industrial discharges, and sewage. Wells must be drilled deeper to tap confined aquifers that contain better-quality water. The deep aquifers are generally protected to a greater extent from surface pollution.

Groundwater supplies usually are inadequate for large urban areas. Consequently, cities typically obtain most of their water from lakes, rivers, or reservoirs and transport it into the cities by underground pipeline or surface aqueducts and canals. Changes in groundwater usage that occur in an area undergoing transition from rural to urban are well illustrated by what happened in the outlying counties of New York City between 1903 and 1965 (Figure 12–18).

When New York City was formed by the consolidation of five counties in 1898, two of its outlying counties, Kings (Brooklyn) and Queens, were mostly suburban or rural. Virtually all of their water was obtained from shallow wells in unconfined glacial aquifers and from deep wells in confined coastal aquifers. Figure 12–18(a) shows that the top of the water table was above sea level everywhere, and reached 35–55 feet above sea level at the highest points.

Rapid settling of the counties occurred in the next 30 years, and groundwater demand increased markedly. By 1936, the water table under Kings and Queens Counties had been drawn down considerably everywhere, reaching 35 feet below sea level in the heaviest use area—Figure 12–18(b). Recharge to the aquifers was now greatly reduced because much of the surface had been paved or covered by buildings. The combination of decreased recharge and increased withdrawal resulted in serious saltwater encroachment in the confined aquifers below, and many wells had to be closed each year. More and more of the area was supplied by water brought to the area by aqueducts from reservoirs upstate.

By 1965 the area was completely urban and most of the wells had been closed—Figure 12–18(c). As a result, recharge now exceeds discharge and the water table is rising steadily, although the area has no present need for the groundwater.

Aquifer Recharge

Adequate groundwater supplies require that the aquifers be recharged continually. The effect of decreasing recharge is well illustrated by the changes in groundwater quality and quantity that have occurred in Florida since rapid urbanization began in the 1950s. Although much of Florida is near sea level, the state has a relatively high central area (maximum elevation around 100 m) running north-south. Limestone aquifers in Florida are recharged from this higher central area, which was rural in the 1950s. The groundwater flows away from this central recharge area to the Gulf of Mexico on the west, the Atlantic Ocean on the east, and the swampy Everglades on the south. With the increasing urbanization of central Florida, especially the development of entertainment complexes in the Orlando area, more

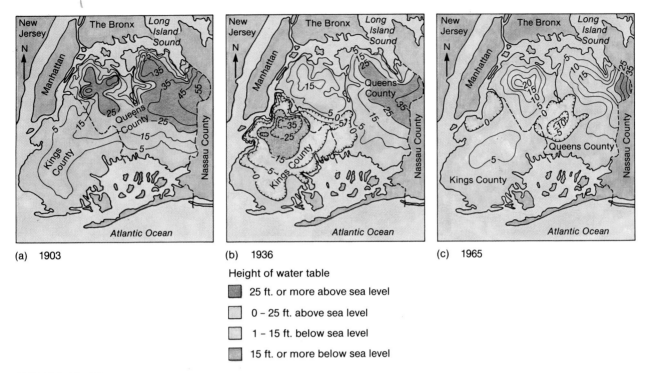

(a) 1903   (b) 1936   (c) 1965

Height of water table

■ 25 ft. or more above sea level

□ 0 - 25 ft. above sea level

□ 1 - 15 ft. below sea level

■ 15 ft. or more below sea level

FIGURE 12–18

Changes in the groundwater resources of Queens County and Kings (Brooklyn) County, New York City, from 1903 to 1965. (a) Groundwater table elevations in 1903. The area was rural and groundwater table elevations were well above sea level everywhere except the industrial area of northwestern Queens County. (b) Groundwater water table elevations in 1936. The area was now more industrialized and densely settled. Deep depressions in the groundwater table occurred in Kings County, and it dropped significantly in all other areas. Saltwater encroachment increased; wells were closed, and the areas were supplied by aqueduct from reservoirs upstate. (c) Groundwater table elevation in 1965. The area was now completely urban. Most wells had been closed and the groundwater table rose a great deal under the area. Virtually the entire area is now supplied with surface water from upstate reservoirs via aqueduct. After New York Water Resources Commission, *An Atlas of Long Island's Water Resources,* 1968, Bulletin 62A, Plate 8, p. 83.

and more of the recharge area was paved over. This decreased recharge, plus the increased groundwater needs of rapidly growing Gulf Coast cities (Tampa, St. Petersburg, and Fort Myers) and Atlantic Coast cities (Palm Beach, Fort Lauderdale, and Miami), has resulted in serious saltwater encroachment into the aquifers in coastal areas.

Many rapidly growing areas are trying to forestall a groundwater crisis by actively increasing recharge. In suburban Nassau and Suffolk Counties on Long Island, New York, laws require construction of re-charge basins in surface sand and gravel as a water-conservation measure (Figure 12–19). Culverts transport the surface water into the recharge basins where the water infiltrates into the ground and eventually reaches the aquifers below.

In Los Angeles, aqueduct water from the Colorado River is pumped *down* special wells to displace the saltwater in coastal aquifers. The Los Angeles area also recharges the aquifers below by controlled flooding of specially prepared areas of permeable ground.

FIGURE 12-19
Recharge basin on eastern Long Island, New York. Rainwater is diverted to the recharge basins where it can recharge the aquifers below. Photo by Nicholas K. Coch.

GEOLOGIC HAZARDS ASSOCIATED WITH GROUNDWATER WITHDRAWAL

Withdrawal of groundwater can be accompanied by various kinds of geologic hazards. We will discuss two of these, collapse and subsidence.

Collapse

Surface collapse is a potentially serious problem in areas underlain by limestone. Solution by ground-water enlarges the rock openings with time, removing support for the surface above. Groundwater that fills the solution cavities provides some support for the surface and may postpone collapse. However, when the groundwater level is drawn down rapidly by domestic or industrial use, support for the surface is reduced, and the surface collapses suddenly to form a sinkhole (Figure 12-20). An additional factor promoting collapse is the increased pressure placed on the surface by the new structures that accompany urbanization (office buildings, roads, homes, and so on).

FIGURE 12-20
Aerial view of sinkhole formation and surface damage in Winter Park, Florida, May 1981. Photo courtesy of R. Deuerling/Florida Geological Survey.

GROUNDWATER POLLUTION

Many areas of the United States have a problem not only with water quantity, but with water *quality* as well. Local contaminants enter the groundwater by infiltration. These contaminants then move downward and laterally within the groundwater flow.

A wide variety of contaminants is affecting the quality of groundwater. Rain and melted snow washing across road surfaces carry into the ground oil, vehicle exhaust products, and salt that is spread to melt ice. Insecticides, nitrate fertilizers, and phosphate fertilizers infiltrate the ground in agricultural areas. Sewage contamination is a threat from both home and municipal treatment systems. Ruptures of underground storage tanks have released chemicals and petroleum products into groundwater supplies. Leaching of landfills and industrial waste piles by rainwater forms contaminants that pollute underlying aquifers. We will consider a few of these situations to illustrate the scope of the problem.

Sewage

In many rural and suburban areas, sewage wastes are directed to individual porous underground tanks called cesspools or septic tanks. Bacteria in these tanks partially break down the sewage, and the liquid effluent is then allowed to infiltrate the ground surface. If treatment is inadequate, or if homes and businesses are located too close together, more sewage enters the soil than can be effectively treated. This results in contamination of unconfined aquifers with harmful bacteria and viruses.

Sewage contamination, as well as all other kinds, is potentially most serious in limestone aquifers. The large, interconnected spaces in many limestone aquifers permit contaminants to travel quickly before they are broken down by soil microorganisms or diluted below dangerous levels. Domestic wastes (sewage) from one home may contaminate the drinking water of a nearby home if the wastes leak into the local aquifer (Figure 1).

Solid Waste

Solid waste is typically disposed of in containment structures called landfills (Figure 2). After a load of waste is dumped, it is covered with sand to minimize odors and animal infestation. However, these unpleasant smells and scavengers are not the major problems with landfills. Rain and snowmelt soaking into the landfill, along with normal decomposition of organic wastes, produces a foul-smelling liquid called *leachate*, along with methane gas. The leachate percolates through the landfill and into the

FIGURE 1
Geologic section showing contamination of groundwater by sewage wastes in an area underlain by limestone. Sewage wastes from the house on left leak into the aquifer below and travel by gravity flow under the house on the right, where contaminated water is drawn upward in the well.

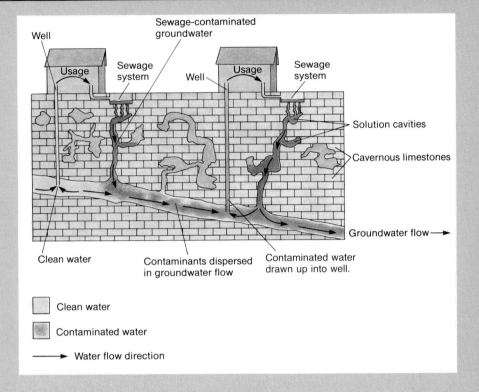

FIGURE 2
Leachate formed in solid-waste landfills can contaminate local surface water and groundwater.

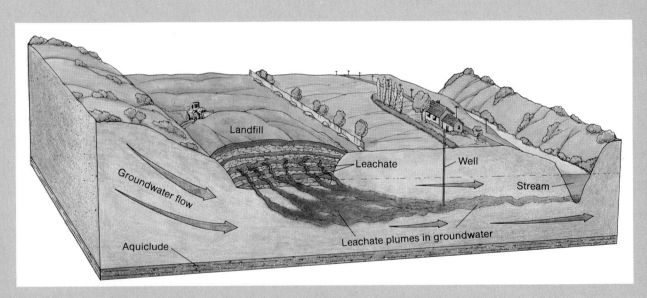

aquifers below. Contamination of aquifers by leachate is a serious problem in densely settled areas that obtain all of their water from the ground, such as Long Island, New York.

Toxic Chemicals

Toxic wastes can enter the ground from disposal ponds, spillage during loading and transport, leakage from storage facilities, burial, or by pumping into injection wells. Injection wells are designed for liquids to be pumped down to a level where they can be safely contained. Many scientists believe that disposal of toxic wastes in deep wells is unsafe because the wastes may migrate from the area through cracks and other openings that are not easily detected beforehand. A number of sources of toxic wastes are shown in Figure 3.

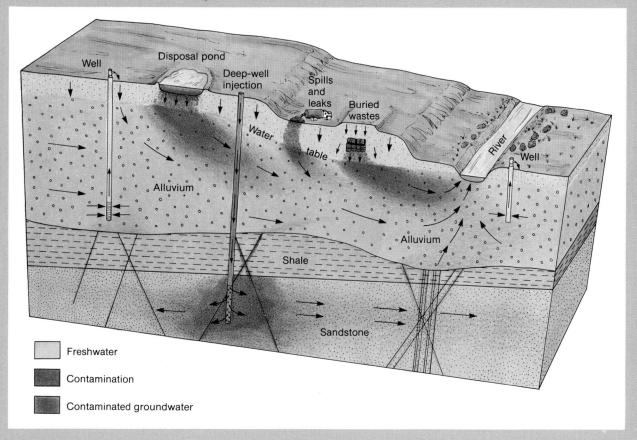

FIGURE 3
Industrial sources of toxic wastes. After American Institute of Professional Geologists, 1983, *Groundwater: Issues and Answers,* p. 16.

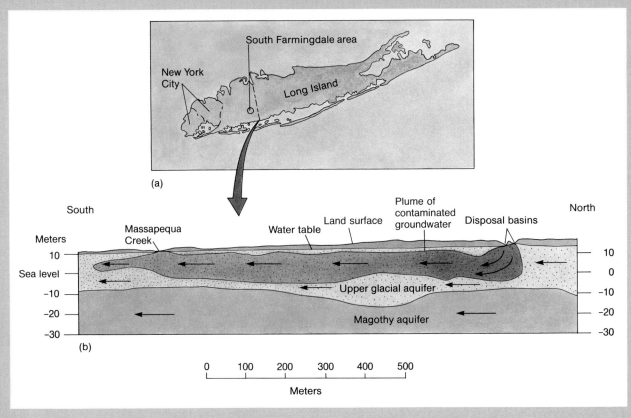

FIGURE 4
Path of groundwater plume contaminated with cadmium and chromium in South Farmingdale,
Long Island, New York. The metal wastes had been disposed of in a surface pit and entered
the groundwater table below. After New York Water Resources Commission, *An Atlas of Long
Island's Water Resources*, 1968, Bulletin 62, Plate 8E, Figure 3, p. 89.

Large-scale industrial pumping of groundwater
can trigger especially serious collapse of the surface.
R. F. Legget in *Cities and Geology* describes a large-
scale collapse in the West Reef area of Johannesburg,
South Africa, which occurred after deep pumping of
groundwater began in 1960. The area is underlain by
thick deposits of dolomite and dolomitic limestone.
Between 1962 and 1966, eight sinkholes larger than
50 m in diameter and deeper than 30 m appeared,
along with many smaller ones. In December 1962, a
large sinkhole developed under a rock-crushing
plant; *the whole plant disappeared* and 29 lives were

lost. In August 1964, surface collapse caused five
deaths as a home dropped 30 m into another suddenly developed sinkhole.

Subsidence

Fluids among the grains in a sedimentary deposit
help to support the surface above. When large volumes of groundwater are removed, the grains come
closer together, resulting in compaction of the sediment layer. The decrease in volume results in a slow
sinking of the surface that is referred to as **subsid-**

In the past, toxic wastes were placed in barrels and buried in landfills and disposal sites. With time, the steel barrels corroded and wastes entered the ground, resulting in severe pollution. One infamous example is the Love Canal area near Buffalo, New York where severe health problems resulted, many homes were condemned, and people were evacuated. In Westhampton on Long Island rupture of underground gasoline storage tanks contaminated the groundwater. New York State now requires that old storage tanks be replaced on a regular basis before corrosion can release their contents into the surrounding area.

In the past, many toxic wastes were disposed of simply by pumping them into pits dug into the ground. The almost inevitable result was groundwater pollution. A good example of this type of groundwater pollution is the disposal of metal-plating wastes in South Farmingdale, New York. This area is on Long Island, whose upper surface is composed entirely of permeable, glacial gravels and sands. The U.S. Geological Survey has documented how this disposal affected the local water supply (Figure 4). Chromium and cadmium were detected as early as 1942 in shallow wells near the disposal pits. By 1962, the body of shallow groundwater contaminated with the metals was about 4200 feet long and about 1000 feet wide and was moving southward at a rate of several hundred feet per year.

ence. Subsidence may result in cracks in the surface and in human structures.

Subsidence is especially serious in coastal areas because it increases the danger of flooding during high tides and especially during tropical storms. Rapidly growing suburban areas south of Houston, Texas, recently have experienced serious problems with subsidence caused by the withdrawal of groundwater and oil. Subsidence in nearby communities like Clear Lake City, which borders the Johnson Space Center, and in coastal areas to the south, has resulted in cracks and breaks in roads and foundations, as well as flooding of structures nearest to tidal waterways (Figure 12–21).

One of the best documented cases of large-scale subsidence in the United States is occurring in the Santa Clara basin south of San Francisco Bay in California (Figure 12–22). The subsidence resulted from excessive pumping of groundwater from basin aquifers. Maximum subsidence in the Santa Clara Valley was 1.18 m between 1960 and 1967, with the greatest subsidence just south of the city of San Jose. Total subsidence measured between 1912 and 1967 was 3.96 m in downtown San Jose.

FIGURE 12–21
Subsidence of land surface along Galveston Bay, south of Clear Lake City, Texas, has increased flooding by tidal waters. The housing complex in the background is now abandoned. Photo by Nicholas K. Coch.

SUMMARY

1. Groundwater accumulates below the ground surface and migrates under the influence of gravity through the pore spaces in sediments and rocks.

2. Pore spaces can be developed at the time of sediment or rock formation or after the rock has been formed.

3. Pore spaces can consist of openings among grains, fractures within the rock, or larger openings resulting from the solution of the rock by groundwater.

4. Permeable sediments or rocks that are capable of holding and transmitting water are called aquifers. Impermeable materials are called aquicludes.

5. The flow rate of groundwater can be predicted from Darcy's law once the cross-sectional area, permeability, length, and inclination of the aquifer are known.

6. Groundwater flow rates are slow, generally only a few centimeters to less than a meter per day.

7. The soil water zone is a thin belt of partially saturated pores just below the surface. The aeration zone is filled partially with water that is in transit to the saturation

zone below. The contact between the aeration zone and the saturation zone is referred to as the water table.

8. The water table follows the contours of the land. If downward-percolating groundwater encounters an aquiclude, it may form a perched water table. Groundwater may reach the surface as a spring on a hillside or at the bottom of a body of water.

9. Withdrawal of groundwater at a rate greater than that of recharge forms a cone of depression on the water table.

10. An unconfined aquifer is one with a free water table whose upper surface is at atmospheric pressure.

11. A confined aquifer is confined above and below by aquicludes and contains groundwater that is under pressure greater than atmospheric pressure. When a well is drilled into a confined aquifer, the water rises in the well under its own pressure to form an artesian well.

12. In humid areas the water table recharges streams (effluent stream) and provides their base flow. In dry regions streams may recharge the water table (influent

FIGURE 12–22

Contour lines connect points of equal surface subsidence resulting from excessive groundwater withdrawal in the Santa Clara basin, California. After Poland, "Land Subsidence in Western U.S.," in Olson and Wallace, eds., *Geologic Hazards and Public Problems,* May 27–28, 1969 conference proceedings, U.S. Geological Survey, p. 77–96.

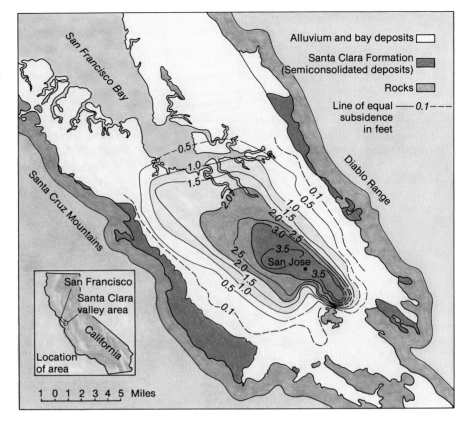

stream), and groundwater may be sufficient to support only a belt of vegetation on the sides of the stream.

13. Perennial streams flow year-round, supplied largely by the water table. Intermittent streams flow part of the year, receiving a portion of their water from the local water table when it is high enough. Ephemeral streams are dry most of the time and contain water only for a short time after it has rained.

14. Local aquifer systems generally tap shallow aquifers in glacial and stream deposits as well as fractured rocks.

15. Coastal aquifers are confined aquifers in which freshwater overlies the saltwater at the seaward end of the aquifer. Penetration of saltwater into these aquifers is saltwater encroachment. Saltwater encroachment increases as freshwater recharge decreases.

16. For every meter the water table is above sea level in an island aquifer, there is 40 times that number of meters of freshwater below that point.

17. Volcanic aquifers transmit water through fractures in basalt. The aquifer can be confined between layers of soil or weathered volcanic ash.

18. Alluvial basin aquifers result from thick layers of stream sediments deposited in a basin bordered by mountains.

19. An oasis occurs in a desert where a break in a confined aquifer allows water to reach the surface, developing artesian springs.

20. Calcium carbonate is dissolved underground by groundwater that contains dissolved carbon dioxide. This solution process proceeds most rapidly just at and below the water table and results in the formation of caves and caverns.

21. Karst topography is an assemblage of landforms characteristic of solution in limestone areas. Karst features include minimal surface drainage, disappearing streams, sinkholes, karst towers, and natural bridges.

22. Calcium carbonate is deposited where groundwater drips into cave openings and the excess carbon dioxide diffuses into the cave atmosphere. Depositional features in caves include stalactites, which grow down from the ceiling, stalagmites, which grow up from the floor, pillars that form as stalactites and stalagmites

grow together, and travertine that is deposited in layers over the cave walls and floors.

23. Groundwater heated by underground magmatic sources can form thermal springs and geysers.

24. Rural areas obtain water mostly from shallow wells drilled into unconfined aquifers. Suburban areas must obtain water from deeper confined aquifers. Urban areas generally must rely on surface water imported into the area by aqueducts or canals.

25. Aquifers are being recharged artificially by using recharge basins and by pumping imported water into the ground.

26. Excessive withdrawal of groundwater in limestone areas has resulted in accelerated collapse of the surface.

27. Excessive withdrawal of groundwater from thick basin aquifers has resulted in subsidence of the surface in many areas.

28. Groundwater supplies are being increasingly contaminated by sewage, landfill leachates, gasoline tank ruptures, road salt, and chemicals that enter the surface and infiltrate downward into the aquifer over time.

REVIEW QUESTIONS

1. How can a rock that had no original porosity develop into a good aquifer at a later time? Give several examples.

2. Clayey sediments have much greater porosities than sandy ones, yet sandy sediments have greater permeabilities. How can this be explained?

3. Describe the shape of a typical water table and account for the vertical fluctuations in the elevation of the water table over the course of a year.

4. What geologic, topographic, and climatic conditions are necessary for the formation of a confined aquifer?

5. What conditions are necessary to form an artesian well?

6. Consider the relationship of groundwater and surface water. What differences exist in this relationship between wet areas and dry areas?

7. What conditions determine whether a perennial, intermittent, or ephemeral stream will develop?

8. How could you tell whether a stream was influent or effluent by looking at it in the field?

9. Describe the conditions that favor the formation of each of these features:
   a. Artesian spring in a coastal bay.
   b. Perched water table.
   c. Natural bridge.
   d. Cone of depression.
   e. Oasis in a desert.

10. How can you determine the depth of freshwater contained in an island aquifer?

11. Describe the chemical conditions that result in both the solution and deposition of calcium carbonate.

12. Over a long period of time, what changes in the surface topography can be expected in a humid area underlain by limestone?

13. Describe the formation of a sinkhole. How has urbanization in areas underlain by limestone increased the incidence of surface collapse?

14. How does excessive groundwater withdrawal from a thick basin aquifer result in subsidence at the surface? Why is this an especially serious problem in coastal areas?

15. What methods have been used to artificially recharge aquifers?

16. Which of these materials would be best to have under a solid-waste landfill? Why?
   a. Fractured granite.
   b. Sandstone.
   c. Limestone with karst topography.
   d. Clay.

FURTHER READINGS

Cohen, P. O., Franke, O. L., and Foxworthy, B. L. 1968. *An atlas of Long Island's water resources.* New York Water Resources Commission Bulletin 62. (A detailed description of water sources, usage, and pollution in a rapidly urbanizing area. Contains many excellent geologic sections).

Ford, R. S. 1978. Ground water—California's priceless resource. *California Geology* 31(2): 27–32. (A good overall review of groundwater, with sections describing its chemical quality and the different types of wells used to obtain groundwater).

Hack, J. T., and Durloo, L. H. 1962. *Geology of Luray Caverns, Virginia.* Virginia Division of Mineral Resources Report of Investigations 3, 43 p. (Describes the formation of features within one of the largest caves in the eastern United States.)

Heath, R. C. 1984. *Ground-Water regions of the United States.* U.S. Geological Survey Water Supply Paper 2242, 78 p. (Describes the geology and types of aquifer systems in all regions of the United States and its possessions. Contains a great number of excellent maps and geologic cross sections.)

Leggett, R. F. 1973. "Hydrogeology of cities." Chapter 4 in *Cities and geology.* New York: McGraw-Hill Book Co., p. 125–76. (Describes how a number of different cities obtain water. Includes a discussion of the aquifer types discussed in this chapter, along with a discussion of surface water supplies.)

Lindorf, D. E., and Cartwright, K. 1977. *Ground-water contamination—problems and remedial actions*. Illinois State Geological Survey Environmental Geology Notes No. 81, 58 p. (Describes a wide variety of contaminants and how they affect the quality of groundwater. Contains numerous case histories).

White, W. B. 1988. *Geomorphology and hydrology of karst terrains*. New York: Oxford Univ. Press, 464 p. (An excellent and detailed description of karst landforms and problems of water supply and pollution in areas underlain by limestone).

13

Glaciers,
Glaciation, and
Climatic Change

North branch of the Trimble Glacier, Alaska Range, Alaska.
Photo courtesy of Austin Post.

laciers today cover only a fraction of the Earth's surface, in contrast to their former extent during the past 1.5 million years, a period of time commonly known as the "Ice Ages." In the popular view, the Ice Ages were times of unrelenting cold when glaciers repeatedly advanced and retreated from polar regions into more temperate latitudes. The true picture is quite different, because this period of time, referred to as the Pleistocene Epoch by geologists, was one of great climatic change, with periods colder than today separated by periods *warmer* than the present.

The climatic changes of the Pleistocene Epoch had worldwide effects. Advancing glaciers eroded loose soil and bedrock and carried them far from their sources. Glacial erosion created dramatic mountain landscapes and carved large basins that subsequently filled with water when the ice melted, such as the Great Lakes. Glacial deposits dammed stream valleys, forcing meltwaters to create new stream systems, such as the Missouri and Ohio, and to greatly enlarge others, such as the Mississippi. Sea level fell and rose repeatedly in response to the formation and melting of glaciers on the land. When sea level was lowered, the shallower parts of the ocean floor emerged, creating intercontinental paths for the migration of plants, animals, and humans.

In areas not covered by ice sheets, climates were cooler than today. The cooler temperatures produced increased rainfall in many presently dry areas, such as the deserts of the American Southwest, resulting in surface water and vegetation on a far greater scale than at present.

GLACIERS

Glaciers are masses of ice that form on land by the compaction and recrystallization of snow. They move downslope or outward in all directions under the pull of gravity.

Where Do Glaciers Form?

Glaciers form wherever there is abundant snowfall and the temperatures are sufficiently cold that snow can accumulate in increasing thickness from year to year. The cold temperatures needed to form glaciers can be found at high elevations as well as higher latitudes. Wherever the winter snowfall exceeds the summer melting in mountains or in low-lying regions, masses of snow can be preserved from year to year.

The lowest elevation of year-round snow is called the **snowline**. The altitude of the snowline is a function of both the ground temperature and the amount of snowfall. In temperate latitudes, mountains that are completely snow covered in winter are only snowcapped at their higher, colder elevations during the summer. In polar latitudes, temperatures may be low enough year-round so that snow covers the ground continually down to sea level.

At any given latitude, the snowline is at a lower elevation on the coastal mountains than on the mountains of the continental interior, which is colder but farther from the coastal source of moisture. Thus, despite warmer temperatures along the coast, the greater snowfall there enables snow to remain year-round at lower altitudes.

How Do Glaciers Form?

The transformation of snowflakes into glacial ice involves melting and recrystallization of snow particles under increasing pressure. Excavation of the higher parts of a glacier shows that the fresh snow at the surface grades downward into a granular snow with a lower porosity, and eventually into ice.

Freshly fallen snow is a porous mass of hexagonal snowflakes—Figure 13–1(a). The snow has great surface area and readily exchanges moisture with the surrounding air. Greater moisture losses occur at the tips of the crystals than across the wide central areas, which results in melting of snowflakes around their edges, rounding them over time—Figure 13–1(b).

The particles also begin to melt at points where they touch each other, because the pressure is greater there. This meltwater, along with any meltwater formed at the surface of the snowfield, moves downward and refreezes around the particles at depth, enlarging them into granular particles called **firn**—Figure 13–1(c). The firn particles form a granular mass of snow with a porosity of about 50%. The porosity of the granular snow decreases under pressure from the snow above—Figure 13–1(d)–(e). This compaction eventually forces the granular particles together, steadily eliminating air

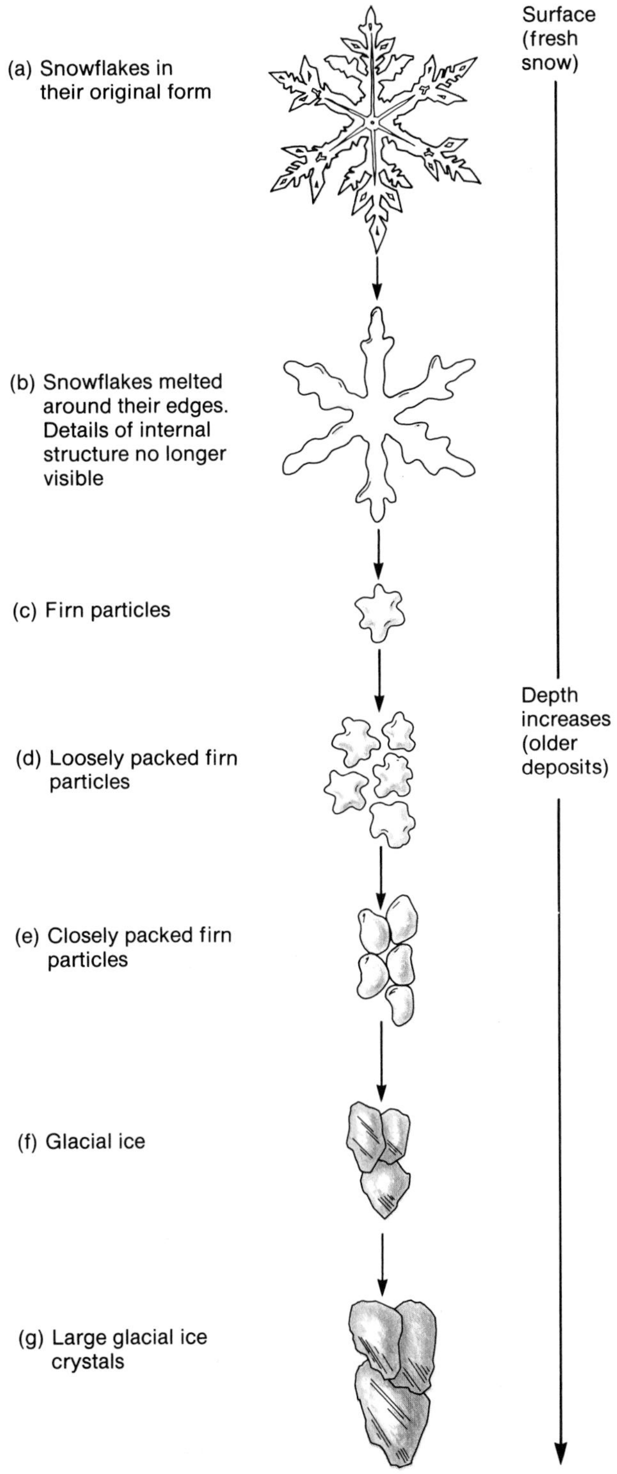

(a) Snowflakes in their original form

Surface (fresh snow)

(b) Snowflakes melted around their edges. Details of internal structure no longer visible

(c) Firn particles

(d) Loosely packed firn particles

Depth increases (older deposits)

(e) Closely packed firn particles

(f) Glacial ice

(g) Large glacial ice crystals

FIGURE 13–1
Transformation of snow into ice.

from pore spaces and forming glacial ice—Figure 13–1(f)–(g). Glacial ice is a solid composed of tightly interlocking ice crystals.

Types of Glaciers

Glaciers may be subdivided into two broad categories. **Mountain glaciers,** or **alpine glaciers,** are confined by mountain walls and have their source above the snowline. In moving downslope, they take the path of least resistance—commonly a preexisting stream valley. They are typically linear in form because they are confined by valley walls. However, they can fan out in an apronlike form as they spread beyond the mountain front. **Continental glaciers** are unconfined glaciers with a domelike form in which the ice flows outward from one or more central areas, with little relation to the underlying topography. Continental glaciers are far larger than mountain glaciers, and they cover all but the highest parts of the ground beneath.

Alpine or Mountain Glaciers. Mountain glaciers form in snow-and-ice-filled rock depressions above the snowline called **cirques.** When a glacier in a cirque grows sufficiently that it can extend down the mountain along preexisting valleys, it is called a **valley glacier** (Figure 13–2). One valley glacier may merge with another that is flowing down an adjacent valley to form a larger **trunk glacier** (Figure 13–2). This is similar to the way in which tributary streams join to form larger streams. However, the ice supplied by each of the valley glaciers is too viscous to mix, and the masses of ice and associated rock debris from each valley glacier flow side by side within the trunk glacier, forming the linear dark bands (rock debris) and light bands (ice) visible on Figure 13–2.

Trunk glaciers may join to form a **piedmont glacier,** a type intermediate between mountain and continental glaciers. Piedmont glaciers are apronlike forms that occupy broad lowlands at the bases of steep mountain slopes and are fed by valley and trunk glaciers (Figure 13–3). The large Malaspina and Bering Glaciers on the coast of Alaska are examples of piedmont glaciers.

Continental Glaciers. Thick continental glaciers covering all or most of the topographic features in a region are called **ice caps** if they are small and **ice** sheets if they are subcontinental or continental in size (Figure 13–4). Ice caps today are located in a number of areas, including Iceland, Baffin Island in Canada, and Spitsbergen in Norway. However, ice sheets exist in only two places—Greenland and Antarctica.

The Greenland ice sheet occupies 80% of Greenland, which is the world's largest island (Figure 13–5). The ice sheet is huge, with an area of 1,726,400 km^2, a volume of about 2,600,000 km^3, and a maximum measured thickness of about 3.3 km. The ice sheet occupies a rock basin whose floor extends to 400 m below sea level. The ice flows radially outward toward the coastline from a high central area, streaming through valleys in the coastal mountains and breaking off into the ocean as masses of ice called **icebergs**.

The Antarctic ice sheet is even larger than that of Greenland; it has an area of 12,530,000 km^2 and a maximum thickness of almost 4.3 km. The Antarctic ice sheet is bordered by more than 1,400,000 km^2 of floating ice. Floating sheets of ice that are connected to and nourished by continental glaciers, and further fed by snowfall and freezing seawater, are called **ice shelves**.

MOVEMENT OF GLACIERS

In Chapter 1, we described an experiment by which eighteenth-century scientists proved that valley glaciers move. This experiment (Focus 1–1 in Chapter 1) showed that ice in the central part of a valley glacier moves faster than ice at the sides. The explanation of *how* glacier ice moves was to come much later, when geologists began to understand more properties of glacial ice.

How Does Ice Move?

The common misconception of ice is that it is a hard, brittle substance that can move only by sliding on its

FIGURE 13–2
Aerial view of several mountain glaciers joining to form a trunk glacier in Alaska. Photo courtesy of Austin Post.

FIGURE 13–3
Aerial view of Malaspina piedmont glacier spreading out beyond the mountain range which nourishes it. Photo courtesy of Austin Post.

FIGURE 13–4
Aerial view of Greenland Ice Sheet. Photo by John Shelton.

FIGURE 13-5

Greenland and its ice sheet. (a) Topographic map, with line of cross section shown in (b). Note that the ice attains an altitude of nearly 3.3 km. Ice streams through the coastal mountains as high-velocity outlet glaciers. (b) Geologic section through Greenland showing the ice sheet and underlying topography. Arrows indicate the theoretical ice-flow direction. Vertical scale is exaggerated for clarity. After R. F. Flint, *Glacial and Quaternary Geology,* © 1971, with permission of John Wiley & Sons, Inc.

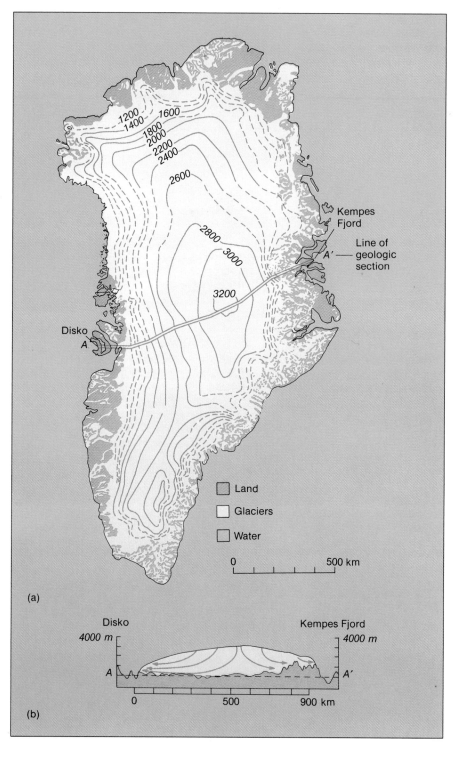

base. However, there is much more to know about ice. The mechanism by which glacial ice moves can be demonstrated with an experiment in which a hole is drilled through a moving glacier (Figure 13–6). After a few weeks, the hole is no longer a straight vertical shaft but has been deformed into a curved feature.

This experiment shows that only a small part of the glacier flow has occurred by sliding along the base, and that most of the movement has been by some type of internal flow. Indeed, the accumulation of from 30 to 60 m of snow, granular snow, and ice provides sufficient pressure to change the characteristics of the lower part of the ice. Although the upper part of the glacier remains brittle, high pressures on the lower part make it flow like a plastic material. Thus, a moving glacier is composed of an upper **brittle zone,** which fractures when stressed, and a lower **plastic zone,** which deforms plastically under stress. The plastic layer is believed to flow by gliding along atomic planes *within the individual crystals*. This process, called **intragranular gliding,** is similar to what happens when a deck of cards is sheared, as shown in Figure 13–7.

Ice also moves by **intergranular slip** and by melting and refreezing. In intergranular slip, the individual ice crystals shift positions as forces are applied. In melting and refreezing, some movement occurs where ice melts at points of greatest pressure and refreezes in areas where pressure is less.

Relative Velocity within Glaciers

The relative velocities within a glacier are similar to those that occur in streams (Chapter 11, Figure 11–5). Friction decreases the velocity of the ice wherever the glacier is in contact with the ground. In the case of valley glaciers, maximum velocities occur in the portion of the ice farthest from the valley walls.

Differences in velocity between adjacent parts of a glacier can cause the upper brittle zone to fracture. These fractures, called **crevasses,** extend downward to the top of the plastic zone. We can illustrate crevasse formation with a few examples from a valley glacier. Crevasse patterns in a valley glacier are shown in Figure 13–8. The ice nearest the valley walls is slowed by friction.

As the faster-moving central ice pulls on the slower-moving ice against the walls, **lateral crevasses** form at the sides of the glacier—Figure 13–8(b). Another type of crevasse forms when the glacier passes over a large obstruction or over an abrupt increase in the slope of the valley

FIGURE 13–6
Motion of ice within a glacier. A vertical drill hole through the glacier is deformed into a curve within a few weeks. Only a portion of glacial movement is by slippage along the underlying rock surface; most movement is internal through deformation and recrystallization. The relative speed of the glacier at different levels is indicated by the length of the arrows.

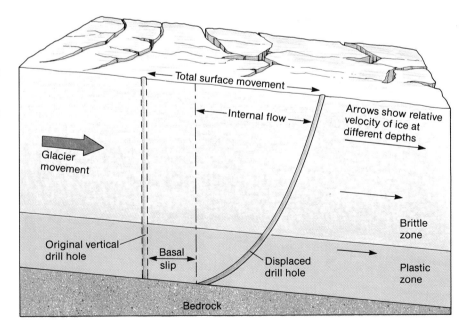

FIGURE 13–7
Glacial ice moves by (a) intragranular slippage within the ice crystals. The movement is analogous to (b) the deformation of a deck of cards by shear force.

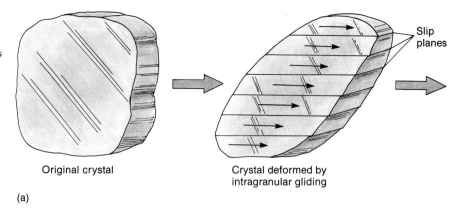

Original crystal

Crystal deformed by intragranular gliding

Slip planes

(a)

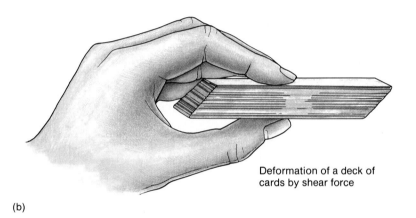

Deformation of a deck of cards by shear force

(b)

floor—Figure 13–8(a). Ice in the plastic zone adjusts by flowing; however, the upper brittle part cannot deform plastically and is ripped apart. After the ice moves over the obstruction (or slope increase) on the valley floor, the crevasses close. Meanwhile, new crevasses form in the brittle ice passing over the obstruction. This results in a belt of **medial crevasses** that cut across the glacier—Figure 13–8(b).

The last type of crevasse occurs at the front of the glacier, where the flow is unconfined. As the ice spreads out, the brittle ice is pulled apart to form a series of radiating **terminal crevasses** across the end of the glacier—Figure 13–8(b). All types of crevasses pose a potential danger to hikers and skiers because they may become covered with a thin "bridge" of freshly fallen snow, concealing a chasm below.

Velocities of Glaciers

Many glaciers move so slowly that no movement is apparent for weeks or months. Others move so rapidly that their motion is apparent within a few days. Why do some glaciers flow so much faster than others? Three major factors determining glacial velocity are *thickness* of the ice, *steepness* of the slope over which the ice is moving, and *cross-sectional area* through which the ice is passing:

☐ The thicker the ice, the greater is the pressure on its base and the faster the ice can move, both by internal flow and, to a lesser degree, by sliding along its base.

☐ Glaciers forming on steep slopes have a greater tendency to flow faster. For example, valley glaciers can move up to 0.3–0.6 m per day, and velocities of 3–6 m per day have been observed at "ice falls," where the glaciers move more rapidly over an abrupt steepening in the valley floor.

☐ Glacial velocity also increases when the ice is forced to flow through a narrow opening. Velocities up to 38 m per day have been observed where the Greenland ice sheet flows through narrow valleys in the coastal mountains of eastern Greenland.

The highest glacial velocities occur in what are called **surging glaciers,** which move several kilometers rather than meters per year. The surging is not continuous but usually is short-lived, from a few months to a few years.

Exactly why glaciers surge is not understood completely. Water can form at the base of a glacier, and a possible cause of surging glaciers is a rapid buildup of water that allows the ice to slip on its bed. To understand how this happens, we must look at how pressure affects the freezing point of water.

At atmospheric pressure, water freezes at 0°C. As pressure on the base of the glacier increases, water freezes at lower and lower temperatures. If the pressure is great enough, the ice at the base of a glacier

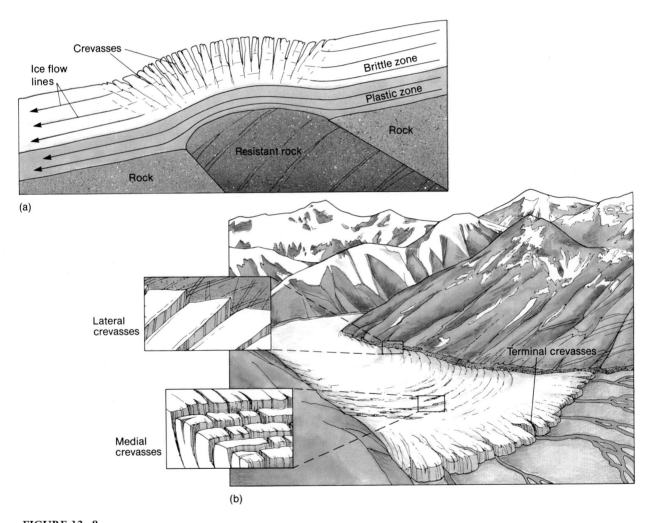

FIGURE 13–8
Valley glacier. (a) Cross-sectional view, showing brittle zone, plastic zone, and crevasse zone. Crevasse zone forms where glacier passes over a steepening of the rock floor. (b) Map view, showing formation of different types of crevasses described in text.

may melt, forming water locally at the same temperature as the surrounding ice. The overlying glacier may move over this water layer, subsequently slowing once again if the pressure drops and the basal water refreezes. This mechanism is important in the process of glacial erosion (a topic discussed later in this chapter); it may be important in the surging of glaciers.

In 1969, a hole drilled 2164 m to the base of the Antarctic ice sheet encountered water under great pressure. The water rushed up the drill pipe, freezing quickly as the pressure was released. Radar probing of the ice sheet in 1978 clearly showed that portions of it sit on a layer of basal meltwater. This water is derived partially from pressure-melting of the basal ice and partially from heat from the Earth's interior. Such accumulations of basal meltwater could provide the lubricating mechanism for a glacial surge of massive proportions. If part or all of the Antarctic ice sheet were to surge into the surrounding ocean, the large volume of water displaced would have disastrous effects on the world's coastal cities.

Economy of a Glacier

Glacier movement can be likened to a continuously moving conveyor belt. As long as there is net accumulation of snow above the snowline, glacial ice is produced there and progresses toward the edge, or **terminus** of the glacier, where it melts. Even though the glacier is melting back in the warmer temperatures at the terminus, new ice is still moving into that area. In winter, snow may cover the glacier well below the snowline, and the front of the glacier may advance. In summer, the snow cover melts, a loss of ice below the snowline occurs, and the glacier may retreat. Additional ice losses occur when the glacier terminus is adjacent to the ocean or other body of water, and the glacier breaks off in pieces that float away as icebergs.

The balance between ice accumulation above the snowline and ice loss below the snowline is expressed as the **glacial economy:**

$$\begin{array}{ccc} \text{Economy of} & = & \text{Accumulation} & - & \text{Melting of} \\ \text{a glacier} & & \text{of snow above} & & \text{ice and} \\ & & \text{snowline} & & \text{snow below} \\ & & & & \text{snowline} \end{array}$$

If the glacier economy is *positive,* the accumulation is greater than the loss, and the terminus of the glacier advances. If the economy of the glacier is *negative,* the loss exceeds the supply, and the terminus of the glacier may retreat. In a negative economy, the ice below the snowline thins and the terminus melts back, *even though ice is continually flowing outward to the edges of the glacier.* If the accumulation and melting are balanced, the position of the ice front remains stationary.

GLACIAL EROSION

Glaciers are extremely effective erosional agents because of the high viscosity of ice, the great pressure

FIGURE 13–9
Glacially polished and striated dolomite outcrop in Wisconsin. Photo by Nicholas K. Coch.

FIGURE 13–10
Mount Henry cirque, Glacier National Park, Glacier County, Montana. Photo courtesy of P. Carrara/U.S. Geological Survey.

that can develop at the base of a glacier, and the rock particles frozen into the ice.

Before discussing erosion, let us introduce two orientation terms used in reference to glaciers: **up-ice** and **down-ice**. Up-ice is the direction from which the ice is flowing and is analogous to the term *upstream*. Similarly, down-ice is the direction in which the glacier is headed, and corresponds to *downstream*.

Mechanisms of Glacial Erosion

The cold climate at the edges of a glacier increases the efficiency of glacial erosion. Extensive frost wedging (see Chapter 9) in these cold climates loosens rock particles at the surface, making them more erodible.

The freezing and melting mechanism described in connection with surging glaciers also is important in glacial erosion. Glaciers exert great pressure on rocks extending upward into the ice. These high pressures may result in melting of the ice on the up-ice side of the rocks. The meltwater can travel short distances around the other side of the rocks, where reduced pressure causes it to refreeze and attach the rocks to the glacier. As the glacier moves past, it carries away part of the rocks, a process called **plucking**. Plucking

is even more effective when the surface rocks have been previously fractured by frost wedging.

Considerable glacial erosion is performed by particles carried at the edges of the glacier. These embedded particles act like sandpaper, wearing away rock by **abrasion**. Small particles carried within the ice cut elongate scratches into the rock surface called **striations**. Abrasion of the underlying rock by glacially transported particles of all sizes gives the rock a polished and striated surface that is characteristic of glacial erosion (Figure 13–9).

Glacial erosion triggers other processes that further accelerate erosion. For example, erosion of the sides of a valley by a glacier may undercut the slope and result in mass movements such as rockfalls or rockslides. Material falling onto the glacier surface as a result of mass movements may be transported on the surface or fall into crevasses to be carried within the ice.

Erosional Features

Glacial erosion modifies the previous topography greatly and can result in dramatic landscapes. Some erosional features, such as sculptured rock outcrops and polished and striated surfaces, are formed by both valley and continental glaciers. However, the most spectacular erosional features are made by val-

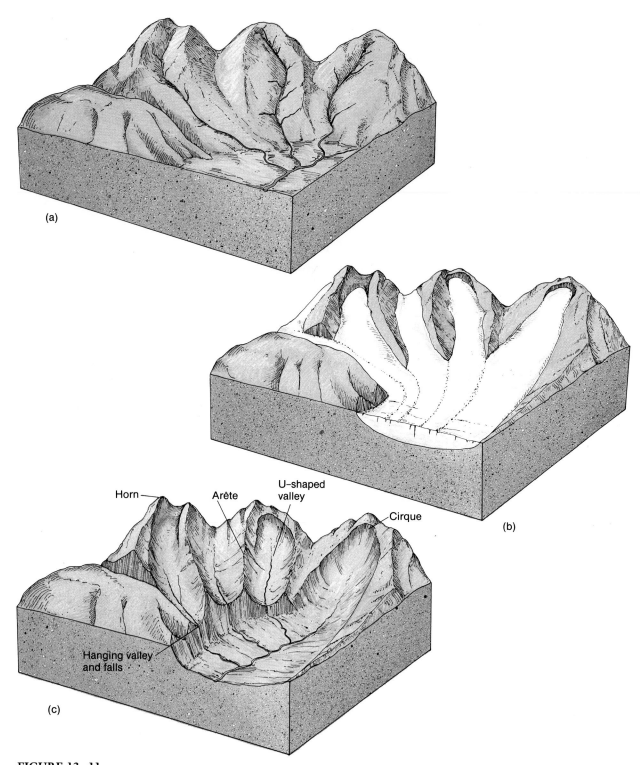

FIGURE 13–11
Evolution of alpine glacial features. (a) Preglacial mountain topography. (b) Mountain area during glacial episode. (c) After several glacial episodes, enlargement of cirque basins forms a variety of alpine glacial features.

ley glaciers in mountain ranges such as the Rocky Mountains and the Alps.

Cirques and Associated Features. Snow accumulating in depressions above the snowline is subject to daily and seasonal warming at the top and outer edges of the snow. As the snow melts, it produces water, which penetrates fractures in the rock below. Subsequent freezing of this water causes it to expand in volume about 9%; over time, this fractures the rock. Meltwater freezes the fractured rock to the glacier, and as the glacier moves downslope it removes the material, thus deepening the glacial basin. By repetition of freezing and thawing cycles, the underlying rock develops into an amphitheater-shaped basin called a **cirque** (Figure 13–10) The cirque is bounded upslope by a steep cliff called a **headwall**. If snow accumulation is sufficient, a glacier may form within the cirque basin and extend downslope along a preexisting valley as a valley glacier.

The successive enlargement of cirques over time produces dramatic landscapes in glaciated mountains (Figure 13–11). The crevasse between the cirque and the headwall is called a **bergschrund;** it allows any meltwater produced during warm spells to penetrate the rocks in the headwall area of the cirque. This water freezes at night or during seasonal cold intervals, bonding the glacier in the cirque to the headwall. Subsequent movement of the glacier plucks out many rock fragments, causing the headwall to recede and enlarging the cirque.

Enlargement of cirques produces the variety of erosional features found in glaciated mountains (Figure 13–11). Two cirques enlarging toward each other produce a narrow ridge between them that is called an **arête**. Two cirques eroding toward each other from opposite sides of a mountain cut a notch into the divide between their headwalls, forming a gap called a **col**. Cirques eroding headward on all sides of a mountain produce an isolated peak called a **horn**. The best known example of this latter feature is the Matterhorn of Switzerland.

Valley Glaciation. Glaciers moving out of cirques and down preexisting valleys erode the valleys, tending to straighten them and modify their cross sections. Stream valleys typically have a V-shaped cross section that results from stream erosion at the base of the valley and mass movements of the material on the slope above. A glacier moving down a stream valley modifies this cross section considerably because the viscous ice molds itself to the valley and directly erodes the valley wall far above the stream channel (Figure 13–12). The plucking and abrasion by the

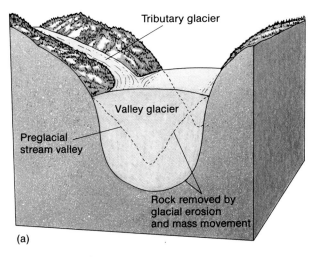

(a)

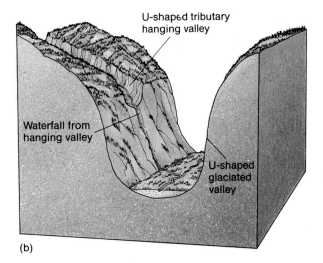

(b)

FIGURE 13–12
Cross-sectional views showing the evolution of a V-shaped stream valley into a U-shaped glacial valley. (a) Valley and tributary glacier at maximum glaciation. (b) U-shaped main and tributary valleys resulting from glaciation. Differential glacial erosion results in the tributary valley remaining at a higher level as a hanging valley.

FIGURE 13–13
U-shaped hanging valley in Yosemite National Park, California. The waterfall (Bridalveil Fall)
is formed when Bridalveil Creek plunges 189 m from the hanging valley onto the Yosemite
Valley floor. Photo courtesy of Julia A. Thomas/U.S. Geological Survey.

valley glacier widens the valley and creates the
U-shaped valley that is characteristic of valley glaciation.

Where a tributary glacier flows into a main valley
glacier, the upper surfaces of the two meet at the
same elevation. However, there may be a considerable difference between the altitude of their *floors,* as
shown in Figure 13–12. This difference results because one glacier may be thicker or is flowing over a
rock that is more easily eroded; thus it erodes its
valley to a greater depth than the other. After both
glaciers have melted back, the tributary valley is left
elevated as a **hanging valley** above the level of the
main valley. A postglacial stream flowing through
the tributary valley plunges as a waterfall into the
main valley below. Yosemite National Park in California (Figure 13–13) has several outstanding ex-

amples of U-shaped valleys and hanging valleys.
Deeply eroded glacial valleys that subsequently become inundated by seawater due to a postglacial rise
in sea level are called **fjords;** they are common in
Scandinavia.

Sculptured Rock. Glaciers overriding exposures of
rock abrade, striate, and polish the up-ice side and
pluck out an irregular surface on the down-ice side.
The result is an asymmetric rock exposure called
roche moutonnée (from the French, "sheep rock")
with the gentle slope facing the direction from which
the ice came (Figure 13–14). Most of these asymmetric rock exposures are only a few meters in
height, width, and breadth; however, some are as big
as hills.

FIGURE 13–14
Glacially polished and eroded diabase outcrop, Greenfield, Massachusetts. The ice overrode the outcrop from left to right, smoothing the up-ice side and plucking the down-ice side of the outcrop (right). The result is an asymmetric rock exposure called roche moutonnée. Photo by Nicholas K. Coch.

GLACIAL DEPOSITION

Glacier ice is a much more powerful erosive and transporting force than water or air because its greater viscosity enables it to carry much larger particles than other common forms of sedimentary transport. In addition, the high viscosity of glacial ice enables it to carry sediment on its surface as well as within and below the ice. This is in marked contrast to the way that sediment particles are carried in air and water.

Sediment carried by a glacier can be deposited in two distinct ways. One is deposition directly by the ice as it moves over the land surface. The other is deposition as the debris-laden ice melts, forming streams that transport rock debris across the ground in front of the glacier. The characteristics of deposits resulting from each mode are quite different.

Drift is the general term for all types of glacial deposits. **Unstratified drift** and **stratified drift** have different characteristics and make up different types of morphological features. We will now consider these two types of deposits in more detail, along with the landforms composed of each type of drift.

Unstratified Drift

Unstratified drift deposited directly by a glacier is called **till**. Till is poorly sorted, generally lacks strat-

ification, and contains angular particles (Figure 13–15). The viscosity of the ice is the reason for this; ice is extremely viscous, so glaciers support and carry everything from boulders to clay. As a conse-

FIGURE 13–15
Glacial till at Montauk Point, New York. The till is poorly sorted and unstratified. Photo by Nicholas K. Coch.

quence, they are very poor sorters of sediments. In addition, the particles make little contact with each other, so they are poorly rounded. Till is closest in character to the deposits made by the viscous mudflows discussed in Chapter 10.

The sizes of the particles in till and their mineralogic composition reflect the type of material over which the glacier passed. Till commonly contains rock fragments, called **erratics,** that have been carried far from their source and deposited on rock of different composition. Among the more interesting erratics are the high-quality diamonds found in Wisconsin, Illinois, Indiana, Ohio, and New York. These diamonds are believed to have a source in east-central Canada, although the specific locations are unknown. The great viscosity of ice allows it to carry erratic rocks of very large size. For example, the Madison Boulder, a block of granite near Conway, New Hampshire, measures 11 × 12 × 27 m and weighs 4700 metric tons. Some tills contain erratics

from a known and unique source area. These erratics can be used to determine the direction of ice movement by drawing a straight line on a map between the source area and the location where the fragments were deposited.

Moraines. The general term **moraine** is used to describe a wide variety of landforms that are deposited by ice at the edges of glaciers (sides, bottoms, and ends). Most moraines are made of till deposited by ice flow, but some may be constructed partially of stratified drift deposited by meltwaters at the edges of glaciers. This type of moraine will be discussed later in this chapter.

Valley glaciers deposit several different types of moraines when the ice melts (Figure 13–16). Material carried at the sides of the glacier can be deposited as **lateral moraines** along the valley walls. Lateral moraines from two adjoining glaciers form a **medial moraine**. The existence and character of me-

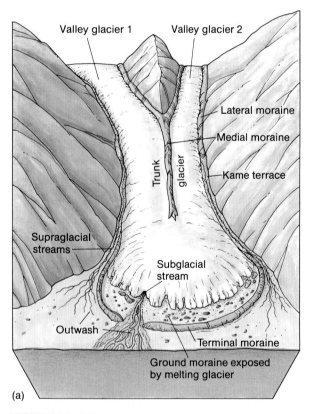

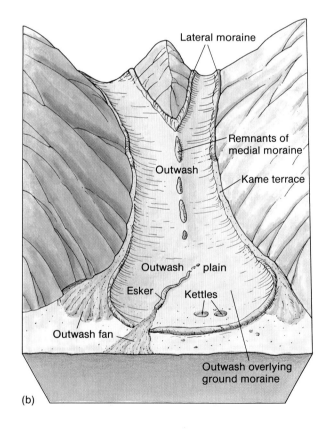

FIGURE 13–16

Trunk glacier. (a) Features on a melting trunk glacier system. (b) Resulting landforms and deposits when deglaciation is complete.

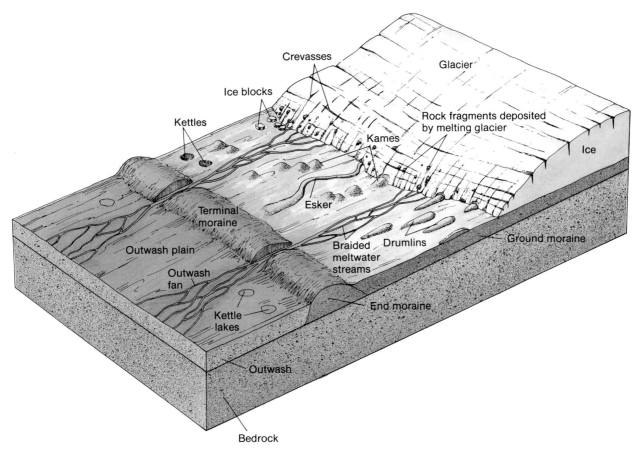

FIGURE 13–17
Landforms and sediment deposits resulting from continental glaciation.

dial moraines are proof that merging glaciers move side by side as separate ice streams for considerable distances without mixing.

In most cases, moraines are composed of till that is plastered onto the land surface by actively flowing ice. Three types of moraines occur in association with both alpine and continental glaciers and are distinguished by their shape and location with respect to the terminus of the glacier. These moraines, along with other features to be discussed later in this chapter, are shown in Figure 13–17.

Smooth-to-gently undulating sheets of till are called a **ground moraine**. Ridgelike deposits formed along the terminus of a glacier are called **end moraines**. The end moraine that forms at the farthest point to which the glacier advances is called a **terminal moraine**. The terminus of a glacier may melt

back, reach equilibrium, and deposit another moraine; an end moraine deposited by a glacier that melts back from a terminal moraine is called a **recessional moraine**. A glacier has only *one* terminal moraine but may have *any number* of recessional moraines.

Drumlins. The plastic ice at the base of a continental glacier can mold the underlying ground moraine into streamlined forms in the same way that flowing water forms sand into ripples or wind forms sand into dunes. Streamlined hills, molded by glaciers from ground moraine, are called **drumlins** (Figures 13–17 and 13–18). Some are composed entirely of till, whereas others are made up of till plastered on a bedrock core. In an aerial view, they have a "teardrop" shape and are elongated parallel to the direc-

FIGURE 13–18
Aerial view of drumlin in Sodus, New York. The ice that formed this drumlin moved from the high, wide end toward the narrow, streamlined tip. Photo courtesy of Ward's Natural Science Establishment, Inc.

tion of glacier movement. Drumlins have a steeper slope facing the direction from which the ice came. (Note that this is the reverse of the asymmetric sculptured rock forms—roche moutonnée—described earlier in this chapter.) Drumlins are generally 1 to 2 km long, 400 to 600 m wide, and 5 to 50 m high. Large numbers of these landforms occur in drumlin fields in Massachusetts, Wisconsin, and New York State.

Stratified Drift

Stratified drift is deposited by water running from the melting glacier. Some melting occurs daily or seasonally on many glaciers with positive economies, and this accounts for the formation of the small patches of stratified drift found in some deposits of till. However, most stratified drift deposits are laid down when the glacial economy is either at equilib-

FIGURE 13–19
Glacial outwash at town of Montauk, New York. The sequence is well stratified and sorting within any given layer is good. Differences in the grain sizes between layers result from fluctuations in meltwater discharge. Photo by Nicholas K. Coch.

rium (the terminus of the glacier is neither advancing nor retreating), or especially when the glacial economy is negative and the terminus of the glacier is melting back. The front of the ice thins as it melts back, decreasing the pressure on the lower portion of the ice. The reduced pressure eliminates the plastic zone and the ice becomes brittle and fractures into many crevasses. The crevasses increase the surface area of the ice exposed to the air, and this accelerates melting. The meltwater washes rock particles from the ice and carries them as its stream load.

Meltwater is a very effective sorting agent because it quickly drops particles that are too large for the water's kinetic energy to move. Gravel-sized particles usually are deposited close to the edge of the glacier, whereas the sand-sized particles can be deposited over a considerable distance beyond the terminus. The finer silt and clay particles are transported in suspension and may be deposited in lakes at the glacial margin or transported for great distances beyond the glacial margin. The result of meltwater activity is a sheet of stratified drift called **outwash** that extends out from the edges of the glacier (Figure 13–19).

The following landforms are composed of stratified drift.

Kames. Moundlike hills of stratified drift are called **kames**. Some kames are composed of sediment deposited by meltwater in crevasses and other openings in or on the surface of stagnating glaciers (Figures 13–17, 13–20). When the ice melts away, the stratified drift is left in the form of isolated or semiiso-

FIGURE 13–20
Formation of kames. (a) Stratified drift is washed into intersecting fractures on a glacier and accumulates there. (b) Melting of ice exposes rounded hill of outwash called a kame.

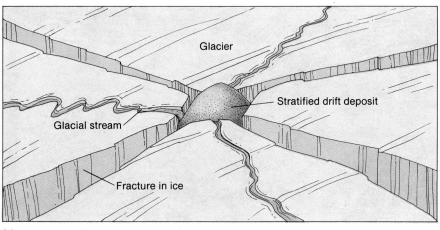

(a)

(b)

FIGURE 13–21
Formation of kame terraces at the sides of a valley glacier. (a) Stratified drift is washed into the spaces between the sides of a melting glacier and the valley walls. (b) Stratified drift is preserved as a kame terrace after deglaciation.

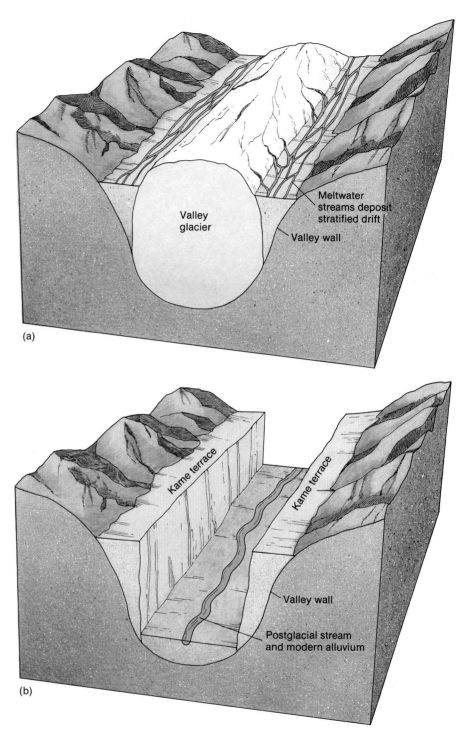

Meltwater streams deposit stratified drift

Valley glacier

Valley wall

(a)

Kame terrace

Kame terrace

Valley wall

Postglacial stream and modern alluvium

(b)

lated mounds. Another type of kame consists of mounds of stratified drift built along the edges of the glacier. When the glacier melts back, the deposits are left as a line of kames marking the former edge of the glacier. The terminal moraine that crosses Long Island, New York was formed in this way.

Kame Terrace. As a valley glacier melts back, the ice thins and melts away from the valley walls. Meltwater streams deposit stratified drift between the stagnating glacier and the valley walls. When the ice melts back through the valley, the deposit is exposed as a flat-topped **kame terrace** (Figure 13–21).

Eskers. Subglacial stream activity forms eskers. An **esker** is a long, narrow, and often sinuous ridge deposited by glacial streams (Figure 13–22). Some eskers form from sediment deposited in open channels on the glacier surface. However, most eskers are believed to be deposited by meltwater streams flowing through tunnels at the base of the stagnating glacier (Figure 13–17). Sediment deposits are laid down on the tunnel floor and accumulate upward, whereas the ice roof or tunnel walls are widened by melting. When the covering ice melts back, it exposes the tunnel deposit as an esker.

Outwash Plains. The outwash from a valley glacier is confined between the valley walls, forming a long, narrow outwash deposit. In contrast, the outwash from a continental glacier or ice cap can extend for great distances along, and tens of kilometers away from, the glacial terminus. Such extensive plains underlain by outwash are called **outwash plains** (Figure 13–17).

Large chunks of ice can separate from the glacier (Figure 13–17) and become buried by outwash. If the climate warms subsequently, the buried masses of ice melt and the surface subsides to form a depression referred to as a **kettle hole**. If these basins later fill with water they are called **kettle lakes**. Kettles and kettle lakes are common features on many outwash plains (Figure 13–23).

Proglacial Lakes. Lakes fed by meltwater that accumulates in low areas on outwash plains or in stream valleys dammed by glacial deposits are called **progla-**

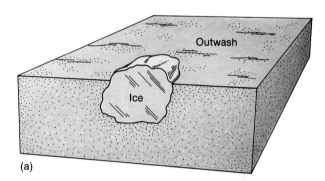

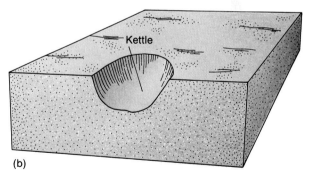

FIGURE 13–23
Formation of a kettle and kettle lake. (a) An isolated mass of ice is left behind a retreating glacier and is covered with outwash. (b) The ice subsequently melts in the postglacial climate, causing the surface above to collapse, forming a depression called a kettle. If the kettle subsequently fills with water, a kettle lake is formed.

FIGURE 13–22
Aerial view of esker emerging from a tunnel (lower center) at the end of La Perouse Glacier, Glacier Bay, Alaska. The dark color of the stagnant ice results from a covering of rock and sediment particles that melted out of the ice. Photo by Terraphotographics/BPS.

cial lakes. Proglacial lakes range in size from small kettle lakes to those that formerly occupied portions of several states. For example, proglacial Lake Hitchcock filled the Connecticut River valley in New England during the last retreat of glacial ice, about 13,000 years ago. Lake Hitchcock was almost 480 km long and up to 16 km wide, had a maximum depth of about 61 m, and extended from Lyme, New Hampshire southward to Rocky Hill, Connecticut.

The lake formed when meltwater backed up between the retreating ice front to the north and a dam of stratified drift at Rocky Hill in the south. Sediment supplied by streams formed deltas at the edges of the lake and deposited rhythmically laminated sediments called *varves* far out into the lake waters (Chapter 6). Eventually the dam was overtopped with meltwater, breached, and the lake drained.

DEVELOPMENT OF THE GLACIAL THEORY

The earliest descriptions of glacial deposits attributed them to the "great flood" recorded in the Bible because it was believed that this was the only way that such huge boulders could have been deposited. A theory for widespread glaciation had yet to be envisioned, for such a theory required a scientist to examine both drift deposits and similar materials being deposited by glaciers today. (This is the uniformitarian approach—"the present is the key to the past"—that we described first in Chapter 1.)

In 1829, a Swiss engineer named J. Venetz first postulated that glaciers once had been more extensive than they are today. He believed that the erratic blocks found in northern Europe and Switzerland had been deposited originally by glaciers. These observations inspired Jean de Charpentier to conduct a field study of active glaciers, during which he became convinced that past glacial activity could account for the features previously attributed to the Biblical flood.

In 1837, de Charpentier showed the results of his study to the respected Swiss naturalist Louis Agassiz. That same year, Agassiz presented detailed evidence for glaciation to a meeting of the Swiss Academy of Natural Sciences. Agassiz's stature as a scientist and his persuasive arguments that glaciers formerly covered large areas of northern Europe and the Alps did much to advance the theory of glaciation. His 1840 book, *Études sur les Glaciers* ("Studies of Glaciers"), contained excellent illustrations that enabled other geologists to understand the deposits of active glaciers and thereby identify the deposits of former glaciers in their own areas.

The first mention of the glacial theory in America was by Edward Hitchcock in a study of the geology of Massachusetts in 1841. Within five years, Hitchcock became the first American geologist to endorse the theory. Subsequent work in many other areas of the world refined and expanded the glacial theory, and it is accepted universally today.

ICE AGES

When we think of ice ages, we naturally tend to think of the most recent ice age—the Pleistocene Epoch. However, there is clear evidence that there were several other ice ages, some as far back as the Precambrian Era.

Pre-Pleistocene Glaciations

The evidence for these ancient glaciations is minimal because of the incomplete preservation of their sedimentary records. The evidence that does exist comes from rock equivalents of modern types of glacial sediments and from polished and striated rock surfaces overlain by rock of a known age. The most characteristic glacial rock is one composed of poorly sorted and unstratified drift and is called **tillite**. We must be careful in assigning a glacial origin to all such rocks because they may have been deposited by other processes, notably mudflows (see Chapter 10). However, wherever rocks similar to glacial till are widespread *and* overlie a polished and striated surface, we can reasonably infer that they are tillites. Some tillites grade laterally into rhythmically stratified shales and siltstones which may represent the rock equivalents of varve deposits in proglacial lakes.

Evidence for the oldest glaciation comes from rocks of Precambrian age. Rocks exposed in Ontario, Canada, record an episode of continental glaciation that occurred about 2.2 billion years ago. Tillites of a younger Precambrian age, about 700 million years, have been found on all continents except Antarctica.

On all the continents of the Southern Hemisphere, certain late Paleozoic to mid-Mesozoic rocks contain distinct tillite sequences. The oldest tillites (Devonian) are in South America, whereas those on the other continents range in age from Pennsylvanian to Permian.

Glaciation on a continental scale has clearly occurred a number of times during Earth history. Later in this chapter we shall examine its cause.

Pleistocene Glaciations

As studies of glacial sediments continued it became apparent that there were several glacial episodes separated by several times when the climate was as warm as today. The warm periods within the Pleistocene Epoch are referred to as **interglacials**. The evidence for glacial-interglacial fluctuations in climate came from the principle of superposition that we described in Chapter 8. Layers of glacial sediments (till) were interbedded with interglacial layers (peat, weathered sediments, soils, and such) a number of times. Field studies suggested at least four major glacial episodes, separated by at least three major interglacial periods within the Pleistocene (Table 13–1).

Our present picture of the Pleistocene Epoch is a time of great and frequent climatic change. At any one place, climates were colder during glacial periods and warmer during interglacial periods. The last and best-documented glacial advance, the Wisconsinan,

TABLE 13–1
Major glaciation subdivisions of the Pleistocene.[1]

Glacial Stages	Interglacial Stages
Wisconsinan (180,000)	Recent (15,000)
Illinoian (550,000)	Sangamon (400,000)
Kansan (1,400,000)	Yarmouth (900,000)
Nebraskan (2,000,000)	Aftonian (1,700,000)

[1]Based on the North American continental sedimentary record. The climatic record deduced from fossil animals and plants in ocean sediments indicates that Pleistocene climatic changes were more complex than indicated by the rather incomplete continental sedimentary record. Approximate age in years before the present is given in parentheses for the beginning of each stage.

illustrates the growth and retreat of continental ice sheets through the coalescing of glaciers from individual areas of accumulation called **ice centers**. Large-scale Wisconsinan continental glaciation started when average annual temperatures dropped about 5°C worldwide, and the climate grew wetter. This cooling resulted in lowering of the snowline, allowing glaciers to expand and occupy areas at lower altitudes, and at lower latitudes year-round.

In North America, massive snow accumulations formed ice centers west of the Hudson Bay region and in Labrador. These ice centers were nourished by moist ocean air from the Atlantic Ocean and the Gulf of Mexico. Cold air above the growing glaciers condensed the moisture, causing it to precipitate as snow. Ice flow from these centers was radial in all directions. Coalescing of the ice from the Hudson Bay and Labrador ice centers formed the massive **Laurentide ice sheet,** which covered most of Canada and eventually extended deep into the northern United States (Figure 13–24). South of the growing Laurentide ice sheet, valley glaciers and local ice caps began forming in highlands such as those in eastern Canada and in the White Mountains of New Hampshire. Ice from these local centers probably was incorporated into the Laurentide ice sheet as it grew southward.

Meanwhile, in the Canadian Rocky Mountains to the west, valley glaciers grew and coalesced into piedmont glaciers and ice caps. These ice masses eventually formed the **Cordilleran ice sheet,** which spread westward toward the Pacific Ocean and eastward beyond the Rocky Mountains, merging in places with the Laurentide ice sheet to form a nearly continuous mass of ice extending across Canada and the northern United States.

The Laurentide and Cordilleran ice sheets are believed to have merged only during the peak of glaciation. At other times there was an ice-free corridor between them. Anthropologists believe that this corridor probably was the pathway south for Asiatic peoples who populated North America during the late Pleistocene Epoch. The pattern of glaciation in Europe was similar; local ice caps occurred in the British Isles and on the Alps. The ice sheet growing over Scandinavia merged eastward with an ice sheet developing in the USSR to from a massive ice sheet that eventually covered northern Europe and part of northern Asia (Figure 13–24).

FIGURE 13–24

Maximum extent of Pleistocene glaciers in the Northern Hemisphere during the last glacial advance. Arrows denote local flow directions. Dashed line marks the approximate boundary between the Laurentide ice sheet on the east and the Cordilleran ice sheet on the west. Polygonal patterns indicate areas of floating ice. Sea level shown is 100 m lower than today. After R. F. Flint and B. Skinner, *Physical Geology,* 2d ed., © 1971, used with permission of John Wiley & Sons, Inc.

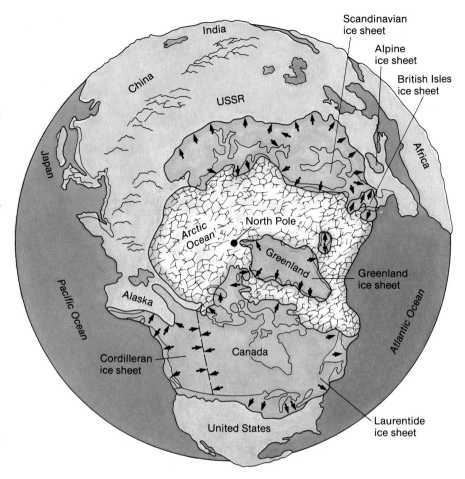

These continental ice sheets advanced and retreated in response to climatic fluctuations within the latest glacial episode, the Wisconsinan. Each glacial advance was accompanied by a drop in worldwide sea level as more and more moisture was used to form continental glaciers. About 15,000 to 18,000 years ago (depending on location), the ice sheets began to thin and melt back from their terminal positions. By 12,000 years ago, the Laurentide ice sheet had melted back from the United States into southern Canada. About 7000 years ago, the Laurentide ice sheet ceased to exist and was represented by remnant ice centers; sea level had risen to within a few meters of its present position and the climate was almost as warm as it is today.

Unglaciated Areas During the Pleistocene

Glaciation on a continental scale is clearly the most spectacular aspect of the Pleistocene Epoch. How-ever, only about 30% of the Earth's land surface was glaciated during the Pleistocene, and we should consider what was going on in the *unglaciated* areas at and beyond the limits of Pleistocene glaciation.

Periglacial Climates. Cold belts called **periglacial climates,** with characteristic structures and landforms, extended many kilometers in front of continental ice sheets. We mentioned earlier that frost wedging loosened particles from rock outcrops in front of the glacier so they could be picked up more readily as they were overridden by the ice. Frost wedging and other periglacial processes formed topographic features far south of the maximum extent of Pleistocene ice sheets. One type of periglacial feature is **block fields**—flat or gently sloping areas covered with a continuous veneer of large, angular blocks, up to 10 m in length. Block fields usually occur on high, flat-topped mountains or plateaus above the timberline in both temperate and cold lat-

THE GLACIATION OF HAWAII

Very few think of glaciers when they visit the island of Hawaii and walk along the palm-fringed subtropical beaches or snorkel across the offshore reefs. However, in the winter, many enjoy skiing on the snow cover of Mauna Kea, highest (4206 m) of the five massive shield volcanoes that form the island of Hawaii. In the Pleistocene Epoch, temperatures on Hawaii were sufficiently colder that moisture-laden winds from the east could nourish glaciers on Mauna Kea.

Geologist Stephen C. Porter, of the University of Washington, has now determined the number, magnitude, and chronology of glaciations on Mauna Kea (see Further Readings). Some of the volcanic units have characteristics indicating that they erupted *under* a thick ice cap. During these subglacial eruptions, the quick release of ponded water formed mudflows that transported volcanic-glacial debris down the mountain. In other cases, layers of drift are separated by lava flows. The drift of the latest glaciation forms steep, sharp-crested end moraines near the summit (Figure 1).

Porter determined that four glacial drifts were interstratified with volcanic rocks. The oldest glaciation occurred between 278,000 and 122,400 years ago; the younger ones are dated more precisely. The last ice cap expansion on Mauna Kea ended 29,500 years ago and the ice cap had melted from the summit before 9080 years ago.

FIGURE 1
Glacial end moraines near the summit of Mauna Kea, Hawaii. Photo by Nicholas K. Coch.

itudes. They form when intensive frost wedging breaks up the underlying bedrock, forming a surficial layer of rock debris. Relict block fields have been reported as far south as North Carolina, well beyond the limits of Pleistocene glaciation.

Frost-wedged rocks can accumulate with snow and ice to form features called **rock glaciers**. This accumulation of rock rubble flows downslope with a form similar to that of a glacier. Melting of the internal ice and snow results in cessation of flow and a deposition of rock material. Active rock glaciers form today near the snowline at cliffs where frost wedging is most active.

Periglacial climates also affected the regolith. Long-continued freezing of the ground results in permafrost (see Chapter 10). The upper part of the permafrost thaws in the summer and then refreezes in the winter, resulting in the formation of surface features such as bands or stripes of rock debris and polygonally fractured soil (Figure 13–25). These features are known collectively as **patterned ground** and can be seen forming today in high latitudes.

As surface sediments are frozen, they contract and may develop vertical fractures a meter or more in depth and initially only a few millimeters wide. During the next warm period, meltwater may reach into the cracks and refreeze when it hits the colder ground below. As the ground refreezes, the crack forms again. In the next thawing period, another thin layer of ice is added to the progressively widening wedge of ice. After a great number of freeze-thaw cycles, the wedge-shaped ice mass in the frozen soil may become a meter or more wide at the top. Melting of the ice by postglacial thawing enables the cavity to be filled with washed-in sediment, forming an **ice-wedge cast** (Figure 13–26). These structures are useful to geologists mapping the extent of periglacial belts around former ice-sheet margins.

Loess. A deposit of predominantly silt-sized particles eroded and deposited by wind is called **loess**. Winds blowing across outwash plains preferentially removed the finer particles, which were deposited as sheets of loess extending far from the edge of the glacier. Loess sheets may reach thicknesses of 35 m near the sources of glacial sediment. However, the individual layers of loess become thinner and the component particles become finer in a direction away from the glacial source. The fossil shells of former air-breathing snails found in loess deposits are of a type suggesting cooler climates than those of today. Carbon-14 dates (see Chapter 8) obtained from these snail shells can be used to determine the age of the loess deposits. Not all loess is of glacial origin, and extensive loess deposits also may occur downwind of major deserts, such as the Gobi Desert in China (see Chapter 14).

FIGURE 13–25
Aerial view of patterned ground in Barrow District, Alaska. The polygons are 7 to 15 m in diameter. Photo courtesy of T. L. Péwé, U.S. Geological Survey.

FIGURE 13–26
Ice-wedge cast. Photo by Nicholas K. Coch.

Pluvial Lakes. The presently semiarid basins of the western and southwestern United States had much cooler and wetter climates during Pleistocene glacial advances. Such a cooler and wetter climate is called a **pluvial climate**. The pluvial climates occurred when warmer air moving into the region from the south encountered the cold air associated with the advancing glacier. The moisture in the air condensed as rain, filling up the formerly dry basins and forming extensive lakes. The increased cloud cover in pluvial climates reduced evaporation greatly, enabling the lakes to grow in size during glacial advances.

Lakes formed during the heavier rainfall and reduced evaporation of glacial advances are called **pluvial lakes**. One of the greatest pluvial lakes was Lake Bonneville, which covered over 50,000 km² of Utah, Nevada, and Idaho and reached a maximum depth over 330 m. The Great Salt Lake of Utah is a remnant of Lake Bonneville; the former shorelines of the old lake are visible on the sides of the Wasatch Mountains outside Salt Lake City (see Figure 13–27).

Pluvial lakes are quite different in origin from the proglacial lakes discussed earlier in this chapter because meltwater played little part in their origin. Further, pluvial lakes grew when glaciers were advancing. In contrast, proglacial lakes grow during warmer periods as glaciers melt and release water.

Changes in Sea Level. Glaciers are part of the hydrologic cycle, so any fluctuation in their size must affect water distribution on the Earth's surface. During a time of active glaciation, moisture evaporated from the oceans falls on the continent as snow, nourishing the expanding glaciers. As more of the Earth's water is stored on the continents as glacial ice, there is a reduction in runoff into the ocean, and a progressive lowering of sea level occurs. It is estimated that Pleistocene sea levels may have been as much as 100 m lower than today's. At a glacial maximum, the Atlantic shoreline lay about 150 km east of the present site of New York City, with the intervening coastal plain forested by spruce and pine, and its population including mammoths, mastodons, and other extinct mammals.

The drop in sea level during glacial advances changed the base level of streams so that streams accelerated their downward erosion, excavating their valleys. During interglacials, sea levels were higher than present because the large volume of water from melting glaciers was added to the oceans, raising levels worldwide. As sea level rose, so did the base level, and streams began to deposit sediment within their valleys. Evidence of sea levels higher than the present occurs on the Atlantic and Gulf coastal plains in a series of coastal features extending up to an altitude of over 40 m above present sea level.

FIGURE 13–27
Lake Bonneville shorelines in the Wasatch Mountains, Utah. Photo by Stephen Trimble.

CAUSES OF ICE AGES

Determining the origin of ice ages is challenging because not only must we have a mechanism for cooling the Earth's surface so that continental glaciation may begin, but we also must account for the numerous cyclical climatic changes indicated by the Pleistocene sediment and fossil record.

Cooling of the Earth

There is evidence of an overall decrease in temperature at middle latitudes between the Late Cretaceous Period and the onset of Pleistocene continental glaciation about 1 million years ago. Evidence for this change comes from studies of fossil land plants and land organisms. Temperature began to decrease rapidly by the start of the Oligocene Epoch (3.5 million years ago), falling 8°C between then and the onset of continental glaciation. A number of hypotheses have

been proposed to account for this cooling trend; we will evaluate each as to whether it can account for the scale and areal range of cooling indicated by field evidence.

Mountain Building. By the Late Cenozoic, mountain building had significantly increased the Earth's relief. The rising landmasses could lower temperatures significantly in continental interiors by acting as barriers to the moderating ocean winds and possibly by interfering with the poleward transfer of heat from the Equator. The cooler elevated highlands would be new sites for valley glaciation. However, it is believed that such mountain building could account for only about 3°C of the observed temperature drop. In addition, most continental ice sheets accumulated on lowland areas far from mountains.

Volcanic Dust. It has been proposed that increasing amounts of volcanic dust during the Cenozoic Era partially blocked the influx of solar energy, thus cooling the Earth's surface. In addition, the particles of dust could have served as nuclei for the condensation of atmospheric water vapor. However, when we look at the Earth's history over much longer stretches of time, we find that sporadic periods of volcanism rarely coincided in time with periods of major glaciations.

Decrease in Solar Radiation. Small variations in solar energy have been noted during historic time. For example, the period from 1450 to 1850 was one both of minimal sunspot activity and apparently of decreased total solar radiation output. Temperatures fell about 1°C in eastern Europe at its coldest (1700), and the snowline elevation dropped significantly throughout the world. This time period is known as the "Little Ice Age."

Although there is other evidence of small-scale reduction in solar energy in historic time, there is no evidence that there were any significant variations in the past or that variations of so small a magnitude would be sufficient in themselves to trigger glacial episodes.

Latitudinal Changes in Continents. Whenever continental masses were located in polar areas with access to oceanic moisture, the potential for extensive continental glaciation increased. At the present time there is ice at both the North and South Poles. However, glaciation is not possible at the North Pole because it is only a mass of floating ice. The South Pole, in contrast, is covered by a continental ice sheet built up on land. The plate tectonics model (see Chapter 1) provides the mechanism for moving continents into colder latitudes. When a continent such as Antarctica was moved to a polar position, it would have both a cold climate and an access to oceanic moisture. These are the necessary conditions for initiating a continental ice sheet.

As ice sheets began to grow on a polar continent they would cool the Earth further by increasing its reflectivity, because more solar radiation is reflected back into space from light-colored surfaces such as snow than from dark-colored ones. As the ice sheet grew, reflectivity would have increased and temperatures would have dropped even further. This "feedback" mechanism would continually decrease global temperatures, ultimately triggering valley glaciation and then continental glaciation in temperate latitudes by the Pleistocene Epoch. This is believed to be the major factor in triggering the latest ice age. We will return to this subject after we consider the causes of cyclicity of glacial climates.

Cyclical Fluctuations in Pleistocene Climates

The hypotheses that we have just reviewed may be relevant in varying degrees to cooling of the Earth, but none can explain the *cyclicity* of Pleistocene climates. The most promising theory to explain cyclic climatic changes is the **Milankovitch theory,** named after its originator and strongest advocate. The Milankovitch theory attributes cyclical climatic changes to three periodic differences in the position of the Earth relative to the Sun.

The first factor is the change in shape of the Earth's orbit, which varies from circular to elliptical and back to circular about every 92,000 years. This results in the nearest and farthest distances to the Sun varying by as much as 3 million miles over the 92,000-year cycle. The second factor is the periodic changes in the tilt of the Earth's rotational axis relative to its plane of rotation around the Sun. The tilt may vary from 21.8° to 24.4° over a cycle of 40,000 years. The third factor is the periodic shifting of the Earth's axis as a result of gravitational attraction by the Sun and Moon (the period is about 21,000 years).

According to the Milankovitch theory, extensive glaciations could begin when the combinations of the three periodic factors we have described produced colder summers in the Northern Hemisphere. When the summers were cold enough so that only a portion of the previous summer's snows melted, snow would accumulate from year to year and the snow cover would eventually be thick enough to form glaciers.

However, although the Milankovitch effect has occurred regularly over geologic time, there is no evidence that glaciations did occur in the past in the regular intervals predicted by the Milankovitch theory. This suggests that the Milankovitch effect does not *cause* ice ages but results in cyclical glacial and interglacial activity once *other* factors have cooled the Earth sufficiently for extensive glaciation to occur.

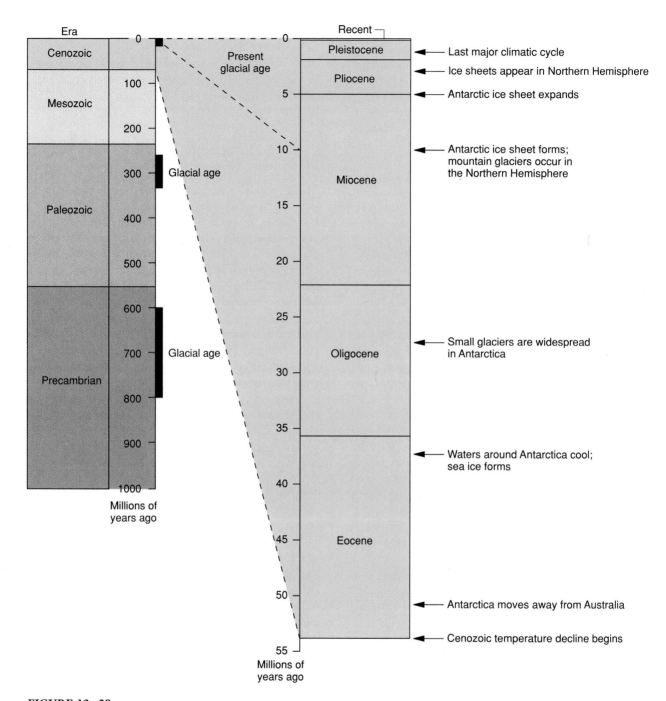

FIGURE 13-28
Climate over the last billion years. In left-hand column, note the much longer duration of pre-Pleistocene glaciations. After Figure 49, p. 190 in *Ice Ages: Solving the Mystery,* by John Imbrie and Katherine Palmer Imbrie, 1979, Enslow Publishers, Short Hills, New Jersey.

ORIGIN OF ICE AGES

Both the origin and cyclicity of ice ages are controversial topics. However, enough data is now available to present a working hypothesis to explain both the causes of ice ages and the cyclical glacial-interglacial climatic changes that occur within them. Although we will speak specifically here of the late Cenozoic ice ages, the same mechanism also could explain the earlier glacial ages recorded in the rock record. The working hypothesis that we will propose involves aspects of both the plate tectonics and Milankovitch theories.

The separation of the continent of Antarctica from Australia and its migration to the South Pole by the Miocene Epoch would have provided both the cold temperatures and the source of moisture required for the growth of continental ice sheets on Antarctica. As ice accumulated in Antarctica it decreased the reflection of solar radiation, cooling the climate, and thus favoring the growth of more glaciers. Geologic evidence indicates that major glaciers existed in west Antarctica by the late Miocene or early Pleistocene, in east Antarctica by the Pliocene, and in the coastal mountains of Alaska by the late Miocene. By the Pleistocene Epoch, the Earth had cooled sufficiently so that continental glaciers began to move into middle latitudes. At this stage, the Milankovitch effect took over and produced the cyclical climatic changes that occurred within the late Cenozoic glacial age. A chronology of glaciations during the last billion years is summarized in Figure 13–28.

FUTURE CLIMATE

How will our climate change in the future? Extrapolations of future climate based on past climatic trends are unclear. How much time do we have before we reach the warmest temperatures of the present interglacial period? Or have we passed that temperature high, and are headed now toward a progressively colder climate? At this point we must look beyond natural factors affecting climate and consider how humans may be influencing climate change.

Some scientists feel that the increasing amounts of particulate matter in the air from manufacturing and burning of fossil fuels will reflect incoming solar radiation back into space, thus cooling the Earth.

Other scientists feel that our increasing use of fossil fuels is increasing steadily the amount of carbon dioxide, CO_2, in the atmosphere, which will lead to increasingly warmer climates. CO_2 in the atmosphere allows solar radiation to pass through but blocks the heat reflected from the Earth's surface, preventing it from escaping back into space. This mechanism has been called the **greenhouse effect** for its similarity to heat buildup from solar radiation in a glass-walled greenhouse.

The greenhouse effect could increase atmospheric temperatures 3°C by the year 2050. This would create conditions for a very different climatic distribution than at present. The worst scenarios for the greenhouse effect are frightening. Present agricultural areas could become semiarid or desert regions, sea level could rise appreciably to flood coastal areas, freshwater supplies could be contaminated by seawater, and lakes could dry up. It is ironic that the human race, which has evolved amid the great temperature changes of the late Cenozoic, now may have the ability to determine the course of climatic change by widespread and accelerated pollution.

SUMMARY

1. Glaciers are masses of ice that form on land by the compaction and recrystallization of snow and move downslope or outward in all directions under the pressure of their own weight.
2. Glaciers form wherever there is a year-round accumulation of snow. The lowest altitude of year-round snow cover on a glacier is called the snowline.
3. As snow accumulates it undergoes a transformation from crystalline snowflakes, to granular snow, to glacial ice crystals.
4. Valley glaciers are confined and restricted to mountains. Ice caps and ice sheets are unconfined and are the largest forms of glaciers.
5. Valley glaciers join to form trunk glaciers. Trunk glaciers can combine to form piedmont glaciers that flow away from the mountains where they originated.
6. Ice sheets exist today only in Greenland and Antarctica.
7. Glaciers move by a combination of slippage along their bases and internal flow resulting from intragranular slippage and recrystallization.
8. A cover of about 30–60 m of snow and ice is sufficient to make the ice below flow plastically. Glaciers

are comprised of an upper brittle zone and a lower plastic zone.

9. Crevasses form in the brittle zone of a glacier whenever tensional stresses pull the ice apart. Different types of crevasses are found at the edges of glaciers, in their centers, and at their ends.

10. Glaciers flow faster when slopes are steeper, when their thickness increases, and wherever they are forced to flow through a narrower opening.

11. Glacier economy is a measure of the balance of snow accumulation above the snowline versus loss of snow and ice below the snowline. Glaciers with positive economies are advancing; those with negative economies are melting back. If accumulation and melting are in balance, ice is being delivered to the terminus as fast as it is melting back and the front of the glacier remains stationary.

12. Glaciers erode bedrock by plucking and abrasion. The end result of glacial erosion is a polished and striated rock surface.

13. Mountain glaciers erode bowl-shaped depressions above the snowline. Two glaciers eroding toward each other produce an arête. When the arête is breached a col is produced. A number of cirques eroding toward a mountain peak can produce a horn.

14. Glacial drift is divided into unstratified drift (poorly sorted with angular particles) and stratified drift (better sorted and contains rounded particles).

15. Erratics are rock fragments that are carried far from their sources and are found on rock of a different composition.

16. Moraines are glacial deposits formed at the edges of the ice. They form at the sides of valley glaciers (lateral moraines), within the glacier where two valley glaciers meet to form a trunk glacier (medial moraine), under the glacier (ground moraine) and at the terminus of the glacier (end moraines).

17. Drumlins are streamlined hills of till formed by continental ice sheets.

18. Kames, kame terraces, eskers, and outwash plains are formed from stratified drift. Kames form as crevasse fillings in the ice. Eskers form in subglacial tunnels. Kame terraces form as valley fills along the side of a stagnating valley glacier. Outwash plains form as meltwater streams carry rock debris from the glacier across the land in front of the edge of the ice.

19. Proglacial lakes form when glacier meltwater accumulates behind a morainal dam.

20. The Pleistocene ice age is divided into glacial periods that had climates colder than today's and interglacial periods when the climate was warmer than it is today.

21. Glaciations occurred several times in the Earth's history, as far back as the Precambrian Era.

22. Continental glaciers form through coalescence of glaciers from individual accumulation areas called ice centers.

23. The Laurentide ice sheet covered most of Canada and extended deep into the northern United States. The Cordilleran ice sheet formed in the Rocky Mountains; it flowed westward toward the Pacific Ocean and eastward toward the Laurentide ice sheet.

24. Unglaciated areas during the Pleistocene still were affected by the cyclic climatic changes. Periglacial climates were characterized by block fields, rock glaciers, and areas of patterned ground. Loess was formed as strong winds removed silt from outwash deposits and carried it downwind to be deposited as sheets of loess.

25. Pluvial climates were wetter than those of today and correlated with times of glacial advances. Pluvial lakes formed in presently dry areas when wetter climates and greater cloud cover resulted in greater rainfall.

26. During glacial periods sea level dropped considerably. This had a significant effect on stream deposition and erosion.

27. The Milankovitch theory attributes climatic changes to periodic differences in the position of the Earth relative to the Sun. Factors in the effect include the eccentricity of Earth's orbit, periodic changes in the inclination of Earth's rotational axis relative to its plane of rotation around the Sun, and the periodic shifting of the Earth's axis as a result of gravitational attraction by the Sun and Moon.

28. The last ice age started when Antarctica drifted to a position over the South Pole. Growth of glaciers on Antarctica gradually increased the reflectivity, cooling the climate further. The increased cooling of the Earth's atmosphere with time enabled valley glaciers to move into mid-latitudes, followed by continental glaciers. At this stage, the Milankovitch effect took over, producing the cyclical changes that occurred within the late Cenozoic glacial age.

REVIEW QUESTIONS

1. What conditions are necessary for the formation of glaciers?

2. What factors control the altitude of the snowline?

3. What transformations occur as snowflakes are converted into glacial ice? What processes are involved?

4. What are the differences among valley, trunk, and piedmont glaciers?

5. Describe the mechanisms by which glaciers move. What factors determine the speed of glaciers?

6. How and where do crevasses form? How deep into a glacier do they extend?
7. Describe the internal structure of a glacier and account for the differences between surface and basal ice.
8. What conditions are necessary for a surging glacier?
9. What is the economy of a glacier? Describe the behavior of glaciers having positive, negative, and zero economies.
10. How do glaciers directly erode bedrock? What role do periglacial climates have in increasing the efficiency of glacial erosion?
11. Describe the mechanisms that result in the formation of a:
    a. Cirque
    b. Arête
    c. Col
    d. Horn.
12. Contrast the mechanisms that result in the deposition of (a) unstratified drift and (b) stratified drift.
13. Contrast the characteristics of till and outwash and account for the differences.
14. How are drumlins formed? What can they tell us about the movement of glaciers?
15. Describe differences in the formation of these types of moraines:
    a. Lateral
    b. Medial
    c. Ground
    d. Terminal.
16. What are the differences between pluvial and proglacial lakes?
17. How do we identify the evidence of past glaciations in the rock record? How does the record of pre-Pleistocene glaciations compare to that of the last glacial age? Why is this so?
18. Describe loess and give the conditions for its formation. Where is it forming today?
19. Describe the formation of the ice sheets that covered the northern United States.
20. What are periglacial features? Describe three different types of periglacial features and describe their origin.
21. What effects did Pleistocene sea level changes have on geologic processes on the continents?
22. How would you distinguish between glacial and interglacial sediments in the geologic record?
23. What relationship exists between plate tectonics and the triggering of worldwide glaciation?
24. How can we best account for the glacial-interglacial cyclicity of temperature changes in the Pleistocene?

FURTHER READINGS

Ashley, G. A. 1987. A facies model for temperate continental glaciers. *Journal of Geological Education* 35(4): 208–16. (A good discussion of sediment types associated with continental glaciers.)

Eliot, John L. 1987. Glaciers on the move. *National Geographic* 171(1): 107–19. (Excellent, well-illustrated treatment of valley and surging glaciers.)

Fairbridge, R. W. 1974. Glacial grooves and periglacial features in the Saharan Ordovician. In D. R. Coates, ed. *Glacial geomorphology*. Binghamton, NY: SUNY at Binghamton, Publications in Geomorphology, p. 315–27. (Describes features associated with a pre-Quaternary glaciation.)

Flint, R. F. 1977. *Glacial and Quaternary geology*. New York: John Wiley & Sons, 892 p. (A classic and exhaustive description of virtually all aspects of glacial and Quaternary geology on a worldwide scale.)

Porter, Steven C. 1979. Quaternary stratigraphy and chronology of Mauna Kea, Hawaii: A 380,000 year record of mid-Pacific volcanism and ice-cap glaciation: Summary. *Geological Society of America Bulletin* 90: 609–11. (A description of the multiple glaciation of a Hawaiian volcano in the Pleistocene. The deposits on the volcano flank consist of interbedded glacial and volcanic deposits.)

Sharp, R. P. 1988. *Living ice—understanding glaciers and glaciation*. New York: Cambridge Univ. Press, 248 p. (A detailed description of glaciers written in a nontechnical and engaging style, with abundant diagrams and photographs.)

U.S. Geological Survey. 1981. *Geologic story of Cape Cod, Massachusetts*. U.S. Govt. Printing Office no. 350–776, 23 p. (A well-illustrated discussion of how continental glaciation created the landforms in this area. Many sequential maps showing changes in the area with time.)

14

Dry Regions and Wind Activity

Great Sand Dunes National Monument, Colorado. Courtesy
of Wards Natural Science Establishment, Inc.

When the word *desert* is mentioned, most people visualize a hot, windy, dry area of rolling sand dunes, perhaps with Arab horsemen or camels gathered around a palm-fringed oasis. Such hot, sandy deserts certainly exist—the Sahara in Africa is one—but most deserts are quite different in appearance and characteristics. For example, many deserts are rocky rather than sandy. Many have sparse, scrubby vegetation; where a seasonal rain occurs, the shrubs leaf out and short-lived annuals cover the ground with blossoms. Some deserts are cold and never warm above freezing.

WHAT ARE DESERTS?

Deserts are more properly thought of as areas of low precipitation and high evaporation, regardless of their temperature or sediment type. **Deserts** are defined as land areas that receive less than 25 cm of precipitation a year; they usually are bordered by **semiarid areas** having precipitation of less than 50 cm per year.

Most of the geologic processes active in humid areas, such as weathering, streams, and wind, also are important in deserts. However, the low precipitation and reduced vegetation in deserts change the relative rates for each of these processes when compared to humid climates. When you travel through a desert and see mostly dried up stream channels, it is easy to assume that stream erosion and deposition are unimportant. However, short but intense periods of streamflow cause extensive erosion and deposition. Another striking feature of deserts is the sharp outline of many rock formations. This is due to the reduced role of chemical weathering, which is a consequence of the low precipitation and sparse vegetation. This reduced weathering results in the absence of thick soils that smooth the topography in more humid climates.

A major misconception about deserts is that they are always hot. If we define deserts on the basis of yearly precipitation, we must include the *cold* deserts that form in polar areas because of the low precipitation there; these are called **polar deserts**. Despite the low precipitation, some parts are covered with snow a good deal of the time because evaporation is so low that snow remains. Some parts of polar deserts have stark, rugged landscapes similar to those of warm deserts (Figure 14–1). In addition, wind plays a major part in the erosion and transport of material in cold deserts, just as it does in warm ones. In this chapter we will treat in detail only the warm deserts.

Two physical principles are important in creating the dry conditions in which deserts form. The principles are: (1) *cold air is denser than hot air,* and (2) *cold air can hold less moisture than hot air.* Therefore, hot air rises and cold air sinks, and cold air is "drier" than hot air. We will use both of these concepts later in this chapter to explain the formation and areal distribution of different types of deserts.

TYPES OF WARM DESERTS

Warm deserts are of three major types. **Climatic deserts** are produced by the distribution of moisture, which is controlled by global atmospheric circulation. **Topographic deserts** form where mountainous masses block the flow of moisture-laden oceanic air into the continental interiors. **Coastal deserts** are caused by cooler oceanic waters, which reduce the amount of moisture falling on the adjacent coast.

Climatic Deserts

A map of the world's climatic zones shows that most deserts exist in two tropical belts from latitudes 10° to 30° north and latitudes 10° to 30° south (Figure 14–2). The largest deserts on the Earth, such as the Sahara in North Africa and the Arabian desert in the Middle East, are located in these latitudes. Why are these latitudes so conducive to the formation of deserts? To explain the formation of climatic deserts we must look briefly at the atmospheric circulation patterns on the Earth's surface (Figure 14–3). Atmospheric circulation from equator to pole is di-

FIGURE 14–1
Polar desert, Wright Valley, Dry Valleys, Antarctica. Photo by Terraphotographics/BPS.

vided into more localized circulation cells that govern the prevailing wind patterns in corresponding latitudinal belts.

The greatest concentration of solar energy occurs in the equatorial region. Here, air is heated and rises to higher altitudes, where it cools. This cooling results in condensation of the water in the air and generates the torrential rains of the equatorial regions. The now-drier air moves away to the north and south from the Equator and begins to descend to the surface around latitudes 30° north and 30° south (Figure 14–3). This air warms as it descends closer to the warm ground, making the air mass capable of holding additional moisture. Thus it releases very little of its moisture as precipitation. The resulting warm, dry winds that blow across the land surface provide conditions for the formation of the great climatic deserts of latitudes 30° north and 30° south (Figure 14–2).

The air between latitudes 30° north and 30° south does not move back to the Equator in a straight north-south line but rather in a northeast-to-southwest direction in the Northern Hemisphere and a southeast-to-northwest direction in the Southern Hemisphere. This deflection of the winds into a curved path is caused by the Earth's rotation and is known as the **Coriolis effect**. As a result of the Coriolis effect, a moving object in the Northern Hemisphere is deflected to *the right of the path along which it is moving,* and a moving object in the Southern Hemisphere is deflected to its *left.*

Winds are named for the direction *from* which they blow, so the prevailing winds in the 0°–30° north latitudinal belt are called the *northeast trade winds* (Figure 14–4). Similarly, the winds moving equatorward from latitude 30° south veer to their left, generating the *southeast trade winds,* which are the prevailing winds in this southern latitude belt.

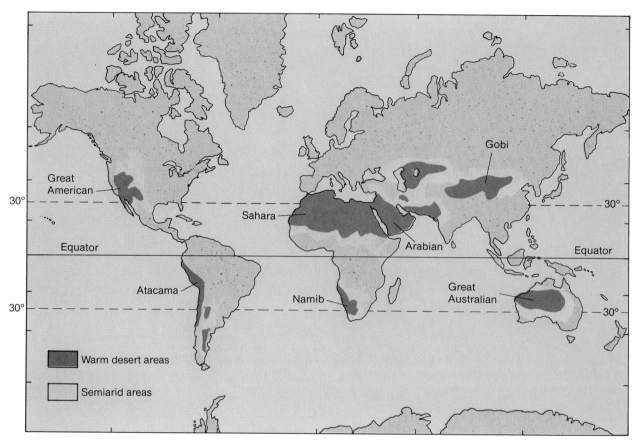

FIGURE 14–2
Areal distribution of deserts and semiarid areas.

Topographic Deserts

Topographic deserts are formed in areas where mountains bar the transfer of moisture from oceanic areas. The deserts of the western and southwestern United States are good examples of topographic deserts. At these latitudes, the dominant winds are from the west and are called the *prevailing westerlies*.

Moisture-laden oceanic air driven inland by the prevailing westerlies rises over the Coast Ranges of California (Figure 14–5). As the air mass rises over the coastal areas, it cools and releases excess moisture as precipitation. Additional moisture is lost as the air mass is pushed even higher to clear the top of the Sierra Nevada. The dry air then descends into the area east of the Sierra Nevada, becoming progressively warmer as it loses altitude. This warm, dry air dehydrates the land, forming the deserts of the American West and Southwest.

The Gobi Desert is the largest in Asia (Figure 14–2) and is located in a similar belt of prevailing westerly winds at a similar latitude. Very little moisture reaches the Gobi Desert because of the shielding effect of The Himalayas and the Tibetan Plateau to the south, and because it is so far inland from oceanic sources of moisture.

Coastal Deserts

One type of desert actually forms along an oceanic coast. These coastal deserts typically are long and narrow and are found on the west sides of continents in the Southern Hemisphere. At first it might seem unusual that a desert could form along an ocean, but it is the significant *temperature difference* between the coastal land and the adjacent ocean currents that brings about the dry conditions under which coastal deserts form.

A cold ocean surface current circles the Antarctic continent in a clockwise movement, driven by prevailing westerly winds (Figure 14–6). This current is deflected up the western coasts of the continents by the Coriolis effect (each current is deflected to its left in the Southern Hemisphere). If the land bordering the cool current is low-lying and hot (as is the case in all Southern Hemisphere continents), conditions are favorable for the formation of a coastal desert. The cold surface water chills the air above the currents, making the air incapable of holding much water. As this cool, dry coastal air moves eastward over the hot, low-lying land, it warms and *absorbs* moisture rather than releasing precipitation. The Atacama of Chile and the Namib of southwestern Africa are good examples of coastal deserts.

CHARACTERISTICS OF DESERTS

Deserts are characterized by extremes of temperatures, precipitation, vegetation, weathering and soils, and stream activity.

Temperatures

Deserts may be very hot in the daytime but become quite cool at night. Why are there such temperature extremes even in "warm" deserts? The lack of vegetation results in more exposed ground area to heat up during the day. At night, the reduced cloud cover in most dry regions allows the ground to radiate back much of its heat, causing temperatures to plummet. Surface temperatures can reach very high values by the middle of the day. One of the highest temperatures ever recorded was 58°C (136.4°F) *in the shade* at Al ' Azīzīyah in the Libyan part of the Sahara

FIGURE 14–3
Air circulation patterns resulting in the formation of climatic deserts near latitudes 30° north and 30° south (only the northern circulation cell is shown).

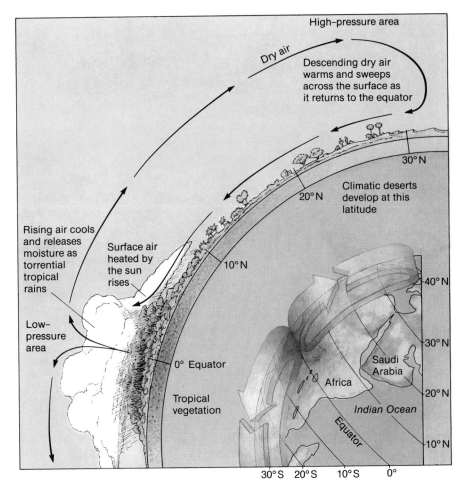

FIGURE 14–4
Generalized atmospheric circulation patterns and major wind systems on the earth's surface. The major wind systems are shown in the global part of the diagram. The diagrams to the right are vertical cross sections showing the circulation-cell pattern typical of each latitudinal wind belt.

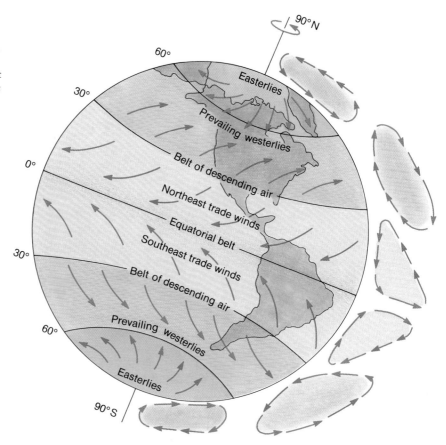

Desert. H. F. Garner provided a fascinating account of the temperature extremes that can be encountered in deserts:

> The writer will not soon forget arising at 3:00 A.M. in the Algerian Sahara in early September. The bucket of water for washing presented a thin film of ice, and a heavy wool sweater and leather jacket were comfortable while riding in an open jeep. By 9:30 A.M., the air temperature was nearing 85°F; by 11:30 A.M., a pocket thermometer registered over 105°F, and at 2:00 P.M. the same thermometer registered 127°F. What passed for a local pub in a nearby oasis cooled its beer by wrapping the bottles in wet sacking and laying them in the sunshine. Evaporation occurs at an almost unbelievable rate under such circumstances, and effective ground moisture levels are fantastically low. The foregoing account amounts to a record of a daily temperature variation of at least 95°F. The beer was delicious.[1]

[1]From *The Origin of Landscapes: A synthesis of Geomorphology* by H. F. Garner. Copyright © 1974 by Oxford University Press, Inc. Reprinted by permission.

Precipitation

The high daytime temperatures in many deserts affect the amount of precipitation that can reach the ground. Precipitation formed by the condensation of moisture in the air at higher altitudes may be evaporated back into the atmosphere as it falls into lower, warmer air. As the air cools at night, the small amount of water in desert air may condense as dew on the rock and sediment surfaces. This condensation occurs in all but the most arid deserts; it enables limited chemical weathering to take place.

Deserts have a wide range of average rainfall, and the amount of rain within any one desert may vary considerably from place to place, even for adjacent areas. For example, the average yearly precipitation at Furnace Creek Station, on the floor of Death Valley in California, is about 4 cm. However, precipitation in the adjacent mountains is much higher. The greater precipitation in desert highlands results in greater weathering and erosion.

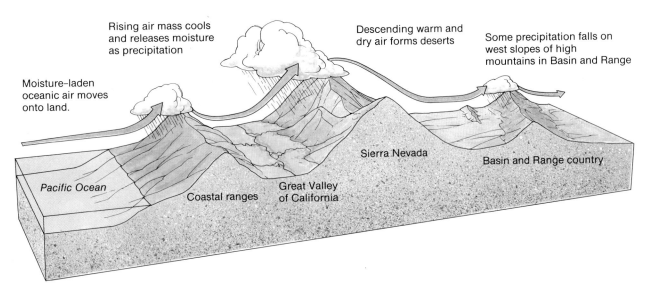

West

Additional precipitation occurs
as air rises higher over the
Sierra Nevada further inland

East

Rising air mass cools
and releases moisture
as precipitation

Descending warm and
dry air forms deserts

Some precipitation falls on
west slopes of high
mountains in Basin and Range

Moisture–laden
oceanic air moves
onto land.

Sierra Nevada

Basin and Range country

Pacific Ocean

Coastal ranges

Great Valley
of California

FIGURE 14–5

Topographic profile from the Pacific Ocean off California eastward into the Basin and Range
country of Nevada, showing the conditions that favor development of topographic deserts in
the American West and Southwest.

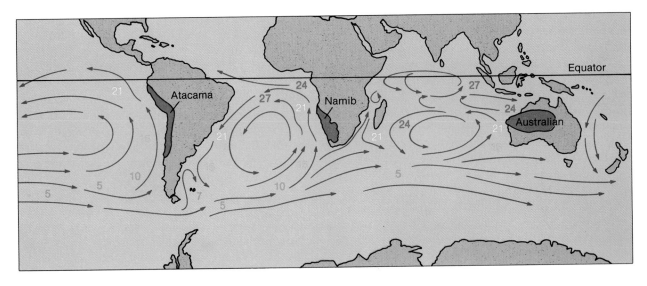

Equator

Atacama

Namib

Australian

FIGURE 14–6

Major ocean surface currents and temperatures in the Southern Hemisphere. Temperatures
(°C) along the currents are shown for a number of latitudinal positions. The cold currents
circling the south polar region are driven up the western sides of the continents in the
Southern Hemisphere, thus providing the cool coastal air masses that form coastal deserts such
as the Atacama Desert (South America), the Namib Desert (southwestern Africa), and the
western Australian deserts. After H. F. Garner, *The Origin of Landscapes: A Synthesis of
Geomorphology,* © 1974 by Oxford University Press; used with permission.

Vegetation

Deserts exhibit a wide range of vegetation densities and plant types which show remarkable evolutionary adaptations for survival in a dry climate. For example, some types of plants have very long roots so that they can tap a deep water table between rainfalls. Other types of vegetation can absorb great amounts of water and have tight outer surfaces that minimize moisture loss. Cactus roots are very shallow and laterally extensive so that they can catch any moisture on the ground surface. Trees, common in humid climates, are rare in deserts because they evaporate too much water through their leaves. When present, they usually are located along the channels of intermittent streams or places where groundwater is near the surface.

Weathering and Soils

In dry regions, physical weathering is more visible than chemical weathering. Especially important are the processes of salt crystal growth, daily temperature changes, and frost wedging that were described in Chapter 9. Physically shattered rock fragments produced by these three mechanisms litter many desert floors, and in places these fragments can be reassembled into larger cobbles and boulders.

Chemical weathering is greatly reduced in deserts because of the low rainfall and scant vegetation. Chemical weathering occurs largely in the short periods when the ground is wet after a flood, in the shaded area around vegetation and between boulders (the shade reduces evaporation), and as a result of the dew that forms on rock surfaces when the desert air cools at night. One indication of chemical weathering is the thin coating, colored dark red, brown, or black, that forms on rock surfaces after long exposure to desert air. These coatings are called **desert varnish** and are composed of oxides of iron, with traces of manganese oxides and silica (Figure 14–7).

The origin of desert varnish is not completely understood. Some believe that it is formed by solution of iron and manganese from within the rock particles, after which the dissolved material is deposited on the outside of the particle as the water evaporates. However, the presence of desert varnish on quartzite and other rock types that do not contain appreciable amounts of iron and manganese suggests that this

FIGURE 14–7
Desert varnish on rock fragments in Death Valley, California. The broken rock shows the purplish-black surface coating characteristic of desert varnish. Photo courtesy of Robert Q. Oaks, Jr.

mechanism may not form all desert varnish. Others believe that desert varnish is of biochemical origin, produced by microorganisms that live in the outer 1 mm of the rock surfaces. These microorganisms produce the desert varnish by biochemical action, using the elements in the rock and the dew that coats rock surfaces at night.

The soils developed in deserts usually are thin and patchy due to both decreased chemical weathering and increased effectiveness of wind and water erosion in the absence of vegetation. Desert soils lack organic accumulations, and contain generally angular rock fragments derived from physical weathering. Caliche soils, rich in soluble minerals, are common in deserts (see Chapter 9).

Drainage

Most of the drainage in desert areas is by ephemeral streams, which carry water only for a small part of the year. Precipitation that falls on steep and largely bare desert slopes usually exceeds the ground's infiltration capacity. This is especially true on areas underlain by rock fragments, on mudflow/debris-flow deposits, and on impermeable rocks. If the rainfall is heavy and relatively short in duration, water can quickly fill and spill over stream channels. The water then moves rapidly across the ground as a shallow sheetflow. Some of this surface flow infiltrates into the ground and recharges the groundwater reservoir below. If the zone of saturation is recharged sufficiently, intermittent streams can maintain their flow between rainfalls. However, these intermittent streams will dry up if the water table drops below the elevation of their channels (see Chapter 12).

Streams in some deserts never reach the sea because the desert basins are ringed by mountains. This creates **internal drainage,** which causes the physical and chemical load of the stream to be deposited within the desert.

If stream discharge is great enough to overcome losses by infiltration into the ground and evaporation into the air, excess water will accumulate in low areas on the desert floors as **playa lakes** (Figure 14–8).

These shallow and broad lakes may persist for days or weeks after stream activity. When the lakes dry up, their dissolved load is deposited as a bed of evaporites.

Some deserts have external drainage and are traversed by through-flowing streams that receive most of their discharge from areas outside of the desert that have higher precipitation. Examples of such perennial streams are the Nile River in Egypt and the Colorado River in the western United States. These streams are able to maintain their flow because their discharge is greater than the combined water loss by infiltration into the ground and evaporation into the dry desert air. External drainage normally results in the physical and chemical load of the stream being deposited outside of the desert. However, human

FIGURE 14–8
Playa surface with salt deposits (white), Death Valley, California. Photo courtesy of Robert Q. Oaks, Jr.

FIGURE 14–9
Arroyo with steep walls and sediment-covered flat floor, showing transported boulders. Arroyo de Corucjuata, Michoacán, Mexico. Photo courtesy of K. Segerstrom/ U.S. Geological Survey.

FIGURE 14–10
Pediment in Granite Mountains, Mojave Desert, California. Photo courtesy of Donald O. Doehring.

modifications on both the Colorado and the Nile have resulted in changes in normal depositional patterns. For example, most of the physical loads of both rivers now are trapped behind dams constructed across them. This has decreased delta growth at the river mouths.

STREAM ACTIVITY

When you visit a desert, the work of the wind may be apparent in sandblasted surfaces, sand dunes, or even a dust storm, and it may seem that wind is the most important factor in the evolution of the landforms.

Look more carefully and you will see many gullies made by ephemeral streams. Desert streams may experience very high discharges during summer thunderstorms or in the spring when snow on adjacent mountains starts to melt. During these high-discharge events, the ephemeral stream gullies become filled with sediment carried in raging streams, mudflows, or debris flows. The type of activity dominating any high-discharge event depends on the ratio of water to debris and the type of material available to be incorporated in the flow. We now know that streams and mass movements are far more important than wind activity in forming desert landscapes.

We should remember also that many present-day deserts were more humid in Pleistocene pluvial climates than they are today. A major proportion of desert features were formed during the increased chemical weathering, stream activity, and mass movements of past pluvial climates. This relict landscape is being modified today only by infrequent stream activity and the work of the wind.

Stream Erosion

Stream erosion, as well as mass movements, forms a wide range of erosional features in deserts. A stream or a mudflow carves a steep-sided channel called an *arroyo* or "dry wash" in the United States and a *wadi* in the Middle East (Figure 14–9).

Another desert feature is the **pediment**. Pediments are bedrock surfaces that slope very gently (less than 5°) away from the base of mountains. The origin of pediments is controversial. Most geologists agree that pediments are erosional features developed on bedrock and that the formation of the pediment is associated with the erosional retreat of the mountain front, which extends the pediment toward the mountains over time (Figure 14–10). The disagreement is over *how* the erosion is carried out.

FIGURE 14–11
Buttes in Monument Valley, Arizona. Photo © Norman R. Thompson, 1990.

FIGURE 14–12
Alluvial fans merge to form a bajada in Death Valley, California. Photo courtesy of Robert Q. Oaks, Jr.

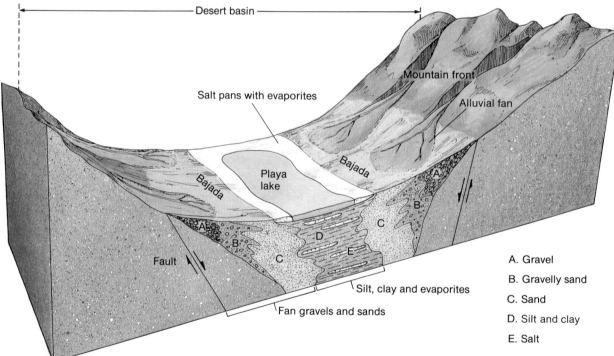

A. Gravel
B. Gravelly sand
C. Sand
D. Silt and clay
E. Salt

FIGURE 14–13
Geologic section showing the sediment facies deposited in a desert basin that has internal drainage. Alluvial-fan sediments become finer basinward where they are interbedded with evaporites derived from the desiccation of playa lakes.

Some attribute pediments to lateral erosion by streams in channels, or to sheetflows. For example, pediments along the Book Cliffs and southward along the base of the high plateaus in Utah are capped by stream gravels that fill stream scours 2–3 m deep. The scours are cut into the readily erodible underlying shale.

However, it is harder to understand how stream-channel migration or sheetflooding can cut a pediment into harder rocks, such as granite. Geologist

John Moss, working in the granite pediments in the Usery Mountains east of Phoenix, found that pediments there were underlain by crumbly granite extending downward to an irregular hard granite surface below. From this evidence, he suggested a two-stage mechanism for pediment formation, starting with deep weathering in a formerly wetter climate followed by removal of the weathered debris by streams and sheetflow in the subsequent dry climate.

When resistant rocks such as sandstone, limestone, and lava are underlain by easily eroded rocks such as shale, topographic features such as mesas are formed. A **mesa** is a broad, flat-topped erosional remnant bounded on the sides by steep slopes and cliffs. The mesa forms as the weaker underlying rocks are weathered and then eroded by mass movements and streams. The resistant rock on top forms a cap that prevents erosion. A **butte** is similar to a mesa, but it is smaller (Figure 14–11).

Stream Deposition

The bases of mountains in deserts are mantled with sediment supplied by ephemeral streams (see Chapter 12). **Alluvial fans** are fanlike stream and mudflow deposits that extend basinward from the mouths of many canyons (Figure 14–12). Alluvial fans vary in size from hundreds of meters to tens of kilometers in length. The fan surface is concave upward, with slopes of 5° to 10° near the apex of the fans at the mountain front to less than 1° at the basin edge of the fan. The fans form as streams and mudflows traveling down mountain slopes drop much of their sediment at the base of the mountain, where the gradient decreases and the flow spreads out of the channel and across the land. Deposition also is aided as the flow sinks into permeable sediments and as it evaporates into the hot, dry desert air. The merging of adjacent alluvial fans forms a sloping depositional surface called a **bajada** (Figure 14–12).

Grain size in alluvial fans decreases basinward. In the central part of the basin, finer alluvial fan sediments interfinger with evaporites derived from the evaporation of playa lakes occupying the lower part of the desert floors (Figure 14–13). Such alluvial basin sediments can reach thicknesses of thousands of meters when the adjacent mountains are being uplifted or the basins are subsiding.

EROSIONAL CYCLE IN DESERTS

The topography in deserts undergoes a series of evolutionary changes, depicted in Figure 14–14. Stream erosion and pediment formation on several sides of a mountain range create breaks in the mountains and isolate erosional remnants called **inselbergs**. If the drainage is internal, the physical and chemical loads of the streams are deposited within the basin. Although the end stage of erosion in both humid and dry regions is a surface of low relief, the development of those surfaces is quite different. In a humid area, the highlands are eroded and the material is removed from the area; the low relief of the land surface therefore is *erosional*. In a dry area, the highlands also are worn down, but they are buried in their own debris because of the internal drainage; much of the resulting land surface therefore is *depositional*.

For an interesting view of spectacular desert erosion in the American Southwest, refer to Focus 14–1.

WIND ACTIVITY

Although wind is not the dominant force in shaping desert landscapes, it is certainly an important one. In fact, wind reaches its maximum effectiveness under desert conditions of dry soil, low vegetation, and frequent winds. Much of what we know about desert wind activity started with the work of R. A. Bagnold, who conducted exhaustive studies in North Africa as a British officer before and during World War II. His work, *The Physics of Blown Sand and Desert Dunes,* was published in 1941. Bagnold's observations have been referred to time and again, and they form the basis of our discussion of the subject.

EROSION AND TRANSPORT BY WIND

Wind readily erodes sand, silt, and clay particles. Once eroded, the coarser and finer particles are transported in different ways.

FIGURE 14–14

Evolution of desert landscapes, a sequence occurring when mountains and basins are stable and the drainage is internal. (a) In this example, the basins and intervening mountains were formed originally by faulting. Over time, the uplifted blocks underwent erosion and alluvial fans built out from the mountain front. The mountain masses were progressively dissected, and temporary playa lakes formed in low areas after rainfall. (b) Relief in the area decreased as the mountains were eroded. Alluvial fans joined to form a bajada along the mountain front. Pediments grew more prominent, and the thickness of the sediments in the basin increased. (c) The late stage of erosion has low relief. Pediments grow as the mountain front is eroded back, and pediments from adjacent mountains have joined to isolate mountain remnants as inselbergs. Basin fill thickness has increased substantially and is beginning to bury the mountains.

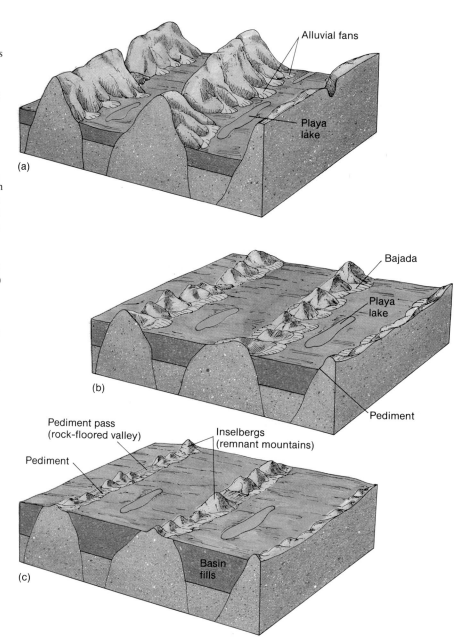

Sand-sized Particles

Wind blowing across the desert may move quite rapidly at a meter or so above the ground. However, as with flowing water, the velocity decreases toward the ground surface, where the effects of frictional drag are the greatest. The air closest to the ground, a zone equal in thickness to about 1/30 of the predominant particle size on the ground, is relatively motionless.

Only those grains that protrude above this thin zone of "dead" air can be eroded. As the speed of the wind increases, a critical velocity or **threshold velocity** is reached in which some of the sand grains start to move in leaps like the saltating particles at the bottom of a stream.

Laboratory experiments indicate that once some of the grains start moving at the threshold velocity, grains will continue to move, even if the velocity

THE HOODOOS OF BRYCE CANYON NATIONAL PARK

Bryce Canyon, situated at the eastern edge of the Paunsaugunt Plateau in southern Utah, is one of our most colorful National Parks. It is at a high altitude (2300–2800 m) in a near-desert area which receives less than 5 cm of rainfall a year. The temperature extremes, steep gradients of streams draining eastward off the plateau, sparse vegetation, and weakness of the underlying poorly consolidated sedimentary rocks all maximize the erosional potential of geologic processes. Consequently, desert weathering, stream activity, and mass movements have carved the edge of the plateau into a mass of spires, curtains, arches, and grottoes.

The geologic story of Bryce Canyon begins in the Tertiary Period when streams deposited sediments that would become a series of varicolored limestones, calcareous sandstone, sandstone, shale, and conglomerate into the lakes and inland seas that occupied this area. Subsequent faulting and uplift formed the plateau and a series of intersecting vertical fractures that cut through it. This set the stage for evolution of the park landscapes.

Rainfall and water from snowmelt infiltrated through the fracture systems and began to chemically weather the rocks. Soluble layers were dissolved and others were decomposed to form clay and iron oxides. The plateau's high altitude enabled frost wedging to play a significant part in rock weathering. Weathered debris was removed from the area by high-gradient streams and mudflows resulting from infrequent rainfalls and snowmelt. Vertical masses of the rock grew isolated from each other as weathering and erosion proceeded along the joint surfaces (Figure 2). Differential weathering between beds created variations in the shape and width of these vertical structures.

Erosion along two intersecting sets of vertical fractures produced spire-like *hoodoos* (Figure 2); erosion along parallel sets of joints produced *curtains*. Where the bases of the curtains were breached by erosion, *arches* formed. As weathering isolates archlike features, *grottoes* form on the face of retreating cliffs.

The Bryce Canyon landscape "renews" itself continually. As hoodoos and curtains are eroded away at the eastern edge of the plateau, new features are being eroded from the retreating cliff face farther to the west (Figure 2).

---

drops below the velocity needed to *start* grain movement. This is because once the grains are elevated into the faster-moving air above, their momentum increases. This increased momentum is transferred to other grains when the saltating grains hit the surface once again. This sets off a chain reaction and soon there is a carpet of sand moving along the ground, bouncing off rocks and other objects and setting other grains into motion. As some grains imbed themselves in the sandy surface, they can push other grains laterally along the surface in a type of grain movement called **surface creep**.

FIGURE 1
Aerial view of Bryce Canyon National Park showing a wide variety of erosional forms. Photo by Nicholas K. Coch.

Silt and Clay-sized Particles

Wind may start movement of sand-sized particles before it can put the finer particles in motion. Finer-grained particles are more difficult to erode because their fine size makes them more cohesive, and they do not extend above the motionless air nearest the ground surface, making it harder for the wind to move them.

How are these particles eventually set into motion? Movement usually begins when the surface is disrupted by a plow, truck, large animal, or off-road

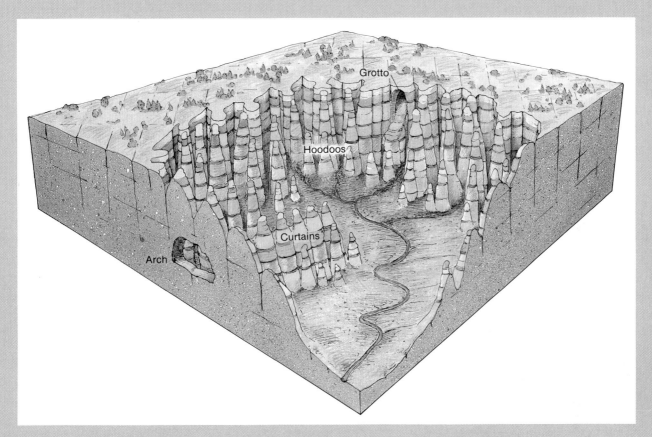

FIGURE 2
Geological development of the erosional forms in Bryce Canyon National Park.

An estimate of the average rate of cliff retreat in this area is 2 feet per 100 years. Bryce Canyon National Park is unusual, for it is one of the few where significant changes can be seen in the landscape over a lifetime. A variety of geologic processes have acted together to produce an ever-changing landscape composed of widely varied features.

vehicle, all of which are trailed by clouds of dust as they break the ground surface, allowing the wind to remove the fine material. The surface also can be broken by saltating sand grains or rolling gravel, both of which throw the finer-grained material into the higher-velocity air above the ground. Tornado-like "dust devils" also stir debris that disturbs the ground surface and suspends the fine material in the air.

The low weight of the fine particles enables them to be carried in the atmosphere for great distances. Fine particles derived from the "Dust Bowl" in the southwestern United States in the 1930s fell on snow in New England and most likely far into the Atlantic Ocean. Since then, examination of fine dust collected on the West Indies island of Barbados showed that it contained mineral and biologic components which were transported from Europe and Africa, thousands of miles away.

WIND EROSIONAL FEATURES

The erosion of loose particles by wind is called **deflation**. Air has only 1/1000 the density of water, so wind usually can transport only sand and smaller-sized particles. However, coarser particles can be transported in special circumstances; on Racetrack Playa in Death Valley, wind apparently transports rock fragments by sliding them across the wet, slick playa surface.

Wind is an excellent sorter of sediment. Consider what happens in the deflation of desert braided-stream deposits or mudflow deposits that include particles ranging in size from cobbles to clay (Figure 14–15). Sand particles eroded from the surface travel downwind by saltation and may be deposited as dunes. Finer particles are lifted higher into the atmosphere and may travel as dust storms for great distances before being deposited beyond the edges of the desert as **loess**. (Loess is discussed in Chapter 13.) As the original deposit is deflated progressively, only coarser particles remain. Eventually, coarse rock particles called **desert pavement** cover the surface and act like armor, protecting the underlying sediments from further erosion (Figure 14–16). Sheet-flows may contribute to the formation of desert pavement, washing out some or all of the finer particles, producing a surface concentration of rock fragments.

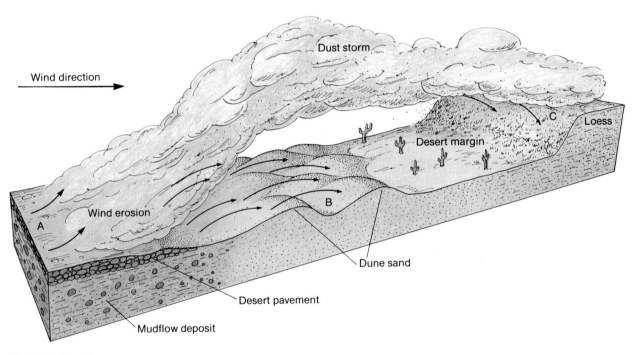

FIGURE 14–15
Deflation of a poorly sorted desert sediment deposit (at A) results in formation of three separate sediment accumulations: desert pavement, dune sand, and loess. Each has a different average size but each is better sorted than the original deposit. A, coarse rock material is concentrated as desert pavement as wind erodes the finer sand and silt from the mudflow. B, sand particles accumulate downwind to form dunes. C, the finest particles are suspended in the wind as dust and deposited at or beyond the desert margins.

FIGURE 14—16
Desert pavement, Mojave Desert, California. The pavement consists of angular rock fragments, many of which are coated with desert varnish. Photo courtesy of Donald O. Doehring.

FIGURE 14—17
Ventifact. The surface of the boulder has been abraded and shaped by windblown sand to form facets that are separated by sharp edges. Photo by Nicholas K. Coch.

FIGURE 14—18
Granite outcrop etched and pitted by the abrasive action of windblown sand. Atacaina Province, Chile. Photo courtesy of K. Segerstrom, U.S. Geological Survey.

Sand grains in saltation can have considerable erosive force. The impact from the grains polishes exposed rock surfaces to form faceted, shiny rock particles called **ventifacts** (Figure 14—17). The abrasive nature of saltating particles within 2 m of the ground surface requires that structures such as wooden telephone poles be shielded with metal collars at the base to prevent them from being weakened by sandblasting. Sand grains in saltation also can erode the bottoms of rock outcrops preferentially, forming pedestal-shaped rocks (Figure 14—18).

WIND DEPOSITION: DUNES

Materials eroded by the wind are deposited in various ways. Fine material may be deposited as loess sheets far from the site of original deflation. Sand-sized particles are deposited closer by, in the form of sheets and mounds of sand behind obstacles and as streamlined sand masses known as **dunes**.

Dune Formation

The only things needed to form dunes are dry granular material and wind. In most cases this material is quartz sand, but dunes can be built of any granular material available. For example, gypsum forms the dunes at White Sands National Monument in New Mexico and calcite makes up the Pleistocene dunes on Bermuda. In fact, Bagnold describes the formation of small dunes composed of ice crystals!

Dunes begin to form when moving sand particles encounter an obstacle. The presence of an obstruction (Figure 14–19) forces the air stream to deflect around the object, creating zones of quiet air, or wind shadows, both immediately before the object (windward) and behind the object (leeward). Sedi-

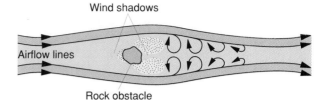

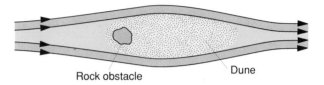

(a) Sand accumulates in wind shadows in front of and behind obstacle

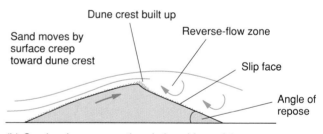

(b) Sand drifts behind obstacle have coalesced into a dune

FIGURE 14–19
Aerial views showing the accumulation of sand in wind shadows around an obstacle. The accumulations form a mound, which may develop into a dune.

ment carried over the object settles in the wind shadows and behind it, building a deposit of sand. Consequently, this sand pile will grow above the height of the obstacle, becoming slightly higher at the

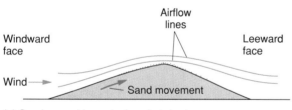

(a) Sand mound forms in the wind shadow

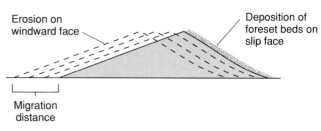

(b) Sand grains move up the windward face of the dune and accumulate at the crest

(c) When the crest of the dune is built up to a steeper angle than the angle of repose, the sand avalanches and forms a foreset bed along the slip face of the dune

(d) Sand dune moves progressively downwind

FIGURE 14–20
Side views showing the formation and migration of dunes.

downwind end. As the sand pile grows higher, it itself becomes an even larger obstacle to airflow, causing even more sand to accumulate on the downwind side. The pile grows upward with a steadily increasing angle on the leeward edge. When a minimum height of about 30 cm is reached and the angle of the leeward face reaches about 34°, the angle of repose for dry sand, a streamlined body called a **dune** is formed.

Dune Migration

As the dune continues to build, it also starts to migrate downwind—Figure 14–20(a). Saltating sand grains landing on the windward slope kick some of the sand grains into the air, pushing other grains up the slope. The grains moving up the slope build the dune to a steeper angle as the sand grains are trapped in the quiet air on the other side of the dune crest and accumulate on the upper part of the leeward slope—Figure 14–20(b).

When the crest of the dune grows steeper than the angle of repose, the slope collapses and the sand grains slide down the steep leeward slope, or **slip face**—Figure 14–20(c). Each of these collapses gen-erates a separate foreset bed along the slip face. The dune advances downwind as wind erodes the sand on the windward face and deposits it as foreset beds on the slip face. The end result of this process is the formation of a streamlined dune with a gently sloping windward face and a steeper-sloping slip face that is inclined downwind—Figure 14–20(d).

Dune Types

There are several types of sand dunes. Each has a distinct shape and occurs under different conditions of vegetation, sand supply, wind velocity, variability of wind direction, and the type of surface over which the dune is moving.

Barchan Dunes. The **barchan dune** is a crescent, with the convex curve upwind and the concave curve and tips pointing downwind. It has a concave slip face that is curved downwind (Figure 14–21). Barchans can migrate as isolated dunes across bare rock or gravel surfaces. The sand moves up the windward slopes and cascades down the slip face and out along the tips of the barchan, moving the dune progressively downwind. Barchans are best developed

FIGURE 14–21
Aerial view of barchan dune in Tule Wash, west of Salton Sea, California. Photo by John Shelton.

where there is a limited supply of sand, little vegetation, and a constant wind direction.

Longitudinal Dunes. Elongate dunes whose long axes parallel the major wind direction are called **longitudinal dunes**. They form in areas with a large supply of sand and high wind velocity and are common in the central areas of great sandy deserts such as the Sahara. These dunes can reach heights of 100–200 m and may range in length from 400 m to over 100 km. The crests of the dunes are sinuous, with curved slip faces on one of the sides (Figure 14–22). Longitudinal dunes are formed by steady winds blowing from one direction, which alternate with brief storm winds at a small angle to this direction.

Transverse Dunes. Dunes that form at right angles to the dominant wind direction in areas of large sediment supply and moderate winds are called **transverse dunes** (Figure 14–23). Transverse desert dunes may grade into barchans on the edges of the dune field, where there is less sand. Another type is found along oceanic coasts and the shorelines of major lakes. The abundant sand made available by coastal erosion and transport is reworked directly by

FIGURE 14–22
Aerial view of longitudinal dune field in Australia. Photo courtesy of Victor Goldsmith.

FIGURE 14–23
Transverse dunes at Padre Island National Seashore, Texas. These dunes form when sand blown from the shore of the Gulf of Mexico (out of picture, to the right) accumulates inland. The height of the slip face on the leeward side of these dunes is about 2 m. Photo by Nicholas K. Coch.

FIGURE 14–24
Parabolic dunes along the North Carolina coast. Photo courtesy of Victor Goldsmith.

onshore winds (blowing from the water onto the land) into irregular and elongate transverse dunes paralleling the coast. Such dunes may form in equilibrium with coastal vegetation; as the dunes grow in height, the shrubs and grass also grow upward, keeping pace with the growth of the dune.

Parabolic Dunes. Although they resemble the crescent shape of barchan dunes in aerial view, the tips of **parabolic dunes** point *upwind* (Figure 14–24). One type of parabolic dune forms from vegetated transverse dunes in coastal areas that have strong onshore winds. Such a dune starts to form from transverse coastal dunes where the local vegetation has been removed by natural or human activity. The local absence of vegetation allows the dunes to be eroded by onshore winds, forming a depression called a **deflation basin** on the windward side. The basin

subsequently becomes elongated parallel to the wind by deflation. The eroded sand piles higher and higher on the convex downwind dune crest, avalanching down slip faces that are inclined landward (downwind). At the same time that the central part of the dune is being extended, the ends of the dune are kept intact by any vegetation on the sides. Advancing slip faces of such parabolic dunes can bury inland forests, structures, and roads.

Characteristics of Dune Deposits. Sand particles transported by wind have distinctive characteristics. They are typically fine-to-medium-grained sands, are well sorted and rounded, and may have a "frosted" grain surface—Figure 6–6(a). Dune sands are fine grained and well sorted because the low density and low viscosity of air severely limit particle sizes that can be carried. Saltating particles frequently collide during wind transport, and these impacts are all the stronger because the intervening air provides less of a cushioning effect than does water. The innumerable impacts among sand particles in dunes wear away the rough edges, rounding the particles. Under a microscope, the sand grain surfaces exhibit numerous scars which collectively make the surface appear frosted. The beds of cross-stratified sand in dunes tend to be thicker than those in water-laid sediments because these dunes generally are higher than the subaqueous dunes formed by water transport.

DUNE DEPOSITS IN THE GEOLOGIC RECORD

Dune deposits in the geologic record usually are distinctive in their white color and in the great thick-

FIGURE 14–25
Tower of cross-bedded Navaho Sandstone in Zion National Park, Utah. Photo by Nicholas K. Coch.

nesses of their cross-beds. The white color reflects the high percentage of quartz in most dunes, whereas the cross-bed thickness results from the height of the dunes that generated the cross-bedding. Two famous wind-deposited sandstones in the Colorado Plateau region are the Coconino Sandstone of Permian age, exposed in the Grand Canyon (see Chapter 8) and the Navaho Sandstone (Figure 14–25) of Jurassic age which makes up the magnificent towers of Zion National Park, Utah. Both of these rock layers are very resistant to erosion and stand out as cliffs.

Ancient wind deposits can be used to determine the prevailing wind direction, or **paleowind,** during the time when the beds accumulated. The paleowind is determined from the direction of inclination of the foreset beds; these beds are inclined in a downwind direction (Figure 14–20). For example, if the foreset beds in a dune deposit are inclined westward, the wind that formed them must have blown from the east. Analysis of paleowind directions in rocks from many places indicates wind directions that are incompatible with the *present* location of these places. This suggests that the continents were located in a different latitudinal position when the windblown sands originally were deposited, and this provides additional evidence for the theory of plate tectonics introduced in Chapter 1.

SUMMARY

1. Deserts are land areas having less than 25 cm annual precipitation and evaporation rates that greatly exceed precipitation rates in most cases.
2. Deserts often are bordered by semiarid areas having less than 50 cm of annual precipitation.
3. Many of the largest deserts on Earth, such as the Sahara and the Arabian, form in latitudes 10° to 30° north of the Equator and in a similar zone south of the Equator, where global atmospheric patterns result in warm, dry air moving Equatorward across the surface.
4. Other deserts form in the centers of continents far from oceanic sources of moisture (e.g., Gobi Desert); where mountains bar the inland passage of moisture-laden ocean air (e.g., the American Southwest); and in coastal areas where cold currents flow offshore (e.g., Atacama and Namib Deserts).
5. Deserts are characterized by wide daily temperature fluctuations and low annual precipitation. Precipita-

tion generally occurs in a few widely spaced rainfalls separated by dry periods. Additional moisture may be provided where snowmelt from adjacent mountains runs into the desert area.

6. Sparse desert vegetation consists of specialized plants that are able to obtain and store the scant available water. The lack of vegetation and the large exposed land area increase the efficiency of both water and wind erosion.

7. In deserts, physical weathering is more important than chemical weathering. Desert soils are thin, discontinuous, and enriched in soluble minerals.

8. Desert stream activity is particularly effective because the sparse vegetation cannot break the force of the rain or anchor the soil. Large quantities of sediments are transported from highlands down onto the desert floors by braided streams, mudflows, and debris flows.

9. Some of the stream discharge, or groundwater flow, or both, reaches the lowest part of the basin, forming temporary lakes from which soluble salts are deposited upon evaporation.

10. Desert stream activity erodes the highlands slowly, depositing sediment in alluvial fans and on the floor of the basin. In most cases the drainage does not extend outside the desert basin, and the bordering mountains are buried slowly in the sediments derived from their erosion.

11. Wind activity reaches its maximum effectiveness in deserts. Sand, silt, and clay are deflated by the wind, leaving behind a layer of rock particles which forms "desert pavement" and inhibits further deflation.

12. The silt and clay can be carried in suspension far from the source to form deposits of loess at the edges of the desert and beyond.

13. The sand moves along the surface by saltation and surface creep, and it can be molded into streamlined forms called dunes. Different types of dunes are formed under different combinations of sand supply, wind strength, consistency of wind direction, and amount of vegetation.

14. Dune sediments typically are fine-to-medium grained, well sorted, rounded sand which may have frosted grain surfaces.

15. Paleowind directions can be ascertained from dune deposits by determining the direction in which the foreset beds are inclined. (The foreset beds of dunes are always inclined in a downwind direction.)

REVIEW QUESTIONS

1. Describe the conditions that form different types of deserts, and give an example of each desert type.
2. Why is there such an extreme change in daily temperatures in deserts?
3. How does the vegetation found in deserts affect weathering, stream erosion, and wind deflation?
4. How is chemical weathering carried out in desert areas? Give an example of a chemical weathering product.
5. How do desert soils differ from the soils characteristic of humid areas? Why?
6. Contrast the deposits formed by alluvial fans and by playa lakes.
7. Describe how the landscape evolves in areas characterized by internal drainage. How does this differ from areas where the drainage is external?
8. Describe the formation of the following desert features:
   a. Desert varnish
   b. Pediments
   c. Bajadas
   d. Ventifacts.
9. Why are fine-grained sediments so difficult to erode by wind? Under what conditions are they eroded more easily?
10. Why is wind the most effective sediment-sorting agent? Describe what types of deposits are formed from the extensive deflation of a desert mudflow.
11. Describe the formation of a dune.
12. How do the foreset beds of a dune form? What information can they provide to a geologist studying dunes in the rock record?
13. What conditions favor the formation of these dune types?
   a. Barchan
   b. Longitudinal
   c. Transverse
   d. Parabolic.
14. Describe the characteristics of each type of dune deposit. Give the reason for each.
15. An ancient rock layer of wind origin, exposed in an area of westerly winds, has foreset beds inclined in a westward direction. What information can be derived from this observation?

FURTHER READINGS

Bagnold, R. A. 1941. *The physics of blown sand and desert dunes*. New York: Methuen, 265 p. (The best overall reference on wind activity and dune formation.)

Doehring, D. O., ed. 1977. *Geomorphology in arid regions*. Binghamton, NY: SUNY Publications in Geomorphology, 272 p. (A collection of papers on a wide variety of desert features, processes, and deposits.)

McGinnies, W. G., Goldman, B. J., and Paylore, P., eds. 1968. *Deserts of the world: An appraisal of research into their physical and biological environments*. Tucson: Univ. of Arizona Press, 788 p. (An excellent overall reference on both geologic and biologic aspects of deserts. A good place to look first for data on a specific desert.)

McKee, E. D., ed. 1979, *A study of global sand seas*. Washington: U.S. Geological Survey Professional Paper 1052, 429 p. (An up-to-date and lavishly illustrated collection of articles about sandy deserts, dunes, and ancient dune deposits.)

Hunt, C. B. 1975. *Death Valley—geology, ecology, and archeology*. Univ. of California Press, 23 p. (A very readable, nontechnical treatment of a number of aspects of this famous desert area.)

15
Coastal Zones

McClure's Beach, Point Reyes, California. Photo by Stephen Trimble.

Coastal zones, where the land meets the sea, are areas of great beauty and diversity. Coastal vistas in the United States include high rock cliffs along the Pacific and New England coasts; sandy shorelines on the Mid-Atlantic and Gulf Coasts; colorful coral reefs in southern Florida; and the high volcanic cliffs and coral reefs that characterize much of the shoreline in the Hawaiian Islands. Many of the same features and processes found along the edges of the ocean also are found along the coasts of large lakes.

COASTAL PROCESSES

A wide variety of processes are active in coastal zones. Most are related to wave activity, but currents generated by the rise and fall of the tides also are important in erosion and deposition within the coastal zone.

Waves

Coastal erosion and deposition are largely a result of wave activity. Waves form where winds blow across the ocean surface. On a clear and windless day at the beach, large waves may be crashing onto the shore and boats offshore are rocking back and forth. Such wave activity in the absence of winds seems puzzling until we realize that it is caused by faraway storms. The turbulent and gusty winds in a storm center exert a frictional drag on the water surface. The transfer of wind energy to the water forms ripples that grow into waves with time. Waves produced by winds in the storm center are irregular and appear to move in several directions, all of which are close to that of the predominant wind. As the waves move away from the direct influence of the wind, they become longer and lower and are referred to as **swells**.

Symmetrical deep-water swells are described by the terms shown in Figure 15–1. The highest part of the wave is the **crest,** and the lowest part is the **trough;** the vertical distance between them is the **wave height**. The horizontal distance between any two similar points (e.g., two crests or two troughs) on successive waves is the **wavelength,** represented by the Greek letter lambda λ. An imaginary line drawn along a **wave crest** is called the **wave front**. The direction in which the wave is moving is the **wave normal,** which is a line perpendicular to the wave fronts. The time (in seconds) for one wavelength to pass a given point is called the **wave period**.

The height and period of waves depend on the speed of the wind, its duration, and the **fetch,** or distance, over which the wind blows. For example, if wind velocity and duration are the same, waves will be higher on large lakes than on small ones because the fetch on the larger lakes is greater. Swells moving away from a storm center can travel thousands of kilometers across deep water before the energy of the wave is released when it crashes onto the coast. During the migration of swells, wave heights may drop as much as 50% in the first 1600 km, but the loss of energy is negligible.

Waves are a mechanism by which energy is transferred along the water surface. Although the wave moves across the surface, the water itself does not. Instead, water particles in a wave have an orbital motion (Figure 15–2). You can experience this motion yourself if you float seaward of the breaking waves along the coast. *As long as there is no local current or wind,* you will only bob up and down in a circular motion, similar to the ball in Figure 15–2, as the waves pass through. After one wavelength has passed through, a floating object returns to its initial position. It is important to reiterate that it is the *wave front* that travels over distances, and not the water.

Experimental studies and field observations show that the orbital motion of the water decreases with depth. Notice in Figure 15–3 that the diameter of the circular orbit at the surface is equal to the wave height. If you dive a few feet, you will continue to move in an orbital path as waves pass overhead, but the diameter of the orbit will be less than at the surface. Orbital motion ceases entirely at a depth equal of about one-half the wavelength of the waves above (Figure 15–3).

This bottom limit of orbital water movement is referred to as the **wave base**. Above the wave base is a zone of agitated water, whereas the water below it is not influenced by wave motion. In Figure 15–3, the depth of the water is greater than the wave base, and thus the ocean bottom is undisturbed by the passage of the waves. However, if waves with a longer wavelength move across the area, the wave base may reach the bottom, and bottom materials will be reworked by the agitated water. Such changes

FIGURE 15–1
Terms used to describe waves.

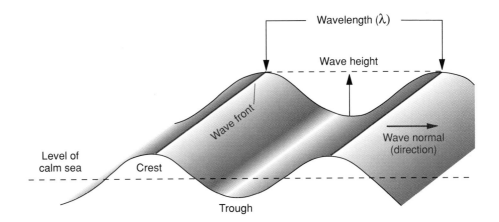

in wave base are important in the erosion and deposition of nearshore sediments. For example, waves from distant storms may have wavelengths as great as 600 m, corresponding to a wave base of 300 m. Waves of this magnitude can stir the bottom sediments far seaward of the coast.

When a wave reaches shallow water, so that the wave base reaches the seafloor, the wave begins to "feel the bottom." The upper part of the wave continues to move landward, whereas the lower part is retarded by friction along the bottom. From this point landward, the character of the waves changes markedly. Wavelength decreases, and the waves get higher and more asymmetrical (Figure 15–4). The zone where these changes occur is called the **breaker zone**. Eventually, the oversteepened waves topple over within the **surf zone,** causing an actual forward movement of the water within the wave. It is in the surf zone that wave energy carries out erosion and deposition. Within the surf zone the energy stored in the wave is put to work for erosion and deposition along the shoreline.

Wave Refraction

A wave approaching a coast does not slow uniformly along its front, because the bottom is not uniformly deep. The part of the wave that passes over a shallow area, such as around a headland where the land juts into the ocean, slows first (Figure 15–5). The parts of the wave that are still in deeper water continue to move landward without decreasing their speed. This

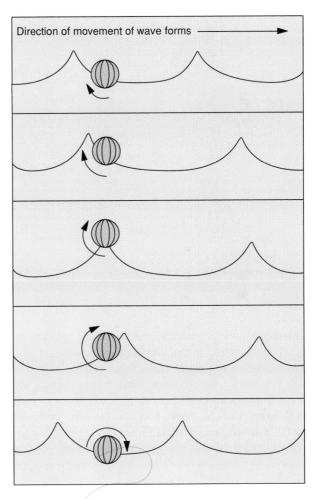

FIGURE 15–2
Motion of ball within a wave. As the waveforms move from left to right, the ball makes a circular orbit, returning to its original position after one wavelength has passed.

COASTAL PROCESSES | 399

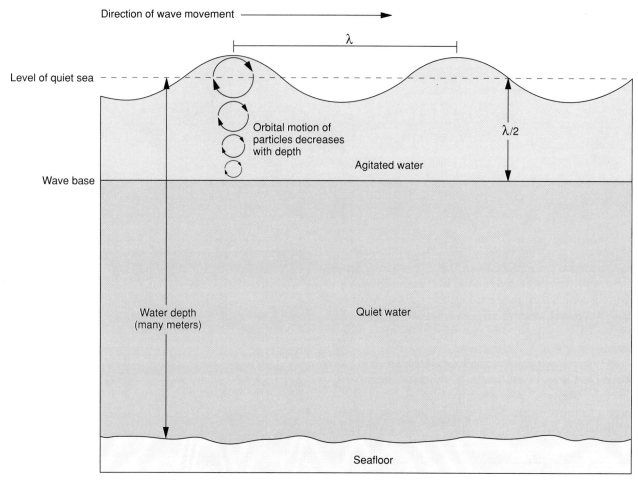

Direction of wave movement

λ

Level of quiet sea

λ/2

Orbital motion of
particles decreases
with depth

Agitated water

Wave base

Water depth
(many meters)

Quiet water

Seafloor

FIGURE 15–3
Water motion within a wave. The orbital motion of the water particles becomes minimal at a
depth equal to one-half of the wavelength. This is called the wave base.

differential wave velocity makes the wave fronts bend, or **refract,** as the waves advance toward the shore (Figure 15–5). **Wave refraction** results in the concentration of wave energy on protruding headlands. Equal segments of the deep-water wave shown in this diagram (A–B, B–C, C–D, and so on) have equal amounts of energy. However, as the waves refract, more of this energy is dissipated across the intervening bays, as shown by the pattern in Figure 15–5. The net result of continued wave refraction is for the waves to erode headlands and to smooth the shoreline by depositing sand in bays between the headlands.

Longshore Drift

When waves break at an angle to the shoreline, large volumes of sediment are moved parallel to the coast. The process involved can be shown by isolating a sand grain and observing its movement in response to waves approaching the shoreline at an angle (Figure 15–6). Onrushing waves move the sand grain up the beach parallel to the wave normals (A–A'). The retreating water moves the grain down the steepest slope of the beach face (A'–B). The sand grain has now moved a small distance along the shoreline (A–B) through this pair of movements.

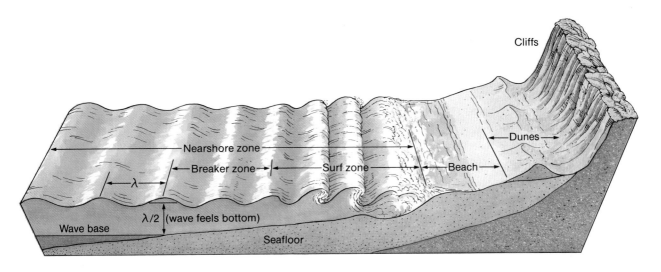

FIGURE 15–4
Cross section showing the changes in waves as they approach the coast. When the water depth equals half the wavelength, the waves begin to break. The wave height increases, the wavelength decreases, and the wave becomes asymmetrical. The waves eventually tumble over in the surf zone.

FIGURE 15–5
Aerial view of wave refraction along an irregular shoreline. Wave refraction concentrates the energy in the wave on the headlands and disperses it over the bays. As the wave fronts enter shallower water, they are refracted progressively so that they become more parallel to the bottom contours.

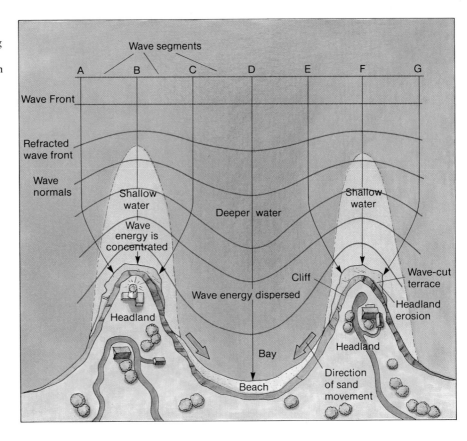

FIGURE 15-6

Wave fronts advancing at an angle to the shoreline move particles of sediment along the shoreline.

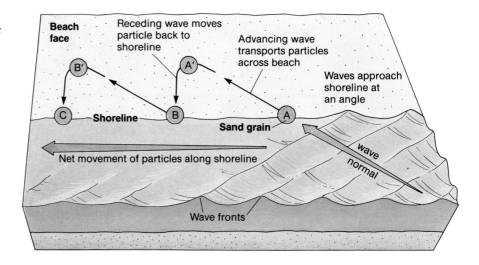

The grain at *B* then is picked up by another incoming wave, repeating the process and moving the grain even farther along the shoreline. This process of carrying sediment along the shoreline by waves and currents is called **longshore drift**.

Longshore and Rip Currents

Waves approaching a coast at an angle generate a **longshore current** that flows parallel to the coast. The longshore current, along with the wave mechanism shown in Figure 15-6, transports sedimentary particles along the shoreline and is important in the formation of coastal depositional features. As waves pile up water nearshore, some of the water eventually returns to the ocean as a current perpendicular to the shoreline called a **rip current**. In many cases, rip currents carry suspended sediment seaward and can be recognized as "plumes" of turbid water extending beyond the surf zone. Rip currents can carry unwary swimmers out into deeper water as well. The currents are powerful, but they dissipate offshore.

FIGURE 15-7

Aerial view of Nauset Inlet on Cape Cod, Massachusetts. Tidal currents moving through this inlet build a flood tidal delta (right) and ebb tidal delta (left). Photo courtesy of S. P. Leatherman.

Tidal Currents

Wind-driven waves and the wave-induced currents just described are the major agents of erosion along oceanic shorelines. However, currents induced by tidal forces become much more important in the sheltered bodies of water between the mainland and offshore islands and at openings in the shoreline called **tidal inlets** (Figure 15–7). Oceanic and bay waters are exchanged through inlets during cycles of high and low tides. Tidal currents moving into the coastal zone as the tidal level rises are called **flood currents**; those moving out into the ocean as the tidal level drops are called **ebb currents**.

Tidal flows are essential for the support of life in the marshes, in the bays behind the oceanic shoreline, and in the nearshore parts of the ocean. Flood currents bring oxygenated ocean water into the coastal areas behind the oceanic shoreline. Ebb currents transport nutrients and organic particles seaward, providing the materials necessary for oceanic life.

Tidal currents redistribute sediment and build a variety of depositional features in coastal zones. For example, large-scale submarine ripples and dunes produced by tidal erosion and deposition are exposed at low tide in many estuaries. In Chapters 6 and 14 we showed how cross-bedding can be used to determine current direction in water-deposited and wind-deposited sediments; the foreset beds of the deposits are always inclined downcurrent. However, in tidal deposits, the flood and ebb currents reverse several times a day. Consequently, in many tidal deposits, cross-bedding directions in successive layers may show differences of current directions of as much as 180°. Such **herringbone cross-bedding** (named after the material pattern) is diagnostic of tidal environments in the geologic record (Figure 15–8).

FIGURE 15–8
Herringbone cross-stratification in the Cambrian Tapeats Sandstone in Grand Canyon National Park, Arizona. The reversals of current directions in successive beds indicate this facies of the Tapeats Sandstone was deposited in a tidal environment. Photo by Nicholas K. Coch.

COASTAL EROSION

Erosion and deposition occur together on all shorelines. However, the relative role of each in forming shoreline features varies from coast to coast. Wherever wave erosion removes more material than is being deposited, *net erosion* occurs, and the coast is cut back. Wherever more sediment is being deposited than eroded, *net deposition* occurs, and the shore builds seaward. Wave erosion is most effective along low-lying sandy coasts. The noncohesive nature of the shore materials, along with the low altitudes along the coast, make them vulnerable to erosion. Storm waves are much higher than normal, so they can erode higher areas inland and quickly remove considerable quantities of sediment. These storm waves wash sand through low points into the areas behind the shoreline in a process known as **overwash** (Figure 15–9).

Some coasts are composed of high bluffs built of sediments. Wave erosion of the base of these cliffs undermines them, and results in a failure of the cliff by slumping. Some cliffs show no apparent change for a few years and then retreat landward tens of meters by slumping in one storm. As a result of this episodic erosion, cliff-retreat rates usually are given as *average* recession rates, such as 0.3 m/year. What happens to the material in the slumped block as it is subjected to wave reworking? The finer material in the slumped mass is either moved along the shoreline (sand) or transported seaward (silt and clay). The coarser material remains behind, forming a submarine **wave-built terrace,** which slopes gently seaward from the base of the cliff (Figure 15–10).

Wave erosion is effective also along rocky coasts, although its rate is much slower than along sandy coasts. Chemical and physical weathering processes described in Chapter 9 weaken the rocks on the cliff face and facilitate wave erosion. For example, freezing water or the growth of salt crystals in rock crevices can exert considerable pressure along minute fractures in the rocks, widening these fractures. Waves breaking along the coast force water under great pressure into the widened rock crevices. This

FIGURE 15–9
Breaching of dunes at Folly Island, South Carolina by Hurricane Hugo, 1989. Sand eroded from the dunes was deposited as washover fans on the marsh behind the dunes. The conical piles of sand are part of an attempt at dune restoration after the storm. Photo by Nicholas K. Coch.

FIGURE 15–10
Aerial view of wave-built terrace developed in front of cliffs of glacial deposits at Montauk Point, New York. The wave fronts are refracting around the terraces in front of each headland. Photo by Nicholas K. Coch.

wedges out blocks of rock that fall into the surf. The rock debris is picked up by the waves and thrown repeatedly against the cliff, eroding a **wave-cut notch** at the base of the cliff. This may undermine the cliff sufficiently to cause a rockfall and an accumulation of talus at the base. The talus acts as an armor for the cliff, protecting it from further erosion until the talus pile is removed by wave erosion. Smaller particles eroded from the front of the talus are moved back and forth by the surf, eroding a flat, gently sloping **wave-cut bench** in the rock layers seaward of the cliff. This wave-cut bench may be exposed at low tide (Figure 15–11).

So far we have been talking about rocky and sandy coasts that have uniform resistance to weathering and erosion. In reality, many coastal cliffs contain rocks that differ in resistance to wave erosion because of their mineralogy or because of fractures that cut across them. Waves erode holes in the flanks of a headland. Where holes from opposite sides meet in the center, they create **sea arches**. Where wave erosion isolates masses of rock from the shoreline, **sea stacks** are formed (Figure 15–12). Landslides may remove a section of rock cliff to make a **sea cave**.

As cliffs recede and the wave-built terrace or wave-cut bench extends further landward, an equilibrium begins to develop. The widening terrace exerts increasing frictional drag on the waves and reduces their energy. The lower wave energy is sufficient to move sediment but not to erode the cliff. Therefore a temporary equilibrium is reached in which the sediment can accumulate as beaches against the cliffs. A subsequent storm may produce waves high enough to strip the beach away and erode the cliffs again. After a number of storms, a more permanent equilibrium develops in which cliff erosion is reduced and beaches are preserved for longer periods of time.

FIGURE 15–11
Aerial view of wave-cut terrace at Bolinas Point, California. Photo by John Shelton.

FIGURE 15–12
Sea stack at Davenport Beach, California. Photo by Nicholas K. Coch.

COASTAL DEPOSITION

Coastal sediments come from three major sources: rivers, reworking of nearshore sediments, and erosion of materials exposed along the shore. Large volumes of sediment are supplied to the shoreline where major rivers, such as the Mississippi, form deltas (see Chapter 11). However, postglacial rise in sea level has decreased stream gradients, trapping most of the stream bed load within the seaward parts of rivers. Only the fine particles can be carried in suspension into the ocean.

Sediments seaward of shorelines also are a source of at least some coastal sands. During episodes of continental glaciation, when sea level was much lower, streams flowed across much of the area that is presently submerged and deposited vast volumes of sediment. Storms are moving some of this material toward the shoreline, but at present geologists are not sure how much material is being moved. Material eroded from the shoreline provides most coastal sediments. Debris may come from rocky cliffs, beach and dune sand, bluffs made of older sediment, or coral reefs in the case of tropical shorelines.

The topography of the offshore area determines what happens to this material as it is moved along the coast by waves and longshore drift. For example, the **nearshore zone** along much of the Atlantic and Gulf coasts is gently sloping and unbroken; sediment can move considerable distances along the shore unless it is interrupted by river mouths or engineering structures. In contrast, sediment movement along parts of the southern California coastline is much more localized because the coastline there is composed partially of submarine canyons that extend into the nearshore area. Stream sediment entering the coastal zone in these areas is moved along by longshore drift until it intersects the heads of one of these canyons. The material then moves seaward in the canyon and is deposited in deep basins offshore.

Coastal sediments are deposited in a number of ways. Newly deposited beach material exposed by a falling tide may be blown by wind into dunes many meters above sea level. Particles moving with shoreline currents may be deposited where there is a decrease in velocity or an increase in water depth. For example, particles of sediment may be swept into tidal inlets by flood currents and deposited in the relatively quiet waters of bays (see Figure 15–7).

Biologic activity also is important in the deposition of coastal sediments. Plants slow currents, causing deposition of their bed load and a part of their suspended load. Other types of organisms, such as shellfish, live on organic material suspended within the water. They draw in water, filter the organic particles, and excrete the associated sedimentary particles as fecal pellets.

COASTAL ZONES

The coastal zone includes the area from where waves first break offshore to the limit of tidal action inland. Within the coastal zone are three major environments—estuaries and bays, tidal wetlands, and barrier islands and beaches. These grade into one another (Figure 15–13).

Estuaries

All rivers that drain coastal zones become **estuaries** near the coast (Figure 15–13). They are tidally influenced and contain water ranging from brackish to salty. The water in an estuary flows seaward on ebb tide and landward on flood tide. Wide, elongate estuaries parallel to the coast are called *bays, sounds,* or *lagoons,* depending on their geographic location. Most estuaries are former river valleys perpendicular to the coast that were submerged by the worldwide postglacial rise in sea level (New York Bay, Chesapeake Bay, San Francisco Bay, and Puget Sound). Some estuaries in higher latitudes are glacially deepened valleys called **fjords** which were drowned by the postglacial rise in sea level. Fjords differ from other types of estuaries in having sheer rock walls that often extend far below sea level.

Tidal wetlands are those portions of the coastal zone that are covered by waters at high tide. **Tidal flats** are unvegetated areas which are exposed fully during low tide. Tidal flats grade upslope into **tidal marshes,** which are partially or completely vegetated areas that may be covered to different degrees during the tidal cycle. Wetlands are accumulation sites for fine-grained sediments and organic material. Extensive wetlands occur around the periphery of estuaries, bays, and sounds. In many coastal areas, wetlands are being filled and built upon (Boston, New York, Norfolk, and San Francisco).

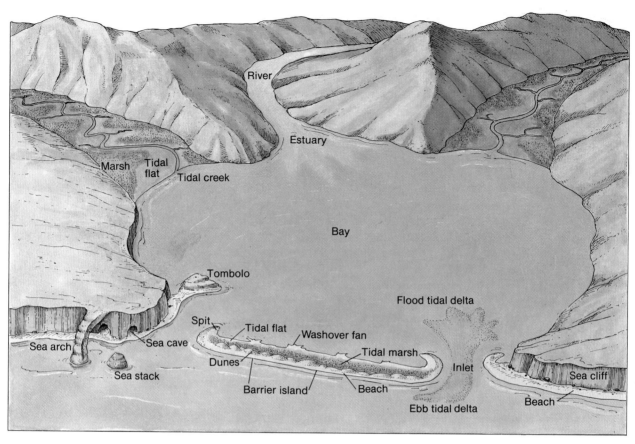

FIGURE 15–13
Features of coastal zones mentioned in text.

Elimination of wetlands causes severe environmental problems. Wetlands provide sheltered areas where fish and other marine life can spawn and where the juvenile stages of marine organisms can grow before moving into the oceans. The large quantities of organic particles and nutrients produced in wetlands by decay of plant and animal tissues are swept seaward to provide nourishment for life in the oceans. These areas also reduce the danger of flooding in coastal storms by absorbing large volumes of water and then releasing it slowly. Recently it has been shown that wetlands have a considerable ability to remove pollutants from water, and they may have great potential for use in sewage treatment.

Beaches and Barrier Islands

Accumulations of coarse sediment along parts of the coastal zone exposed to wave action are called **beaches**. The beach extends from the low-water line landward to where there is a change in morphology, type of sediment, or vegetative cover. In general, beaches extend inland to rock or sediment cliffs, dunes, or the seaward limit of permanent vegetation.

Beaches occur in a variety of forms (Figure 15–13). Small beaches are formed along concave portions of cliffed shorelines. More extensive beaches are formed along **spits,** which are elongate sand bodies attached to land and built into the water by sediment eroded from headlands or sandy islands. Beaches also form on **tombolos,** which are islands connected to each other or to the mainland by spits.

The widest and most extensive beaches are found along **barrier islands**—elongate bodies of sand which are parallel to, but separated from, the mainland by a bay, sound, or lagoon. Barrier islands differ from spits because they are much longer and wider and are not attached to land. However, spits can form at the ends of barrier islands as sand is transported into inlets by tidal currents. Barrier islands make up most of the Atlantic and Gulf Coast shorelines of the United States. Several major U.S. cities are located on barrier islands, including Galveston, Texas; Miami Beach, Florida; and Atlantic City, New Jersey.

The origin of barrier islands is somewhat controversial and several possibilities for barrier island formation have been proposed. Some barrier islands can form when a spit grows out from a headland to enclose a bay. They also may form as rising sea level submerges coastal dunes. The drowned dunes form the nucleus for accumulation of nearshore sediment that ultimately forms a barrier island. However, most barrier islands probably formed as waves eroded shallow-water sediment and piled up the sand as linear bodies. These shoals gradually built up to sea level to become barrier islands. The islands became higher as winds blew the dry sand into dune systems.

Barrier islands are composed of a wide variety of sedimentary deposits. Large changes in grain size occur between adjacent areas, reflecting changes in both the intensity and agent of transport. For example, coarse sand, gravel, and broken shells may be deposited in the surf zone at the same time that wind is depositing fine sand in dunes a few tens of meters from the shore. As you walk from the ocean toward the bay across a barrier island, you will traverse the following sedimentary environments: surf zone, beach, dunes, vegetated area behind the dunes, tidal marsh, tidal flat and bay. In no other depositional system are sediment facies of such diverse character deposited so close to one another.

COASTAL STORMS

In recent years, coastal geologists have come to appreciate how much of the change seen in shorelines can occur in a few short-lived storm events. In the long, quiet periods between storms, coastal processes merely redistribute sediment and restore the shoreline.

The Nature of Coastal Storms

A cyclonic storm or hurricane is a storm system having low pressure in the center, surrounded by winds that blow in a counterclockwise path. As the storm

TABLE 15–1
Factors determining the severity of coastal storms.

Factor	Effect
Wind velocity	The higher the wind velocity, the greater the damage.
Storm surge height	The higher the storm surge, the greater the damage.
Location relative to center of storm	Viewed from the storm center and in the direction of storm movement, areas to the right are hit harder because the velocity of the counterclockwise storm winds (viewed from the top) is added to the forward velocity of the storm-center movement.
Coastal shape	Concave shoreline sections sustain more damage because the water is driven into a confined area by the advancing storm, thus increasing wave height.
Storm center movement	The slower the storm moves, the greater the damage. The worst possible situation is a storm that stalls along a coast.
Nature of coast	Rocky coasts are the least disturbed. Cliffed sedimentary coasts can retreat by slumping. Damage is most severe on low-lying barrier island shorelines because they are easily overwashed by storm waves.
Previous storm damage	A coast weakened by even a minor previous storm will be subject to proportionally greater damage in a subsequent storm.
Human activity	With increased development, property damage increases and more floating debris becomes available to knock down other structures.

approaches a coast, the low pressure at the center causes the sea surface below it to rise. In addition, its winds act as a bulldozer, pushing water ahead of it and raising the ocean surface further. Winds raise huge waves on the surface and tides may raise the level of the water even more. Thus it is apparent that water levels during a storm can be much higher than mean sea level. The rise in the water surface that results from all of these factors is called the **storm surge**. Storm surges routinely rise 2–3 m and can go beyond 12 m in a hurricane. Willard Bascom, in *Waves and Beaches,* describes the effects of the most deadly storm surge in American history:

> Probably the most famous example of a storm surge was the Galveston, Texas "flood" of 1900. On that occasion a hurricane with winds of one hundred and twenty miles per hour raised the water level along the shore of the Gulf of Mexico fifteen feet above the usual two-foot tidal range. The storm waves,

FIGURE 15–14
Aerial view of Nauset Spit on Cape Cod, Massachusetts after a major blizzard in 1978. Extensive overwashing of sand has formed washover fans along the length of the spit. The spit was almost breached during the storm where it is attached to the mainland (top of photo). Photo courtesy of S. P. Leatherman.

probably another 25 feet high, rode in atop the storm and demolished the city; some five thousand people drowned.[1]

[1]From *Waves and Beaches* by Willard Bascom. © 1980 by Willard Bascom. © 1964 by Educational Services Incorporated. Reprinted by permission of Doubleday & Co.

The effect that a given storm has on a coast is related not only to the strength of the storm but to a number of other independent factors, summarized in Table 15–1.

FIGURE 15–15
Seasonal changes on the beach at Boomer Beach, La Jolla, California. (a) Winter. (b) Summer. Photos by John Shelton.

(a)

(b)

Coastal Erosion and Deposition During Storms

High waves associated with storms erode large quantities of sediment from beach and dune areas. This material can be shifted rapidly along the shore by accelerated longshore currents, or it may be removed to offshore areas. On a low-lying barrier island, storm waves may cut into beach and dune deposits and wash the sediment through low areas in the dunes onto the back side (mainland-facing) of the island (Figure 15–14). **Washover fans** are lobe-shaped deposits of sand eroded from the ocean side of a shoreline and deposited inland (Figure 15–9). Erosion of the front side of a barrier island and deposition of washover fans on the back side results in the migration of the barrier island toward the mainland with time, a process called **barrier rollover**.

If a storm is particularly severe, a new inlet may breach the barrier island, allowing sediment to be deposited into the bay as a **tidal delta** (Figure 15–7).

Coastal Recovery After Storms

If you have a favorite beach, you may have noticed that it looks very different in winter and summer. In winter, beaches are generally narrower, steeper sloping, and covered by coarser sediment—Figure 15–15(a). This is because storms are more frequent in the winter, and sand is removed more quickly than it can be supplied to the beach by streams or offshore sources. In the summer, storms are infrequent and wave height is generally lower; these conditions promote beach growth.

It has long been known that sand stripped from a beach is eventually returned, to some degree. However, only recently have we been able to understand *how* this process occurs. Detailed studies of Lake Michigan beaches after storms have shown that sand is returned to the beaches through formation of an offshore ridge of sand and its migration landward (Figure 15–16). During this landward migration, the ridge is separated from the beach by a water-filled trough called a *runnel*. The ridge shown in Figure

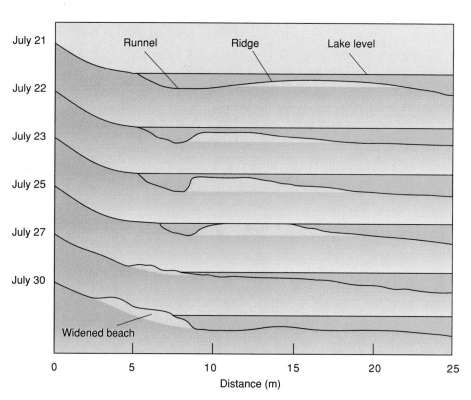

FIGURE 15–16
Natural beach replenishment after a storm on Lake Michigan. An offshore ridge migrates gradually landward and eventually reaches the beach, increasing its width. During its landward migration, the ridge is separated from the beach face by a water-filled runnel. After Davis and Fox, "Natural Beach Restoration on Lake Michigan," *Journal of Sedimentary Petrology* 42:416, Fig. 4. Modified with permission of Society of Economic Paleontologists and Mineralogists.

FIGURE 15–17
Underwater view of coral reef, Florida Keys. Photo by Nicholas K. Coch.

15–16 has "welded" itself to the storm beach and restored its width in little more than a week.

CARBONATE SHORELINES

Most American shorelines fit into one of the coastal types that we already have described. However, the shoreline is quite different in many parts of the Hawaiian Islands, the American Virgin Islands, and in all of South Florida. Although these shorelines have many of the same shore features that we have discussed so far (beaches, dunes, inlets, and such), they are made up largely of deposits of carbonate-secreting organisms such as calcareous algae and corals. These organisms comprise massive linear, wave-resistant structures known as **coral reefs** (Figure 15–17).

Coral Reefs

Coral reefs require very specific conditions for development. The water must be warm, clear (no sediment), shallow (so that light penetrates below) and agitated (so that food particles may be brought to the stationary corals). Corals are animals, but algae (plants) live within the tissues of the coral. It is believed that algal photosynthesis results in the depo-

sition of the calcium carbonate that constitutes the stony structure in which corals live. Algae associated with the corals require sunlight for growth, and coral growth is limited to the upper 40 m of the ocean, the depth to which sunlight penetrates. Clear water is essential both for light penetration and because sediment particles in the water would cover the corals and kill them. In several parts of the world today, coral reefs are threatened not only by pollution but by turbid water resulting from blasting through the coral for navigation channels or harbors and by dredging to create marinas and housing developments.

Wherever favorable conditions occur, coral reefs can continue to grow outward and upward, even if sea level rises. The reef-making organisms grow fastest in the direction of open water or in the direction facing the prevailing winds, because these conditions provide the oxygenated waters and food required for coral growth. Consequently, most reefs are wider and better developed on windward coasts. The skeletal debris eroded from coral reefs by waves is transported by tidal and wind-driven coastal currents and is built into the same type of features—spits, tidal deltas, and beaches—that are found along sandy shorelines in higher latitudes.

Tropical storms can severely damage coral reefs. Only the uppermost part of the reef mass contains live corals, and this is the area most affected by storm waves. Waves and wave-driven debris destroy a considerable part of a reef in a major storm. However, shortly after a storm, new coral growth starts, and soon a layer of live coral covers the reef once again. It is essential to preserve coral reefs, not only for their natural beauty and place in the oceanic ecosystem, but also because these massive structures break the force of storm waves and thus provide essential hurricane protection to coastal cities such as Miami Beach and Miami in Florida.

Types of Coral Reefs

There are three basic types of coral reefs; two are shown in Figure 15–18.

A **barrier reef** is a linear feature parallel to a coast and separated from it by a wide lagoon or bay. Barrier reefs form where the offshore slope is very gentle (Figure 15–18). Natural channels between coral reef segments permit the exchange of waters between the bay and the ocean. All reefs in southern Florida are of this type. The longest barrier reefs in the world are along the east coast of Yucatán (Mexico) and Belize (Central America), and The Great Barrier Reef along the east coast of Australia (Figure 15–19).

Fringing reefs are built right against the coast and have their greatest growth rates where the surf is strongest. This type of reef is more common along coasts with steep offshore slopes (Figure 15–18). Fringing reefs are common on volcanic islands such as Hawaii and in the American Virgin Islands.

The third type of reef is an **atoll**. An atoll is a circular reef surrounding a lagoon in which there may be an island. In 1842, Charles Darwin first proposed an origin for these structures, based on his field observations. Other theories for formation of atolls have been proposed, but Darwin's theory still fits most of the facts. Darwin proposed that atolls form from the subsidence of volcanic islands and their associated fringing reefs (Figure 15–20).

The fringing reef continues to grow upward as the volcano subsides—Figure 15–20(a). When most of the island has subsided, a barrier reef is formed, encircling the island—Figure 15–20(b). After all of the original volcanic island has subsided, a lagoon takes its place as the circular reef pattern of the atoll forms. Storms breaking on the reefs of the atoll pile material above sea level as islands on the reef—Figure 15–20(c).

Although many coral atolls are uninhabited, some of the larger ones are suitable for habitation. If Darwin's subsidence theory is correct, there should be volcanic rock under the carbonate rocks of an atoll. Support for Darwin's theory came shortly after World War II, when deep drilling on the Pacific Ocean atolls of Eniwetok and Bikini revealed volcanic rock under the coral limestone of the atolls.

CHANGING SEA LEVEL

Until now, we have referred to sea level as if it were constant, moving small distances up and down with

FIGURE 15–18
Geologic section across an oceanic island, showing the reef types that develop as a function of offshore slope. A fringing reef forms on steep slopes, and a barrier reef forms on gentle slopes.

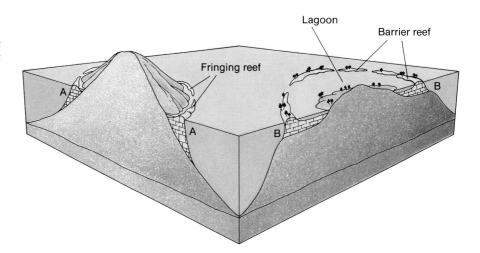

FIGURE 15-19
Aerial view of The Great Barrier Reef. Photo courtesy of Australian Tourist Commission.

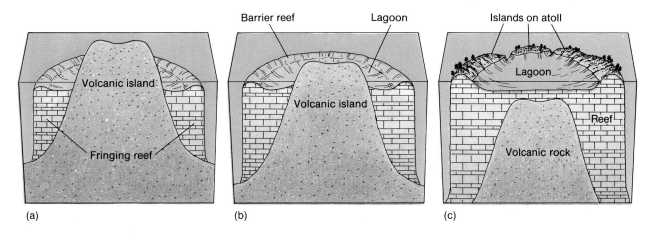

(a) (b) (c)

FIGURE 15-20
Formation of an atoll, according to Charles Darwin. (a) A fringing reef develops on the edges of a subsiding volcanic island. (b) The reef continues to grow upward, keeping pace with the island's subsidence. The fringing reef evolves into a barrier reef, with development of a lagoon. (c) After the volcanic island subsides below sea level, storms pile material upon the remnants. The barrier reef forms the circular atoll around a central lagoon.

the tides, and rising higher during the short periods of a storm surge. However, we now know that long-term changes in sea level are occurring. Such long-term sea-level changes have occurred time and again in the Earth's history, as recorded in the sedimentary rock record. The present changes in sea level are more alarming because, for the first time, humans and their cities are threatened.

A steady rise in sea level is occurring today as a result of postglacial melting. This meltwater is being added to the oceans and is raising sea level. Changes in the level of the ocean surface resulting from increasing or decreasing the amount of water in the ocean are called **eustatic** changes in sea level.

The "greenhouse effect" discussed in Chapter 13 is resulting also in an accelerated rise in sea level. At the present time there is controversy over how fast sea level will rise because of the global warming resulting from the greenhouse effect. If this effect is real (and some scientists feel that it is not), several major coastal cities may be inundated early in the next century.

Rising sea level is creating a landward migration of the shoreline wherever the crust is stable (or subsiding). Detailed studies of Atlantic and Gulf coastal barrier islands have shown that net long-term erosion is occurring along those shorelines. Geologist Chris Kraft of the University of Delaware has pro-

FIGURE 15–21
Location of past and future shoreline positions on the central Atlantic Coast. Projection of present processes of coastal change and rates of rise in sea level indicate that major population centers on the coastal plain may be submerged in as little as 10,000 years. After J. C. Kraft, 1973, *Coastal Geomorphology,* Fig. 22., p. 352.; redrawn with permission of Donald R. Coates.

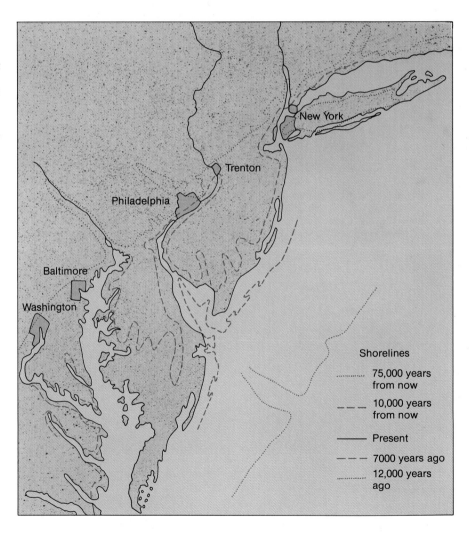

Shorelines
.... 75,000 years from now
– – – 10,000 years from now
——— Present
– – 7000 years ago
.... 12,000 years ago

FIGURE 15–22
Elevated beach ridges (dark arcuate lines at center of photo) at Marble Point, McMurdo Sound, Victoria Land, Antarctica. These ridges demonstrate that the land has risen since the disappearance of the most recent Pleistocene ice sheet. Photo courtesy of W. B. Hamilton/U.S. Geological Survey.

jected the present rate of sea-level rise (Figure 15–21). Such projections indicate that many of the great coastal cities could be submerged eventually.

Kraft's projection does not include the accelerated sea level rise attributed to the greenhouse effect. When we add that to the postglacial rise, we see that coastal flooding could be even higher and come earlier than predicted in Figure 15–21.

In some places, the shoreline is not migrating inland as sea level rises because the *land* is rising at a faster rate. Tectonic forces may result in this elevation (or submergence) of coastal areas. For example, consider what is now happening in areas formerly overlain by continental glaciers during the Pleistocene Epoch. The great thicknesses of these ice sheets depressed the crust. As the glaciers withdrew, the land surface began to rebound to its original level. This rebound is a slow process that is still occurring in many previously glaciated areas, such as Canada and Scandinavia. In areas where such rebounding is occurring, beaches formed along Pleistocene shorelines have been uplifted. These areas show a number of former beaches in a "stairstep" arrangement (Figure 15–22).

Another type of tectonic change is caused by earthquake activity. For example, in the 1964 Alaska earthquake, parts of the coast either were warped upward or subsided beneath the sea.

The combined effect of tectonic and eustatic changes in sea level makes it difficult to determine the magnitude of change in many areas. For example, subsidence in the central Gulf Coast area makes

sea level rise faster, whereas elevation along the Oregon coast makes it seem as though sea level is not rising at all. In both cases, however, there is an actual rise in the ocean surface.

COASTAL RESOURCES

The coastal zone is the ultimate repository for most of the minerals and organic material carried by streams. Several important minerals are reworked by coastal waves and currents into economically significant deposits of titanium, zircon, diamonds, tin, and iron.

Large amounts of petroleum and natural gas were formed from the deposition of organic material in ancient coastal environments such as swamps and deltas. Phosphate rock, essential for fertilizers, is believed to have formed in ancient nearshore coastal environments where phosphate-rich bottom waters were forced to the surface. It is now being mined from deposits in Florida and North Carolina. Deposits of sand and gravel concentrated by ancient surf action are mined for use in highway fill, concrete, and the making of high-quality optical glass.

HUMAN EFFECTS ON COASTAL ZONES

Increased coastal development in the face of accelerating beach erosion has led to the use of engineering structures in an attempt to prevent erosion and save peoples' homes (see Focus 15–1). In discussing these structures, we will use the terms updrift and downdrift to refer to geographic positions along a shoreline in relation to the direction of transport of sediment: *updrift* is the direction from which the sediment is coming, and *downdrift* is the direction toward which it is moving. (This usage is similar to the terms *upstream* and *downstream* in stream studies.)

Structures built within the coastal zone alter the dynamic equilibrium along coasts by cutting off the sand supply needed to nourish beaches downdrift. Walls of rock, concrete, or wood, called **groins,** are commonly built out at right angles to the shoreline in an attempt to trap sand and nourish the beach (Figure 15–23). The beaches updrift grow out by net deposition. However, beaches downdrift of the

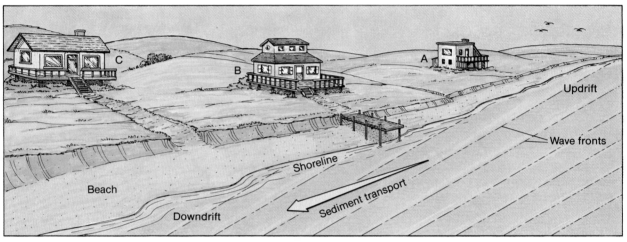

(a)

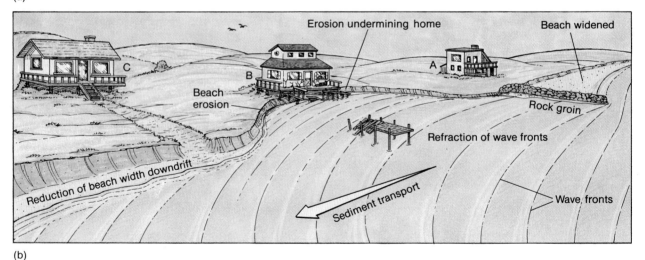

(b)

FIGURE 15–23
Effect of groin construction on coastal erosion and deposition. (a) Conditions prior to groin construction. (b) Homeowner A decides to widen the beach by building a rock groin into the surf at the edge of his property. The beach widens at A by sand deposition updrift of the groin. However, the groin has prevented the beach downdrift from being replenished, and net erosion occurs there, undermining the home at B.

418 | COASTAL ZONES

Coastal erosion is not a new problem. The rock record shows that the edges of the continent have repeatedly been submerged by the ocean, and erosion has been a reality for at least 3.5 billion years. Beach erosion has become an increasingly prevalent problem because people are building more and more on shorelines and are unaware of the closely interrelated dynamics affecting the balance between erosion and deposition on a coast. This increase is a function of several factors: a rise in sea level brought about by postglacial melting and the greenhouse effect; storms and hurricanes; and erosion and weakening of the coast by human modifications.

None of our coasts is being spared. Beaches in New York City are being stripped by the construction of groins, jetties, and stabilized inlets to the east on Long Island. Miami Beach must be restored periodically by sand

FIGURE 1
Extensive beach and cliff erosion at Malibu, California. Photo by Nicholas K. Coch.

pumped from offshore. Erosion is rampant in Louisiana, which has **40%** of the nation's wetlands, because the land is subsiding as sea level is rising. Many areas of the California coastline are being eroded because the river-supplied sand that normally replenishes them is being trapped in flood-impoundment reservoirs inland (Figure 1).

The effects of sea-level rise are dramatic. On the low-lying, gently sloping coasts of the Atlantic and Gulf of Mexico, a 1 m rise in sea level can result in a 100 m landward migration of the shoreline. This means that even elevated homes located on the beach will be destroyed within a hundred years, as waves erode away their supporting pilings.

How does a barrier island react to a rise in sea level? In a natural situation, the ocean side of the island is cut back and this material is moved across the island in storms to make washover fans on the bay side. In this model, the island width remains the same and the entire barrier island migrates landward. This barrier rollover preserves the barrier islands as the sea level rises. Present attempts to use seawalls to prevent the front of barrier islands from being cut back will lead only to their eventual destruction, because the walls also prevent deposition of washover fans on the bay side. However, it is hard to convince a coastal landowner to let his home erode away so that the barrier island at that place can be preserved by rollover.

FIGURE 2
A pile of boats washed from their moorings onto land behind the Isle of Palms, South Carolina during Hurricane Hugo, 1989. The damage from this destruction alone is estimated at $40–50 million. Photo by Nicholas K. Coch.

Storms and hurricanes always have exacted a toll on coastal areas, but the damage will only grow worse in the future as a result of human activities. On the night of September 22, 1989 we had a good look at the kind of destruction that a powerful hurricane can do to a developed coast when Hurricane Hugo made landfall just north of Charleston, South Carolina. The eye of the storm was 50 km across with winds to 217 kph (135 mph) and a wave surge of 3.5–6 m. Hugo was to prove the costliest storm in American history. Winds knocked down forests, smashed houses, picked up loose de-

FIGURE 3
Damage by wave surge and high wind from Hurricane Hugo on Sullivans Island, South Carolina. The ocean is at lower part of photo. Many homes in the oceanward row were destroyed. Photo by Nicholas K. Coch.

FIGURE 4

Breaching of a barrier island and
formation of a storm inlet at
Pawleys Island, South Carolina
during Hurricane Hugo. The ocean
is at the bottom of the photo.
Photo by Nicholas K. Coch.

bris and used it to damage structures, and washed boats onto land (Figure
2). The wave surge washed over barrier islands, removing seawalls, roads,
utility poles, and many structures (Figure 3). Where barrier islands were
narrow, low-lying, and unprotected by dunes, they were breached and inlets
were created (Figure 4).

Human modifications of coasts are causing new erosion problems. The
elimination of wetlands for housing developments has removed their natural
flood-protection properties, and storm-swollen estuaries now flood the barrier island from the bay side as a storm moves inland. As dunes are removed
for home construction, the most effective barrier to storm waters is lost.
Groins built to trap sand and widen beaches updrift have caused serious
problems downdrift (Figure 5).

Coastal development also will increase the loss of property and life in
storms. Warning systems have dramatically reduced loss of life in hurricanes,
but property loss has increased each year because of the rapid development
of our shorelines. For example, the preliminary estimates of insured property damage from Hurricane Hugo on the U.S. mainland are $3–4 *billion*.

The total damage may be two or more times that. Many states now require that new homes be built on pilings 3–5 m above sea level in order to minimize loss in coastal flooding. Unfortunately, this may be of little use if storm-eroded debris batters down the supports!

Are America's beaches all washed up? No, but the *houses* on many of them are. Serious problems can be minimized only when we understand the dynamics of shorelines and the inevitable consequences of sea-level rise. Zoning can be used to prevent building (or overbuilding) in coastal areas that are subject to erosion. Many geologists believe that instead of making expensive and futile attempts (in the long-range view) to preserve eroding shorelines, more of them should be cleared and returned to public use. Examples of such public coastal resources are the national seashores as well as numerous city, county, and state parks.

National seashore areas are kept in as natural a state as possible for all to enjoy. Storms will cause damage to roads, parking lots, bathhouses, and concessions, but these can all be rebuilt inexpensively, compared to the costs of restoring a heavily built coast. Once we understand the way coastal processes work and the geologic changes that will occur, we can preserve and use our coastal areas safely and to great benefit well into the future. This should be our goal.

FIGURE 15–24
Aerial view of stabilized tidal inlet, Southampton, New York. The direction of longshore drift
is from right to left. The beaches updrift of the inlet are wide, whereas those on the downdrift
side (left) are being eroded. Photo by Nicholas K. Coch.

groin are eroded because the groin traps the sand
that would have replenished them.

Sediment transport along a coast is disturbed also
by the construction of long rock walls called **jetties**
on either side of tidal inlets. These structures serve to
prevent sand from filling the inlet, thus maintaining
boat access between the ocean and the bay. Construction of jetties causes beach erosion downdrift in
a manner similar to groins. In addition, most of the
sedimentary particles that are able to pass the updrift
jetty do not make it across to the downdrift side
because they are swept into the inlet by tidal currents
and are deposited on tidal deltas in the bay (Figure
15–24).

Another method used to prevent beach erosion is
the construction of **seawalls** parallel to the shore.
These massive and expensive structures do prevent
erosion of the beach behind the seawall for several
years. However, they eventually accelerate beach ero-

sion. The erosion occurs because the wave energy is
reflected off a vertical wall, rather than being dissipated across a beach. The reflected energy is directed
downward toward the bottom, causing erosion of
the beach in front of the seawall. This results in a
narrowed beach, useable only at low tide. Eventually, the seawall is undermined and fails, and a new
seawall must be built in a more landward position.
Seawalls and other similar structures also prevent the
normal erosion of materials from cliffs and beaches
which are essential to maintaining the longshore-
drift system.

Other artificial methods of beach restoration, such
as rebuilding dunes with sand dredged from bays or
offshore, have had mixed results. Engineering structures built to "control" beach erosion have proved
quite costly to implement and maintain. In many
cases, an erosion problem solved in one area has produced a new problem downdrift.

SUMMARY

1. Coastal zones are areas of rapid geologic change located within the area of tidal influence.
2. Wind blowing across the ocean surface generates swells that move away from the storm center. When the waves reach water depths equal to one half of their wavelengths, they decrease in wavelength, increase in height, lose their symmetry, and eventually collapse in the surf zone.
3. Wave fronts approaching the shore at an angle move sand grains up and down the beach face, gradually moving large volumes of material along the coast in the process of longshore drift.
4. Waves approaching an irregular coastline are refracted when part of the wave enters a water depth less than one half of its wavelength. The refracted waves concentrate the energy on headlands and disperse it across the intervening bays. The effect of wave refraction over time is to gradually straighten the shoreline.
5. Wave erosion along rocky shorelines is accomplished by the force of the waves driving water under great pressure into rock crevices. Rock debris carried by the waves erodes wave-cut notches and benches. Waves undermine sediment cliffs and cause failure by slumping. Differential erosion forms features such as sea arches, caves, and stacks.
6. A variety of currents move sediment particles in the coastal zone. The process of longshore drift, as well as the longshore current, transports sediment along the beach. Rip currents move sediment out from the beach into the nearshore waters. Flood and ebb tidal currents move sediments in and out of tidal inlets and erode and deposit materials in the tidal channels, flats, and marshes of the estuarine portion of the coastal zone.
7. Coastal zones are composed of (a) estuaries and bays, (b) wetlands made up of tidal marshes, tidal flats, and tidal channels, (c) and beaches and barrier islands.
8. Storms make significant changes in sediment type and morphology along the coast. Their effects are greatest along low-lying coasts and along barrier island–bay shoreline segments.
9. Storms erode the front side of barrier islands and wash the material inland to form washover fans along the bay. When a storm cuts through a barrier island it forms an inlet and tidal delta. These storm-cut inlets usually close as a result of spit growth across their mouth.
10. Storm damage is a function of the height of the storm surge, the tidal stage, the shape of the coast, the wind velocity, speed of movement of the storm, the height and composition of the coast, and damage caused by a previous storm or by engineering structures.
11. Tropical low-lying shorelines commonly are composed of calcium carbonate secreted by organisms such as calcareous algae and corals. The calcium carbonate is broken apart by wave action into sediment particles.
12. Coral reefs are wave-resistant structures that form in warm, clear, and agitated ocean waters. Fringing reefs form on steeply sloping offshore areas, such as the sides of a volcano. Barrier reef shorelines form on gently sloping offshore areas, and the world's greatest reefs are of this type. Atolls are circular reefs that originally formed as fringing reefs around a subsiding volcano.
13. Sea level has been rising as a result of postglacial melting of ice and the greenhouse effect. The resulting landward migration of the shoreline is most pronounced in areas where the crust is stable or subsiding. The rising sea level is causing serious beach erosion and loss of coastal structures. Barrier islands roll over and migrate landward as sea level rises.
14. Human activities have caused increasingly serious problems along shorelines. Engineering structures such as groins, jetties, and seawalls have proven costly and of limited long-term help in solving the problems for which they were designed. In addition, many of these engineering structures have caused increased beach erosion downdrift.
15. Titanium, zircon, diamonds, tin, and iron are being recovered from ancient coastal placer deposits. Phosphate, limestone, petroleum, and natural gas are other resources that are being obtained from coastal deposits.

REVIEW QUESTIONS

1. Describe the differences among longshore, rip, and tidal currents.
2. What conditions are necessary for an effective longshore drift system along a coast?
3. Describe the evolution of an irregular rocky shoreline over time, and account for the changes.
4. All rivers entering the coastal zone are transformed into estuaries. What are the differences between rivers and estuaries?
5. What happens to the material eroded from a beach in a storm? How does the beach recover afterward?
6. Some storms cause far less damage than others. What factors govern the damage that can result from a storm?

7. What conditions result in the formation of different types of reefs?

8. A severe hurricane is approaching a barrier island–bay coast. What changes to the barrier island would you expect as a result of this storm?

9. We know that sea level is rising worldwide, and yet there are shorelines where this is not apparent. Why?

10. Many coastal localities have laws mandating that all new homes must be built on pilings 3–5 m above sea level to prevent flood damage. Do you think this is an effective method?

11. An east-west shoreline faces the ocean to the south. The sediment source is in the east. A home owner builds a groin out into the surf to trap sand and extend his beach. Describe the probable result of this action.

12. Describe the response of a barrier island to a rise in sea level. What does this imply about the future of homes built close to the beach?

13. Coastal flooding during storms results from ocean flooding prior to and during the storm, and flooding from the bay seaward after the storm has moved inland. How is this so?

14. Formulate a plan for coastal development and management in light of the inevitable changes that most geologists expect will occur along shorelines in the coming years.

FURTHER READINGS

American Geological Institute. 1981. Old solutions fail to solve beach problems. *Geotimes* December:18–21. (A discussion of the role of engineering structures and the rise in sea level with respect to beach erosion.)

Davis, R. A. 1986. *Oceanography–An introduction to the marine environment*. Dubuque, IA: William C. Brown. (Refer to chapters 11, "Estuaries and Related Environments"; 13, "Barrier Island Complexes"; 14, "Rocky Coasts"; and 15, "Reefs." Provides a very readable discussion of most of the topics in the present book at a slightly higher level.)

Dolan, R., Hayden, B., and Lins, H. 1980. Barrier islands. *American Scientist* 68:16–25. (Describes the formation and characteristics of barrier islands. Includes a discussion of the environmental aspects of development in coastal zones.)

Leatherman, S. P. 1979. *Barrier island handbook*. College Park, MD: University of Maryland Geography Department, 109 p. (A readable discussion of barrier island environments and dynamics.)

Lundgren, L. L. 1986. "The Coast and Coastal Management." Chap. 11 in *Environmental geology*. Englewood Cliffs, NJ: Prentice-Hall. (A very good discussion of coastal erosion and how the coastal environment can be managed.)

McCormick, L., Pilkey, O. H., Jr., Neal, W. J., and Pilkey, O. H., Sr. 1984. *Living with Long Island's South Shore*. Durham, NC: Duke Univ. Press, 157 p. (Describes many aspects of this shoreline segment; useful to homeowners. One of several similar publications giving details on specific shoreline areas in Louisiana, North Carolina, South Carolina, and Texas.)

U.S. Army Corps of Engineers, North Central Division. 1978. *Help yourself—A discussion of erosion problems on the Great Lakes and alternative methods of shore protection*. General Information Pamphlet, 25 p. (A detailed, well-illustrated treatment of major structural methods used in coastal stabilization.)

16
Oceans

View from the U.S. Space Shuttle *Columbia,* 300 km above the coast of the northeastern United States. New Jersey is at far left and southern Canada is at far right. Visible at left are Long Island, Cape Cod, and Nantucket Island and Martha's Vineyard (next to Cape Cod). The portion of ocean at right shows Georges Bank, one of the world's most productive fishing areas. Photo courtesy of National Aeronautics and Space Administration.

tudies of continental geology began in the 1700s, and by the end of the nineteenth century we had a fairly detailed picture of continental composition and topography. In contrast, modern oceanographic studies began on December 21, 1872, when a British research ship, the H.M.S. *Challenger,* set out from Portsmouth, England. Three and a half years and 68,890 miles later, the *Challenger* returned after conducting oceanographic measurements in all the oceans except the Arctic. The data that *Challenger* scientists gathered—on water depths and currents; on the physical, chemical, and biological aspects of the water; and on the nature of bottom sediments and life—were published in 50 volumes. These provided our first detailed information about the oceans.

Not much more data was obtained until the 1950s, in spite of the fact that oceans cover 72% of the Earth's surface. For example, only three decades ago, we still thought that ocean basins were relatively featureless and were submerged parts of continents. Today, we know differently: far from being flat-bottomed, the oceans contain the longest mountain ranges on Earth and mountains having relief greater than any on the continents. In addition, we now know that currents not only move along the surface of the seas but sweep along deep parts of the ocean floor as well, and that the ocean floors are not covered by a uniform sediment type but by a wide variety of sedimentary facies.

Oceanography is the study of the physical, chemical, biologic, and geologic aspects of oceans. In this chapter we will look first at the topography on the seafloor; then at the nature and behavior of the water filling the ocean basins; and finally at the sedimentary facies deposited on the seafloor. Most of the evidence now used in support of the plate tectonics model has come from recent studies of the ocean floors, so we will conclude by discussing the origin of ocean features in light of the plate tectonics theory.

EXPLORING THE OCEAN FLOORS

Oceanographers today can study the oceans and ocean basins in ways that nineteenth-century scien-

FIGURE 16–1

Precision depth recording (PDR). A sound impulse generated by the research vessel is reflected off the ocean bottom and recorded by receivers on the vessel. The depth of the water is calculated by measuring the delay time between the impulse generation and return of the signal, and knowing the velocity of sound in water.

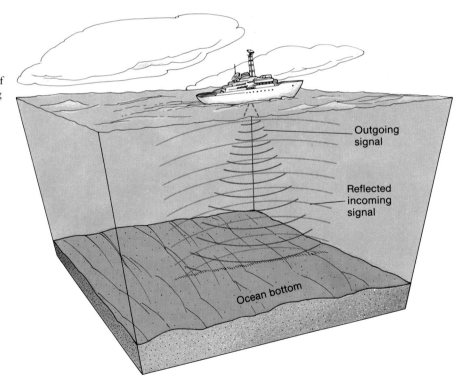

Outgoing signal

Reflected incoming signal

Ocean bottom

FIGURE 16–2

Profiles of the ocean floor. (a) Precision depth recorder (PDR) record. The vertical axis is exaggerated nine times compared with the horizontal axis to show small details. (b) A drawing of the same profile without the vertical exaggeration. Photo by Woods Hole Oceanographic Institute.

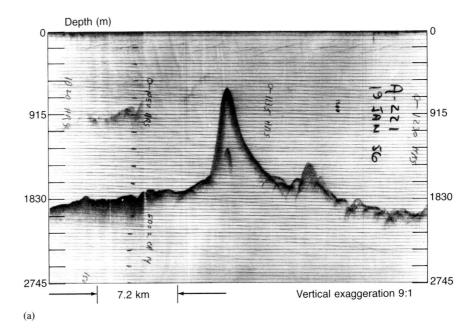

Depth (m)

915

1830

2745

7.2 km   Vertical exaggeration 9:1

(a)

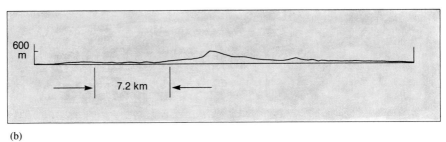

600 m

7.2 km

(b)

tists could only dream of. Changes in the techniques of measuring water depths and collecting samples of sediments illustrate the extent of these advances.

Measuring the Depth of the Oceans

Scientists on the H.M.S. *Challenger* obtained the first accurate ocean depth readings. They simply lowered a weighted line and recorded the depth when the weight hit the bottom. The process was long and tedious. In deep water, it took as much as an hour and a half for the weight to reach the bottom.

Today, water depths can be obtained rapidly and under adverse weather conditions by using an electronic device called a **precision depth recorder (PDR)**. In using a PDR, the research vessel generates a sound which is reflected (echoed) from the ocean floor to a sensor on the ship. The depth is calculated according to the time it takes for the sound signal to reach the seafloor and be reflected (Figure 16–1). The results of many PDR analyses provide a continuous profile of the bottom topography, as shown in Figure 16–2.

Like all graphs, PDR recordings are scaled—for example, in Figure 16–2, each division of the vertical scale (for water depth) represents 100 m, and the increment shown on the horizontal scale (for distance) represents 7.2 km. However, these scales have been deliberately selected to exaggerate the vertical scale out-of-proportion to the horizontal scale, so that very subtle changes in relief can be detected. This **vertical exaggeration** of the ocean-bottom relief makes it appear greater than in reality, but aids in analysis.

Collecting Samples of the Seafloor

When weighted lines were used to measure water depth, the bottom of the weight often was coated with sticky tar or wax so that it would bring up a small sample of the sediment from the bottom. Larger samples were collected by dredging— lowering a device to scoop up whatever lay on the bottom. Dredging still is used because it collects large samples relatively inexpensively. However, dredge samples are not useful for detailed studies because the procedure tends to mix the surface layers.

To collect samples of material that lie below the surface of the ocean floor, oceanographers use a weighted cylindrical coring tube called a *piston corer*. This device cuts through the sediment much like an apple corer cuts through an apple, obtaining a cylindrical sample. The sediment sample rises in the tube as it cuts deeper below the bottom. A piston within the tube creates suction that holds the sediment core within the tube as it is withdrawn from the bottom. Modern piston corers can recover undisturbed cores up to 30 m long (Figure 16–3). Piston core samples reveal changes in the depositional conditions of the ocean floor with time; they have provided the detailed climatic record discussed in Chapter 13.

Today scientists can descend into even the deepest parts of the oceans in small research submarines. They can observe features directly or take photographs of them. Most submersibles also are equipped with mechanical arms which can be maneuvered by the scientists to select geological and biological samples.

Piston cores are limited to approximately the upper 30 m of ocean floor, so the Deep Sea Drilling Project (DSDP) was initiated to obtain sediment and rock samples many times deeper. DSDP adopted many of the standard drilling techniques used in oil fields, but major technological changes had to be made to enable a ship to remain in position and successfully drill a hole below great depths of water. A specially designed ship, the *Glomar Challenger* (Figure 16–4), was used in the drilling program, and many holes went completely through the sediment floors and into the underlying bedrock. Samples from these holes have provided a complete record of the origin of the ocean basins and how they have changed over time.

The Deep Sea Drilling Program obtained cores from 624 sites through 1983, using the *Glomar Challenger*. In 1985 a new Ocean Drilling Program (ODP) continued the work, using the drilling ship *JOIDES Resolution*. As of April 1990, the ODP had completed drilling at 184 sites.

The Rock Sequence Under the Ocean Floors

Information on ocean sediments and the rocks beneath them has been obtained as described above. Additional information on deeper deposits has been obtained from techniques in which research ships use much more powerful sound sources than those used in PDR work. Waves produced by these devices not only are reflected from the seafloor but penetrate

FIGURE 16–3
Layering in a portion of a piston core from the Atlantic Ocean floor east of Martinique in the Windward Islands. Water is very deep at this site, some 5025 m. Depth of the core is shown in centimeters below the seafloor. Calcareous ooze appears down to 220 cm; the sediment between 220–230 cm is terrigenous sand and muddy sand. The section below 230 cm consists of interbeds of the sediments in the two sections above. Photo courtesy of Lamont-Doherty Geological Observatory/Deep Sea Sample Repository.

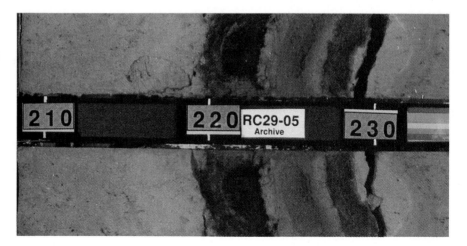

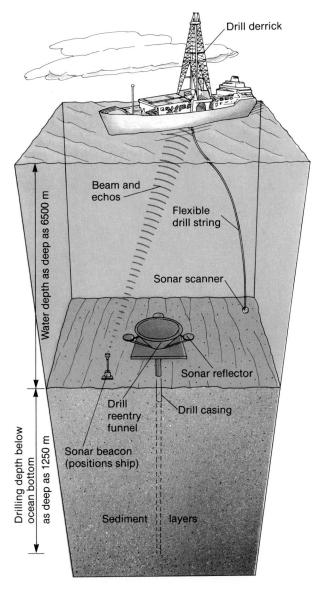

FIGURE 16-4

The *Glomar Challenger* is specially equipped to take long cores of ocean bottom sediments and underlying rocks in great depths of water. The vessel is positioned by sonar over the drill site and is maintained in that position by powerful thrusters on the sides and ends of the ship (sea and wind conditions are sensed electronically and computers activate the appropriate thrusters). The drill string may be removed from the hole and replaced with guidance from a sonar scanner, which seeks the sonar reflectors positioned around the hole.

below it and are reflected by the various layers. This technique can produce a "cross section" that shows both the stratification and the structure of sediment rocks beneath the ocean floor.

Observations at many sites have determined that the ocean floor contains a distinctive set of layers. The rock equivalent of this sequence, as found in the geologic record, is called the **ophiolite suite** (Figure 16–5). The base of the ophiolite suite is ultramafic rock, such as dunite and peridotite. The ultramafic rocks pass upward into mafic intrusives (gabbro) and closely spaced parallel basaltic dikes called **sheeted dikes**. These are overlain in turn by pillow basalt, interlayered baked sediments and basalt flows, and finally by shale, which represents consolidated former oceanic sediments.

SEAFLOOR TOPOGRAPHY

New research tools described above have given us a detailed picture of ocean-floor topography. We now know that the floor is quite rugged in some areas, but so flat in others that it is necessary to use vertical exaggeration to show many of the features. Oceanic islands and mid-oceanic ridges rise high above the ocean floors, whereas oceanic trenches extend far below the average depth of the sea. In fact, the maximum relief on the ocean floor—the distance between its highest and lowest points—is greater than the relief on the continents. If Mount Everest, the highest mountain on the continents (altitude 8848 m), were placed in the deepest oceanic trench (depth −11,022 m), its top would not come close to sea level.

Figure 16–6 is a map of the North Atlantic Ocean basin showing major topographic features. Figure 16–7 is a simplified and vertically exaggerated profile of the seafloor off the east coast of North America. As we discuss ocean-floor topography, we frequently will refer to these figures.

Continental Shelves

Every continent is surrounded by a **continental shelf**—a gently sloping surface that extends seaward from the shoreline (Figures 16–6 and 16–7). Continental shelves have very gentle slopes, typically with about 1 m vertical drop over a horizontal dis-

FIGURE 16–5
Geologic section through the ocean
floor, showing the ophiolite suite.

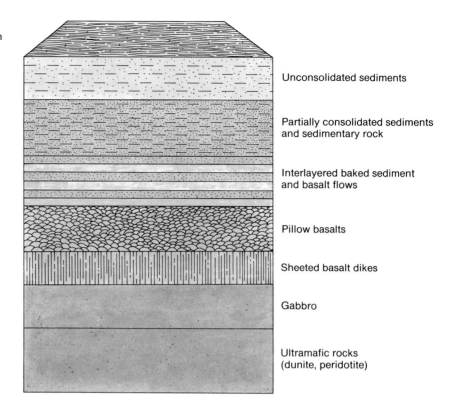

Unconsolidated sediments

Partially consolidated sediments
and sedimentary rock

Interlayered baked sediment
and basalt flows

Pillow basalts

Sheeted basalt dikes

Gabbro

Ultramafic rocks
(dunite, peridotite)

tance of 500 m (a gradient of 1:500). Continental shelves are the shallowest parts of oceans, with water depths at their outer edges ranging from about 35 m to about 240 m. The widths of the shelves also are highly variable. The stable east coast of North America has a wide shelf, whereas that of the tectonically active west coast is much narrower and is broken by faults into a series of elongate basins.

Because the continental shelves are the shallowest parts of the ocean basins, they were affected most severely by the large-scale changes in sea level associated with Pleistocene glaciation. When continental glaciers reached their maximum extent during the Pleistocene, sea level may have been 100–130 m lower than today. The shoreline would have been far out on most continental shelves, where much of what is now beneath water would have been exposed to the air.

In colder climates, continental glaciers moved across the areas of the exposed shelf, leaving an irregular glacial topography similar to that seen on the continents. In warmer latitudes, sediments were deposited far out on the shelves by streams which

flowed across wide coastal plains. We can trace the courses of these ancient streams by PDR data, which outline elongate depressions called **shelf valleys** (Figure 16–7). Many shelf valleys, such as the Hudson Shelf Valley off New York City (Figure 16–8), connect with the mouths of modern rivers. This indicates that the Hudson River at one time flowed across what is now the continental shelf.

The topography and sediments found on continental shelves in the middle to high latitudes are largely the result of stream or glacial deposition, and they have been modified only recently by ocean processes. The continental shelves are the only part of the ocean basins for which this can be said; they also are the only part which can be thought of as submerged portions of the continents.

Continental Slopes and Submarine Canyons

The outer edge of the continental shelf terminates at the **shelf break,** which marks the beginning of a more steeply sloping region called the **continental slope** (Figures 16–7 and 16–8). Gradients on the

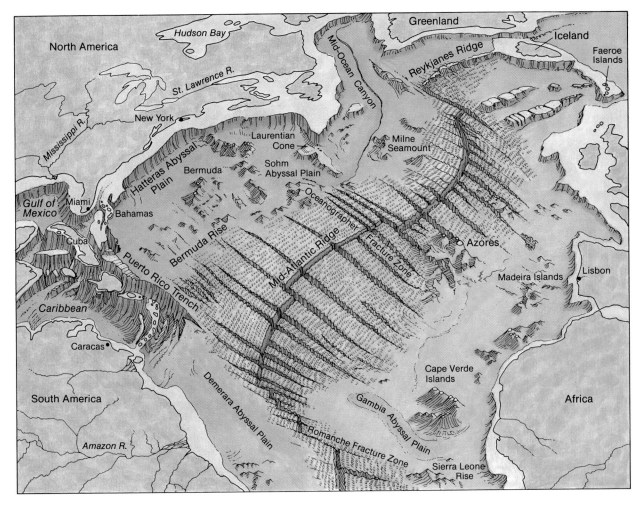

FIGURE 16–6
Topographic features on the floor of the Atlantic Ocean. From a painting by Heinrich Berann,
courtesy of Aluminum Company of America.

continental slopes range from 1:2 to 1:40. One consequence of this steeper gradient is that sediments tend to be unstable and move downslope into deeper water if disturbed.

The continental slopes are incised deeply in places by steep, V-shaped valleys called **submarine canyons**. They extend across the shelf and down the slope onto the deep seafloor below. Major submarine canyons sometimes have tributary canyons which join the main valley, as with modern stream systems on land. Some submarine canyons, such as those off southern California, have their heads almost in the surf zone, but others begin only at the shelf break. Some, such as the submarine Hudson Canyon (Figure 16–8), obviously were linked at one time to modern streams. The Hudson Canyon extends from its shelf valley shown in Figure 16–8, which issues from the mouth of the present Hudson River Estuary. However, other canyons have no counterparts among modern rivers. Originally they may have extended from ancient rivers whose channels subsequently were buried by continental shelf sediments.

The upper parts of submarine canyons could have been cut by stream erosion at the same time as the shelf valleys, during periods of low sea level during the Pleistocene Epoch. At such times the Hudson River would have flowed across the continental shelf and deposited its sediment near the upper edge of the continental slope. However, the lower parts of the canyons could not have been carved by streams because they are found at depths far lower than any exposed by glacial lowering of sea level. They must have been formed by some type of submarine erosion process. Most oceanographers believe that the deeper parts of the submarine canyons were eroded by dense, sediment-laden currents that moved down the continental slope from their initial sites of deposition. (We will examine this topic later in the chapter.)

Continental Rises

At the base of the continental slope there is an area of gentle gradient (average 1:150) called the **continental rise**. It marks the transition from continental slope to the deep seafloor (Figures 16–6 and 16–7). The smooth surfaces of the continental rises are broken by channels, which appear to be extensions of

submarine canyons. Recent geologic studies have shown that the continental rise is a depositional feature built in part from sediments that have been transported down the continental slopes. The sediment-laden currents that carve the submarine canyons deposit much of their sediment at the base of the continental slope as submarine fans. This process is very similar to the formation of a series of alluvial fans at the base of a mountain range in dry areas (see Chapter 14).

Abyssal Plains and Abyssal Hills

The continental rises gradually grade into the flattest part of the ocean basins, regions called **abyssal plains,** where gradients are typically as low as 1:1000. The abyssal plains are covered by sediment from two sources: siliciclastic sediment derived and transported from the continents into the depths by

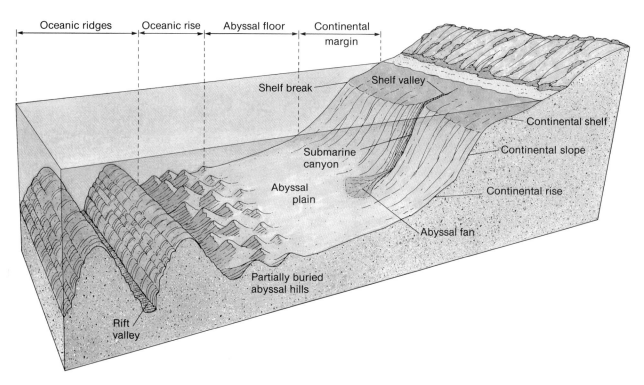

FIGURE 16–7
Generalized profile showing the submarine topography across the western part of the North Atlantic Ocean. The profile has a high vertical exaggeration so that subtle features can be seen more clearly.

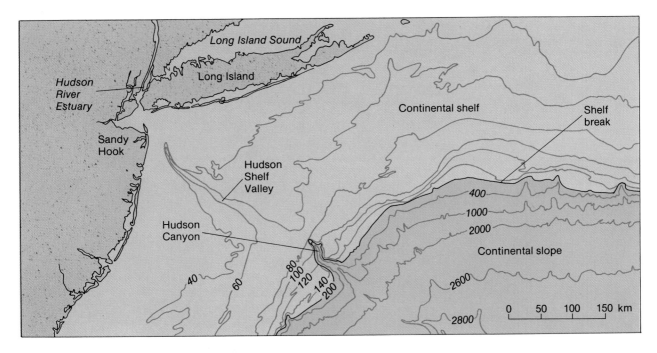

FIGURE 16-8
Contour map of continental shelf and slope south of Long Island, New York. Note the Hudson Shelf Valley that cuts across the shelf between the Hudson River Estuary and the Hudson Canyon.

currents, and biogenic and chemically precipitated sediment that forms within the ocean.

The flatness of the abyssal plains is broken by **abyssal hills,** which rise a few meters to hundreds of meters above the plains and range up to 10 km in diameter. Deep sea drilling has shown that abyssal hills are extinct basaltic volcanoes capped by a thin layer of sediment. The hills are less abundant near continental rises, possibly because they have been buried completely by continental sediment.

Oceanic Ridges and Fracture Systems

Large linear mountain ranges called **ocean ridges** rise from the depths of every ocean basin and are interconnected in a worldwide network over 60,000 km long. In the Atlantic Ocean the ridge (Mid-Atlantic Ridge) is at the center of the basin, whereas in the Pacific Ocean the ridge (East Pacific Rise) is near the eastern margin (Figure 16–9). Segments of the ridges are given geographic names for purposes of identification. Some of the segments are very

long—the Mid-Atlantic Ridge, for example, extends for about 21,000 km—but others, such as the Galápagos Rise, are short.

The Mid-Atlantic Ridge shows the features typical of oceanic ridges. Over most of its length, it is below sea level, but in some places individual peaks form islands such as Iceland in the North Atlantic, the Azores, and Saint Helena and Tristan da Cunha in the South Atlantic (some are shown in Figure 16–6). At the edges of the ridge, the ocean floor gradually rises above the abyssal plain, and seafloor relief becomes more rugged toward the central part of the ridge. At the very crest is a deep linear depression called a **central rift valley,** which is flanked by steep rift mountains. The central valley is a divergent plate boundary where two plates are moving away from each other. This area is characterized by current earthquake activity and basaltic volcanism.

Look at the North Atlantic in Figure 16–6. Notice that the Reykjanes Ridge on which Iceland is located is not a continuous ridge but is offset by a fracture that runs perpendicular to two ridge segments. By "offset," geologists mean that the two

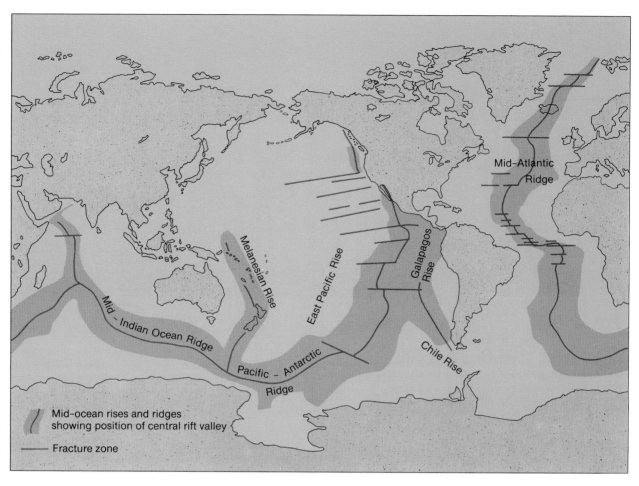

FIGURE 16–9
The interconnections among the worldwide network of ocean ridge systems. The ridge systems
are offset by fracture zones.

ridge segments were once continuous, but they later broke apart and moved away from each other while remaining parallel. In many parts of the oceanic ridge network, similar fractures cut across the rift valleys and offset them. These fractures extend the width of the ridges and most or all of the adjacent abyssal plain. In places they come close to the continental slopes, and some geologists believe that certain fractures can be traced onto the continents. Individual fractures can be traced for thousands of kilometers, and many of the larger ones have been named (Figure 16–6).

Examination of the rocks on islands rising from the ridges and the results of deep sea drilling have told us a great deal about the composition and struc-

ture of ocean ridges. They are composed of a distinctive type of basaltic lava called **MORB (mid-ocean ridge basalt),** which is different from basalt extruded on the continents. Core samples of ocean ridge basalts reveal intense fracturing and dynamic metamorphic recrystallization (see Chapter 7) near fracture zones, but metamorphism does not play a major role in the formation of the ridges. Radiometric dating of basaltic rocks collected from various parts of the worldwide ridge system shows that the basalts nearest the ridge axes are the youngest, and the farther the basalt is from the axis, the older it is. Ridge crests are covered by a thin veneer of sediments; the sediment blanket thickens away from the axes and is thickest over the oldest basalts.

FIGURE 16–10

Formation of oceanic ridges and seafloor spreading. (a) Rifting forms the central valley and adjacent rift mountains and provides an avenue for the eruption of mid-ocean ridge basalt (MORB). (b) As rifting proceeds, old MORB is carried away from the ridge crest on the spreading plates as new MORB is being erupted at the ridge crest. (c) As the process continues, the ocean widens. The oldest lavas are found at the margins of the ocean, the youngest at the ridge crests.

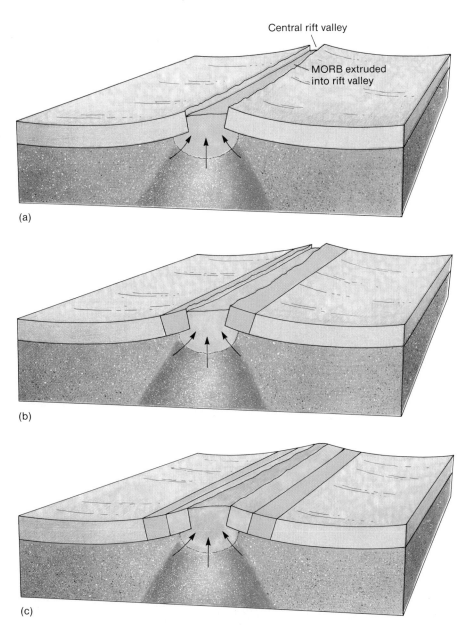

We will now consider what all these observations tell us about the origin and evolution of oceanic ridges.

Origin of Oceanic Ridges—Seafloor Spreading.

According to the plate tectonics model, ocean ridges are divergent plate boundaries where new ocean crust is being formed (Figure 16–10). The ridges form when strong tensional forces are applied to the lithosphere from below, causing breaking along the rift valleys. As the lithosphere is split into what eventually will become two separate plates, MORB rises into the rift zone and accumulates to form the ridge. As rifting continues, more lava rises to the surface and the older lavas are carried away from the rift zone in conveyor-belt fashion. The freshly extruded basalt forms the ridges, but as the older lavas are carried away from the zone of rifting, they cool, contract, and become denser. As they become denser, they sink and form the basement upon which the abyssal plain sediments accumulate.

This, according to plate tectonicists, is how ocean basins form. As a continent is broken apart, the region underlain by MORB grows larger, resulting in a basin which soon is filled by ocean waters. This process of ocean growth is called **seafloor spreading,** and ocean ridges commonly are referred to as **spreading centers**. Spreading is continuing at the Mid-Atlantic Ridge; as a result, Europe and North America are continuing to move farther apart. The process is relatively rapid as geologic processes go: during a typical college semester, New York City and London move about one cm away from each other!

Oceanic Trenches, Island Arcs, and Marginal Seas

The deepest parts of the oceans are found near the ocean margins in narrow, arc-shaped depressions called **oceanic trenches** (Figure 16–6). The greatest water depth recorded so far (−11,022 m) was measured in the Challenger Deep of the Marianas Trench in the western central Pacific. The trenches are mostly oriented so that their convex sides face the deep ocean whereas their concave sides face the nearest continent (Figure 16–11).

On the continental (concave) side of most trenches is an **island arc**—a curved string of islands characterized by active volcanism and frequent earth-

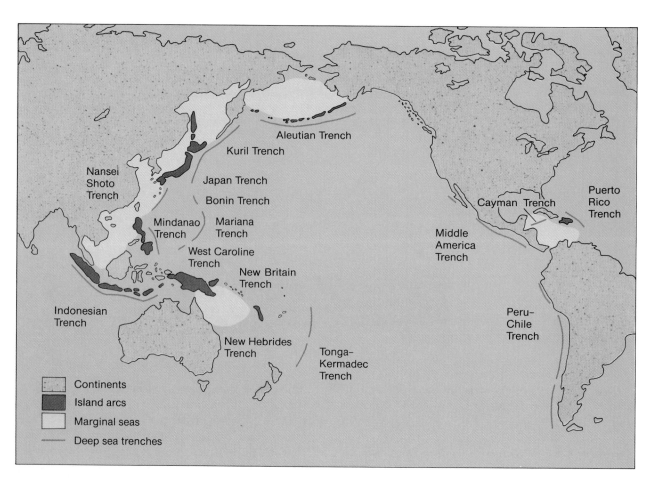

FIGURE 16–11
Location of oceanic trenches, island arcs, and marginal seas in the Pacific Ocean.

quakes. For example, the Aleutian, Japanese, and Philippine island systems in the Pacific Ocean are island arcs bordered by trenches. Despite its great size, the Atlantic Ocean has only two island arcs, the Greater Antilles Arc and the Scotia Arc. The Greater Antilles Arc separates the Atlantic Ocean from the Caribbean Sea and includes the islands of Puerto Rico, Hispaniola, and Cuba (Figure 16–6). In just one instance, a continent borders an oceanic trench. The trench is the Peru-Chile, which lies off the west coast of South America; rather than an island arc, the Andes Mountains lie near the trench.

Volcanism on the island arcs is more varied in composition than volcanism along the ocean ridges, consisting primarily of andesite with lesser amounts of rhyolite and basalt. Thick sequences of volcaniclastic debris flank the island arcs and are interlayered with lavas and tephra. The rocks are deformed by compression and show the effects of regional metamorphism. Debris eroded from island arcs often is deformed as it is being deposited, producing a distinct sedimentary rock called **melange** (Figure 16–12); it has poorly sorted blocks of partially lithified sediment chaotically arranged in a fine-grained matrix such as shale. Melange is found in the area between the trench and the volcanic arc in a mass called an **accretionary wedge**.

Behind (on the concave side of) most island arcs are relatively shallow basins called **marginal seas** or back-arc basins (Figure 16–11). These basins separate the island arc–trench system from the nearest continent. The Caribbean, the Sea of Japan, and the Philippine Sea are three of the larger marginal seas. Island arc–trench systems and their associated marginal seas are more abundant around the margins of the Pacific than in any other ocean.

The shallow marginal seas trap sediment deposited into them and prevent this sediment from reaching the abyssal plains beyond the island arcs. Where arc-trench systems nearly surround an ocean basin such as the Pacific, this severely restricts the amount of continent-derived sediment supplied to the basin. We shall discuss the significance of this later in the chapter.

Origin of Island Arc–Trench Systems by Subduction. In the plate tectonics model, island arc–trench systems are convergent plate boundaries at which oceanic crust is returned to the asthenosphere by the process of subduction, which is illustrated in Figure

FIGURE 16–12
Melange. Photo by Nicholas K. Coch.

FIGURE 16–13
Geologic section through an arc-trench system.

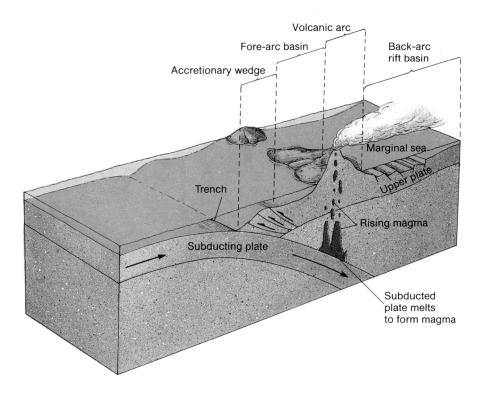

Volcanic arc

Fore-arc basin

Back-arc
rift basin

Accretionary wedge

Marginal sea

Upper plate

Trench

Rising magma

Subducting plate

Subducted
plate melts
to form magma

16–13. During subduction, one lithospheric plate is thrust beneath another. The descending plate drags the overlying plate down to form the trench; this disturbs the accumulated sediment and forms the melange of the accretionary wedge. Heat generated by friction causes melting at depth along the subducted plate, and magma rises to form the volcanic islands that make up the island arc. Continued collision of the two plates causes bending and breaking of rocks along the boundary.

The presence of the ophiolite suite in the rock record on many arcs indicates that material originally from the ocean floor has been added to the arc during the subduction process. The ophiolites are trans-

FIGURE 16–14
The process of obduction. Material is scraped from the descending plate and plastered onto the overlying plate.

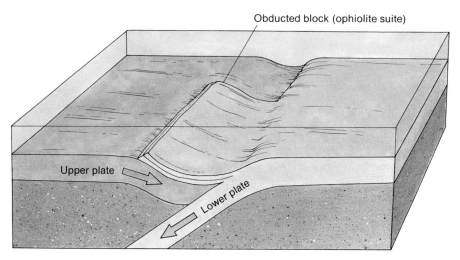

Obducted block (ophiolite suite)

Upper plate

Lower plate

ported into the island arc by a process called **obduction**. During obduction, the overriding plate acts like the blade of a carpenter's plane, literally scraping a section from the subducting plate (Figure 16–14). The detached (obducted) segment then is plastered onto the upper plate as the ophiolite suite.

Basaltic Island Chains

Chains of oceanic islands composed of basaltic shield volcanoes are called **"Hawaiian-type" islands** after their most famous example. These oceanic mountains differ from the oceanic ridges and the island arcs previously discussed, because they are associated with neither central rift valleys nor oceanic trenches.

The Hawaiian islands show systematic changes along their length, which provide a clue to their origin (Figure 16–15). The southernmost major is-

land, Hawaii, has active volcanism and its landscapes generally show little change from a fresh volcanic terrain. The next island to the north, Maui, has a dormant volcano and a landscape that has been modified considerably more by surficial processes. The northernmost island, Kauai, has extinct volcanoes and a landscape that has been deeply carved by weathering and erosion into one of the most beautiful in the world.

To geologists, such changes in weathering and erosion indicate a difference in relative age among the islands, with the northernmost islands being older than the southernmost. Indeed, radiometric dating of volcanic rocks shows that the Hawaiian Islands become progressively older the farther north they are from the area of active volcanism on Hawaii. Furthermore, the basalts in the islands of the Midway Group, which is north of the Hawaiian Islands,

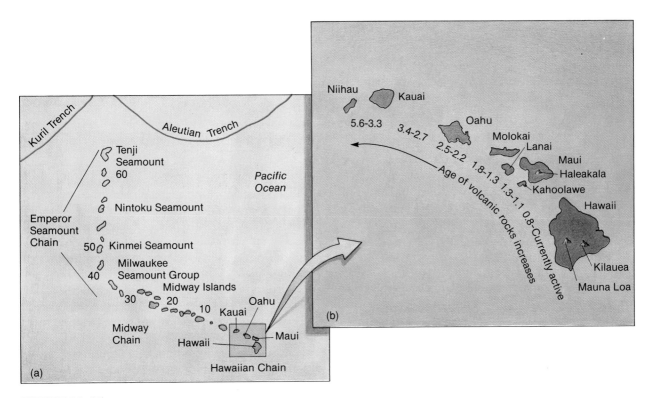

FIGURE 16–15
Ages of volcanism (in millions of years before the present) in the Hawaiian Islands, Midway Islands, and the Emperor Seamount chain. Ages on right-hand map are those of Hawaiian Islands that are above sea level. Ages on left-hand map are from submerged basaltic volcanoes that underlie atolls and coral islands (Midway Island Chain) or seamounts (Emperor Seamount Chain). Ages of volcanic rocks derived from I. McDougall, "Potassium-Argon Ages from Lavas of the Hawaiian Islands," *Geological Society of America Bulletin* 75:107–28, 1964.

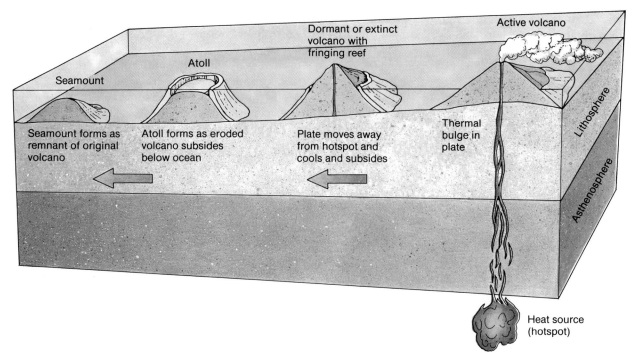

Seamount

Atoll

Dormant or extinct
volcano with
fringing reef

Active volcano

Seamount forms as
remnant of original
volcano

Atoll forms as eroded
volcano subsides
below ocean

Plate moves away
from hotspot and
cools and subsides

Thermal
bulge in
plate

Lithosphere

Asthenosphere

Heat source
(hotspot)

FIGURE 16–16

Origin of Hawaiian-type islands. Volcanic islands form over a hotspot and are moved away from it by continual plate movement. As the islands move away from the hotspot, they cool, and the rock density increases, resulting in subsidence of the island. This subsidence plus surficial erosion reduces the islands in height more and more with time. Fringing reefs grow upward around the subsiding islands to form atolls. When the islands eventually sink below the surface, they evolve into seamounts.

are older than the Hawaiian basalts, and the Emperor Seamount Chain, which is farthest north, is the oldest of all.

Origin of Hawaiian-Type Islands—Hotspots in the Mantle. Hawaiian-type islands occur *within* a plate rather than at a divergent margin (oceanic ridges) or at a convergent margin (island arcs). How can we have volcanic activity within a plate? How does it lead to the formation of islands that show a steady increase in age away from the active volcanic center?

The answer lies in the presence of an area of high heat flow below the asthenosphere. Such areas are called **hotspots** and cause melting in the mantle. This forms basaltic magma, which moves upward through the lithosphere (Figure 16–16) and is extruded first on the ocean floor above. As more and more lava is extruded, the mass rises above the ocean surface as a Hawaiian-type island.

This explains how a single volcano forms, but it does not explain how a chain of volcanoes is created. The answer is that the hotspot mechanism must be combined with plate movement to produce multiple volcanoes. As soon as a volcanic island is formed, the moving plate carries it away from the hotspot and a new volcano begins to form on the ocean floor above the hotspot. That island in turn is moved away from the hotspot, and another island begins to form there; and so on (Figure 16–16). Long-term plate movement results in a long train of volcanoes whose age increases away from the hotspot, just as the ages of the Hawaiian Island–Emperor Seamount Chain increase toward the north.

Seamounts and Guyots

Features that resemble submerged volcanoes and that rise at least 1 km above the seafloor are called

seamounts. Seamounts are particularly abundant along the ocean fractures that offset the ocean ridges. They are extinct basaltic volcanoes that apparently never managed to build their cones high enough to reach sea level.

Submarine hills that resemble seamounts in size and shape, but have flat tops rather than jagged peaks, are called **guyots** (pronounced gee-'yo). Shallow-water fossils collected from the sediment on guyots and the similarity of their flat tops to wave-cut terraces (see Chapter 15) suggest that guyots once were volcanic islands that were eroded by wave action, planed off at sea level, and then slowly subsided beneath the surface of the sea. Atolls form where coral reefs can maintain their growth, whereas the oceanic islands subside.

SEAWATER

The water that fills the ocean basins differs significantly from place to place in biologic content, salinity, concentration of dissolved gases, temperature, density, and concentration of suspended particles. A large volume of ocean water with its own distinctive characteristics is called a **water mass.** Water masses move through the ocean basins much like the hot and cold air masses that move across the Earth's surface.

Salinity

The most notable characteristic of seawater is its **salinity**—the concentration of dissolved solids. It is this dissolved load that gives seawater a salty taste and makes it unfit for human consumption. Much of this load comes from chemical weathering of earth materials, which frees ions that are carried away by streams and usually reach the ocean. Roughly half of the naturally occurring elements have been detected in solution in seawater, including the only slightly soluble, rare elements such as gold and silver. However, the most abundant ions are the most soluble ones, and seven of these account for more than 99% of the ocean's dissolved load, as shown in Table 16–1. It is easy to see from this table why the common evaporite minerals are halite (NaCl) and gypsum ($CaSO_4 \cdot 2H_2O$).

Salinity is generally expressed in parts per thousand (ppt) of dissolved ions in water. The average salinity of surface ocean water is 35 ppt (3.5%), but salinity varies from area to area. Low salinities occur in humid tropical areas where high rainfall dilutes the surface waters and in polar regions where melting ice from glaciers has the same effect. Higher salinities occur in the subtropical desert latitudes where high evaporation rates exceed rainfall and dissolved ions are concentrated to a larger degree. The highest salinities occur in hot, very dry areas such as the Persian Gulf and the Red Sea. In the shallower parts of these areas, the waters may be **hypersaline,** containing more than 40 ppt of dissolved ions. Coastal waters in temperate climates may have salinities lower than ocean waters as a result of the addition of freshwater by rivers.

Dissolved Gases

All the gases of the atmosphere are found in ocean waters, but oxygen and carbon dioxide are probably the most important for living creatures and for ocean sedimentation. Both gases are present in the upper

TABLE 16–1
The most abundant ions dissolved in seawater.

Ion	Formula	Percent of All Dissolved Ions
Chlorine	Cl^-	55.04
Sodium	Na^+	30.61
Sulfate	$(SO_4)^{2-}$	7.68
Magnesium	Mg^{2+}	3.69
Calcium	Ca^{2+}	1.16
Potassium	K^+	1.10
Bicarbonate	$(HCO_3)^-$	0.41
Total		99.69

few meters of the ocean, where exchange and mixing with the air are greatest.

The **photic zone** is the upper few tens of meters in which plant life flourishes because of the penetration of sunlight. Through a series of chemical reactions called photosynthesis, plants use the Sun's energy to manufacture their own tissues from carbon dioxide. Oxygen is released by the reactions. All but the very top layer of the photic zone is thereby depleted of carbon dioxide and enriched with oxygen. Below the photic zone, no plants exist to remove carbon dioxide from the water, but dissolved oxygen is extracted by animals. Some oxygen also is consumed by the decay of organic debris falling toward the seafloor. As a result, dissolved oxygen tends to decrease with depth. If it were not for oceanic circulation, most of the bottom waters would be devoid of oxygen and hence of most life forms.

Temperature

Ocean water temperatures vary with latitude and depth. As would be expected, surface-water temperatures are highest in equatorial regions and decrease steadily toward the poles. A typical profile through the ocean at a middle latitude can be divided into three zones on the basis of temperature (Figure 16–17). The uppermost zone, extending to depths

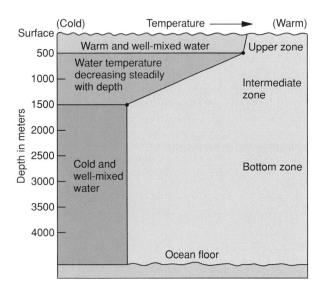

FIGURE 16–17
Generalized variation of temperature with depth in the oceans.

of about 500 m, is the warmest and is characterized also by a relatively homogeneous temperature. The warmth comes from the absorption of heat from the Sun, and the temperature homogeneity is the result of the waves and currents that act to mix the water thoroughly.

The bottom layer is homogenized by deep currents but is extremely cold, often just barely above 0°C, the freezing point of freshwater. In the intermediate zone, temperatures decrease steadily with depth between the upper and lower zones.

Density

The density (mass per unit volume) of seawater depends on several factors, including temperature, pressure, salinity, and the amount of suspended sediment. Increases in salinity and amount of suspended sediment increase the mass of the water, thereby increasing its density. Water expands (increases in volume) when heated, so warm tropical water is less dense than cold Arctic or Antarctic water.

Density is a controlling factor in oceanic circulation because the movement of some water masses is initiated when dense water sinks beneath less-dense water. These movements are examined below.

OCEANIC CIRCULATION

Ocean waters are constantly moving, due to several agents. We already have seen how winds produce waves and longshore currents (Chapter 15) and how the Sun and Moon help generate tidal currents (Chapter 2). In addition, there are well-established global oceanic circulation patterns in which water masses move both along the surface and at depth. These water movements—called surface currents and deep currents—play important roles in sedimentation, climate modification, and the production of marine food resources.

Surface Currents

The Earth's atmosphere has a well-established circulation pattern (see Chapter 14). The prevailing wind system in an area exerts a force on the surface waters in the oceans, which sets them into motion to produce **surface currents**. Wind is the basic driving

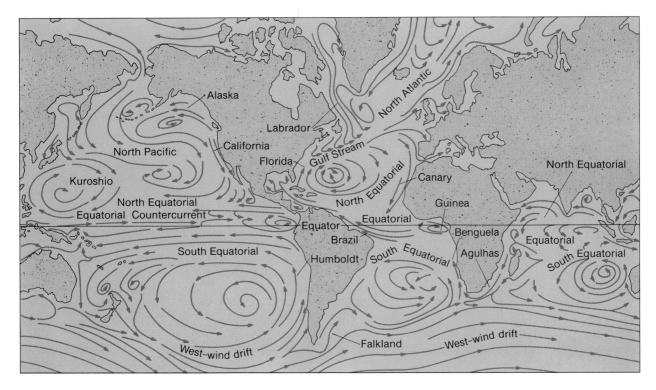

FIGURE 16–18

Map of the world's major surface currents. The currents in the Northern Hemisphere curve clockwise, while those in the Southern Hemisphere turn counterclockwise as a result of the Coriolis effect.

force, but the actual direction of the currents is influenced by the Coriolis effect (discussed for air movement in Chapter 14) and by deflection around large landmasses. Water masses can move great distances, as shown by the major surface currents in Figure 16–18. We might expect surface motions to be restricted to the shallower depths of the oceans, but there is evidence that some surface currents may extend downward to depths of 2000 or 3000 m.

The importance of these currents to humans is illustrated by one of the best known—the Gulf Stream of the North Atlantic. Surface waters in the North Atlantic are driven southward into equatorial regions by the northeast trade winds but then are deflected northward to move along the east coast of North America as the Gulf Stream. The Gulf Stream transfers equatorial heat energy to higher latitudes. This warmth moderates what otherwise would be extremely severe climates in such places as Iceland, the British Isles, and Scandinavia.

Deep Currents

Oceanographers have long known about and mapped surface currents, but they became aware of **deep currents** only recently. Deep currents are quite different from those on the surface at the same place. For example, off North Carolina, the Gulf Stream flows northward, while under it a much colder current flows southward.

Surface currents are basically horizontal movements of water generated by wind. In contrast, deep currents have significant vertical movement as well as horizontal and are generated by the differences in density of water masses. Water density is controlled by temperature, salinity, and turbidity.

Hot and Cold Currents. Cold water at the poles sinks because of its density and flows along the bottom toward the warmer equatorial regions. In the Atlantic Ocean section shown in Figure 16–19,

FIGURE 16–19

Vertical north-south generalized section through the Atlantic Ocean, showing deep current circulation resulting from density differences among the water masses. The section has a significant vertical exaggeration so that the currents can be shown more clearly. ABW = Antarctic Bottom Water; AIW = Antarctic Intermediate Water; NADW = North Atlantic Deep Water; MW = Mediterranean Water (saline). After G. Neumann and W. J. Pierson, Jr., 1966, *Principles of Physical Oceanography*, Prentice-Hall, p. 466, with permission of the authors.

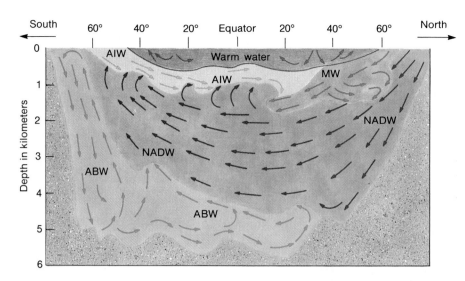

these movements are the North Atlantic Deep Water mass, which originates in the north polar region, and the Antarctic Bottom Water mass, which originates in the south polar region. This circulation mixes the ocean waters and forms a homogeneously cold region near the seafloor, as described earlier.

Saline Currents. The difference in density between highly saline and normal marine waters is responsible for the density currents that control water circulation

in the Mediterranean Sea (Figure 16–20). The Mediterranean is situated in a semiarid region, and evaporation has produced waters that are more saline (and hence, denser) than those of the Atlantic Ocean. At the Strait of Gibraltar, where the two bodies are connected, relatively light-density (normal salinity) Atlantic Ocean water flows along the surface into the Mediterranean, while denser (more saline) Mediterranean water flows beneath it into the Atlantic. The Mediterranean water sinks to about

FIGURE 16–20

Density current movement between the Mediterranean Sea and Atlantic Ocean. Relatively cool and normally saline Atlantic water flows along the surface into the Mediterranean. The water is warmed, evaporation increases, and the water becomes more saline (therefore denser) and sinks. This warm, saline Mediterranean water flows back into the Atlantic as a distinct density current which sinks to about 1000 m in the Atlantic.

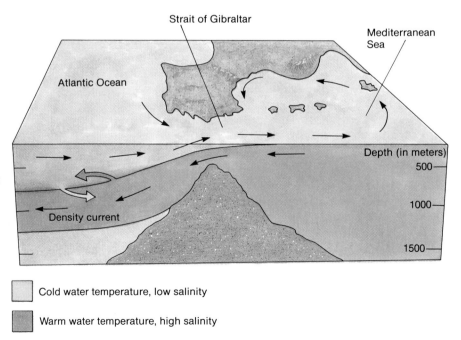

1000 m in the Atlantic and forms a distinct water mass (see Figure 16–20) before being dissipated gradually. This circulation pattern effectively brings about a complete exchange in the waters of the Mediterranean Sea in approximately 75 years.

Turbidity Currents. Sediments deposited near the upper part of the continental slope become unstable when they are piled more steeply than the angle of repose (see Chapter 10) or when they are disturbed by vibrations, such as in an earthquake. They then move down the slope as a **turbidity current,** a cloudy mass of water and suspended sediment that sinks because its sediments make it denser than its surroundings (Figure 16–21). Turbidity currents are important mechanisms for carrying sediment from the continental margins into the deep oceans. They are the source of much of the sediment that has built the continental rises, covered some of the abyssal hills, and spread out onto the abyssal plains. In addition, abrasion of the seafloor by particles suspended in the turbidity currents may carve submarine canyons and their extensions across the continental rises.

OCEANIC SEDIMENTATION

Recent advances in oceanographic techniques have given us a detailed picture of sedimentation on the ocean floors. We now know that the ocean floors are covered by a wide variety of sediments that reflect differences in climate, terrestrial sediment supply, organic productivity, current patterns, and the tectonic framework of adjacent continental margins. Marine geologists recognize three major classes of oceanic sediments: terrigenous, biogenic, and authigenic.

Terrigenous Sediments

Sediments derived from the erosion of continents or oceanic islands are called **terrigenous sediments**. Rivers bring sediment to the ocean where it is redistributed by waves and currents. Terrigenous sediments in lower latitudes generally are fine grained because of the prevalence of chemical weathering in those areas. In contrast, terrigenous sediments in high latitudes are coarser because of the dominance of physical weathering and the input of glacial debris. Volcanic islands supply large amounts of tephra

FIGURE 16–21
Turbidity currents and the deposition of turbidites. Slumping of sediment on the upper continental slope results in formation of a dense suspension of sediment and water which flows down the continental slope and across the abyssal floor. Particles settling from the turbidity current produce a turbidite layer—a graded bed in which the average particle size in the sediment decreases from the base to the top. The turbidite layer overlies a layer of pelagic ooze deposited under quiet water conditions.

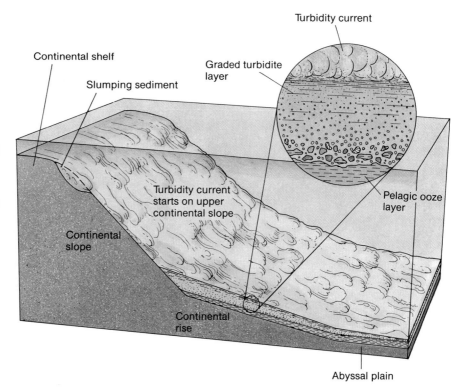

to the ocean; the deposits are thickest on the seafloor downwind and downcurrent of the islands. Many volcanic islands in low latitudes are surrounded by coral reefs (see Chapter 15), and wave erosion supplies broken reef debris to the ocean floor.

Relatively coarse-grained terrigenous sediment can be transported for considerable distances within the ocean basins by a variety of mechanisms. In high latitudes, icebergs break off the edges of glaciers and float out to sea, where they eventually melt. As they melt, the poorly sorted sediment they carry is deposited on the ocean floor. Deposition of sediment from melting icebergs is called **ice rafting**.

Both coarse- and fine-grained terrigenous sediments can be transported far out from the continents by turbidity currents. The deposits from turbidity currents grade upward from coarse material at the base to fine-grained material at the top and are called **turbidites** (see the detail inset in Figure 16–21). Each turbidite bed generally is overlain by a layer of very fine-grained material, which is deposited from the waters above in the time interval between turbidity-current flows from the continental margins. Each turbidite bed becomes finer and thinner oceanward.

Winds blow fine-grained sediment far out across the oceans. Surface and deep currents can carry this sediment, and fine particles from other sources such as rivers, great distances from where they entered the ocean. Eventually, fine-grained particles settle slowly through the water column and are deposited on the ocean floor. Deposits of organic and inorganic particles that settled through the water are called **pelagic** deposits. Marine geologists also refer to these deposits as **pelagic oozes** in reference to the consistency of the sediment when it is extruded from piston cores. Oxidation of iron in pelagic deposits produces a reddish-brown color, and consequently the pelagic oozes also are called **brown clays**.

Biogenic Sediments

Sediments formed from accumulations of marine-organism skeletons on the seafloor are called **biogenic sediments**. These are a significant part of the sediment in many parts of the oceans, and are either calcareous ($CaCO_3$) or siliceous (SiO_2) material. **Calcareous oozes** are extremely fine-grained sediments composed of the skeletons, skeletal parts, and shells of calcareous organisms (Figure 16–22). They are abundant in the South Pacific, Indian Ocean, and Atlantic Ocean, except for the colder latitudes.

The abundance of calcareous forms in surface waters is not necessarily reflected in the bottom sedi-

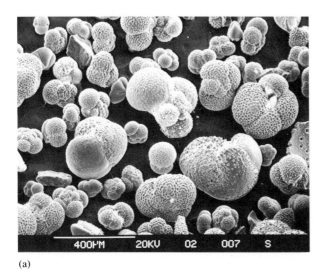

(a)

(b)

FIGURE 16–22
Scanning electron photomicrograph of biogenic ooze, showing (a) calcareous and (b) siliceous forms. Scale indicated on photograph. Scanning electron microscopy by Dee Breger, courtesy of Lamont-Doherty Geological Observatory.

ments below. The calcium carbonate from which these organisms build their skeletons becomes more soluble as pressure increases and water gets colder. Both increased pressure and colder water are characteristic of deep ocean regions, so when an organism dies and its skeleton sinks, the calcium carbonate is dissolved gradually as it passes through the cold, deep waters. At a depth known as the **calcium carbonate compensation depth (CCCD)**, pressures and temperatures are such that almost all calcareous material is dissolved before it reaches the seafloor. The CCCD varies somewhat from ocean to ocean but is generally between 3.5 and 4.2 km. Thus, in areas of deeper water, even though a large community of calcareous-shelled organisms lives at the surface, calcareous biogenic sediments may be completely absent from the sedimentary record accumulating on the seafloor below.

Siliceous oozes are fine-grained biogenic sediments composed of siliceous organisms (Figure 16–22). Siliceous oozes occur in great volumes in the colder waters at high latitudes, partly because a large amount of nutrients is contained in this water and because the shells of calcareous organisms dissolve in the cold water, enriching the pelagic ooze in the siliceous organisms.

Authigenic Sediments

Perhaps the most intriguing ocean deposits are the **authigenic sediments,** which form on the ocean floor through chemical reactions between ions dissolved in seawater and materials on the bottom. Authigenic sediments are most abundant in areas where terrigenous sedimentation rates are relatively low. The most common authigenic deposits are **manganese nodules,** which are rounded masses rich in manganese that cover wide areas of the seafloor (Figure 16–23). Similar manganese-rich materials form coatings on rocks exposed on the ocean bottom. In addition to the manganese, significant concentrations of iron, nickel, and cobalt are found in both the nodules and the coatings.

Manganese nodules apparently form by precipitation from solution, with preferential deposition on hard objects already on the sea bottom. The manganese is deposited in several stages, as indicated by the series of concentric layers observed when the nodules are cut open. Growth rates appear to be highly vari-

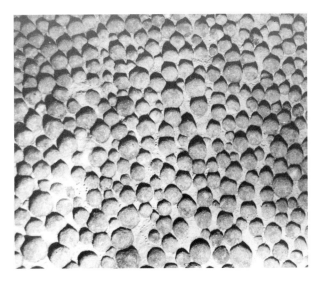

FIGURE 16–23
Field of manganese nodules on the ocean floor. Photo courtesy of Lawrence Sullivan/Lamont-Doherty Geological Observatory.

able and are reported by researchers to range from less than 1 mm to more than 100 mm per million years.

Oceanic Sedimentary Facies

The type of sediment found on the ocean floor at any place reflects the relative input of terrigenous, biogenic, and authigenic sediments. Description of a great number of bottom-sediment samples has enabled marine geologists to map the distribution of sediment facies on the ocean floor (Figure 16–24). As we have come to learn more about the origin of oceanic sedimentary facies, we have been able to use uniformitarian principles to reconstruct the paleogeography of ancient oceans from their sedimentary rock record.

THE AGE AND ORIGIN OF THE OCEANS

Ocean basins long were thought to be among the most ancient Earth features, but dating of rock samples from the Deep Sea Drilling Project showed that the basins are much younger than assumed. The oldest fossils found in sediments are of Jurassic age (roughly 180 million years old). Dating of basaltic

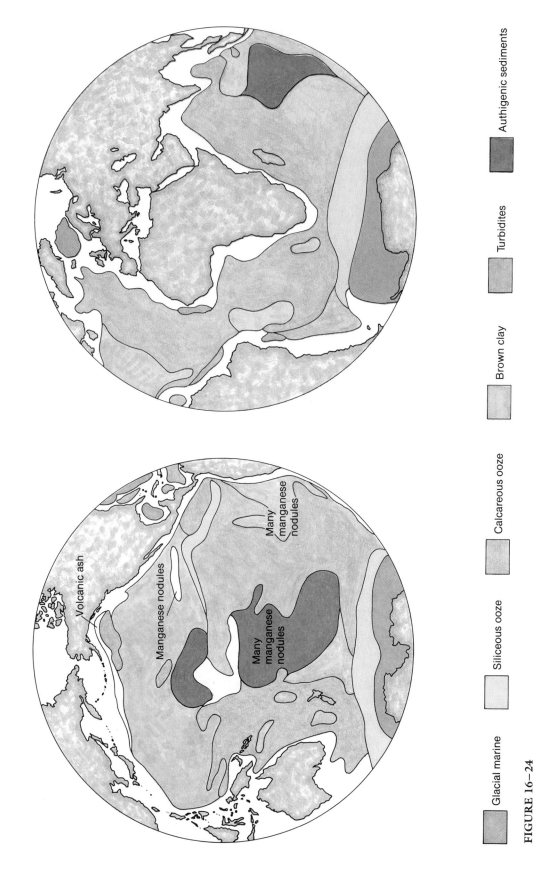

FIGURE 16-24

Geographic distribution of major sediment facies in the oceans. After map from *Submarine Geology*, 3d ed., Fig. 14.3, by F. P. Shepard. Copyright 1973 by Harper & Row, Publishers, Inc. Used by permission of Harper & Row, Publishers, Inc.

Glacial marine   Siliceous ooze   Calcareous ooze   Brown clay   Turbidites   Authigenic sediments

Volcanic ash

Manganese nodules

Many manganese nodules

Many manganese nodules

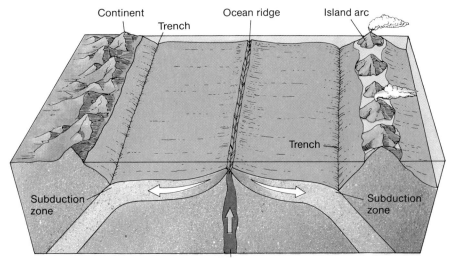

FIGURE 16–25
The plate tectonics model for the origin of ocean basins. Seafloor spreading involves rifting and extrusion of basalt at the ocean ridges to produce new ocean floor materials. The cool, older ocean floor is subducted into the asthenosphere at ocean trenches.

Continent   Ocean ridge   Island arc
Trench
Trench
Subduction zone   Subduction zone
Hot basaltic magma

rocks that underlie the sediments has produced similar results. Basalts at the ocean ridges and on Hawaiian-type islands where eruptions are continuing today are understandably young, but the oldest basalts ever recovered from the seafloor also are only Jurassic. It seems that the ocean features that cover more than 70% of the surface of the Earth have been here for only 180 million of the 4.6 billion years—4%—of the planet's history.

In view of these facts, how are oceans formed? Answering this question involves asking two others: Where did the water in the oceans come from? How did the present basins form?

As to the origin of ocean waters, there is general agreement among geologists. During the early stages of Earth history, volcanic eruptions brought water vapor into the atmosphere, just as they do now. When the surface of this planet cooled sufficiently, most of the water vapor condensed on the surface to form the hydrosphere. Then, as now, water filled whatever depressions were available. Soluble ions released by weathering of rocks were carried into the ocean basins, creating the saline waters that we have today. As more ions were added to the oceans over billions of years, other ions were being removed to form chemical rocks. Thus an equilibrium evolved, and the salinity of the oceans is believed to have remained essentially constant since they were formed.

As to the origin of ocean basins, the plate tectonics theory provides an acceptable explanation. As explained earlier in this chapter, an ocean basin forms when a continent breaks apart and forms a rift into which existing ocean waters migrate. The ocean grows by seafloor spreading (Figure 16–25) as the plates move away from the spreading center at the oceanic ridge. The growing plates extend away from the spreading center and are eventually consumed in a subduction zone under a continent or island arc. Only the plate tectonics model, built as it was upon data obtained during oceanographic research, seems to explain all the features of the ocean basins satisfactorily. In the next chapters we will consider the topographic features of both continents and oceans in greater detail.

OCEANIC RESOURCES

Oceanic food resources, such as fish, shellfish, crustaceans, and certain varieties of marine plants called algae, have been used since the beginning of human development. The continental shelves bordering the oceans provide over 90% of our marine food resources although they occupy only 8% of the area of the oceans. This is because light penetration and the concentration of nutrients from the land is greatest on the continental shelves. In contrast, most of the deep ocean is barren except for certain parts of the ocean ridge system where hot springs containing hydrogen sulfide erupt (see Focus 16–1).

Unfortunately, this 8% of the oceans is receiving nearly 100% of oceanic pollution. The most dra-

BLACK SMOKERS IN THE OASIS OF THE DEEP

P rior to 1979, the deep waters of the oceans were considered to be devoid of life. After all, how could life be sustained far below the photic zone and in temperatures close to freezing? This picture changed quickly in April 1979, when excited scientists in the research submarine *Alvin* reported a startling landscape at a depth of 2.5 km on a portion of the Galápagos Rift System in the east-central Pacific Ocean (Figure 1). The scientists described a spectacle of geology and biology that had never been seen before (Figure 2). Jets of hot (350°C), dark water were erupting from chimneylike forms which now are called **black smokers**. The seafloor was not barren but covered by a variety of barnacles, clams, crabs, tube worms, and other animals, many of which had never been described before.

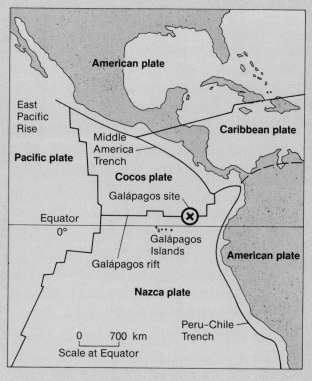

FIGURE 1
East-central Pacific Ocean region, showing Galápagos rift study area.

How could such concentrations of life survive in such cold, deep, dark waters? The answer lies in the relationship between the bottom water and the rock below. Oceanic ridges are where MORB is being extruded onto the ocean floor. The extruded lava cools quickly into pillowed lavas and flows, and they are commonly fractured. Cold water sinks into these fractures and is heated by the hot rock below the seafloor. This hot water dissolves mineral matter from the heated rocks. The hot waters, laden with minerals and gas (hydrogen sulfide), emerge through fractures and vents on the seafloor.

The heated, sulfur-rich waters emerging from the vents (Figure 3) are what make these active rift areas an "oasis of life" in the otherwise barren oceanic deeps. Some of the bacteria, which normally are present in the ocean, can obtain their energy from the **chemical energy** in hydrogen sulfide, and they multiply in great numbers in areas where this warm, gas-rich water emerges from the seafloor. The bacteria are food for larvae of larger organisms drifting by in the current. The bottom is covered with a variety of animals that feed on larvae and other small organisms and which grow to extraordinary sizes. For example, huge clams, 10 by 30 cm, grow in great numbers around the warm-water vents.

FIGURE 2
Vent area, showing fractured basalt and a variety of life forms, including giant clams. Photo by Carl Wirsen/ Woods Hole Oceanographic Institute.

FIGURE 3
Hot, dark water emerging as jets from black smokers on the ocean floor. Photo courtesy of Dr. Fred N. Speiss/Scripps Institute of Oceanography.

The scientists had discovered a whole new ecosystem, one based on chemical energy extracted from hydrogen sulfide rather than on the energy from sunlight captured through photosynthesis. This showed that sunlight is not always the primary source of energy for life.

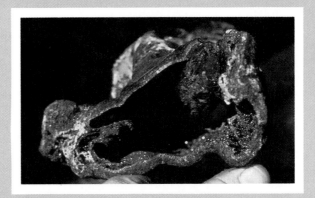

FIGURE 4
Cross section of black smoker mineral deposit, showing various mineral layers. Photo by Rod Catanach/Woods Hole Oceanographic Institute.

Another spectacular sight was the precipitation of metals in mounds and chimneys around the vents as the hot, mineral-rich waters are rapidly cooled by mixing with the near-freezing bottom waters. The mechanical arm on the *Alvin* obtained samples of the crusts making up the black smokers. The material was found to have a dull exterior composed of zinc sulfide and a bright golden interior composed of iron sulfide (Figure 4). In some areas, the metallic crusts were interlayered with sediments, suggesting that the metal precipitation is an intermittent event.

The researchers suggest that the crust at the Galápagos Rift breaks apart and is the site of extensive basalt extrusion about every 10,000 years. As the new oceanic crust cools, cracks develop through which seawater moves into the hot rocks below, eventually reappearing at the surface as hot, mineral-rich waters which precipitate metals in contact with the icy bottom waters. Pelagic sediment is deposited between eruptive events, and the newly formed oceanic crust shifts away slowly from the ridge crest by plate movement. After a period of time, basaltic eruptions resume on the ridge crest at the Galápagos Rift, and the cycle starts again.

matic example is the dumping of liquid wastes into coastal waters and into the streams that feed them. Continued degradation of ocean waters poses a serious threat to all levels of the marine food chain and to points far from where pollutants are introduced into the ocean. The surface and deep currents that we have discussed are proving to be effective in dispersing pollutants throughout the ocean basins.

Seawater is being desalinized to provide drinking water in some places. Although operational, it is very expensive because energy is needed to heat the seawater until it evaporates, producing freshwater vapor and a residue of dissolved salts. So far this process is practical only where there are limited sources of freshwater and a plentiful supply of energy, as in the Middle East.

The oceans are potentially a source of incredibly large amounts of raw materials if we can learn to extract the elements dissolved in seawater and mine the manganese nodules on the seafloor. Only a tiny percentage of the substances found in seawater currently are extracted from it (for example, common salt, NaCl) because of the enormous expense involved. Low concentrations of most elements in seawater require that great quantities of water be processed at an enormous expenditure of energy and money. Technology is now being developed to collect manganese nodules from the deep seafloor, but jurisdiction over these and other oceanic resources is controversial. In developing the new technology of oceanic resource recovery in a little-known ecosystem, we must be careful not to create more environmental problems.

SUMMARY

1. Our evolving knowledge of the oceans shows them to be far different from our view of only 35 years ago. Far from being the oldest and most unchanging features on Earth, we now know that they are among the youngest, and that they are dynamic—growing in some areas, shrinking in others. Their floors contain the longest and highest mountain chains on Earth and a wide variety of sedimentary facies.

2. Continents are surrounded by gently sloping surfaces covered by shallow water (the continental shelves) that pass seaward into more steeply sloping regions called continental slopes. The deep, flatter parts of the oceans (the abyssal plains) contain small hills that ap-

pear to be extinct volcanoes and large linear mountain ridges called ocean ridges. The ocean ridges are spreading centers where MORB (mid-ocean ridge basalt) is extruded to form new oceanic crust. The ridges are offset but not completely truncated by crosscutting oceanic fracture systems.

3. The deepest parts of the oceans, the oceanic trenches, are found close to continental margins and are associated with either volcanic island arcs or continental mountains, such as the Andes. Oceanic trenches represent subduction zones where oceanic crust descends into the mantle. Island arc–trench systems are separated from the continents by shallow seas called marginal seas or back-arc basins.

4. Basaltic islands may occur singly as seamounts, in submerged flat-topped mountains called guyots, or in large clusters of shield volcanoes as Hawaiian-type islands. Hawaiian-type islands form over a hotspot and migrate away from it on a moving plate. With time, this generates a chain of islands extending away from the hotspot.

5. Seawater varies in temperature, density, and composition from place to place. Average seawater contains 35 parts per thousand of dissolved solids in the form of ions, of which sodium and chlorine are by far the most abundant. All the gases in the atmosphere are found in solution in the oceans, particularly oxygen and carbon dioxide.

6. The density of seawater increases as salinity and suspended sediments increase and as temperature decreases. Water masses may rise or sink past others, depending on their relative densities.

7. Oceanic water is in a constant state of circulation involving both surface and deep waters. Surface water movement is driven by the world's prevailing winds, whereas the deep currents result from density adjustments. Cold, highly saline, and highly turbid waters initiate oceanic circulation in some areas by sinking beneath lighter water of more normal salinity.

8. The ocean floors are underlain by a layered section, called an ophiolite suite, composed from top to bottom of unconsolidated sediment, interlayered sedimentary rocks and basaltic lava flows, pillow basalts, sheeted dikes, gabbros, and ultramafic rocks.

9. Sediment on the seafloor comes from a continental source (terrigenous), from biologic activity (biogenic), or from chemical precipitation (authigenic).

10. Coarse terrigenous sediments reach their oceanic sites of deposition by stream transport, ice rafting, and turbidity currents.

11. Finer particles, some blown into the ocean by winds, are carried in suspension by surface and deep currents for long distances. Eventually, these particles settle out of suspension and are deposited as pelagic ooze.

12. Biogenic sediments consist of fine-grained calcareous and siliceous skeletal remains of near-surface organisms that die and sink to the bottom, accumulating as calcareous and siliceous oozes. Calcareous organisms tend to dissolve in the deep, cold waters of the oceans at the calcium carbonate compensation depth (CCCD). Therefore, fossilized calcareous shells are not found as abundantly in the ocean sediment as their present abundance in surface waters would suggest.

13. Authigenic sediments such as manganese nodules form by precipitation in areas of the ocean basins where the rate of sedimentation is known to be relatively low.

14. Dating of sediments and rocks of the ocean basins shows that they are young—no older than Jurassic. According to the plate tectonics model, ocean basins form when continental crust is rifted and MORB is extruded into the depression to form new oceanic crust. The oceanic plates move away from the spreading center at the oceanic ridge and eventually are subducted back into the mantle. Existing ocean water fills the growing depression to form a new ocean.

15. Oceans are potentially a vast reservoir of food, mineral, and energy resources if we can develop the technology necessary to extract what we need cheaply and without damaging the fragile, and little understood, oceanic environment.

REVIEW QUESTIONS

1. How was the age of the ocean basins determined? What is the geologic significance of the findings?
2. How do the rocks under the sediments of the continental shelf and the oceanic ridges differ? Why?
3. Which oceanic topographic feature is most affected by climatic change? Why?
4. What contrasts are there in continental shelf sediments in high and low latitudes? Why?
5. Describe the formation of a turbidity current and the resulting sedimentary deposit.
6. Describe the formation of the continental rise. Can you think of one or more continental analogs?
7. What are the topographic and geologic features of oceanic ridges? How do they form?
8. What processes are active at a subduction zone? What are the geologic products?
9. What systematic changes are found among the Hawaiian Islands? How can we account for them?
10. Describe the temperature changes observed with depth in general in the oceans, and account for any differences you see.
11. What factors cause differences in salinity in different parts of the ocean?

12. What factors cause differences in density in seawater?
13. Contrast the factors that drive surface and deep currents.
14. Describe the formation of three different kinds of deep currents. Give an example.
15. Describe the conditions in which an abundance of calcareous organisms at the surface would not be reflected in a deposit of calcareous ooze on the seafloor.
16. What is the origin of brown clay deposits in the oceans?
17. Describe the origin of the ocean basins according to the plate tectonics model.

FURTHER READINGS

Ballard, R. D. and Grassle, F. L. 1979. Return to the oases of the deep. *National Geographic* 156(5):689–705. (Describes the features found on the Galápagos Rift vent site. Includes spectacular pictures of organisms, black smokers, and mineral deposits.)

California Division of Mines and Geology. 1987. A thumbnail sketch of California's marine geology. *California Geology* 40(8):171–80. (An overview of the morphology, processes, and resources associated with an active plate margin.)

Davis, R. A., Jr. 1986. *Oceanography—An introduction to the marine environment.* Dubuque, IA: William C. Brown, 431 p. (A readable description of many aspects of the ocean).

Heezen, B. C., and Hollister, C. D. 1971. *The face of the deep.* New York: Oxford Univ. Press, 659 p. (A collection of outstanding bottom photographs of animals and surface features on the ocean floor. Accompanying text explains the origin of the features).

Kennett, J. P. 1982. *Marine geology.* Englewood Cliffs, NJ: Prentice-Hall, 813 p. (An exhaustive treatment of aspects of marine geology and geophysics at an advanced level.)

U.S. Geological Survey. 1980. *Conference on continental margin mass wasting and Pleistocene sea level changes.* Circular 961, 133 p. (A collection of short, well-illustrated papers on widely varied aspects of continental shelf, slope, and rise environments.)

Woods Hole Oceanographic Institution. 1987. Changing climate and the oceans. *Oceanus* 29(4), 98 p. (A collection of papers showing the oceanic evidence for climatic change.)

Woods Hole Oceanographic Institution. 1988. DSV *Alvin*—25 years of discovery. *Oceanus* 31(4). (A collection of papers resulting from research carried out on the deep-sea submersible *Alvin*. Papers include a description of the Mid-Atlantic Ridge, p. 34–40, and life at the Galápagos Rise, p. 41–46.)

17
Deformation of Rocks

Folded Paleozoic strata near Borah Peak, Idaho. Photo by
John Shelton.

We know that continents move about on the surface of the Earth and collide with one another. Ocean basins open and close. Plates grind past one another in transform faults or are subducted. What happens to rocks when they are squeezed between colliding plates or are part of a plate that is splitting apart? They are bent as shown in this chapter's opening photograph, or they are broken as in Figure 17–1. In this chapter, we will examine how rocks respond to the enormously powerful forces of moving plates and will show how analysis of deformed rock helped geologists develop the plate tectonics model.

Sedimentary rocks are deposited in horizontal beds. Features such as the folds in the opening photograph and faults in Figure 17–1 show that the original horizontality and continuity of the beds have been disturbed—we say that they have been *deformed.* Folds and faults are called **geologic structures,** and the study of deformed rocks and deformation is called *structural geology.*

HOW DO ROCKS DEFORM?

Sedimentary rocks are solid and hard; they shatter and splinter when hit with a hammer. We also would expect natural forces to break them, and Figure 17–1 shows that this does happen. The breaks are called **faults** when the blocks of rock that they separate have moved from their original positions. The movement may be only a few centimeters, or may be hundreds of kilometers, as in the case of the famous San Andreas fault in California.

Rocks do not always break during deformation. The opening photograph, for example, shows **folds,** rocks that bent and buckled but did not break. They have acted plastically and flowed like modeling clay rather than breaking as we might expect rock to do. How can this happen? What determines whether a rock will bend or break, or in geologic terms, be folded or faulted?

Folding Versus Faulting

Events in our everyday lives help to show that *time* and the *intensity of the deforming forces* determine how a rock deforms. For example, a sharp karate chop can

break a board, and a hard-hit baseball will shatter a window. Intense and rapidly applied forces cause things around our homes to break, and do the same to rocks. In contrast, a wooden shelf bends and stays bent if heavily loaded for a long time. Old panes of window glass become thicker at the bottom because gravity causes the ions in the glass to move downward. Even the brittle ice of a glacier becomes plastic and flows when placed under great pressure. Wood, glass, and ice all break when subjected to deforming forces that are strong and rapidly applied. However, if weaker forces are applied over long periods of time, these same materials respond by flowing.

Rocks thus may break during some kinds of deformation and flow and be folded in others. It may take weeks for wood to bend, centuries for glass to flow, and millions of years for rocks to fold. However, the most spectacular example of the several possible types of behavior is provided by Silly Putty in a matter of minutes—see Focus 17–1.

For a rock or any other material to be folded, its atoms must change position relative to one another. This causes a slow flow of matter which results in the new, deformed shape. The processes by which this flow occurs in rocks are exactly the same as those by which the ice in a glacier flows, as described in Chapter 13.

Factors Controlling Deformation

Laboratory studies help us understand how folds develop, and what conditions are needed before the same rocks will be faulted. We can place samples of different rocks in a hydraulic press, where we can vary temperature and lithostatic pressure (the pressure due to progressive burial). This enables us to recreate conditions at any depth in the crust and to predict how rocks will deform at different levels. Directed pressures cause most deformation in the Earth; we simulate these by pressing a piston against the sample (Figure 17–2). These experiments show that a rock's response to deforming forces is controlled by (1) its composition and texture, (2) the nature and intensity of the directed pressures, and (3) the temperature and lithostatic pressure.

Heat and Pressure. The undeformed cylinder of marble shown at the left in Figure 17–2 can be shortened by applying strong directed pressure, as

FIGURE 17–1
Faults in sandstones in California. The layers of rock have been broken and shifted as indicated by the arrows. Photo by Allan Ludman.

FIGURE 17–2
Experimental deformation of marble. A marble cylinder (left) is deformed by compression in a hydraulic press. The deformed cylinders shown have been subjected to the same directed pressure, but different lithostatic pressures as indicated. The center specimen has deformed by brittle faulting, and the specimen on the right by ductile folding. Photo courtesy of M. S. Paterson.

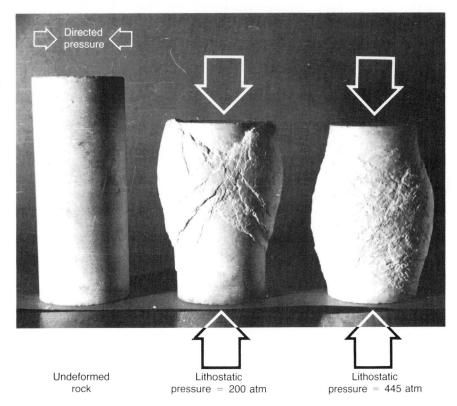

Directed pressure

Undeformed rock

Lithostatic pressure = 200 atm

Lithostatic pressure = 445 atm

indicated by the arrows. Whether the shortening will be caused by faulting (center sample) or folding (right sample) depends on the temperature and lithostatic pressure. At low lithostatic pressures, most rocks break easily; faults like the cracks in the deformed center sample develop, and the rock changes shape by shifting along the cracks. If lithostatic pressure increases, the same amount of directed pressure will cause the rock to flow plastically, as shown by the sample at the right. Note that there are

PRINCIPLES OF ROCK DEFORMATION REVEALED BY SILLY PUTTY

nyone who has played with Silly Putty, a popular toy, knows that it behaves weirdly as it stretches, bounces, and breaks. Surprisingly, insights into how rocks deform can be gained from a short session with Silly Putty.

1. The role of temperature—when Silly Putty is first taken from its egg-shaped container, it feels stiff and can be stretched or bent only with some effort. Under these circumstances, it often will break instead of bending. After being in your warm hands for a few minutes, however, it becomes pliable and stretches easily. Rocks behave the same way! When they are relatively cold (as in the upper part of the crust) it is difficult to flex them, but as they become hotter (when they are buried deeply in the crust) they bend and flow easily.

2. The role of time—hold Silly Putty in both hands and pull gently for a few seconds. It will stretch easily, eventually becoming a fine filament. Reform the original ball and then pull more rapidly. The Silly Putty stretches a bit and then snaps. Rocks again follow this behavior. A moderate force applied over a short period of time will result in brittle behavior, whereas lengthy application of a gentle force results in a more extended flow, even though the total force (force/second × number of seconds) may be less in the brittle case.

no faults in this sample. The rock changed shape by flowing rather than by breaking and slipping. The result in nature is a fold.

Other experiments show that heat has a similar effect: cold rocks tend to break and hot ones tend to flow. This shows that position within the Earth plays a major role in determining the geologic structures that can form. Rocks in the upper parts of the crust develop faults much more readily than those in the lower crust or mantle because they are colder and under less lithostatic pressure. The same rocks that are broken and faulted in crustal conditions would deform by flowing and folding in the mantle where temperatures and pressures are greater.

Stress. Lithostatic and directed pressures acting on a rock are resisted by the rock's internal cohesive forces—the chemical bonds that lock ions and atoms in mineral structures and the cement or interlocking texture that holds the grains in place. The term **stress** is used to describe the magnitude of the external forces acting to deform rocks. The amount of stress depends not only on the intensity of the applied pressures but also on the surface area over which they act (Figure 17–3). The greater the area over which the pressure is spread, the smaller the stress will be. Conversely, the smaller the area on which the pressure is focused, the greater the stress will be. Deformation occurs only when stress overcomes a rock's internal cohesiveness.

Rocks are affected by three kinds of stress, which differ in their directional properties: compressional, tensional, and shear (Figure 17–4). **Compressional stresses** occur when a rock is squeezed, and the opposing forces are directed *toward* one another—Figure 17–4(a). **Tensional stresses** de-

FIGURE 17–3
The relationship among force, surface area, and stress. The football player in (a) causes less stress on the player he has tackled than the player in (b) because his weight is spread over a larger surface area.

velop when a rock is pulled apart, with the forces directed *away* from one another—Figure 17–4(b). Both compressional and tensional stresses develop perpendicular to the deforming forces. **Shear stresses** are generated along planes *parallel to* the deforming forces—Figure 17–4(c), where rocks are twisted in opposite directions. (This is exactly what happens when paper is caught between the blades of a pair of scissors or shears.)

Strain. Rocks change their shape by faulting or folding in response to stress. The amount of change is called **strain** and usually is measured by the amount of change in the original length or width of the rock. Thus, stress indicates the *intensity of the deforming forces,* whereas strain reflects the *response of the rock* to those forces. The way in which any rock deforms can be described in terms of its stress and strain.

Behavior of Rocks During Deformation

When rocks deform, they do so in three different ways, depending on the relationship between the stresses applied and the resulting strain. The different types of behavior are called elastic, ductile, and brittle deformation.

Elastic Deformation. A weight suspended from a spring or rubber band causes **elastic deformation** of the spring or rubber (Figure 17–5). If a small weight is hung from a spring, the spring stretches. If twice the weight is hung from the spring, twice the elongation will occur; tripling the weight triples the elongation; and so forth. In elastic deformation, stress is directly proportional to strain.

Elastic deformation is *temporary.* When the weights are removed, the spring returns to its original length. Once the stresses are relaxed, the deformed material returns to its undeformed state. This is why we cannot show a picture of elastically deformed rocks. Figure 17–5 shows elastic behavior caused by tensional stresses, but the same behavior can result from compression—the spring would become shorter and then return to its original length.

Ductile Deformation. A permanent kind of deformation that results from the flow of material is called **ductile deformation.** Up to a point, a spring will

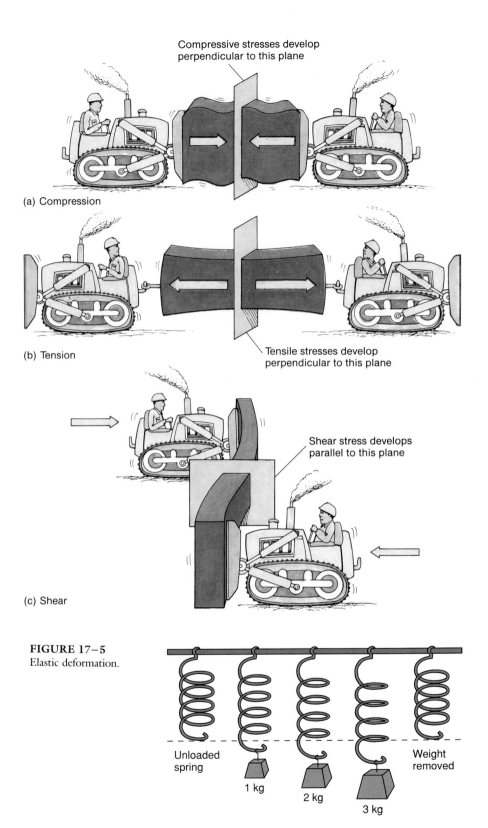

FIGURE 17–4
Types of stress.

Compressive stresses develop
perpendicular to this plane

(a) Compression

Tensile stresses develop
perpendicular to this plane

(b) Tension

Shear stress develops
parallel to this plane

(c) Shear

FIGURE 17–5
Elastic deformation.

Unloaded
spring

1 kg

2 kg

3 kg

Weight
removed

FIGURE 17–6
Ductile deformation.

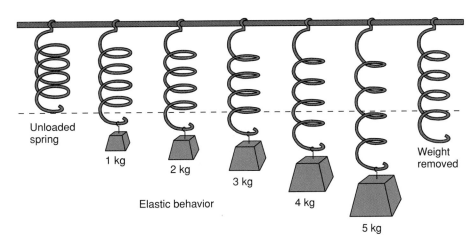

Unloaded spring

1 kg

2 kg

3 kg

Elastic behavior

4 kg

5 kg

Weight removed

Ductile behavior

behave elastically as weights are loaded on. However, if too much weight is added, the metal stretches out of shape. When the weights are removed, the spring cannot return to its original length; unlike elastic deformation, this change in length is permanent (Figure 17–6). Ductile deformation of rocks results in folds like those shown in the opening photograph of this chapter.

Brittle Deformation. During brittle deformation, deforming forces overcome a material's cohesiveness and it breaks like the spring shown in Figure 17–7. All rocks have a maximum strain beyond which their cohesive forces can no longer hold them together. When a rock experiences more strain than this limit, it breaks. Brittle behavior also results in permanent geologic structures, such as the faults shown in Figure 17–1.

Faulting and Folding of Rocks. Experiments show that most rocks experience all three types of behavior, depending on the amount of stress and the temperature/pressure conditions. Rocks generally behave elastically under low-stress conditions, but as stress increases they experience ductile deformation and eventually brittle deformation—they break. The springs in Figures 17–5, 17–6, and 17–7 follow this sequence and so do most rocks.

Some rocks deform much more easily than others. Rocks that change shape under small stresses are called **incompetent rocks,** and those that deform only under enormous stresses are called **competent**

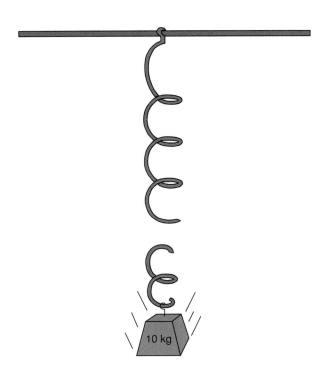

10 kg

FIGURE 17–7
Brittle deformation.

rocks. Whether a rock is competent or incompetent may depend on the type of stress affecting it. Some rocks resist changes under great tensional or compressional stress, but deform easily when subjected to small shear stress. Some rocks, however, are incompetent under all types of stress; these include limestone, dolostone, gypsum, and rock salt.

FIGURE 17–8

Salt domes. Schematic diagram
showing extreme ductile
deformation of halite. Overlying
beds have been removed for clarity.
With increased pressure from
overlying rock (left to right), the
ductile behavior of the halite
increases and the salt domes become
larger.

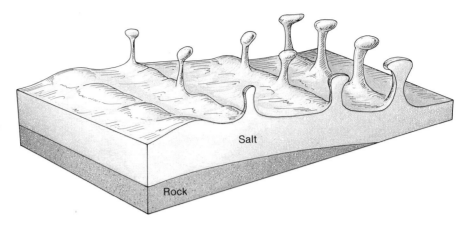

A rock's composition and texture play important roles in controlling the way it deforms. Rocks composed of silicate minerals generally behave more competently than nonsilicate rocks such as limestone and marble because of the greater strength of bonds in silicate minerals. Igneous and metamorphic rocks with interlocking grains are more competent than clastic sedimentary rocks because their interlocking grains provide a strong framework that is difficult to overcome.

Rock salt is one of the least competent rock types, and requires very little stress before it begins to deform ductilely. In some areas, the weight of overlying rocks is enough to make halite behave with extreme ductility. The salt flows like a highly viscous liquid and forms weird, grotesquely bulbous projections called **salt domes** (Figure 17–8). The upward movement of the salt is caused by both the pressure of the overlying strata and the fact that salt has a lower specific gravity than most other rocks and therefore has a tendency to rise.

EXAMINATION OF DEFORMED ROCKS IN THE FIELD

Comparisons of *naturally deformed* rocks with those deformed in the laboratory can help us learn about the conditions of stress and strain within the Earth. Geologic structures let us reconstruct the deformation history of the Earth much as a rock's texture lets us reconstruct the processes by which it formed. We just have to know what to look for.

What Can We See?

Elastic deformation is temporary, and leaves no permanent geologic structures to indicate that it has taken place. However, in geologic terms, "temporary" can mean thousands of years, and large-scale elastic behavior associated with the most recent ice ages can be detected by very careful observation and measurement. The melting of vast continental glaciers 10,000 years ago removed a large mass of ice from the continents, comparable to the removal of weights from a spring. Just as a spring returns to its original unloaded position, the continents are now rising very slowly in the process of **isostatic rebound** (Figure 17–9). We can tell that this is going on by noticing ancient shoreline features, which are now uplifted to elevations many feet above present sea level.

The return to the original state takes a fraction of a second for a spring, but thousands of years for a continent. The difference is due to the fact that a rebounding spring must push aside only the air between its coils, but a continent can rebound only when the asthenosphere beneath it flows slowly back to its initial position. The weight of the ice sheets pressed the lithosphere downward and caused the less-rigid asthenosphere to flow outward, away from the depressed continent. When the ice melted, the pressure on the asthenosphere was lessened and mantle material flowed slowly back to its original position as the unloaded lithosphere rose. The amount of uplift is shown in Figure 17–9 by the contour lines, each of which represents a particular amount of rebound. The greatest isostatic rebound in North America is centered near Hudson Bay in Canada. Presumably, this is where the greatest weight of ice was concentrated, and where elastic strain was greatest.

Ductile and brittle deformation are much easier to recognize because they leave folds and faults as records of their activity. Folds and faults are distributed throughout the world, and dating shows that they have occurred throughout geologic time.

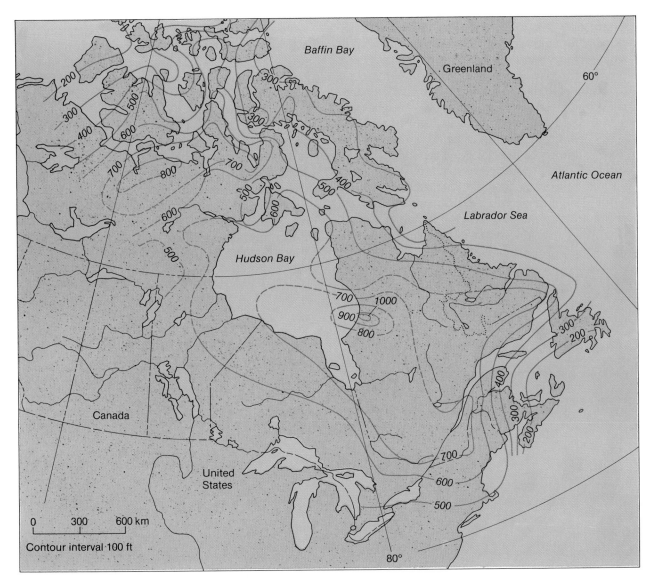

FIGURE 17–9
Isostatic rebound of North America following glacial melting. The Pleistocene ice sheet that advanced over the United States was centered in the vicinity of Hudson Bay. Melting of the ice sheet removed a large mass from the crust, and the Earth is now rising in response to the unloading. Contours indicate the amount of uplift in feet above sea level since the area was deglaciated. After Energy, Mines and Resources Canada, *National Atlas of Canada,* MCR 1128, 1972. Reproduced with the permission of the Minister of Supply and Services Canada, 1990.

Determining the Orientation of Deformed Rocks

Sedimentary rocks generally are deposited in horizontal beds. The degree to which these beds were changed by folding or faulting from their original horizontal position is a measure of the amount of deformation. There is no principle of original horizontality for igneous rocks, however, so that folds in batholiths may well be impossible to detect. Therefore, for the sake of simplicity, we will confine our discussion of folding and faulting to the sedimentary rocks in which these structures are best displayed.

In Chapter 1 we mentioned that one of the difficulties in studying geologic features is the problem of *scale*. This problem often arises in field studies of deformed rocks because some folds are so large that it is not immediately apparent to the geologist that a fold is present (Figure 17–10). It usually is necessary to measure the orientation of bedding at many outcrops in order to determine that a large-scale fold is present.

Geologists use two terms to describe the three-dimensional orientation of a planar feature such as a bed: strike and dip (Figure 17–11). The **strike** of a bed is the compass direction of a horizontal line drawn on the bed. In Figure 17–11, the surface of the water is perfectly horizontal; as a result, the intersection of the water and the bedding plane is the same as the strike line. By placing a compass next to this line, its direction may be measured relative to north or south. Thus, a strike of N53°E (read as "north 53 east") means that the horizontal line drawn on the bedding plane trends 53° east of north, as shown in Figure 17–11.

The **dip** of the bed shown in Figure 17–11 is the angle between the bedding plane and the horizontal. Dip tells us how much the bed has been changed from its original horizontal position. The amount of inclination of a bedding plane (the dip angle) is measured at right angles to the strike direction, but for a bed striking at N53°E there are two directions in which this bed could dip: to the northwest or to the southeast. The bed on the left side of Figure 17–11

FIGURE 17–10
A geologist faces the problem of scale. Folds may be so large that it is difficult to recognize them. The geologist shown here sees only a small outcrop of nearly horizontal strata, and is unaware of its relationship to the large fold.

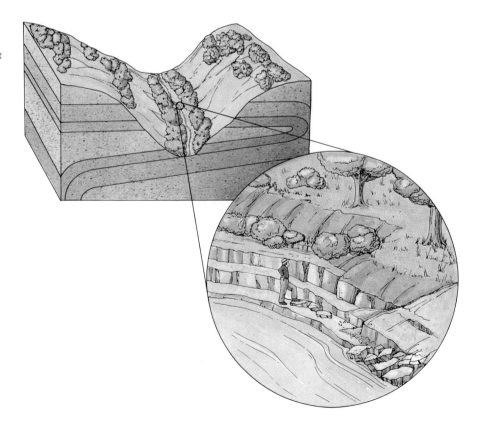

FIGURE 17-11

Strike and dip. The horizontal water surface defines the strike line on both outcrops of the folded layer. Note that since the strike lines are parallel, the two outcrops have the same strike. However, they dip in opposite directions.

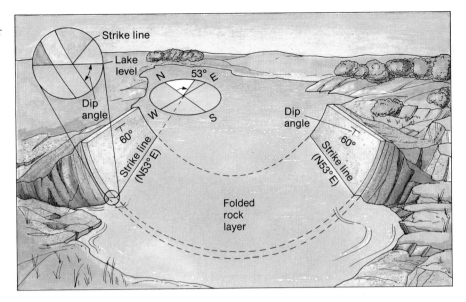

has a dip of 60° to the southeast, and its complete attitude would be written as N53°E, 60°SE. The bed on the right side of Figure 17-11 has exactly the same *strike*, but a dip of 60°*NW* (N53°E, 60°NW). A perfectly horizontal bed has a dip of 0°, a perfectly vertical bed a dip of 90°.

The symbols used in Figure 17-11 are used on maps to indicate strikes and dips. The long bar points in the direction of strike; the short bar points in the dip direction; and the amount of dip angle is indicated. The attitude of any planar feature can be recorded using strike and dip. Field geologists studying structures in deformed rocks record the strike and dip of such planar features as fault planes and foliations, as well as beds of rock.

FOLDS

Each fold is a permanent record of ductile deformation, and a careful study of folds often enables geologists to reconstruct the compressional stresses active during deformation.

Anatomy of a Fold

We begin our study of folds with a brief anatomy lesson. Each fold consists of two sides, called **limbs,** which join at a line called the **fold axis**—Figure 17-12(b). Consider a sequence of sedimentary layers that have been folded as in Figure 17-12. All of the axes for a particular fold lie on a single surface called the **axial plane**. In most cases, the axial plane forms at right angles to the compressional stresses that caused the folding. The following descriptions of different kinds of folds are easy to understand if you remember these three terms.

Types of Folds

Folds fall into two basic categories. Folds in which the limbs dip toward the fold axis are called **synclines**. Folds in which the limbs dip away from the fold axis are called **anticlines**. Anticlines and synclines are classified further by the attitudes of their axial planes and axes, by the dips of their limbs, and by the angles between their limbs (Figure 17-13).

For example, if *axial planes* are vertical, folds are said to be *upright*—see (d) in Figure 17-13. If their axial planes are horizontal (e), the folds are **recumbent**. Folds whose *axes* are horizontal are said to be **nonplunging** (f); those with axes inclined at some angle are said to be **plunging** (g). It is thus possible to describe a fold as a "gently plunging, upright anticline." Other modifiers are shown in Figure 17-13. An **overturned fold** is one in which one of the limbs has been turned so much that it is upside down.

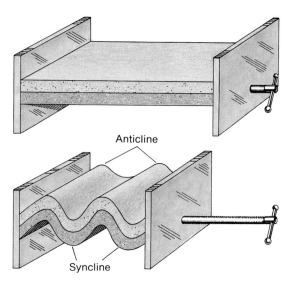

Anticline

Syncline

(a) Folds are produced by compression

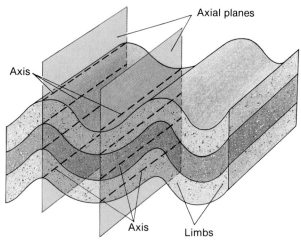

Axial planes

Axis

Axis   Limbs

(b) Anatomical components of folds

FIGURE 17–12
Anatomy of a fold.

There are two special kinds of folds in which rocks dip inward or outward from a single point, rather than toward or away from an axis. In **basins**, the rocks dip toward a central point—Figure 17–13(h). In **domes**, the rocks dip away from the central point—Figure 17–13(i).

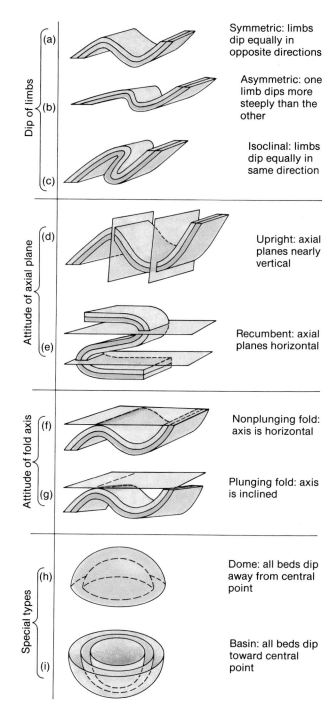

Dip of limbs

(a) Symmetric: limbs dip equally in opposite directions

(b) Asymmetric: one limb dips more steeply than the other

(c) Isoclinal: limbs dip equally in same direction

Attitude of axial plane

(d) Upright: axial planes nearly vertical

(e) Recumbent: axial planes horizontal

Attitude of fold axis

(f) Nonplunging fold: axis is horizontal

(g) Plunging fold: axis is inclined

Special types

(h) Dome: all beds dip away from central point

(i) Basin: all beds dip toward central point

FIGURE 17–13
Classification of folds. The bases for classification are dip of limbs, attitude of axial plane, and attitude of fold axis.

Features Observed in Folded Rocks

Interpretation of the deformation history of rocks is aided by small-scale features that form during folding. Even at a small outcrop, the presence of these features shows that the rock is part of a large-scale fold and may enable geologists to postulate the shape and orientation of the structure.

Cleavage and Foliation. In many folds, closely spaced fractures develop more or less parallel to the axial planes (Figure 17–14). These fractures are called **rock cleavage** and should not be confused with cleavage in minerals. Rock cleavage affects entire beds instead of just a single mineral grain.

We saw in Chapter 7 that regionally metamorphosed rocks often develop a foliation of platy minerals. One of the major agents of regional metamorphism is the directed pressure that is also responsible for folding. Folding may take place without metamorphism, but when regional metamorphism accompanies folding, a foliation forms that parallels the

(a)

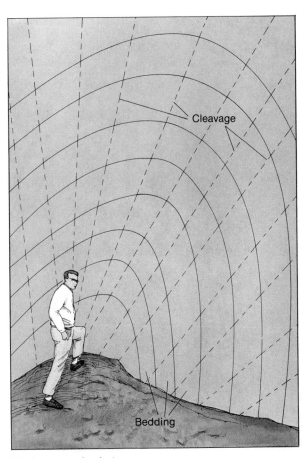

(b) Area seen in photo

FIGURE 17–14
Cleavage in folded rocks. (a) Anticline in slate of the Precambrian Walden Creek Group near Walden, Tennessee showing well-developed axial plane cleavage. The cleavage fans somewhat but is essentially parallel to the axial plane of the fold. (b) Sketch showing relationship between folded beds and axial plane cleavage. Photo by Robert D. Hatcher, Jr.

axial planes of the major folds. This happens because, as with cleavage, foliation forms at right angles to the dominant compressional stress.

Outcrop Patterns of Folded Rock. Folds that we observe in the field often have been subjected to long periods of erosion before we get to study them. Erosion removes much of the evidence that we need to understand folding. It flattens the ground surface, removing the arches of anticlines and the troughs of synclines (Figure 17–15). Indeed, if the rock in the axial region of an anticline is easily eroded, an *erosional trough* may form exactly where we would expect the structural arch of the anticline. Highly variable strikes and dips and the presence of cleavage or foliation in scattered outcrops often are all that we can see to indicate that the rocks have been folded.

However, erosion also can be helpful. When horizontal rocks are eroded, the layers are exposed in a pattern shown in Figure 17–16(a). Erosion of tilted or folded rocks produces a very different pattern. Resistant beds stand above the ground surface as ridges, whereas nonresistant beds are more rapidly weathered and eroded and form lowlands. Linear ridges and valleys indicate that rocks have been tilted or deformed into nonplunging folds—Figure 17–16(b). Zigzag patterns traced out by alternating ridges and valleys tell us that plunging folds are present—Figures 17–16(c). The ridges (or valleys) converge in areas called **fold noses**. Anticlines plunge toward their noses, whereas synclines plunge in the opposite direction (away from their noses).

FAULTS

When stresses cause more strain than a rock can absorb by ductile flow, the rock breaks. The breaks are called **fractures** if they are oriented randomly or **joints** if they are oriented systematically. They are called **faults** only if there is significant *movement* of the rock on opposite sides. Faulting may be produced by compressional, tensional, or shear stresses, and the type of fault that develops depends on the type of stress applied. Once we learn how to recognize different kinds of faults, we can determine what stresses acted on a given area in the geologic past.

How Does Faulting Operate?

Consider an area that is being deformed (Figure 17–17). As stresses build, the rock first undergoes elastic deformation, then ductile deformation, and finally brittle deformation. At the rupture point, the stress that had been accumulated is relieved and a small offset is produced as the blocks move. If the deforming forces continue to operate after the opposite sides of the fault have stopped moving, stress will again build in the rock, only to be released in a later episode of fault movement. When we look at a fault today and measure its total offset, we must re-

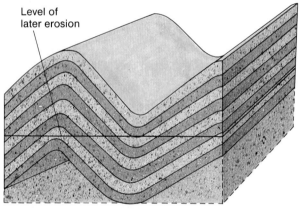

Level of later erosion

(a) Folded strata before erosion

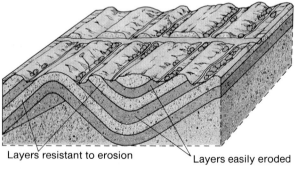

Layers resistant to erosion

Layers easily eroded

(b) Topography after erosion

FIGURE 17–15
Effects of erosion on folds. Erosion removes anticlinal crests and synclinal troughs. Resistant beds form ridges whereas nonresistant layers are eroded to form valleys.

member that the displacement probably did not take place during a single period of movement but rather in a series of small-scale motions over a long time. For example, the San Andreas fault of California exhibits an offset of hundreds of kilometers, but the fault has been active for about 60 million years. The present offset was accomplished by many small movements over that time period.

Faults, like folds, can be observed at several scales. Faults may be seen within the confines of a single outcrop, as shown in Figure 17–1, or many faults may be concentrated in large, regional-scale fault zones.

Anatomy of Faults

Faults consist of the three simple elements shown in Figure 17–18. These are the two blocks of rock on opposite sides of the fault and the fault plane itself. The blocks on opposite sides of an inclined fault plane are given names inherited from the early British coal miners who often encountered faults as they worked underground. The **hanging wall** rests on (or hangs upon) the fault plane, whereas the **footwall** forms the foundation (or footing) upon which the fault plane and the hanging wall rest. (In a vertical fault, there is no hanging wall or footwall.)

Types of Faults

Faults are classified according to the nature of the movement of their blocks (Figure 17–19). However, it is nearly impossible to determine the *absolute* direction of movement. The apparent motion of one block relative to another is the best we can do, but that is enough. The blocks in Figure 17–19(a) may well have moved in opposite directions as shown by the arrows, but both sides could just as well have moved upward (left side more than the right side) or downward (right side more than the left side). The arrows indicate the *relative motion*.

Some faults are called **dip-slip faults** because their movement has been up or down the dip of the fault plane, with little if any horizontal motion. Others are **strike-slip faults,** with most of the displacement in a horizontal direction parallel to the strike of the fault

plane and with little or no vertical offset. More complex motion is found in oblique-slip and rotational faults (Figure 17–19).

Dip-Slip Faults. Dip-slip faults are produced by either tensional or compressional stresses. Dip-slip faults in which the footwall block has moved upward relative to the hanging wall block are called **normal faults**. (There is nothing "abnormal" about other kinds; it is simply that this was the type of fault found most often in the British coal fields, and it was therefore considered to be the "normal" type.) Normal faults form where the Earth's crust is subjected to tension (Figure 17–20).

Complex systems composed of several normal faults that dip alternately toward and away from one another result in alternatingly downthrown and upthrown blocks, as shown in Figure 17–20. In the American Southwest, the downdropped blocks are called **basins** and the uplifted blocks are called **ranges**. (The term "basin and range" should not be confused with the type of fold called a basin shown in (i) of Figure 17–13.) In Europe, the same structures are called a **graben** and a **horst,** respectively. Cliffs that form along the fault plane as the blocks move are called **fault scarps**.

East Africa today is experiencing tensional stresses, as indicated by the extensive East African Rift System, a series of large grabens bounded by horsts (Figure 17–21). One graben in this system is the low area occupied by the Red Sea; another is the Gulf of Aden. A third is responsible for Olduvai Gorge, where so many important fossil remains of ancient humans have been found.

The topography at the crests of the ocean ridges is very similar in form to the East African Rift System, leading geologists to believe that the ocean ridges are the result of large-scale tensional stresses. This, of course, is precisely the meaning of the plate tectonics interpretation of seafloor spreading.

Reverse faults are those in which the footwall block has moved downward relative to the hanging wall block (they have the reverse sense of motion from normal faults). Reverse faults are caused by *compression* (Figure 17–22), and usually are found in association with strongly folded rocks, showing that folding too is a compressional process.

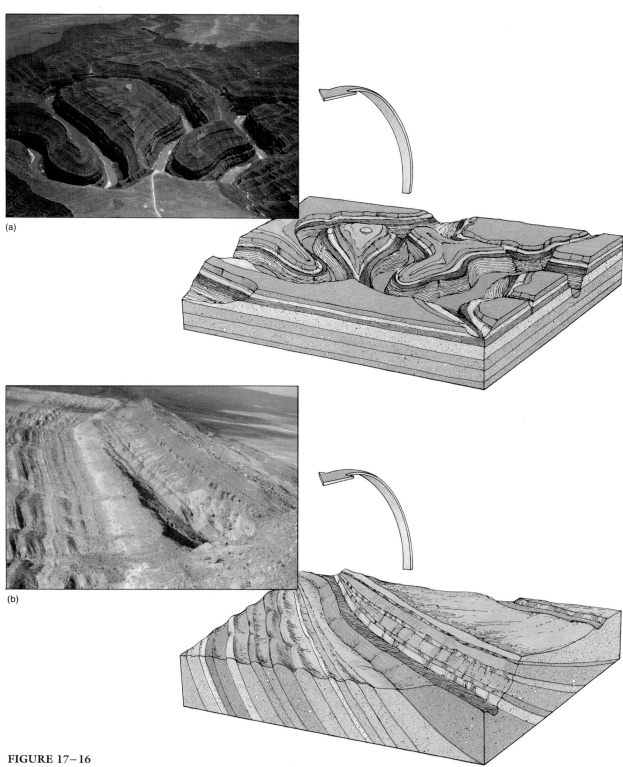

FIGURE 17–16
Outcrop patterns of eroded structures. (a) Pattern developed in horizontal strata (Goosenecks of the San Juan River in Utah). (b) Tilted beds, southeastern Utah.

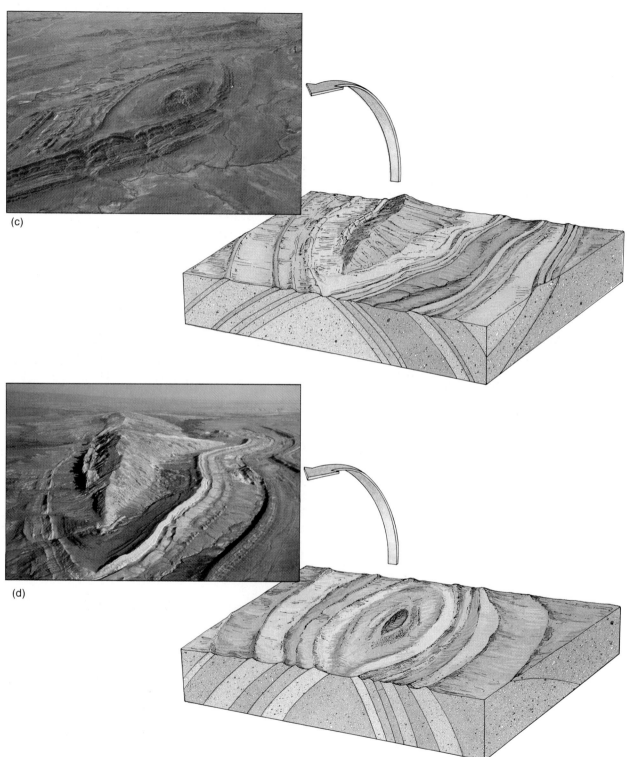

FIGURE 17–16 (continued)
(c) Plunging fold—this anticline plunges toward you (Sheep Mountain Anticline, Wyoming).
(d) Domes (Little Dome, Wyoming). Photos by John Shelton.

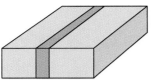
(a) Undeformed rock

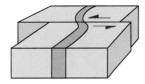
(b) Rock deforms ductilely

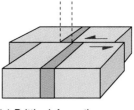

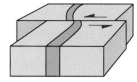
(c) Brittle deformation (faulting)

(d) Ductile deformation begins again as stress continues

FIGURE 17–17
Stepwise development of offset along a fault. Final offset is produced by several small episodes of motion as shown.

FIGURE 17–18
Anatomy of a fault. Arrows show relative sense of displacement.

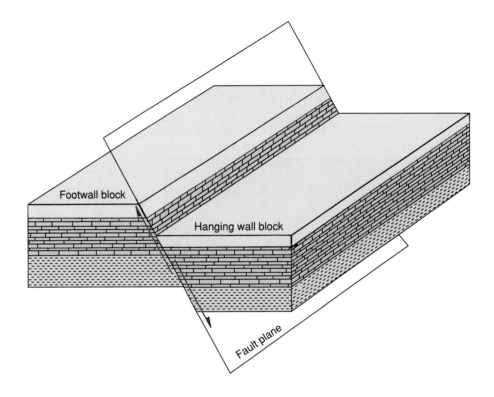

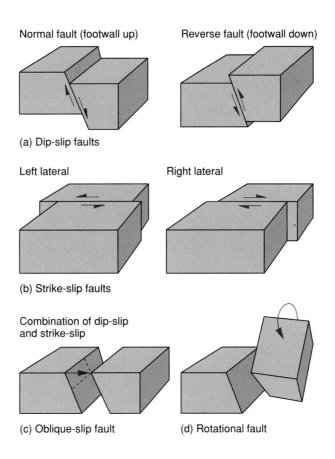

FIGURE 17–19
Classification of faults.

Normal fault (footwall up)

Reverse fault (footwall down)

(a) Dip-slip faults

Left lateral

Right lateral

(b) Strike-slip faults

Combination of dip-slip and strike-slip

(c) Oblique-slip fault

(d) Rotational fault

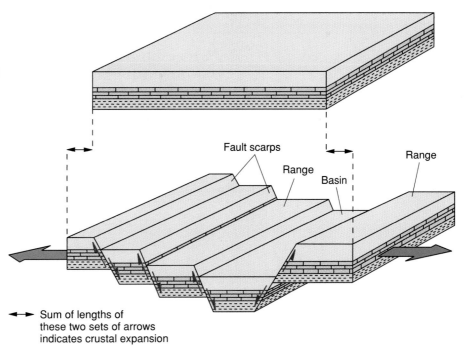

FIGURE 17–20
Crustal expansion is created by tension. The normal faults shown here produce what is called basin-and-range topography in the United States; in Europe, the term *horst* is used for uplifted blocks and *graben* is used for downdropped structures.

Fault scarps

Range

Basin

Range

Sum of lengths of these two sets of arrows indicates crustal expansion

FIGURE 17–21
The East African Rift System. (a)
Olduvai Gorge. (b) Stresses inferred
for the rift system. Photo courtesy
of Patrick Brock.

(a)

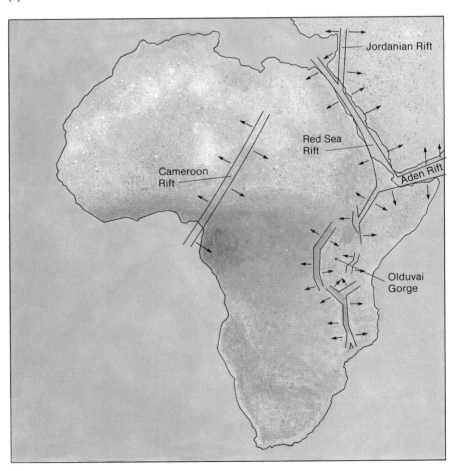

(b)

Thrust faults are a special class of reverse faults in which the fault planes dip very gently, often only a few degrees. Many thrust faults are responsible for the transport of large masses of old rock upward and eventually on top of younger rocks, thus reversing the sequence expected from the principle of superposition. Because thrust faults dip gently, the upward motion is gradual, and rocks may move laterally many kilometers from their original depositional sites in order to move upward a few hundred meters (Figure 17–23).

Thrust faults are found in many mountain ranges, including the Alps, Rockies, and Appalachians. Such mountain systems must therefore have been produced by compression. In the plate tectonics model, subduction zones are simply a special kind of thrust fault, one that forms by intense compression when two plates collide.

Strike-Slip Faults. Displacement during strike-slip faulting is parallel to the strike of the fault, and is thus essentially horizontal. From the viewpoint of a person standing on one of the two fault blocks and facing the fault, motion of the opposite block is either to the left or the right. This provides the classification for strike-slip faults: **right-lateral** or **left-lateral** (Figure 17–19).

Strike-slip faults are the result of shear stresses, and these stresses may arise during major episodes of regional tension or compression. Strike-slip faults thus are found with either normal or reverse faults in structurally complex areas, and are common in deformed regions throughout the world. As a matter of

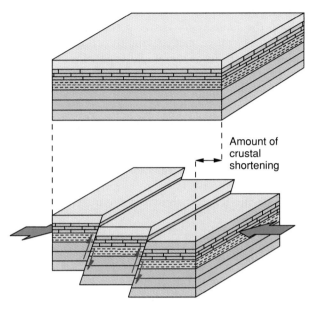

FIGURE 17–22
Crustal shortening is produced by compression, causing these reverse faults.

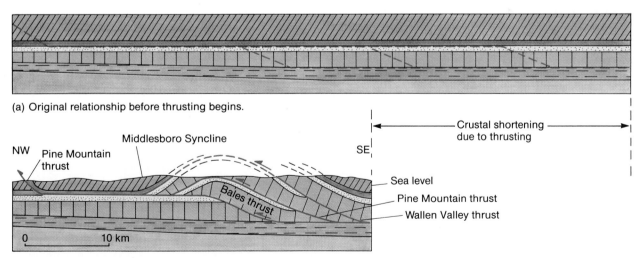

(a) Original relationship before thrusting begins.

(b) After compression, which has produced several thrust faults.

FIGURE 17–23
Thrust faulting in the Southern Appalachians. Modified from Harris and Milici, 1977, U.S. Geological Survey Professional Paper 1018.

fact, strike-slip faults are among the best-known geologic features in the world, after volcanoes. The San Andreas fault, for example, is a right-lateral strike-slip fault familiar to millions of people.

The amount of movement along a strike-slip fault can be measured if we can match once-connected objects that lie on opposite sides of the fault. The offset in faults that are currently active often is revealed by offset of such modern features as streams, fences, and rows of plants in fields (Figure 17–24).

Transform Faults. The plate tectonics model places particular importance on a special class of strike-slip faults that act as neutral plate boundaries. These are the **transform faults** discussed in Chapter 1, along which plates grind past one another. Transform faults are more than just particularly large strike-slip faults—they are a special kind of fault. The difference between a transform fault and a "typical" strike-slip fault lies in the blocks that are moved.

To understand the complexities involved, let us consider the evolution of a transform fault that cuts a midocean ridge (Figure 17–25). At the point of initial rupture, the ridge segments are separated by some distance along the fault, giving an impression of left-lateral offset—Figure 17–25(a). However, each block is itself undergoing tension because of

seafloor spreading. New crustal material forms at the ridge crests on both sides of the fault as rifting opens the ridges wider and wider—Figure 17–25(b). The spreading produces motion along the portion of the transform fault between the two ridge crests that is actually *right-lateral,* exactly the opposite motion from what we first interpreted. Further, everywhere along the fault *except* between the ridge crests, the two blocks move in the same direction, not in opposite directions as we would expect in a fault.

Transform faults connect segments of oceanic ridges, segments of island arcs, or separate pieces of other transform faults. In every case, the transform fault separates blocks on which some kind of motion is in progress—spreading, subduction, or faulting. Transform faults are not limited to the oceans. The San Andreas fault is a transform fault on a continent; it separates segments of an ocean ridge called the East Pacific Rise, as shown in Figure 17–26.

Recognizing Faults in the Field

Some faults, such as those in Figure 17–1, are easily recognized because the fault plane and both fault blocks are exposed to view. Others are not nearly so obvious. Ancient faults are particularly difficult to

FIGURE 17–24
A geologist stands where recently plowed furrows in a California field have been offset by even more recent fault motion. Photo courtesy of U.S. Geological Survey.

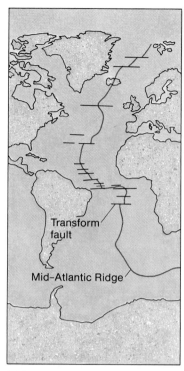

(a) Transform faults in the ocean

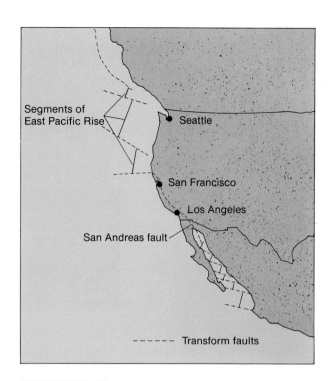

FIGURE 17-26
The San Andreas fault is a transform fault connecting segments of the East Pacific Rise.

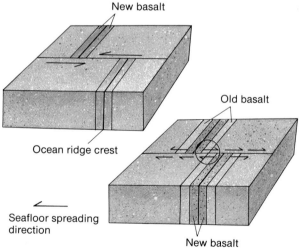

(b) Formation of a transform fault

FIGURE 17-25
Transform faults. (a) Transform faults appear along the Mid-Atlantic Ridge. (b) At left, the ocean ridge forms with irregular fractures due to the Earth's curvature and thickness of the lithospheric plates. The apparent offset is *left lateral*. At right, seafloor spreading continues at each ridge segment. Motion on opposite sides of the transform fault is complex, but between ridge segments *right-lateral* motion occurs.

identify because erosion and deposition may have flattened original fault scarps, and streams may have straightened the offset originally formed in strike-slip motion.

Faults can be recognized by textures produced during the crushing and grinding that takes place when two blocks of rock move past one another. The blocks abrade each other as they move, and angular fragments of the rocks are broken off along the fault plane. Fault motion grinds some of these into extremely fine material, but leaves a few larger pieces. Such crushed and broken rocks are called **fault gouge**. With continued grinding, and as grains reorient and recrystallize in response to increased stress, the fine-grained, foliated *mylonitic* fabric discussed in Chapter 7 is formed in some fault zones.

Finely pulverized material produced in fault zones commonly smooths and polishes the fault plane in much the same way that a jeweler's abrasive powder polishes gemstones. Slightly larger, more-resistant grains are dragged across the polished surface and carve fine grooves. These grooved and polished sur-

FAULTS | 483

FIGURE 17–27
Slickensides produced by faulting of
fine-grained granite, San Gabriel
Mountains, California. The grooves
indicate that most of the fault
motion was horizontal. Photo
courtesy of W. B. Hamilton/U.S.
Geological Survey.

faces are called **slickensides** (Figure 17–27), and are useful in identifying the direction of movement in the fault. The movement of the block must have been parallel to the groove, but *which way* parallel to the groove is not as easily determined. Slickensides thus may tell us that a fault is strike-slip, but not if it was left-lateral or right-lateral. They also demonstrate dip-slip motion, but not whether the motion was normal or reverse.

Faults may have a distinctive topographic expression because gouge and mylonite are generally less resistant to erosion than the unbroken materials lying outside fault zones. Differential erosion removes the rocks along the faults, creating aligned valleys, ponds, lakes and other lowland areas along the strike of the fault (Figure 17–28).

DEFORMATION OF ROCKS AND TECTONIC HYPOTHESES

The first hypotheses concerning the origin of mountains came from studies of deformed rocks. When geologic mapping spread across Europe and then into North America in the 1800s, geologists discovered that the Alps, Appalachians, and Rockies contain some of the most intensely deformed rocks on the continents. The expanding, contracting, and pul-

sating Earth models discussed in Focus 1–2 were proposed to explain the structures found in those mountain ranges. One of the reasons that the plate tectonics model has replaced its competitors is that it alone explains the current and ancient deformation history of the Earth.

Modern Deformation

Today, for the first time, we are able to directly measure deformation *as it is happening* in the oceans as well as on the continents. Modern technology has provided the geologist with strain gauges, instruments that measure the strain built up in rocks of active fault zones. We can measure how rapidly strains build to the breaking point, how long the buildup-rupture cycle illustrated in Figure 17–17 takes, and how often this cycle occurs. Extremely precise surveying instruments help measure the small amounts of displacement associated with individual episodes of faulting. Once we understand modern deformation, we can use principles of uniformitarianism to explain ancient geologic structures.

Studies of earthquakes associated with active faults enable us to determine whether the faults are normal, reverse, or strike-slip. This, in turn, permits us to construct a picture of the large-scale tensional, compressional, and shear stresses acting on the Earth

FIGURE 17–28
Linear depressions and sag ponds along two strands of the San Andreas fault zone, San Luis Obispo County, California. Photo courtesy of R. E. Wallace/U.S. Geological Survey.

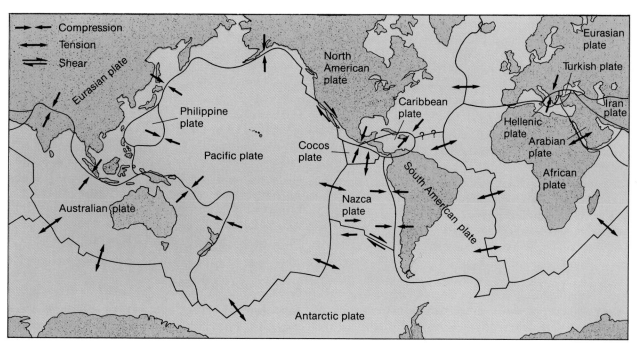

FIGURE 17–29
Stresses in the modern Earth inferred from relative motions of lithosphere plates.

today. Figure 17–29 is a simplified stress map based on relative motions deduced from earthquake studies. The relationship to the plate model shown in Figure 1–6 is obvious.

Figure 17–29 shows that the Earth is presently experiencing tensional and compressional stresses at the same time. This is not compatible with either the expanding or contracting Earth models, and only marginally acceptable in the pulsating Earth model. These models require uniform worldwide stresses, not the combination exhibited by Earth today, and must therefore be invalid for the Earth at this point in time. If we can show that the Earth has behaved the same way in the past, plate tectonics would be the only viable tectonic hypothesis of the five discussed in Focus 1–2.

Interpreting Ancient Stresses

Folds and faults are evidence of Earth's stress history. We saw earlier that folds result from compression, and that the axial planes of most folds form at right angles to the compressional stresses. Fold belts therefore record both the existence of compression in the past and the orientation of the ancient compressional stresses. For example, axial planes of the major folds in the Appalachian Mountains strike roughly northeasterly along the entire mountain system. The maximum compressional stresses during folding must have been at right angles to these axial planes: from the northwest and southeast.

Large-scale normal faults long have been recognized in the American Southwest and along the East Coast of the United States because of large horst-and-graben structures. We already have seen other examples of tension-produced normal faults in the world, so that tension must have operated in the past.

We now see that both tensional and compressional stresses have affected the Earth in the geologic past. The important question is whether they acted *simultaneously,* as they do today.

Unraveling the Record of Ancient Deformation

Relative-age and absolute-age dating methods permit us to date deformation events and to correlate deformations throughout the world. By the early part of the twentieth century, geologists in Western Europe and North America had found many similarities in deformation history. Both continents have been subjected to periods of intense deformation and mountain building, separated by periods of relative inactivity. Each major episode of deformation is called an **orogeny,** and the timing of European and American orogenies proved to be very similar.

For example, rocks of Western Europe experienced two distinct orogenies during the early part of the Paleozoic Era, and similar events took place in the northern Appalachians at almost exactly the same times. Late Mesozoic and early Cenozoic times saw more orogenies in the American West, in the Alps in Europe, the Himalayas in Asia, and the Andes in South America. A picture of simultaneous compressional events throughout the world seemed to emerge.

Recent studies have modified this conclusion. We now know that tension affected eastern North America at the beginning of the Mesozoic Era, at the same time that intense compression affected Japan. Tension and normal faulting continued along the east coast of North America at the same time that the late Mesozoic orogenies mentioned above were beginning in the American West.

The evidence is clear: Earth has behaved in the past much as it does today, with both compression and tension occurring at the same time in different places. Only the plate tectonics model fits this condition.

GEOLOGIC STRUCTURES IN EVERYDAY LIFE

Some geologic structures, particularly faults, can create hazards if they are not recognized when building sites are chosen. As a result, most major construction projects involve preliminary geologic assessment of the local terrain. Other structures have provided valuable deposits of oil and gas.

Petroleum and Natural Gas

Much of the world's supply of petroleum and natural gas comes from places where these precious fuels are trapped and concentrated in geologic structures. The most productive structures are anticlines and salt domes, and these are called **structural traps** in the petroleum industry (Figure 17–30). A brief discussion will show how structural geology, combined with a knowledge of porosity and permeability, can keep our homes warm and automobiles running.

Oil and natural gas are commonly concentrated in the axial areas of anticlines—Figure 17–30(a). They

FIGURE 17–30
Structural traps for oil and gas.

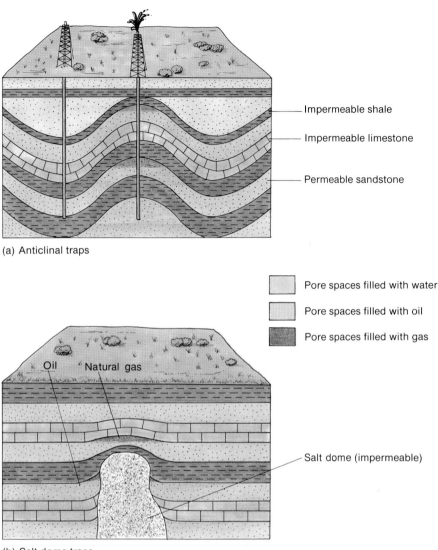

(a) Anticlinal traps

— Impermeable shale

— Impermeable limestone

— Permeable sandstone

Pore spaces filled with water

Pore spaces filled with oil

Pore spaces filled with gas

Oil   Natural gas

— Salt dome (impermeable)

(b) Salt dome traps

do not *form* in these places but *migrate* to them according to the principles of groundwater flow and are trapped at the crests of the anticlines. Rocks in which the oil and gas originally form are called **source rocks;** those in which they are eventually trapped and from which we recover them are called **reservoir rocks**.

Oil and natural gas are fluids, and their behavior in rocks is comparable to that of water: they accumulate in aquifers, and are blocked by aquicludes (Chapter 12). Oil has a lower specific gravity than water, and therefore floats on water, whether in a river, a tank, or when displaced from pore spaces in an aquifer. Gas has an even lower specific gravity. If all three are present, gas will be at the top, oil in the middle, and water at the bottom.

When an aquifer is folded to form anticlines and synclines, water pressure displaces the oil and gas from the lowest areas (the axes of the synclines) and forces them to migrate upward into the axial areas of the anticlines—Figure 17–30(a). If the oil-bearing aquifer is overlain by an aquiclude, the fuels will be trapped at the fold crests. If there is no aquiclude, the oil and gas will escape upward, will not be concentrated, and will be much more difficult to find and recover. It is clear that well drillers must understand local structures before they consider spending millions of dollars to drill an exploratory well.

Salt domes also are responsible for many oil and gas deposits. As salt flows upward during extreme ductile deformation, it heaves the overlying rock upward into an arch and thus produces an anticline where oil and gas can be trapped—Figure 17–30(b). In extreme cases, the salt actually pierces the overlying strata. When this happens, oil and gas displaced by water pressure, as in the case of anticlines, migrate upward in an aquifer until they come into contact with the walls of the salt dome. Rock salt is impermeable and prevents further migration; the oil and gas are trapped against the salt and accumulate there in large enough amounts that they can be recovered profitably.

Geologic Structures as Hazards

Some geologic structures are dangerous, although their danger is not always recognized. It seems intuitively foolish, for example, to build a dam, factory, or home on an active fault, yet many homes, schools, hospitals, and businesses sit directly on the San Andreas fault in California. This fault has been the site of several major earthquakes in the past century, including the destructive earthquake of October 1989, yet building near it and on it continues. We will study this problem in greater detail in our discussion of earthquakes in Chapter 19, but remember that an earthquake is merely the result of displacement along a fault.

Sometimes the problem is deciding whether or not a fault is active. After all, if deforming forces in an area stopped hundreds of millions of years ago, ancient faults may be relatively safe to build on. We must keep in mind, however, that humans view time in terms of decades or centuries, whereas geologic time involves millions of years. Thus, much debate preceded construction of the Indian Point nuclear power plant in southeastern New York State because the location was on the Ramapo fault (yes, there are also faults in the eastern United States). Arguments among geologists about whether the fault was capable of motion came to an abrupt end in 1978 when the fault settled the debate for us. The power plant

had been built by that time, but thankfully the earthquake caused no significant damage.

Even inactive faults can pose a problem for construction because foundations anchored in gouge tend to be less stable than those in unfractured bedrock. Their high porosity and permeability make gouge zones useful as collectors of ores deposited by groundwater and hydrothermal fluids. Unfortunately, these same properties also facilitate seepage of groundwater into construction sites and basements, and the seepage of water out of reservoirs and sewage out of treatment plants.

Small-scale features also can be hazardous. "Falling Rock Zones" on highways often are places where rocks are strongly jointed, fractured, or cleaved. We saw in Chapter 10 that rockfalls and rockslides are aided by fracturing, and that oversteepening of slopes by excavations in areas of dipping rock also can lead to mass movements. Sometimes even small movements can be fatal.

SUMMARY

1. Rocks subjected to intense deforming forces may either deform ductilely by folding, or brittlely by faulting.
2. Deforming forces produce compressional, tensional, and shear stresses as they are resisted by the cohesiveness of rocks. The amount of deformation in response to these forces is called strain.
3. During initial deformation, most rocks behave elastically—the amount of strain is directly proportional to the stress applied. Elastic deformation is temporary; when the forces are relaxed, the rocks return to their original shape.
4. When stresses build to a particular point, which varies for each rock type, rock material flows and deformation becomes permanent. Such behavior is called ductile deformation and is illustrated by folds and salt domes.
5. When a rock's cohesive strength is completely overcome, it breaks. Zones of breakage include random fractures, systematically aligned joints, and faults. Faults are breaks along which motion has taken place.
6. Different types of geologic structures are produced by different types of stress. Folds are the result of compression, and develop so that their axial planes are perpendicular to the dominant compressional stresses. Faults result from all three kinds of stress. Compression causes reverse and thrust faults, and accompanies local crustal contraction. Tension produces normal faults and

is associated with local crustal expansion. Shear stresses cause strike-slip faulting.

7. The Earth currently exhibits a complex pattern of stresses. Tension is taking place at the ocean ridge crests, compression at the island arc–trench systems, and shear in the oceanic fracture zones and major strike-slip faults on the continents.

8. Modern stresses are not uniform throughout the world, which would be expected if the globe were expanding or contracting. Ancient stresses, as recorded by folds and different kinds of faults, similarly indicate nonuniform stresses at any given time in Earth history. Only the plate tectonics model satisfactorily explains the details of these stress fields.

9. Geologic structures, notably anticlines and salt domes, often are places where petroleum and natural gas are concentrated, and are thus sought by exploration geologists. Mineralizing fluids pass readily through fractured rocks, so that fault zones are prospected for valuable mineral deposits.

REVIEW QUESTIONS

1. Examine the faults shown in Figure 17–1. What kind of faults are they? What kind of stress was involved in producing them?

2. Two sedimentary rocks subjected to the same conditions of temperature, pressure, and stress are seen to deform very differently. Explain how this is possible.

3. Which kind of rock would provide the most complete record of deforming forces: competent or incompetent? Explain fully.

4. What changes would you expect in the deformational behavior of a rock as it is buried deeper and deeper in the Earth's crust?

5. An outcrop of slate has a strongly developed cleavage, and this cleavage is tightly folded. From what you know about cleavage, what can you deduce about the deformation history of this rock?

6. Explain how transform faults are different from other strike-slip faults.

7. The movement of ice in a glacier was described in Chapter 13. Describe this movement in terms of rock deformation behavior.

8. Why is it difficult to recognize folds in a batholith?

9. Why is oil found in the axial regions of anticlines but not in the axial regions of synclines?

10. A geologist walking through a new field area finds some places where sedimentary beds are horizontal but are cut by a nearly vertical cleavage. What conclusions can be drawn from these observations?

FURTHER READINGS

Billings, M. P. 1978. *Structural geology*. 3d ed. Englewood Cliffs, NJ: Prentice-Hall. (One of the classic textbooks in structural geology, with descriptive sections on each major structural type.)

Burchfiel, B. C. 1983. The continental crust. *Scientific American* 249(3):130–45. (A description for laypersons of the behavior of the Earth's crust.)

Davis, G. H. 1984. *Structural geology of rocks and regions*. New York: John Wiley & Sons. (An excellent structural geology text with imaginative and entertaining diagrams illustrating how deformation occurs.)

Suppe, J. 1985. *Principles of structural geology*. Englewood Cliffs, NJ: Prentice-Hall. (A higher-level structural geology text with detailed discussions of deformation.)

18

Mountains and Mountain Building

Curtis Glacier on Mt. Shuksan, North Cascade Range, Washington. Photo by Kirkendall/Spring.

People have always considered mountains to be special, often so special as to be religious symbols. Thus, the ancient Greeks thought that their gods lived on Mount Olympus; the Hawaiians believed that Pele, goddess of fire, lived in Kilauea volcano. Noah's ark came to rest on Mount Ararat. Moses received the ten commandments on Mount Sinai.

Mountains are such visible indicators of geologic activity that they are also special to geologists: they challenge us to understand the processes that raise land to such heights. Back in Chapter 1, Focus 1–2 outlined four responses to this challenge—the expanding, contracting, pulsating, and plate tectonics versions of the Earth's behavior. We have seen in earlier chapters that only the plate tectonics model adequately explains most aspects of Earth behavior, structure, and topography. In this chapter we will examine the different kinds of mountains and see whether plate tectonics is equally successful in explaining their origins.

WHAT IS A MOUNTAIN?

A **mountain** is defined as a part of the Earth's crust that stands more than 300 m above the surrounding land surface and has a restricted summit area, steep sides, and considerable bare rock surface. Smaller heights are called hills. Large flat areas that rise prominently above their surroundings but do not have sharp peaks are called **plateaus**.

By this definition, three major physiographic features of the ocean basins are true mountains: ocean ridges, volcanic island arcs, and Hawaiian-type islands. In Chapter 16 we discussed the plate tectonics processes that build mountains in the oceans; in this chapter we will focus on the mountains found on the continents. These differ in several important ways from oceanic mountains, and the differences help us understand more about how the Earth works.

Some mountains are isolated peaks that rise majestically above surrounding lowlands. Most, such as those shown in this chapter's opening photograph, occur in groups called **mountain ranges**. Individual peaks in a mountain range usually form by the same processes and at about the same time as the others. The Cascade Range in the American Northwest, for example, contains mountains such as Saint Helens,

Rainier, Baker, Hood, Shasta, and Lassen Peak—all recently active stratovolcanoes.

Several ranges may be associated in a still larger group called a **mountain system**. For example, the Appalachian Mountain System consists of many ranges, including the Great Smoky Mountains, Blue Ridge, Berkshires, and the Allegheny Mountains. The Rocky Mountain System is even larger, and includes such ranges as the San Juan Mountains, Uinta Mountains, Teton Range, Beartooth Range, and others. Both the Appalachian and Rocky Mountain systems have histories of sedimentation, volcanism, deformation, plutonic activity, and metamorphism that span hundreds of millions of years. A single range within such a mountain system may record only a small part of that complex history.

TYPES OF MOUNTAINS

Earth's most important mountain systems are shown in Figure 18–1. A comparison with Figure 1–6 reveals that some are found at plate margins and others at the centers of plates. Location relative to plate boundaries is important, but it is only one of the ways in which mountains differ from one another. They also may vary in the types of rock of which they are made and the nature and intensity of deformation involved in their formation.

Composition

Mountains contain nearly every type of rock. Some are made entirely of volcanic rock, others of plutonic rocks, and others of a wide variety of sedimentary rocks. Some contain all of these types, along with their metamorphosed equivalents. Compositions that are so varied indicate the operation of very different mountain-building processes. For a tectonic model to be acceptable, it must adequately explain these compositional variations.

Nature of Deformation

Deformation plays an important role in forming many mountains, but is entirely absent from others. Hawaiian basalts are for the most part undeformed, as are volcanic rocks of the northern Cascade Range. In contrast, rocks of the Appalachians, Alps, and

TABLE 18–1
Simplified classification of mountain systems.

Dominant Rock Types	Oceanic Mountains		Continental Mountains	
	Deformed	Undeformed	Deformed	Undeformed
Volcanic	Island arcs; ocean ridges	Hawaiian-type islands; seamounts	Andean-type mountains	Cascade Range-type mountains
Sedimentary + Igneous ± Metamorphic	(None)	Atolls	Fold mountains (e.g., Appalachians) Fault-block mountains (e.g., Sierra Nevada, Basin and Range)	Erosional mountains (e.g., Adirondacks, Colorado Plateau)

Rockies are intensely folded and cut by large-scale thrust faults and strike-slip faults, as described in Chapter 17. Plutons of the Sierra Nevada are up-lifted along major normal faults.

We saw in Chapter 17 that these different structures result from different kinds of deforming forces. Thus, some mountains must form mostly by compression, others by tension, and still others without any sign of deforming stress. Very different mountain-building processes are indicated by such varied deformation histories.

A Classification Scheme for Mountains

Table 18–1 is a simplified classification scheme for mountains based on the criteria described above, and on whether the mountain is in an ocean or on a continent. We already have seen that continents and oceans differ in several ways. We would not expect mountains on the continents to be exactly the same as those in the oceans, and we will summarize these differences at the end of this chapter. One difference stands out prominently, however, and must be mentioned here. All oceanic mountains are *volcanic*. Even coral atolls, composed entirely of reef-building organisms, are able to rise above the ocean because they are built on top of ancient volcanic cones.

How Old Are Mountains?

Table 18–1 does not consider one important factor—time. Not all mountains are the same age. Some, such as the Andes along the western margin of South America, the Himalayas between India and China, and the oceanic ridges, are very young. The forces that produced them are still active and they are growing even as we study them. Others, including the Appalachians and the Ural Mountains in Russia, are much older. The forces that created them stopped millions of years ago, and they are now being leveled by weathering and erosion. The height of a mountain range may be used, as a first working hypothesis, to indicate its age: the higher the mountains, the younger they are.

Intensely deformed and regionally metamorphosed rocks are found in the deeply eroded cores of many of the world's mountain systems. Large flat areas of Precambrian rock that display the same features of deformation and metamorphism are found on every continent (Figure 18–2), and are called **Precambrian shields**. These shields are remnants of Earth's oldest mountains, all that remain of mountain systems that at one time rivaled the Alps and Andes in height. It has taken literally billions of years of erosion to strip away the overlying rocks to reveal the mountain roots. The Precambrian shields show that mountain building is not a new phenomenon, but one that has operated throughout Earth history.

MOUNTAINS ON THE CONTINENTS

It is easier to study continental mountains than those that lie beneath the oceans' surfaces. We can collect samples and measure structures without expensive diving or remote-controlled apparatus, and can much more easily trace the extent of different rock types. However, it is more difficult in some ways to

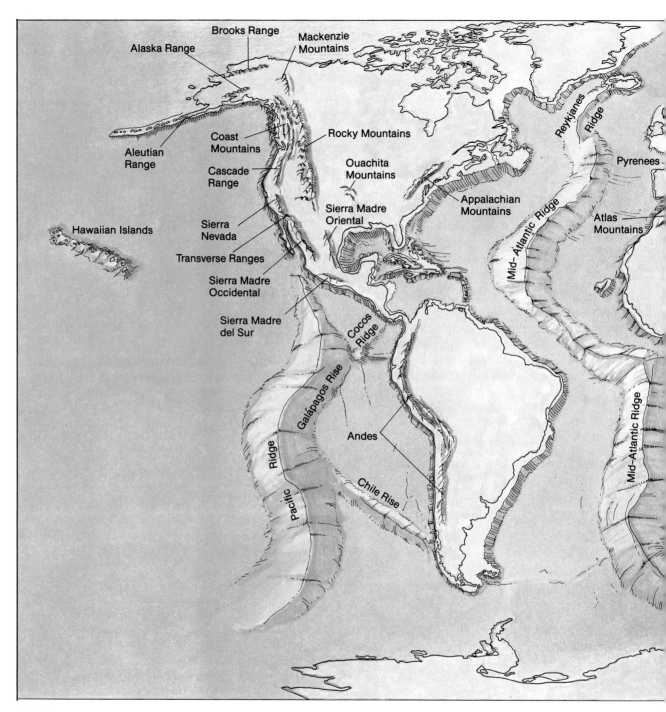

FIGURE 18–1
Earth's major mountain systems. In Asia, the term *Shan* means mountain range.

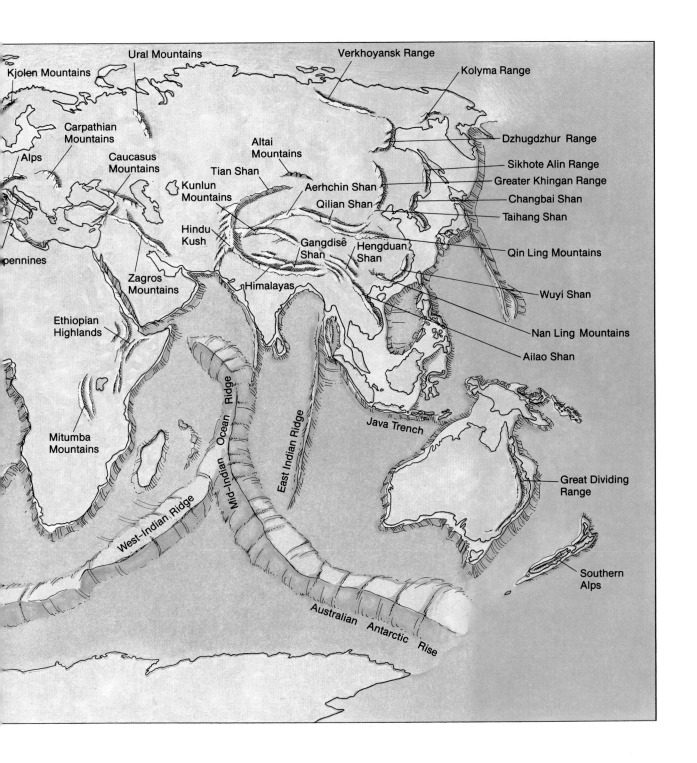

Kjolen Mountains

Ural Mountains

Verkhoyansk Range

Kolyma Range

Carpathian
Mountains

Alps

Caucasus
Mountains

Altai
Mountains

Dzhugdzhur Range

Tian Shan

Aerhchin Shan

Sikhote Alin Range

Kunlun
Mountains

Qilian Shan

Greater Khingan Range

Changbai Shan

Taihang Shan

pennines

Hindu
Kush

Gangdisê
Shan

Hengduan
Shan

Qin Ling Mountains

Zagros
Mountains

Himalayas

Wuyi Shan

Ethiopian
Highlands

Nan Ling Mountains

Ailao Shan

Mitumba
Mountains

Java Trench

Great Dividing
Range

West-Indian Ridge

Mid-Indian Ocean Ridge

East Indian Ridge

Australian Antarctic Rise

Southern
Alps

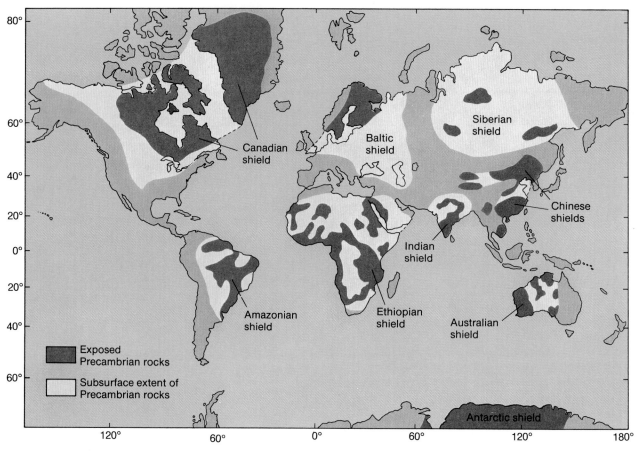

FIGURE 18–2
Distribution of the Precambrian shields. Rocks similar to those of the shields also occur below
the surface, covered by younger strata.

define the processes by which continental mountains
are built. Many of our most famous peaks developed
their shapes from a combination of mountain building and erosion by glaciers and streams. Underwater
mountains are modified little by erosional processes
and their shapes reflect the processes that made
them. For example, the crest area of the Mid-Atlantic
Ridge is clearly the result of the rifting that formed
the ridge.

Some continental mountains are essentially volcanic, and the processes that formed them are comparable to those active in the oceans. Most, however,
are more varied and far more complex. Typical continental mountains are made of many rock types,
including a wide range of volcanic and plutonic igneous rocks; clastic, biogenic, and chemical sedimen-

tary rocks; and their metamorphosed equivalents.
Those continental mountains not formed by simple
volcanism can be grouped into classes based on their
deformational histories: **fold mountains** (compressional), **fault-block mountains** (tensional), and **erosional mountains** (essentially undeformed).

Volcanic Mountains

Vast outpourings of lava on the continents commonly form lava plateaus (Chapter 5) rather than
mountains. Those mountains on the continents that
contain significant amounts of volcanic material are
generally situated at continental margins near an
ocean, and form by subduction of a lithospheric plate
beneath the continent.

FIGURE 18–3
The Villarrica volcano of the central Chilean Andes rises from a basal elevation of about 200 m
to a summit at about 3000 m. Photo courtesy of Dr. Charles R. Stearn.

The Andes Mountain System. The Andes of western
South America illustrate the similarities and differences between the continental mountains and
subduction-related island arcs of the oceans. The
similarities are that the Andes are located at a plate
boundary, they are bounded on one side by an ocean
trench (the Peru-Chile Trench), and they contain
several active stratovolcanoes (Figure 18–3) which
erupt large volumes of andesite, rhyolite, and basalt.
Thick deposits of volcaniclastic sedimentary rock
also are present. These rocks are tightly folded, and
thrust faults are common in some parts of the system.

Unlike island arcs, the Andes contain limestones,
dolostones, and old (Precambrian) gneisses and
schists, and a large batholith composed mostly of
granite occupies a prominent place in the Chile-Peru
part of the system. The similarities outweigh the differences, and the Andes are thought to result from

subduction of the oceanic Nazca plate beneath the
South American plate (Figure 18–4). The differences arise because continental rocks do not behave
the same as oceanic rocks; subduction of an oceanic
plate beneath a continental plate will produce mountains that are slightly different from those caused by
subduction involving two oceanic plates. We will examine reasons for this later in this chapter.

Figure 18–4 explains the present tectonic setting
of the Andes system, but does not begin to show the
complex history that the South American plate underwent prior to the current episode of subduction.
Focus 18–1 fills in some of the details, and shows
why continental mountains are far more difficult to
understand than those of the oceans.

The Cascade Range. The Cascade Range of the
American Northwest is a dominantly volcanic mountain range, but not quite the same as the Andes. The

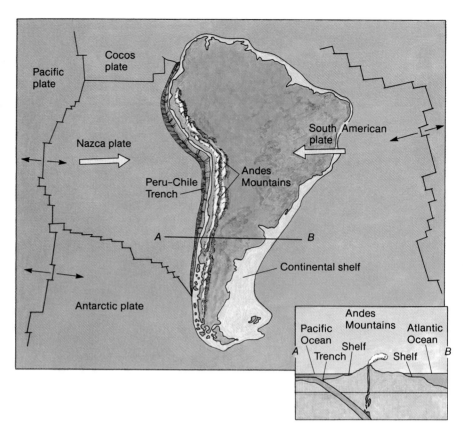

FIGURE 18–4
Plate tectonics model for origin of the Andes mountains. Subduction of the Nazca plate and Antarctic plate beneath South America causes deformation; melting of the subducted plate generates magma to form Andean volcanoes such as those shown in Figure 18–3.

Cascades consist of several andesitic stratovolcanoes in Washington, Oregon, and northern California which have erupted significant volumes of basalt and rhyolite (Figure 18–5). At first glance, this is the only similarity with the Andes or island arcs because the Cascades are well inland from the Pacific Ocean and there is no trench immediately offshore as there is in the Andes.

The presence of andesites, volcaniclastic sedimentary rocks, and structures indicating intense compression in parts of the range suggest that subduction must be involved. Data from earthquake waves (explained in Chapter 19) indicate that there is indeed a subducted slab beneath the Cascades. The Cascade Range results from subduction of the Juan de Fuca plate beneath the North American plate; melting of the subducted slab at depth yielded the magma which erupted to form the volcanoes.

The Olympic Mountains and Coast Ranges that separate the Cascades from the ocean probably represent an intensely deformed **accretionary**

wedge—Figure 18–5(b). The wedge formed as sediment filling the trench was continuously deformed during the subduction process. Note that two very different kinds of mountain ranges—the volcanic Cascades and sedimentary Olympic Mountains and Coast Ranges—are the result of a single tectonic process. Subduction created a trench in which the sediment that formed the Olympic Mountains and Coast Ranges was deposited and deformed, and melting of the subducted slab formed the magmas that produced the Cascades.

Fold Mountains

Most of the world's greatest continental mountains are the result of compressional deformation. The Himalayas, Alps, Ural Mountains, Rockies, and Appalachians, among others, are characterized by tightly folded rocks and owe their elevation to buckling and thrusting produced by enormous compressional forces. Fold mountains are the most complex type of mountain, and each fold mountain system has some unique features not found in the others. To illustrate aspects of fold mountains, we shall first examine the Appalachians, one of the most intensely studied mountain systems in the world.

The Appalachian Mountains. The Appalachians extend along the East Coast of North America from Alabama to Newfoundland, a distance of 3000 km (Figure 18–1). The visible part of the Appalachians is nearly 600 km wide, but sediments of the Atlantic

FIGURE 18–5
Plate tectonics explanation for the Cascade Range. The mechanism is similar to that for the Andes in Figure 18–4, but there is no trench west of the Cascades comparable to the Peru-Chile Trench system because the North American plate and the deformed accretionary wedge strata of the Olympic Mountains lie above the subduction zone.

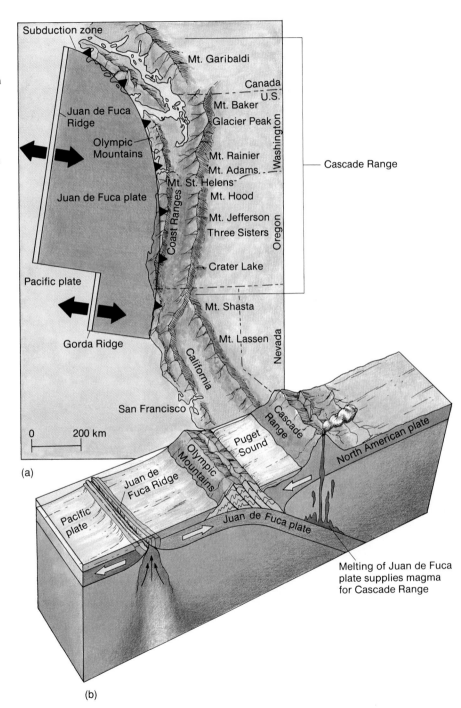

and Gulf Coastal plains cover the system on the east and south, respectively, masking its full extent.

Rock Types. Nearly every type of rock can be found somewhere in the Appalachians. Sedimentary rocks are represented by a wide range of clastic, biogenic, and chemical types, indicative of both marine and terrestrial environments. Igneous rocks span the complete range discussed in Chapter 4, and are present in both plutonic and volcanic varieties. Nearly every grade of regionally metamorphosed and contact-metamorphosed rocks can be found, as well as extensive zones of dynamic metamorphism.

In addition to the wide *variety* of rocks, one of the unique features of fold mountains is the great *thickness* of their sedimentary rocks. Estimated thicknesses of 10,000 to 20,000 m are proposed by researchers in the Appalachians. In one of the first detailed geologic studies in the United States, James Hall showed in 1857 that the thickness of rock units deposited during a given interval of geologic time increased sharply from west to east across New York and New England. This eastward thickening is illustrated in Figure 18–6.

A change in sedimentary facies accompanies the increase in thickness. The thinner sequence to the west consists of quartzose sandstones, limestones, and dolostones. Analysis of the strata and the fossils they contain indicates that they formed in shallow water. The thicker strata to the east are graywackes and shales interbedded with volcanic rocks. Sedimentary features and a scarcity of fossils suggest a

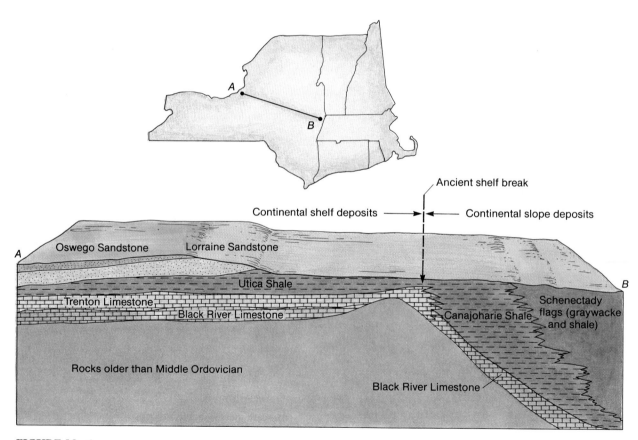

FIGURE 18–6
Thickening and facies changes of Ordovician strata in New York State. After C. O. Dunbar and J. Rodgers, *Principles of Stratigraphy,* © 1957, with permission of John Wiley & Sons, Inc.

EVOLUTION OF THE ANDES MOUNTAIN SYSTEM— OH, NO! HERE COMES ANOTHER PLATE . . .

The Andes was one of the first continental mountain systems for which a plate tectonics model was proposed, principally because of the readily identified relationship among ocean crust, trench, and adjacent volcanic belt. The modern configuration was superimposed on an existing mountain system, one that evolved, also by plate tectonics processes, over a period of hundreds of millions of years. Its history shows how continents grow, and again reinforces a concept that has been obvious from many earlier chapters: oceans and continents are always changing.

Figure 1 shows that subduction and plate collision have been going on at the western edge of South America since the early Paleozoic. The earliest subduction closed an ocean that once separated what is now the eastern part of South America from an isolated continental block labeled "Chilenia" in Figure 1.

Chilenia collided with ancestral South America in the Middle Devonian following subduction-related volcanism and deformation. The enlargement of South America by adding a piece of continental material from some-where to the west as an intervening ocean closed is an example of how continents can grow and oceans disappear with time.

During the Carboniferous, subduction began again at the western edge of the composite Chilenia–South America megacontinent. This led to development of a new volcanic belt on the Chilenian part of the plate. Changes in the steepness of the subducted plate caused the volcanism to shift position and led to sporadic orogenic events accompanied by folding, thrusting, and metamorphism. When the dip of the subducted plate was gentle, volcanicity occurred far inland because the plate had to move farther to reach the depths at which melting occurs; when the dip was steep, volcanism occurred closer to the trench.

Subduction of ocean floor continued from the Carboniferous to the present, but most geologists agree that it was not a single oceanic plate that

deeper water environment for these rocks, and the volcanic rocks resemble those of island arcs. Similar observations have now been made in fold mountains throughout the world.

The rock types and their distribution provide vital information about the origin of fold mountains. The two facies correspond to rocks found today on con-tinental shelves (carbonate-quartzite) and continental slopes (graywacke-shale). We thus can infer that the area now underlain by the Appalachian Mountains was at one time located at the edge of a continent.

Deformation. Intensity and style of deformation vary in the Appalachians, and, as in other fold mountains,

was involved. In some places, particularly in the southern part of the Andes, ancient ocean ridges appear to have been subducted beneath South America. The ancestral Pacific Ocean was segmented into several smaller plates, some of which survive today as the Antarctic, Nazca, and Cocos plates. Others have been subducted back into the mantle.

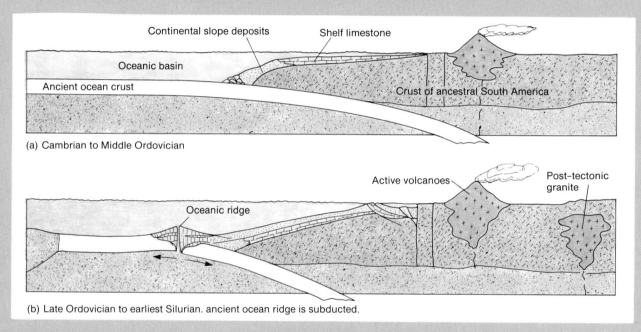

(a) Cambrian to Middle Ordovician

(b) Late Ordovician to earliest Silurian. ancient ocean ridge is subducted.

FIGURE 1
Growth of the South American plate by successive accretion of formerly separate plates. Accretion of the Chilenia plate to South America occurred during Devonian time by subduction of an ancient ocean. Subduction of the ocean west of Chilenia then began, initiating the present orogenic cycle.

the continental shelf deposits are much less deformed than those of the ancient continental slope. The Appalachian Mountain System is divided into four sections, based largely on type of deformation and the age of the rocks present (Figure 18–7):

☐ The **Blue Ridge Province** is a region of Precambrian rocks that have been deformed several times (folded and then refolded) and subjected to high-grade regional metamorphism.

☐ The **Piedmont Province** is occupied by Paleozoic sedimentary and volcanic rocks. It is also an area of intense folding, but one in which metamorphic grade is much more variable. Some rocks are low grade but others are high-grade migmatites.

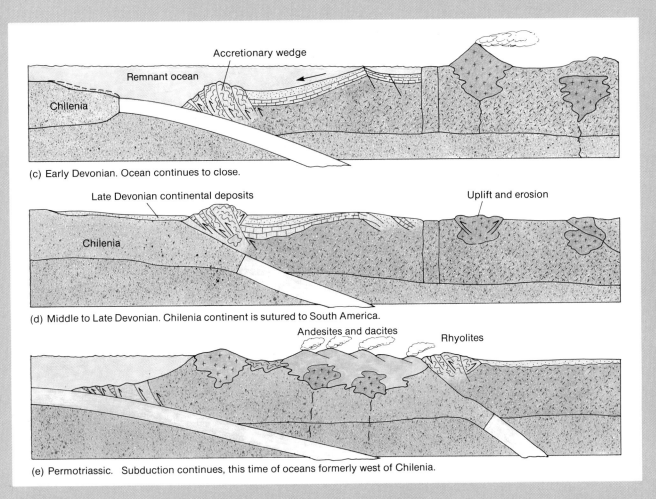

(c) Early Devonian. Ocean continues to close.

Accretionary wedge
Remnant ocean
Chilenia

Late Devonian continental deposits
Uplift and erosion
Chilenia

(d) Middle to Late Devonian. Chilenia continent is sutured to South America.

Andesites and dacites
Rhyolites

(e) Permotriassic. Subduction continues, this time of oceans formerly west of Chilenia.

FIGURE 1
(continued)

□ The **Valley and Ridge Province,** in contrast, contains sedimentary rocks that have been tightly folded but barely metamorphosed.

□ The **Appalachian Plateau Province** lies at the western margin of the Appalachian system. It is composed of flat-lying, totally unmetamorphosed sedimentary rocks. These include the youngest strata of the entire mountain system, ranging in age from the middle to late Paleozoic Era. Surface exposures suggest that these rocks are undeformed or only slightly warped, but data from wells and from seismic profiling (described in Chapter 19) reveal that even these flat-lying rocks have undergone large-scale thrust faulting, as shown in Figure 18–7(b).

MOUNTAINS ON THE CONTINENTS | 503

FIGURE 18-7

Tectonic subdivisions of the
Appalachian Mountains system. (a)
Location. (b) Cross-section showing
differences in rock type and style of
deformation. After Cook and
others, 1979, "Thin-Skinned
Tectonics in the Crystalline
Southern Appalachians," *Geology*
7:563–67.

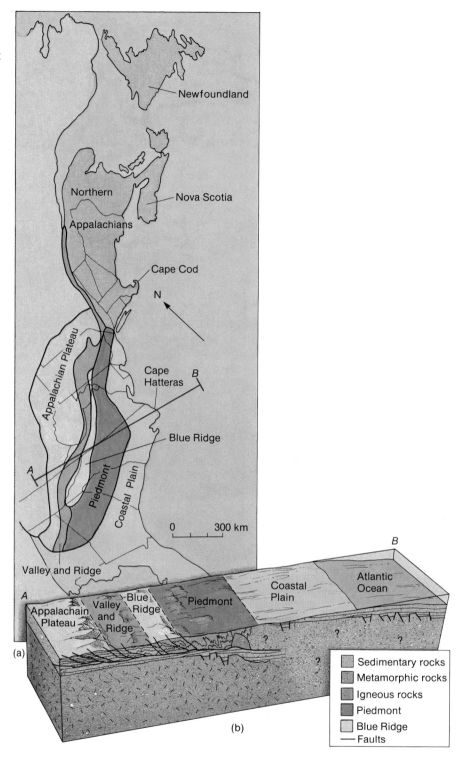

FIGURE 18–8
Ophiolites in the Appalachians. (a)
Obducted ophiolite sequence on the
northwest coast of Newfoundland.
(b) Distribution of ophiolites in the
Appalachians. The linear pattern is
thought to mark a suture between
ancient plates. The entire suite is
preserved in only a few cases; in
many locations, metamorphosed
ultramafic rocks are interpreted to
represent the lower parts of the
suite. Photo (a) courtesy of Dr. W.
S. F. Kidd.

(a)

Indeed, we have learned in the past decade that
thrust faulting is as important as folding in building
many fold mountain belts. Thrusts are now recognized within the Valley and Ridge Province, an area
formerly described as the best example of a fold
mountain system.

The deformation history responsible for the differences among the four provinces consists of several
major mountain-building events called **orogenies**
spread out over nearly 800 million years from the
late Precambrian to almost the end of the Paleozoic
Era. Each orogeny included several episodes of folding, usually accompanied by regional metamorphism, plutonism, and contact metamorphism. In
order for a tectonic model to successfully explain the
origin of the Appalachian Mountains, it must explain
these orogenies.

Ophiolites in the Appalachians. Rocks of the ophiolite
suite have now been identified in the Appalachians,
and in many other fold-mountain systems. Intense
shearing and metamorphism have, in many places,
dismembered the suite and shuffled its components
from their initial positions, but in places like Newfoundland the complete sequence is preserved—
Figure 18–8(a). Complete or partial ophiolite suites

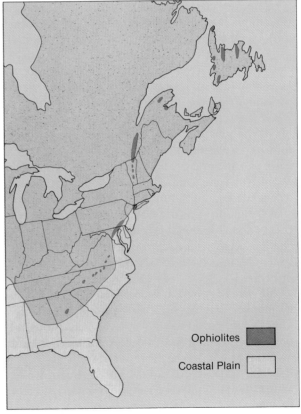

Ophiolites

Coastal Plain

(b)

are now recognized along the entire Appalachian system—Figure 18–8(b). Remember that these rocks are interpreted as pieces of ancient *oceanic* crust. What are they doing on the continents, far from the nearest oceanic crust?

Origin of Fold Mountains. The presence of ophiolites, volcanic rocks similar to those of island arcs, and structures produced by intense compression ALL indicate that plate collision, subduction, and obduction have been involved in the formation of the Appalachians and other fold mountains. We saw earlier that a collision between two oceanic plates creates island arc mountains in the ocean, and that collision between an oceanic and a continental plate produces Andean-type mountains. When two *continental* plates collide, the result is a fold mountain system.

The evolution of the Appalachians involved several steps, each explainable by modern plate tectonics processes, and shown in Figure 18–9. Essentially, the mountain system formed by two processes: (1) breakup of a large continent by rifting and seafloor spreading during late Precambrian time, which formed the ocean whose shelf and slope deposits are preserved in the Appalachians today; and (2) gradual closing of this ocean by subduction throughout the Paleozoic Era, which explains the island arc volcanic rocks and regional metamorphism.

Obduction during the closing of the ocean explains the presence of the ophiolites. Each orogeny records a collision between plates within the ocean basin. By the end of the Paleozoic Era, the ancient ocean had completely disappeared and the Appalachians marked the suture between formerly separate continental plates. That was the end of the Appalachian mountain-building cycle, but not the end of tectonic activity for North America. The Atlantic Ocean had yet to be born, an event discussed in Focus 18–2.

Similar histories are proposed for other fold mountains that are now far from any ocean. The Ural Mountains in the Soviet Union are thought to be a suture joining Asia and Europe, and the Himalayas are believed to be the result of suturing India to Asia. In each of these cases, an ocean that once separated the sutured continents has disappeared. At various times in the histories of these ancient fold mountains, there would have been active subduction zones, volcanoes like those of the Andes, and trenches adjacent to the continents. The formation of Andean-type mountains thus may be simply one step in the formation of a fold-mountain suture. During Silurian and Devonian times, an Andean-type mountain system may have existed along the eastern edge of ancestral North America. The Andean history described in Focus 18–1 is similar to the Paleozoic history of the Appalachians. If the Pacific Ocean were to close, and Asia were to collide with South America, the situation would be similar to what happened in the Appalachian Mountains at the end of the Paleozoic.

The Rocky Mountains. The Rocky Mountains (see Figure 18–1) are another fold (and thrust) system that formed much as the Appalachians did—as a suture between formerly separate plates. However, they are much higher, with many peaks rising over twice the height of the tallest Appalachian peaks. This suggests that the Rockies are a younger mountain system, a hypothesis confirmed by dating of its orogenic events.

As in the Appalachians, the Rockies were affected by mid- and late-Paleozoic orogenies. The climactic event in the Appalachians occurred during Permian time, but deformation continued in the Rockies well into the Tertiary Period. Mountain building in the Rockies thus continued for more than 200 million years after it had stopped in the Appalachians. Erosion has been operating in the Appalachians much longer than in the Rockies and, as a result, Appalachian peaks are far lower than those of the Rocky Mountain System.

The post-orogenic histories of both systems are different. We already have seen that the Appalachian cycle based on the closing of an ancient ocean was followed by the beginning of a new tectonic cycle with the opening and continued spreading of the Atlantic Ocean. Tensional forces are acting today beneath western North America, as we shall see below. These have uplifted local mountain ranges within the Rockies, but have not gone so far as to initiate the next rifting and seafloor-spreading cycle.

Fault-Block Mountains

Continental mountains that form by large-scale normal faulting are called **fault-block mountains**. The

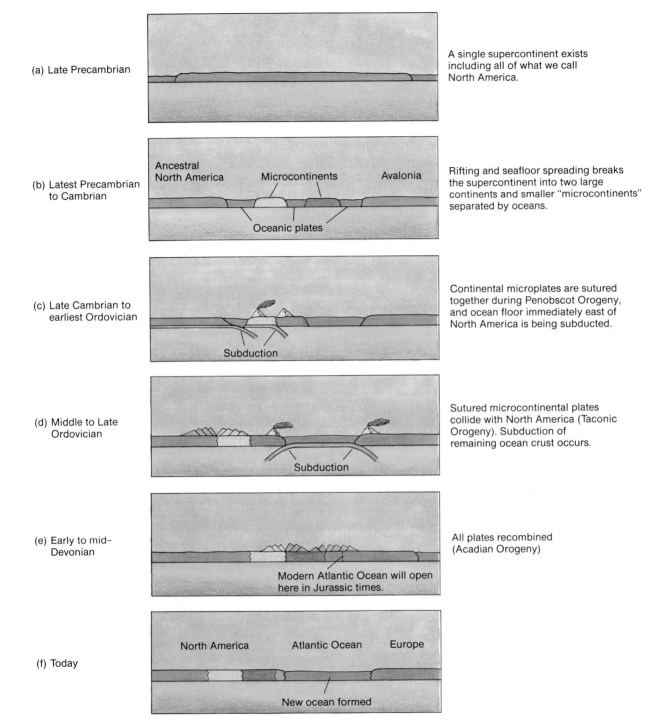

(a) Late Precambrian

A single supercontinent exists including all of what we call North America.

(b) Latest Precambrian to Cambrian

Ancestral North America   Microcontinents   Avalonia

Oceanic plates

Rifting and seafloor spreading breaks the supercontinent into two large continents and smaller "microcontinents" separated by oceans.

(c) Late Cambrian to earliest Ordovician

Subduction

Continental microplates are sutured together during Penobscot Orogeny, and ocean floor immediately east of North America is being subducted.

(d) Middle to Late Ordovician

Subduction

Sutured microcontinental plates collide with North America (Taconic Orogeny). Subduction of remaining ocean crust occurs.

(e) Early to mid-Devonian

Modern Atlantic Ocean will open here in Jurassic times.

All plates recombined (Acadian Orogeny)

(f) Today

North America   Atlantic Ocean   Europe

New ocean formed

FIGURE 18–9
Evolution of the Appalachian Mountains began with (a) a single supercontinent in the Late Precambrian, and progressed through (b) opening of an ocean during the latest Precambrian–Cambrian, through (c), (d), and (e) progressive closing of that ocean by subduction during Paleozoic time, and (f) rifting during the Mesozoic Era to shape the present edge of the North American continent.

"AVALONIA"—A LITTLE BIT OF THE OLD WORLD HERE IN AMERICA

I f Appalachian history is correctly portrayed by Figure 18–9, New England and Maritime Canada contain two sutures created during Paleozoic time (Figure 1). Ancestral North America was first sutured in Ordovician times to a microcontinental plate, perhaps something a bit bigger than the island of Madagascar off the southeast coast of Africa, and then to the larger "Avalonian" continent later in the Paleozoic.

This latter suture no longer welds two large continents together because a second cycle of plate-tectonic activity affected North America after the plate collision near the end of the Paleozoic Era. Seafloor spreading began in Triassic and Jurassic times beneath the supercontinent that had formed during the formation of the Appalachian Mountains, rifting it and creating the modern Atlantic Ocean. The break took place east of the mid-Paleozoic suture, so that part of ancient Europe—the continental plate referred to as "Avalonia"—was left behind on the modern North American continent (Figure 1). Southeastern Newfoundland, parts of Nova Scotia and New Brunswick, southeastern Maine, and the area surrounding Boston belonged to Europe 400 million years ago, rather than to North America.

FIGURE 1

Paleozoic sutures in the northern Appalachians. The western suture formed during the Taconic orogeny in Ordovician times when the unnamed central plate accreted to ancestral North America. The eastern suture marks the boundary between this composite plate and the "Avalonian" plate.

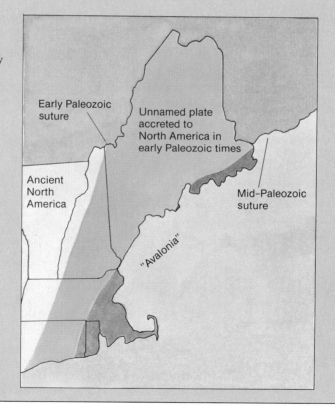

rocks that make up these mountains may be of any type and may have had complex histories prior to the tension that caused the faulting. We will look at two examples of fault-block mountains from the American Southwest—the Sierra Nevada of California, and the Basin and Range of Utah and Nevada.

The Sierra Nevada in California. The Sierra Nevada is a fault-block mountain range formed by the tilting of a single large crustal block (Figure 18–10). This block is nearly 600 km long and 100 km wide, and is capped by mountains rising over 4000 m above sea level. Events prior to normal faulting were those found in fold mountains, including the formation of continental shelf and slope deposits in late Paleozoic and early Mesozoic times. Folding was fol-

lowed by the repeated injection of granitic plutons that formed the huge Sierra Nevada Batholith. The region then became tectonically quiet and was eroded.

During the Pliocene Epoch, after millions of years of orogenic inactivity, normal faulting began along what is now the eastern margin of the Sierras. A steep scarp formed along the east flank of the range where uplift was greatest, and the land sloped more gently to the west, into what is now the Great Valley of California. Erosion of the tilted block produced the rugged topography of the Sierra Nevada that is so strikingly displayed in Yosemite National Park. Mount Whitney, the highest peak in the range, is thus the result of erosion of the uplifted block, and not a volcano like Lassen Peak in northern California.

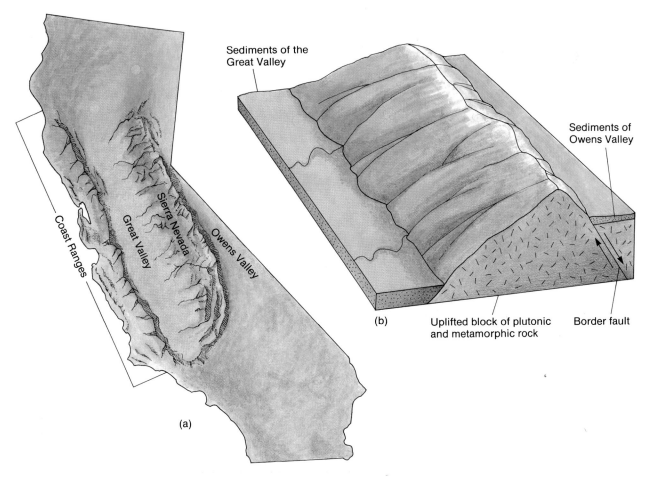

FIGURE 18–10
Sierra Nevada mountains, California. (a) Location. (b) Simplified structure.

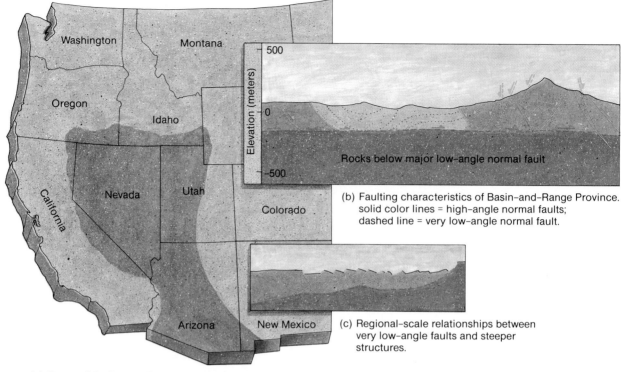

(b) Faulting characteristics of Basin-and-Range Province.
solid color lines = high-angle normal faults;
dashed line = very low-angle normal fault.

(c) Regional-scale relationships between
very low-angle faults and steeper
structures.

(a) Extent of Basin-and-Range Province.

FIGURE 18–11
Cross-section showing structure of the Basin and Range Province, southwestern United States.
Modified from Davis and Lister, 1988, *Detachment Faulting in Continental Extension:
Perspectives from the Southwestern U.S. Cordillera,* Geological Society of America Special Paper
218, p. 133–60.

Basin and Range Province in Utah and Nevada.
The Basin and Range Province is a series of horsts
and grabens typical of areas subjected to tension
(Figure 18–11). A complicated orogenic history
preceded the faulting as in the Sierra Nevada, but
normal faulting is what has caused the present topography.

Tensional stresses that created the Basin and
Range Province were first applied in late Tertiary
times, at about the time that the Sierra Nevada was
uplifted. Deformation of the Basin and Range Province continued into the Quaternary Period and is still
going on today. Volcanism accompanied faulting in
some areas, particularly in northwestern Nevada.
The magmas reached the surface by migrating along
fault planes that penetrated deep into the crust. The
volcanic rocks are a combination of basalts and rhyolites, an association found throughout the world

where continental plates have been subjected to tension. Geologists initially thought that the normal
faults continued steeply downward, but now believe
that the faults curve and become nearly horizontal
with depth, as shown in Figure 18–11(c).

This may not sound very significant, but the
amount of offset on a gently dipping fault needed to
produce uplift like that in the Basin and Range Province is much greater than that needed on steep faults.
We have had to completely revise our estimate of
how much crustal stretching has occurred.

Origin of Fault-Block Mountains. Uplift and tension in the oceans are indicated by the ocean ridges,
but similar forces produce different results on the
continents. This is yet another indication of the differences between continents and oceans. Some geologists suggest that basin-and-range structures form

when a continental lithospheric plate slides over an active ocean ridge. The East Pacific Rise, for example, is now truncated by the San Andreas fault system in the Gulf of California (see Figure 17–26), but at one time an active segment of this ridge may have been located beneath the continent. Fault-block mountains thus may be the surface expression of rifting that has been transmitted through the entire thickness of the North American continental plate.

Erosional (Upwarped) Mountains

Some mountains owe their relief to stream and glacial erosion of a region that is being uplifted without apparent deformation. This gentle uplift is called **epeirogeny** to distinguish it from orogeny, the intense deformation that raises fold mountains.

Epeirogeny is operating now beneath the Colorado Plateau in Colorado, Utah, and Arizona. Extremely rapid uplift has elevated this region to heights thousands of meters above sea level, but a mountainous topography has not developed because of the climate. The region is so arid that neither stream nor glacial dissection has operated as extensively as it has in the Rocky Mountains, but the major through-flowing stream, the Colorado River, has been able to carve the Grand Canyon through the Plateau. Indeed, the spectacular depth of the Grand Canyon is due to the fact that the river has been able to carve its way through the rocks as rapidly as they have been uplifted.

The cause of epeirogeny is not fully understood. It may be related to large-scale compressional or tensional stresses, perhaps to some grand-scale compression involving the North American and Pacific plates.

Dome Mountains

The Black Hills of South Dakota do not fit into any of the categories of continental mountains listed in Table 18–1. They are one of the best examples in the United States of **dome mountains**—mountain ranges with generally circular outlines that cannot be explained easily by the plate tectonics theory.

Dome mountains form by a pistonlike uplift of rocks in a localized area, as shown in Figure 18–12. The upper layers are arched into a domal structure, and erosion strips away the cover rocks to expose the

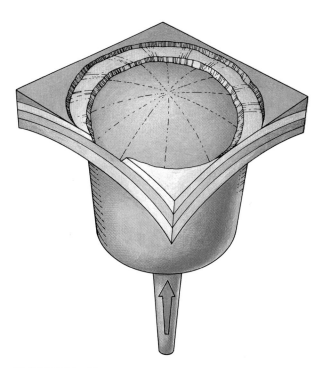

FIGURE 18–12
Dome mountains form by a pistonlike local uplift that arches originally horizontal strata. This type of mountain is thus very different from other types that result from largely lateral movements of crust and mantle.

older units in the core of the dome. In the Black Hills, rocks of Paleozoic and Mesozoic age are arched over the Precambrian core of the dome. It is in the Precambrian core that Mount Rushmore (part of a granitic pluton) and the famous Homestake Gold Mine (in metamorphic and plutonic rocks) are located.

Dome mountains do not fit into the plate tectonics model because they appear to result from dominantly *vertical* forces acting in a localized area in the midst of a lithospheric plate. Other types of mountains form by dominantly *lateral* forces (compression or tension) or by volcanism. In Chapter 20 we will examine models of mountain building that involve vertical forces, to see if they can be reconciled with (or replace) the plate tectonics model.

Mountains, Tectonics, and Vacations

The United States contains most of the types of continental mountains described above, and many are so spectacular that they are part of our National Park

TABLE 18–2
Types of mountains in our national parks.

Type of Mountain	National Park	
Oceanic Mountains		
Volcano produced by mantle hotspot activity	Haleakala	Hawaii Volcanoes
Continental Mountains		
Volcanic mountains	Lassen Volcanic Mount Rainier	Crater Lake Katmai
Fold-thrust mountains	Shenandoah Great Smoky Mtns. Wrangell-St. Elias Rocky Mountain	Olympic Glacier Acadia Redwood
Fault-block mountains	Yosemite Sequoia	Kings Canyon
Erosional (upwarped) mountains	Grand Canyon Bryce Canyon Canyonlands	Arches Zion

System. There is no better way to understand the differences between the types of mountains described in Table 18–1 than to visit these national parks (Table 18–2). The beauty of the parks makes these visits unforgettable, and your enjoyment will be enhanced by understanding something of the processes that have built the mountains and sculpted them. Pamphlets and books that discuss details of the mountains' evolution are available at each of these parks.

COMPARISON OF OCEANIC AND CONTINENTAL MOUNTAINS

Plate tectonics satisfactorily explains most of the types of mountains, but we have seen that basic differences between oceans and continents must be responsible for differences in their mountains. Those differences are more fundamental than can be explained by the lack of erosion in the oceans. You may wish to refer to Chapter 16 for details about the oceanic mountains as we compare them with their relatives on the continents. The differences are inherent in the processes, compositions, and ages of the mountains, and include:

1. *Oceanic mountains are invariably young, but continental mountains show a much wider range of*

age. There are no rocks in any ocean older than Jurassic, and the rocks at the ridge crests today are less than 10,000 years old. The ocean ridges are growing as you read this chapter, as are the Hawaiian Islands and the island arc volcanoes of Japan and the Aleutians. Some continental mountains are still growing, but even young mountains such as the Andes, Cascade Range, and the Himalayas contain rocks that are hundreds of millions of years older than those of the oceanic mountains. In addition, the folding that produced the Appalachians and Ural Mountains ended millions of years before the modern ocean ridges began to spread. Why are the rocks of continental mountains so much older than those of oceanic mountains?

2. *All oceanic mountains are volcanic, but only some of those on the continents are.* Although continental mountains like the Cascades and Andes do contain large amounts of lava and tephra, the great fold mountains are dominantly sedimentary. Why? Is it easier for magma to reach the surface in the oceans than on the continents? Or is there some other explanation?

3. *The greatest mountains on the continents are the result of compression, whereas those of the oceans result from tension.* Why?

4. *Sedimentary rocks are the major constituent of continental fold mountains, but are insignificant in most oceanic mountains.* Why?

There are other important questions that must be answered. If the plate tectonics theory is correct, we must then take the next step and ask: What causes the tension at the ocean ridges? How can a plate move? What is the origin of the forces that thrust one plate beneath another in subduction zones? Why does the subducted plate always seem to be comprised of oceanic material and never of continental material?

We cannot resolve these problems here, because the answers can come only from a detailed knowledge of the interior of the Earth. How can we know that there is an asthenosphere if we have never drilled through the crust, and therefore cannot prove directly that it exists? As a matter of fact, how do we know that Earth contains a crust, mantle, and core?

In the next chapter, we will describe the methods used by geologists to investigate those parts of Earth's interior that we have never seen, and explain the reasoning by which the internal structure shown in Figure 2–6 has been established. With that additional background, we shall tackle the basic questions about plate tectonics in Chapter 20.

ISOSTASY AND MOUNTAIN RANGES

We saw in Chapter 17 that isostatic rebound is occurring in North America as a result of unloading following melting of continental ice sheets. The formation of a fold mountain system such as the Himalayas involves a tremendous crustal thickening—essentially loading the crust more than any glacier could. We would therefore expect isostatic rebound to take place when erosion carves away the mountains. It does, and the discovery that Earth's lithosphere behaves buoyantly eventually led to an understanding of deep-earth processes. This buoyancy is known as **isostasy**. We will briefly follow the trail that led geologists to this important discovery.

Mountains Have Roots: A Geologic Detective Story

In the mid-nineteenth century, British engineers making a topographic map of India ran into unexpected difficulties just south of the foothills of the Himalayas (Figure 18–13). The distance between two points (Kalianpur on the Ganges Plain and

FIGURE 18–13
India, showing locations of the points between which the surveying error was detected.

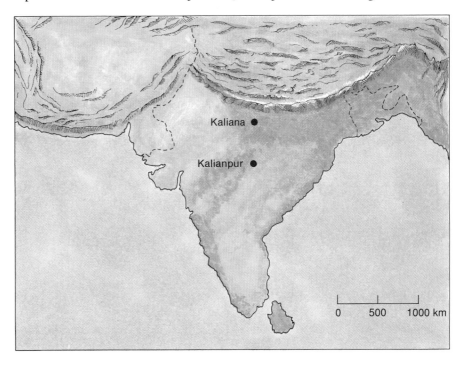

FIGURE 18–14

Gravity as a source of error in the Indian star sighting. (a) The angle measured is between the line of sight and the vertical. (b) If a large mass, such as a mountain range, deflects the plumb bob, an incorrect angle will be used in the calculation.

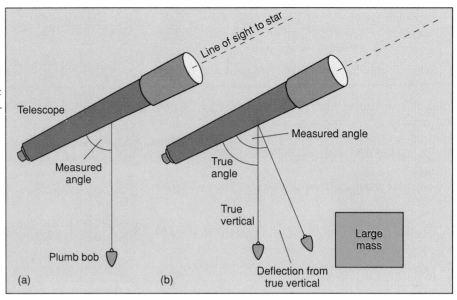

Kaliana in the Himalayan foothills) was calculated by two methods: astronomic sighting on a star and standard surveying techniques. These yielded two slightly different results, off by only 150 m (about 500 feet) in an overall distance of nearly 600 km (375 miles). Surveying, even in the mid-1800s, required better accuracy than that.

An explanation was soon suggested by J. H. Pratt. Star sightings involve measuring the angle between a vertical line and the line of sight to a particular star (Figure 18–14). Pratt suggested that the plumb bob used to determine verticality in the Himalayan foothills might have been attracted by the local mass of the mountains, producing a line that was not quite vertical, and leading to the distance error. This would solve the problem neatly, and Pratt quickly calculated the amount of gravitational attraction expected from mountains the size of the Himalayas. He was surprised to find that the deflection of the plumb bob was only one-third of what it ought to have been. This was no longer just a problem for the surveyors; geologists now had to explain why the mountains exerted far less gravitational pull than expected.

Two hypotheses were proposed to resolve the problem, one by Pratt and the other by G. B. Airy (Figure 18–15). Pratt suggested that the mountains and plains of India are underlain by rocks of different densities. Pratt thought that mountains are made of lower-density rock than lowlands, leading to less mass in the mountains than we would expect based on their size. A lower mass meant that the gravitational attraction by the mountains would be less than expected, accounting for the lower-than-expected gravitational pull. This idea would solve the problem, except for one fact: geologic mapping shows that the plains and mountains are composed of rocks of the same density. Pratt was wrong.

In contrast, Airy believed that mountains and plains are underlain by the same type of rock. He proposed that beneath every mountain there is a **root** of low-density rock extending downward into denser rock (Figure 18–15). This low-density root provides less mass beneath the mountains than at a comparable depth beneath the plains. Net result: a lower gravitational pull by the mountains. Detailed measurements of gravitational attraction near mountains and of the passage of earthquake waves through the Earth indicate that Airy's explanation is correct.

Isostasy: Mountains "Float"

Both Pratt and Airy envisaged a buoyant behavior for the Earth in which light rocks near the surface float on denser rocks below. This buoyancy is called isostasy when applied to Earth processes. Earth makes isostatic adjustments the way a ship does as it is loaded and unloaded (Figure 18–16). A ship loses

FIGURE 18–15

Comparison of the Pratt and Airy explanations for the apparent lack of mass in mountainous regions. (a) Pratt's hypothesis states that mountains are made of lower-density rock than lowlands, so their total mass is less than their size suggests. (b) Airy's hypothesis states that mountains and lowlands are made of the same-density rock, but mountains are underlain by low-density roots, which lowers their total mass compared to the adjacent plains. Pratt's model was disproved when geologists found the same rocks in the plains and mountains.

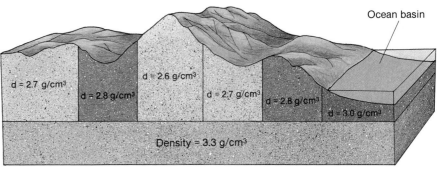

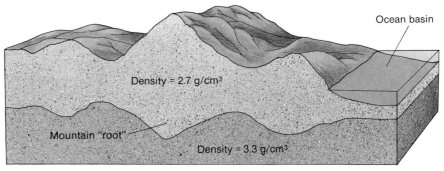

FIGURE 18–16

Isostasy likens a continent to a ship. (a) Loading the ship makes it rest lower in the water; unloading the ship lets it rise in the water. (b) Mountain roots act like the keel of a ship. As mountains are eroded (unloaded), they are buoyed upward isostatically.

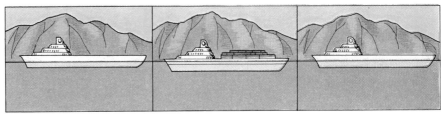

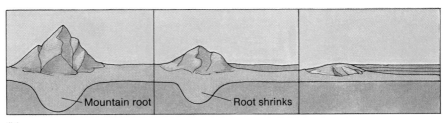

some of its buoyancy when cargo is loaded, and settles lower into the water. As it is unloaded, it rises again. Continents behave in a similar way, but much more slowly, because the asthenosphere flows much more slowly than water. The fluid nature of the upper mantle inferred from the Airy model was the first clue that the asthenosphere existed, but scientists using tectonic theories of the late 1800s did not even suspect such a possibility.

The formation of a mountain is comparable to loading a ship; a root develops and sinks into the mantle. As the mountain is eroded, some of its mass is removed and redistributed on the surface—Figure 18–16(b). The amount of rock removed is considerable; debris shed from the Rocky Mountains forms the Great Plains of the American Midwest and sediment eroded from the Appalachian and Ouachita Mountains forms the Atlantic and Gulf Coastal Plains. If isostatic equilibrium is maintained, the mountain roots must decrease in size to match the mass lost to erosion. As erosion proceeds, the (temporarily) excess buoyancy of the root causes isostatic uplift, raising the partially eroded mountain and shrinking the root. This adjustment keeps stream gradients steep and thus promotes further erosion. This cycle continues until mass is equally distributed between the mountain and adjacent plains, at which point there is no longer any mountain and the root has disappeared.

Isostasy and Plate Tectonics

Isostasy also plays a role in plate tectonics. The process of subduction involves the downward thrusting of relatively low-density crustal rocks into higher density mantle, and strong forces are required to overcome the buoyancy of the subducted slab. The buoyancy of continental crust is so great that it cannot be overcome by these forces. As a result, *continents cannot be subducted.* Ocean crust is denser than average continental crust, has less buoyancy, and can be subducted. Isostasy thus explains why subduction-related continental mountains like the Andes are different from island arc—trench systems of the oceans.

SUMMARY

1. Mountains are topographic features that rise more than 300 m above their surroundings. They occur singly or in groups called mountain ranges and mountain systems.
2. Mountains are found in the oceans and on the continents, at plate boundaries and in the centers of plates.
3. Oceanic mountains are dominantly volcanic, whereas continental mountains contain large volumes of sedimentary, plutonic, and metamorphic rock as well as volcanic material.
4. Some continental mountains, situated near an ocean, are dominantly volcanic, and resemble island arcs. The Andes mountains have many large volcanoes, a deep ocean trench just offshore, and structures comparable to those of the arcs.
5. The Cascade Range contains similar volcanoes, but they are far from the continental margin and lack a trench just offshore. They are the result of subduction beneath North America, and the Olympic Mountains and Coast Ranges that separate the Cascades from the Pacific Ocean are believed to be intensely deformed accretionary wedges associated with that subduction.
6. Nonvolcanic continental mountains are classified by the type of deformation present. The deformation types are fold mountains, fault-block mountains, erosional mountains (essentially not deformed), and dome mountains.
7. Fold mountains are caused by compression, resulting in large folds and thrust faults. These mountains contain very thick sedimentary and volcanic deposits that are more varied than those of the oceanic mountains. Ophiolites also are present in fold mountains, indicating an origin by subduction and eventually by collision between two continental plates.
8. Fold mountains experience deformation in a sporadic fashion; each major episode is called an orogeny, and is generally accompanied by metamorphism and plutonism.
9. This sporadic deformation history is, in many instances, the result of progressive suturing of small plates to a large continental plate.
10. Fault-block mountains form by tensional forces acting on continents. Some result from simple tilting, whereas others involve horst-and-graben structures. Rhyolitic and basaltic volcanism accompany many of the latter type.

11. Epeirogeny is regional upwarping or uplift without deformation; it produces upwarped mountains.
12. Dome mountains apparently form by local pistonlike vertical motion rather than the lateral motion characteristic of most other kinds of mountains. Of all the types of mountains, these are the most difficult to explain with the plate tectonics model.
13. Mountains have roots, downward projections of low-density crustal rock into the upper mantle. Isostasy controls the height of the mountains and the size of their roots. As erosion carves away the mountaintops, uplift and the flow of material in the asthenosphere shrinks their roots.

7. What can the existence of Precambrian shields tell us about how long plate tectonics processes have been operating on Earth?
8. In Chapter 5 we learned that flood basalts are present on nearly every continent. Based on what you have learned about deformation and plate tectonics, suggest an origin for these extensive eruptions.
9. Explain the statement: "Regional metamorphism is the result of plate collision."
10. Explain the difference between subduction and obduction.
11. Both the Pratt and Airy models for isostasy explained the surprisingly low gravitational attraction of the Himalayas. What were the differences between these models? Explain why the Airy model is favored.
12. Show how isostatic uplift keeps mountain peaks at high elevations even while erosion is working to lower them.

REVIEW QUESTIONS

1. Construct a classification scheme for all mountains (continental and oceanic) based on the types of stress applied during their formation.
2. In this scheme, which continental mountains would be grouped with which oceanic mountains? What differences would exist between oceanic and continental mountains in each group?
3. Explain how the elevation of mountain peaks can be used as an indication of the age of the mountains. Then, using information from an atlas or gazetteer for data, list the following mountain systems in order of their age: the Himalayas, Appalachians, Andes, Rockies, Alps, Apennines.
4. Discuss the principal differences between mountains on the continents and those in the oceans. Why do these differences exist?
5. What features would you use to identify a suture between two ancient plates?
6. Volcanic rocks erupt during formation of several different kinds of mountains, and the different types of rock help indicate which kind of mountain-building process is active. What type(s) of mountain would you expect to be associated with:
   a. An andesite-basalt-rhyolite suite?
   b. A basalt-rhyolite suite?
   c. MORB?

FURTHER READINGS

Cameron, Ian. 1984. *Mountains of the gods*. New York: Facts on File Publications, No. 248. (A fascinating look at the peoples who live in what is called the "Roof of the World," the several mountain systems formed by the India-Eurasia collision, and the European explorers who brought back word of the largest mountains on the continents.)

Dietz, Robert. 1972. Geosynclines, mountains, and continent building. *Scientific American* March 1972:30–38. (A discussion linking older concepts of mountain building to the newer plate tectonics theory.)

James, David. 1973. The evolution of the Andes. *Scientific American* August 1973:60–69. (A well-written plate tectonics explanation for the origin of the Andes Mountains.)

Molnar, P., and Tapponier, P. 1977. The collision between India and Eurasia. *Scientific American* April 1977:30–41. (Plate tectonics evolution of several of the world's largest mountain systems.)

Wyllie, Peter. 1976. *The way the Earth works*. New York: John Wiley & Sons, 296 p. (A well-written treatment of the plate tectonics theory as applied to the origin of mountains.)

19

Earthquakes and Seismology

Nimitz Freeway collapse in Oakland, California following the October 17, 1989 earthquake. Photo by UPI/Bettmann Newsphotos.

On Tuesday, October 17, 1989, millions were settling down in front of their TV sets to watch the San Francisco Giants battle the Oakland A's in the third game of the World Series from Candlestick Park in San Francisco. They got more than they bargained for, as the San Andreas fault shifted and a major earthquake struck northern California. For the first time in history, an earthquake was shown live on prime-time television. People worldwide agonized with Californians as buildings and bridges collapsed, flames engulfed a small part of San Francisco, and civil defense workers rescued survivors and prepared for the inevitable aftershocks.

To most of us, an earthquake means San Francisco and Oakland shaken and damaged by the 1989 shock, or the total destruction of Armenian cities and villages in December 1988. Actually, earthquakes occur every day, most so mild that they can be detected only with sensitive instruments. Although we think of earthquakes in terms of destruction, they do have a helpful aspect: it is mainly from the study of energy released during earthquakes that we have been able to infer the structure and composition of the Earth's interior and the processes of plate tectonics. The branch of geology that studies earthquakes is called **seismology**.

In this chapter, we will first look at the nature of earthquakes, the hazards they pose for humans, and steps being taken to lessen their impact. We will then see how seismologists use earthquakes to learn about the Earth's interior.

EARTHQUAKES

An **earthquake** is a shaking of the ground caused by the sudden release of energy during motion along a fault. The energy is transmitted through the Earth in pulses called **seismic waves,** which are analogous to the ripples that radiate outward from a rock dropped into a pond (Figure 19–1). When the waves reach the Earth's surface, they make the ground shake.

The movement can be barely perceptible—a feeling similar to vibrations from a large truck passing

FIGURE 19–1
Generation of waves (a) in a pond; (b) in the Earth.

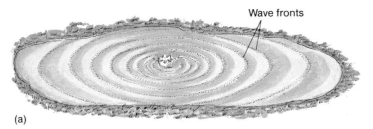

(a)

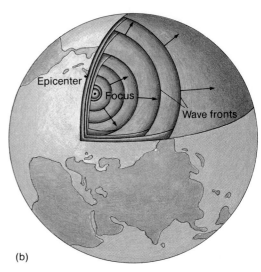

(b)

by—or it can be so violent that it throws us from bed or knocks us from our feet. Ground motion during a 1970 Peruvian earthquake leveled many houses. The 1989 northern California earthquake (called the Loma Prieta earthquake for where it was centered) broke buildings, bridges, highways, and water mains. Photos showing the collapse of the Nimitz Freeway became symbols of how the area had suffered (see the photograph that begins this chapter).

Causes of Earthquakes: The Elastic-Rebound Hypothesis

In 1906, San Francisco suffered an earthquake that caused far more damage than the 1989 shock. Many more buildings collapsed, casualties were much greater, and fires engulfed large portions of the city. After the quake, geologist H. F. Reid made a detailed study of rocks along the San Andreas fault that led him to propose the **elastic-rebound hypothesis** as an explanation for earthquakes (Figure 19–2). According to Reid, deformation along the fault was mostly an accumulation of *elastic* strain, which grew until rupture took place. Beds that had bent elastically prior to faulting—Figure 19–2(b)—returned to an unbent state after movement—Figure 19–2(c). Reid believed that energy is stored in rocks as elastic strain accumulates and is released during rupture. It is this energy that travels through the Earth as seismic waves.

Reid's work helps us understand that fault motion is not continuous, but rather a sporadic, episodic process as shown in Figure 17–17. Fault blocks build strain over a long period of time, and then release it suddenly. This often is referred to as a **stick-slip** process: the blocks *stick* together until enough strain builds to cause rupture so that they can *slip* past one another. Most of the stored energy is released when the blocks rupture, but small quakes called **aftershocks** usually follow the major motion. These occur because fault planes are not smooth (as shown in Figure 19–2) but are rough, irregular surfaces. After the main motion has occurred, the fault may be stuck temporarily where two projections come into contact. When the two blocks overcome this local instability, the energy released produces an aftershock.

The actual site where faulting occurs and from which the energy is released is called the earthquake **focus** (plural: foci). The focus often is many kilometers below the surface. Seismologists locate earthquakes by referring to the point on the Earth's surface directly above the focus; this point is called the earthquake's **epicenter**.

Most earthquakes have shallow foci within 60 km of the surface (Table 19–1), but a few occur at great depths in the mantle. As we saw in Chapter 17, the deeper that rocks are buried, the greater is their tendency to behave ductilely during deformation. Earthquakes are the result of brittle behavior, so most should occur in the upper, more brittle parts of the Earth.

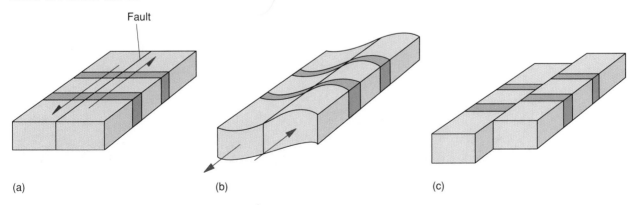

Fault

(a)   (b)   (c)

FIGURE 19–2
The elastic-rebound hypothesis of earthquakes. (a) Unfaulted vertical beds before displacement, showing location of future fault plane. (b) Beds bend elastically as displacement begins along the fault. (c) Rocks rupture and are offset along the fault. Elastic strain is released as seismic waves and the beds return to their initial unbent shape.

TABLE 19–1

Frequency of earthquakes during 1963–1966, based on depth of focus.

Type of Focus	Depth of Focus (km)	Number
Shallow	0–60	10,855
Intermediate	60–300	3492
Transient	300–450	257
Deep	>450	739

Measuring the Strength of an Earthquake

The size or strength of an earthquake depends on how much energy was stored in the deformed rocks before faulting. If a fault undergoes displacement by many small stick-slip increments, relatively small amounts of elastic strain are involved in each rupture; the result is many small-scale earthquakes. However, a single large displacement can occur if elastic strain builds over a long period of time. The release of all the stored energy at once can cause a catastrophic earthquake. Seismologists measure the

FIGURE 19–3

Intensity of the 1811 New Madrid, Missouri earthquake as felt in the eastern United States. Curves represent areas of equal earthquake intensity and are labeled according to the modified Mercalli intensity scale (see Table 19–2). The diagram was constructed by analyzing records of the earthquake damage compiled from newspapers, official reports, and personal diaries. After Nuttli, 1973, "The Mississippi Valley Earthquakes of 1811 and 1812: Intensities, Ground Motion, and Magnitude," *Seismological Society of America Bulletin* 63(1): 227–48; used with permission.

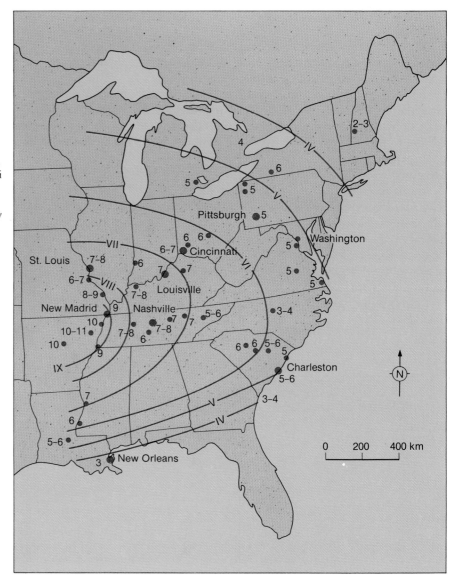

strength of an earthquake in two ways, referring to either the *intensity* or *magnitude* of the event. Intensity is a measure of how much damage the earthquake causes, whereas magnitude is based on the amount of energy released.

Intensity. The **intensity** of an earthquake is a measure of the extent to which human-built structures are damaged. Earthquake damage generally is greatest close to the epicenter and decreases with distance from the fault. For example, Figure 19–3 shows the

TABLE 19–2
Modified Mercalli intensity scale.

Intensity Value	Description
I	Not felt. Marginal and long-period effects of large earthquakes.
II	Felt by persons at rest, on upper floors, or favorably placed.
III	Felt indoors. Hanging objects swing. Vibration like passing of light trucks. Duration estimated. May not be recognized as an earthquake.
IV	Hanging objects swing. Vibration like passing of heavy trucks, or sensation of a jolt like a heavy shell striking the walls. Standing motor cars rock. Windows, dishes, doors rattle. Glasses clink. Crockery clashes. In the upper range of IV, wooden walls and frames creak.
V	Felt outdoors; direction estimated. Sleepers wakened. Liquids disturbed, some spilled. Small unstable objects displaced or upset. Doors swing, close, open. Shutters, pictures move. Pendulum clocks stop, start, change rate.
VI	Felt by all. Many frightened and run outdoors. Persons walk unsteadily. Windows, dishes, glassware broken. Knickknacks, books, etc., off shelves. Pictures off walls. Furniture moved or overturned. Weak plaster and masonry D[1] cracked. Small bells ring (church, school). Trees, bushes shaken visibly, or heard to rustle.
VII	Difficult to stand. Noticed by drivers of motor cars. Hanging objects quiver. Furniture broken. Damage to masonry D, including cracks. Weak chimneys broken at roof line. Fall of plaster, loose bricks, stones, tiles, cornices, also unbraced parapets and architectural ornaments. Some cracks in masonry C[1]. Waves on ponds; water turbid with mud. Small slides and caving in along sand or gravel banks. Large bells ring. Concrete irrigation ditches damaged.
VIII	Steering of motor cars affected. Damage to masonry C; partial collapse. Some damage to masonry B[1]; none to masonry A[1]. Fall of stucco and some masonry walls. Twisting, fall of chimneys, factory stacks, monuments, towers, elevated tanks. Frame houses moved on foundations if not bolted down; loose panel walls thrown out. Decayed pilings broken off. Branches broken from trees. Changes in flow or temperature of springs and wells. Cracks in wet ground and on steep slopes.
IX	General panic. Masonry D destroyed; masonry C heavily damaged, sometimes with complete collapse; masonry B seriously damaged. General damage to foundations. Frame structures, if not bolted, shifted off foundations. Frames cracked. Serious damage to reservoirs. Underground pipes broken. Conspicuous cracks in ground. In alluviated areas, sand and mud ejected, earthquake fountains, sand craters.
X	Most masonry and frame structures destroyed with their foundations. Some well-built wooden structures and bridges destroyed. Serious damage to dams, dikes, embankments. Large landslides. Water thrown on banks of canals, rivers, lakes, etc. Sand and mud shifted horizontally on beaches and flat land. Rails bent slightly.
XI	Rails bent greatly. Underground pipelines completely out of service.
XII	Damage nearly total. Large rock masses displaced. Lines of sight and level distorted. Objects thrown into the air.

[1]Key to 1956 revision prepared by Charles F. Richter, *Elementary Seismology*, W. H. Freeman, San Francisco, 1958, p. 137–38:
Masonry A—Good workmanship, mortar, and design; reinforced, especially laterally, and bound together by using steel, concrete, etc.; designed to resist lateral forces.
Masonry B—Good workmanship and mortar; reinforced, but not designed in detail to resist lateral forces.
Masonry C—Ordinary workmanship and mortar; no extreme weaknesses such as failing to tie in at corners, but neither reinforced nor designed against horizontal forces.
Masonry D—Weak materials, such as adobe; poor mortar; low standards of artisanship; weak horizontally.
Source: After H. O. Wood and F. Neumann, "Modified Mercalli Intensity Scale of 1931," *Seismological Society of America Bulletin* 21(4): 277–88. Used with permission.

intensity of the 1811 New Madrid, Missouri earthquake, among the most powerful ever to have affected our continent. It was felt throughout most of the part of North America that was settled by European colonists at the time. Although barely noticed in New England, damage was greater to the south and west, closer to the epicenter.

The **modified Mercalli intensity scale,** which is used by seismologists and civil defense personnel to describe earthquake intensity, is outlined in Table 19–2. No single value can be given to the *intensity* because, as shown in Figure 19–3, intensity depends on location relative to the epicenter. Furthermore, well-constructed buildings can survive more shaking than those not designed to withstand earthquakes. The modified Mercalli scale takes these differences into account by classifying types of construction materials into the four categories footnoted in Table 19–2.

Magnitude. The **magnitude** of an earthquake is a measure of the energy released at its focus. It is determined by measuring bedrock motion when the energy arrives at the surface in the form of seismic waves. As the waves pass through rock, they cause it to vibrate, and the vibrations are detected and recorded by sensitive instruments called **seismographs**.

The principles by which seismographs operate are shown in Figure 19–4. Seismographs are anchored in bedrock so that they can measure Earth motion directly. Each of the simplified devices shown in Figure 19–4 consists of a rotating drum covered with paper and a pen attached to a mass suspended next to the drum. Thus, the drum is attached to bedrock, but the pen is not. When the Earth moves, the instrument moves with it, but the suspended mass and pen remain motionless. The paper moves along with the Earth and the stationary pen traces out the ground motion in a record called a **seismogram**—Figure 19–4(c). Seismographs can measure either horizontal motion—Figure 19–4(a), or vertical motion—Figure 19–4(b), depending on how the mass is suspended.

Magnitude is calculated from a seismogram by measuring the **amplitude** of the recorded waves. The amplitude is half the distance between crest and trough. A large release of energy causes high amplitudes, and thus large magnitudes. An earthquake's magnitude is described according to the **Richter magnitude scale,** a system named after the American seismologist Charles F. Richter. In this system, magnitudes are expressed as whole numbers and decimals. For example, the "Good Friday" 1964 Alaskan earthquake measured 8.3 on the Richter scale, the 1970 Peruvian earthquake 7.7, and the 1989 Loma Prieta earthquake 7.1. The difference between magnitude and intensity is demonstrated by the fact that the 1988 Armenian earthquake measured "only" 6.9 on the Richter scale but caused far more damage than the higher-magnitude Loma Prieta event (see Focus 19–1).

On the Richter scale an increase of one whole number means a 10-fold increase in wave amplitude and a 32-fold increase in energy release. For example, an earthquake of Richter magnitude 5.5 struck southern California in November, 1979. The 1964 Alaskan earthquake was nearly three magnitude units higher (8.3), indicating that its wave amplitudes were nearly 1000 times greater ($10 \times 10 \times 10$), and its released energy nearly *30,000 times greater* ($32 \times 32 \times 32$) than the southern California quake.

Earthquakes with magnitudes greater than 7.0 are called "great earthquakes" and are catastrophic when they affect heavily populated areas. Those with magnitudes lower than 5.0 generally do not cause much damage to modern buildings, although they can destroy poorly or primitively built structures. The threshold for human detection of an earthquake is about 2.0 on the Richter scale; lower values have too little ground motion to be noticed, and a seismograph must be used to detect them.

Fortunately, the vast majority of earthquakes are of low magnitude, as shown in Table 19–3. The

TABLE 19–3

Frequency of earthquakes (worldwide) of different magnitudes, January 1963–June 1966.

Richter Magnitude	Number
1–2.99	>20,000
3–3.99	1101
4–4.99	9937
5–5.99	3918
6–6.99	299
>7.00	3

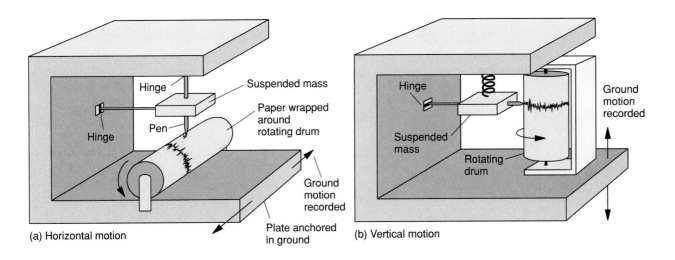

Hinge

Suspended mass

Paper wrapped around rotating drum

Pen

Hinge

Ground motion recorded

Plate anchored in ground

(a) Horizontal motion

Hinge

Ground motion recorded

Suspended mass

Rotating drum

(b) Vertical motion

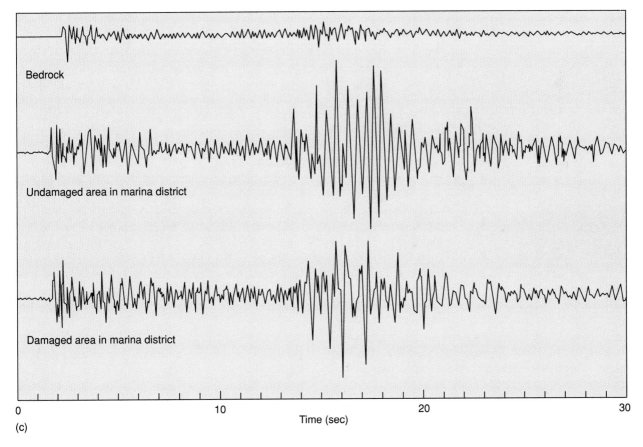

Bedrock

Undamaged area in marina district

Damaged area in marina district

0   10   20   30

Time (sec)

(c)

FIGURE 19–4

Simple seismographs. In both units, the suspended mass remains motionless while the rest of the instrument, anchored to bedrock, moves as the ground moves. A record of the ground's horizontal movement (a) or vertical movement (b) is traced on a drum that is driven by a clock gear. A drawing of a seismogram record of the devastating October 1989 Loma Prieta earthquake (c).

NORTHERN CALIFORNIA AND ARMENIA: A TALE OF TWO EARTHQUAKES

The earthquake that struck northern California on October 17, 1989 came less than a year after the catastrophic shock in Armenia. Many who watched the California event on television feared that they were witnessing a similar disaster, one in which thousands of lives might be lost. Indeed, the magnitude of the Loma Prieta quake was first reported as 6.9, almost exactly the same as the 6.8 Armenian event.

There were other similarities as well. Both earthquakes hit at the worst possible times: during rush hour in California, and at noon in Armenia when people were at work or school. Both occurred on plate boundaries, the Armenian quake taking place in an area where several small plate fragments have been sutured to the Eurasian continent, and the Californian quake occurring on the San Andreas transform fault. Both occurred in seismically active regions. We have mentioned some of the many seismic events in California; Armenia experiences numerous low-magnitude shocks, and has been jolted several times by great earthquakes. Between 1970 and 1983, Armenian deaths numbered 7000 in earthquakes. Earlier shocks were much deadlier; 80,000 died in 1667 alone.

The results of the two earthquakes, however, were very different. The Armenian quake was much deadlier, killing 50,000 to 100,000, *more than a thousand times the casualties in northern California*. Damage to cities was much more extensive as well: two-thirds of Leninakan (population 290,000) was destroyed, along with half of Kirovakan (150,000 inhabitants). Spitak, a city nearly the size of Santa Cruz, California ceased to exist on December 7, 1988. Many asked why California had been so "lucky" this time.

largest earthquakes approach 9.0 on the Richter scale. More powerful earthquakes are unlikely because rocks have finite strengths and can store only a certain amount of strain energy before rupturing.

Earthquake Hazards

Earthquakes kill thousands and cause millions of dollars in damage every year. Most of the damage is caused by ground vibration at the epicenter, but damage may occur hundreds and even thousands of kilometers away. To minimize the effects of earthquakes we must learn how the following factors affect the nature and amount of destruction: different kinds of ground motion, the type of rock or soil, and location relative to the epicenter.

Ground Movement. Although buildings are rigid, they are designed to move a little in response to high winds. For example, the twin towers of the World Trade Center in New York City can sway several centimeters in a gale. However, they can bend only

A combination of geologic and human factors explain why northern California suffered far less than Armenia. The segment of the San Andreas fault that moved is located for the most part in sparsely inhabited areas, much of it in the Santa Cruz Mountains. Santa Cruz and Watsonville were close to the epicenter and were severely damaged, but San Francisco and Oakland are about 100 km away and escaped the most serious damage. In contrast, the fault segment that moved in Armenia passed *directly through* the major cities and towns that it destroyed. If the locked segment upon which San Francisco and Oakland lie had moved, damage would have been far worse.

In addition, California was better prepared for a great earthquake than Armenia. The effects of earthquakes on buildings have been studied extensively in California since 1906. Strict laws require earthquake-resistant construction for new buildings, and many older structures have been reinforced. In contrast, many of the schools and factories that collapsed in Armenia had been built with prefabricated slabs of unreinforced concrete that were attached only weakly to uprights. As one geologist put it during an interview: "Earthquakes don't kill people, buildings do."

Preventive construction measures in California are supplemented by a strong disaster preparedness program. Earthquake drills are held throughout the state, and civil defense teams practice response to simulated disasters. This does not lessen the damage, but vastly improves the rescue effort. Armenian authorities seemed unprepared for an earthquake; rescue and recovery efforts were slow by California standards, and many Armenians faced winter in makeshift housing a year after the earthquake.

so far before breaking. As a result, strong ground motion can destroy a building if it has not been constructed to be earthquake resistant. In general, the taller a building, the greater the potential damage, because a structure's motion is magnified with height. To demonstrate this, make a framework building with an Erector or Tinkertoy building set and then shake the base. The upper floors sway much more than the lower ones.

In a real building, the upper floors may suffer more damage because of shaking than the lower ones. Witnesses to the 1978 earthquake in Mexico City described such great motion in high-rise buildings that the upper floors of adjacent buildings, normally separated by a few meters, actually collided and broke against one another.

Ground motion causes building failure in several other ways. During the disastrous 1755 earthquake in Lisbon, Portugal, ground shaking nearly leveled the city. Massive masonry in churches gave way and fell on worshipers, and palaces and hovels alike crashed to the ground.

FIGURE 19–5
Damage caused by ground shaking during the October 1989 Loma Prieta earthquake. (a) Building damage. (b) Damage in a home in Los Gatos, California. Photo (a) by Wide World Photos; (b) by David LaMarche.

(a)

(b)

When the ground shakes, walls may collapse—Figure 19–5(a), objects may be thrown from shelves—Figure 19–5(b), and some structures may collapse completely. Supposedly well-built concrete-floored structures are susceptible to damage by a phenomenon called **pancaking** (Figure 19–6). These buildings are vulnerable at the points where strong, poured-concrete floors are attached to upright girders. The floors remain intact, but they can detach from the walls; upper floors fall onto those below, causing them to collapse until the entire building has been razed.

FIGURE 19–6
Pancaking. The upper floors of this building, which houses the Ministry of
Telecommunications and Transportation in Mexico City, pancaked during the earthquake of
September 19, 1985. Photo courtesy of M. Celebi/U.S. Geological survey.

Earthquake damage often is more extensive in areas underlain by unconsolidated sediment than in those underlain by bedrock. Sediment lacks the rigidity of bedrock, and as a result undergoes more complex and potentially more damaging motion. This effect can be illustrated with a bowl of gelatin. A sharp tap on the side of the bowl produces only slight vibration in the bowl itself, but the gelatin quivers erratically. Soil and debris lying on bedrock behave the same way, and amplify quake-induced movements. It was no accident that the worst damage to San Francisco in 1989 occurred in the Marina

District where homes are built on landfill. The problems facing Mexico City, one of the most densely populated cities in the world, are worse. The *entire city* is built on the unconsolidated sediments of an ancient lake, so even an epicenter hundreds of kilometers away can cause severe damage.

Liquefaction. Liquefaction, the sudden loss of strength in a mass of water-saturated sediment, was discussed in Chapter 10. Vibrations caused by the passage of seismic waves through sandy deposits is sufficient in some cases to cause liquefaction and extensive damage to structures built on the sediment.

During a 1964 earthquake in Niigata, Japan, most of the poorly constructed buildings in the city were destroyed by ground motion. In addition, several modern buildings designed to be earthquake resistant had to be abandoned, even though they were not destroyed (Figure 19–7). The sediments beneath these buildings liquefied, lost their structural strength, and caused the buildings to rotate as they sank into the ground.

Liquefaction most readily affects sandy sediments that have been rapidly deposited in bodies of water because the sands trap large amounts of water in their pores. Unfortunately, this description fits most material used to claim new land from the sea. The marinas and housing built on land reclaimed from San Francisco Bay are particularly vulnerable to this kind of damage.

Mass Movements. Vibration caused by an earthquake also can trigger large-scale mass movements at great distances from the epicenter. An extreme example of this phenomenon was a disastrous debris avalanche in Peru in 1970, which occurred nearly 100 km from the earthquake's epicenter. Rock, soil, and glacial ice dislodged from the peak of an Andean mountain rushed downslope at velocities of hundreds of kilometers per hour, completely burying the

towns of Yungay and Ranrahirca and 30,000 of their inhabitants. Figure 19–8 shows two views of Yungay; both are peaceful scenes, but their serenity masks the horror of the debris avalanche. The village square shown in Figure 19–8(a) was completely buried by the avalanche, and no trace remains in Figure 19–8(b).

Large-scale slumping is much less violent but much more common. During the 1964 Alaskan earthquake, massive slumps destroyed the Turnagain Heights area of Anchorage, and slump-induced subsidence disrupted life in the downtown part of the city (Figure 19–9).

Tsunamis (Seismic Sea Waves). People fleeing the destruction of the 1755 earthquake in Lisbon, Portugal thought they were safe on new stone wharves along the Tagus River; the wharves were solidly constructed and in the open where no buildings could collapse upon them. If worse came to worst, they felt they could be evacuated from the stricken city by boat. Unfortunately, they were unaware of a phenomenon affecting coastal cities that are subjected to the effects of distant earthquakes, and hundreds were killed when massive walls of water called **tsunamis** swept up the river and crashed over their temporary sanctuary. (The name tsunami means "harbor wave" in Japanese.)It could have been far worse; tens of thousands died when tsunamis slammed into the

FIGURE 19–7
The results of liquefaction. Photo courtesy of U.S. Geological Survey.

FIGURE 19–8
(a) The town plaza of Yungay, Peru, before the 1970 debris avalanche. (b) The same area after the avalanche. Photos courtesy of Lloyd Cluff.

(a)

(b)

eastern coast of the Japanese island of Honshu in 1896, and even more were lost along the coast of Java after the explosive eruption of Krakatau in 1883. Tsunamis are particularly dangerous because they often strike unprepared cities thousands of kilometers from the epicenters of the earthquakes that generate them.

Tsunamis occur when faulting with vertical offset occurs on the ocean floor (Figure 19–10). A seismic wave is produced in the ocean that is similar to some seismic waves in rocks. Tsunamis have amplitudes of only a few meters, but their wavelengths (crest-to-crest distances) may exceed 200 km. This kind of wave is so flat that people on a ship in midocean

FIGURE 19–9
Scarp produced by earthquake-induced subsidence in downtown Anchorage, Alaska. Prior to the earthquake, the sidewalk in front of the stores was at the level of the adjacent street; the earthquake lowered it nearly 4 m. Photo courtesy of U.S. Geological Survey.

would never be aware that a tsunami had passed beneath them, even though tsunamis can travel faster than 500 km/hr. A tsunami slows by friction when it enters shallow water, and "breaks" much like the waves described in Chapter 15. In narrow channels and bays, the water may be funneled into a solid wall up to 20 m high (Figure 19–11). The kinetic energy of such a wave enables it to do extensive damage.

There are some warnings of tsunamis. Just before their arrival, water commonly recedes from the shoreline to far below the low-tide level. This is a warning to get to high ground quickly, but too many curious onlookers take it as an invitation to walk out on the exposed shore to collect stranded fish and shells. All too often they are drowned when the tsunami roars ashore. Another warning: a tsunami is not a single wave, but is part of a wave train that may include many crests and troughs. It often takes hours for the energy to dissipate, and many have returned to their homes after the first wave only to perish in later ones.

Other Earthquake Hazards. Fires caused by earthquakes actually cause more damage and loss of life than ground shaking, liquefaction, and tsunamis. Where homes are made of wood and food is cooked over open fires, the collapse of wooden walls into fire pits can cause disastrous fires that destroy entire cities. Before modern times, when nearly everything was built of wood, fires were the chief fear during earthquakes.

FIGURE 19–10
Origin of tsunamis. Vertical faulting of the ocean floor starts the water movement that eventually becomes a tsunami. (a) Ocean crust with line of impending fault. (b) Sudden vertical displacement of seafloor causes a momentary drop in local sea level. (c) Water rushes into the depression, but overcorrects, locally raising the sea level. (d) Sea level locally oscillates before stabilizing. These oscillations are transmitted as long, low waves that travel thousands of kilometers.

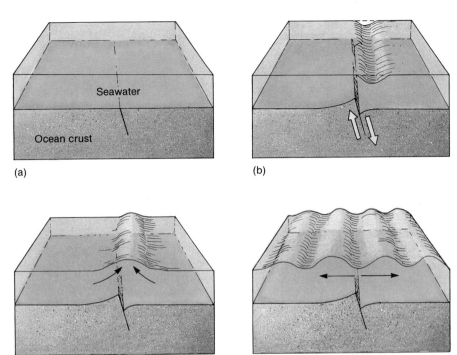

Things have not changed that much, despite our "advanced technology," as shown by the Loma Prieta quake. We use brick, steel, concrete, and glass in our buildings, but wood is still an important construction material and many of the plastics that we use produce toxic fumes when melted. Natural gas pipelines in major urban centers are susceptible to breaking during an earthquake, making our cities vulnerable to fires that start when gas leaks from ruptured mains.

Water mains also break during earthquakes, so our ability to fight fires can be severely diminished.

FIGURE 19–11
A tsunami coming ashore in the mouth of the Wailuku River, Hawaii. One segment of the bridge was destroyed by an earlier wave of the same tsunami. Photo courtesy of Shigeru Ushijama.

During the 1906 San Francisco earthquake, broken water lines made it nearly impossible to fight the widespread fires. Similar problems arose in 1989, and fireboats came to the rescue of the Marina District. Further, highways closed by quake-triggered mass movements or overpass collapse can make evacuation and rescue impossible.

Earthquakes easily destroy dams, leading to disastrous local flooding, and to long-term water supply problems. Surprisingly, several dams near San Francisco are built directly on the San Andreas fault!

Some earthquakes have had profound psychological effects on survivors. The Lisbon earthquake of 1755 lasted only 10 minutes, but struck the Portuguese capital city on the morning of All Saints' Day. Many were crushed in churches when the walls came tumbling down, and the survivors wondered for years why the faithful had been so punished.

Predicting Earthquakes

One of the major problems with earthquakes is that they occur unexpectedly. If we knew that an earthquake was coming, we could evacuate, clear shelves of fragile merchandise, board windows, shut off gas lines, and deploy fire-fighting teams *before* the damage began. We must learn how to predict the time, epicenter, and magnitude of an earthquake far enough in advance to make the necessary arrangements.

We are making progress toward the goal of earthquake prediction. A few hours before it happened, Soviet scientists accurately predicted the time and magnitude of the November 2, 1978 earthquake in the Fergana Valley near the Tadzhikistan-Kirghiz border north of Afghanistan. In February, 1975, Chinese seismologists gave 24-hour warning of a potentially severe quake in the province of Liaoning near North Korea. Despite the numbing cold of a Manchurian winter, people left their homes and spent the night in tents; cars and animals were removed from garages and barns. These precautions paid off and thousands of lives were saved when a major earthquake struck the region, toppling many buildings.

What clues led seismologists to predict these earthquakes? We have learned to recognize a few warning signals; although we do not understand fully how all are related to earthquakes, they have been useful:

1. Groundwater flow sometimes is disrupted just before severe earthquakes. Artesian wells lose pressure and deep wells suddenly run dry.
2. Radon gas is released from the ground along fault zones just before some earthquakes. The Soviet and Chinese scientists who predicted the earthquakes reported that radon emissions nearly doubled before the quakes began.
3. Subtle changes in the electrical conductivity of rocks in fault zones occur before the rocks rupture.
4. Unusual radio static was detected just before the Fergana Valley earthquake.
5. Minor seismic activity may increase and then decrease slightly shortly before a severe earthquake.
6. In some cases the ground surface tilts before a quake.
7. Animals apparently can sense imminent earthquakes, by means as yet unidentified. Many witnesses report that dogs and horses become very restless or frightened minutes or hours before an earthquake strikes. Birds leave the ground to roost in trees, even including such notoriously poor fliers as chickens. Abnormal behavior of cattle and insects also has been documented.

The best chances of predicting an earthquake lie in well-studied seismically active areas. Careful monitoring of the strain accumulating along active faults can tell when the rocks are about to rupture. Strain meters are anchored across the fault zones to record minute increases in the stored elastic strain energy. If we know the rupture strength of the rocks and the rate at which the strain is increasing, we can estimate when faulting will occur.

Understanding the stick-slip mechanism of major faults can help predict earthquakes. When slippage occurs on a fault like the San Andreas, it involves only a small portion of the entire fault. During the 1906 San Francisco earthquake, for example, a segment of the fault nearly 400 km long shifted. The 1989 faulting involved a 50-km segment that lay just south of the piece that moved in 1906. Some segments of the fault move almost continuously, but others are said to be *locked:* movement has occurred north and south of these segments but they have not released their strain. Such segments are prime candidates for major earthquakes.

Unfortunately, San Francisco lies on a locked segment of the San Andreas fault, one that has been storing strain for more than 80 years. The Parkfield segment of the fault between San Francisco and Los Angeles has experienced activity on a fairly regular schedule for many years, but it has been locked and is now overdue for what will probably be a major earthquake.

Although earthquake prediction is still in its infancy, scientists have had some successes, as described. Indeed, the locked segment responsible for the October 1989 Loma Prieta earthquake was iden-

tified by several seismologists, and one even predicted the magnitude of the event (Figure 19–12). Unfortunately, we cannot yet be very specific as to timing, and all that could be said (in 1987) was that a "good probability" of rupture existed within the next 30 years. Other potential earthquake sites are shown in Figure 19–12, along with the probability of occurrence and estimated magnitude.

There also have been failures. The 1979 earthquake in southern California and those in the northern part of the state in January and May of 1980 were completely unexpected. The 1988 Armenian earthquake occurred in a seismically active region near a well-known plate border, but people there also were totally unprepared, as no one had predicted the event.

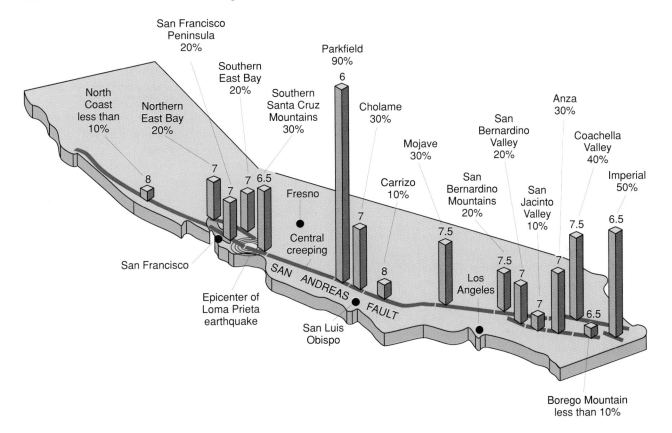

FIGURE 19–12
The probability (in percent) of major earthquakes occurring along the San Andreas fault during the 30-year interval between 1988 and 2018. Estimated earthquake magnitudes are given above each column. This diagram was drawn in 1987, *before* the Loma Prieta event. Note the 90% probability of motion along the Parkfield segment. After U.S. Geological Survey pamphlet, *The Loma Prieta Earthquake of October 17, 1989.*

Can We Prevent Earthquakes?

No, not yet. If earthquake prediction is still a few steps beyond us, the ability to prevent an earthquake is little more than a dream. It may not be an impossible dream, however. Severe earthquakes require storage of large amounts of elastic strain energy in rocks. What if we could find some way to release that energy in small amounts as it accumulates? We would then provide a safety valve that could stop great earthquakes from affecting our cities.

The first step toward earthquake prevention was taken in the spring of 1962, when the Denver area was rocked by hundreds of low-magnitude earthquakes. Denver is situated in a moderately active area, but this level of activity was abnormal. Alarmed by the surge in seismicity near a major city, geologists searched for an explanation.

When they plotted the earthquake epicenters on a map, they were surprised to find them clustered in a circular area surrounding the U.S. Army Rocky Mountain Arsenal. Chemical warfare weapons were manufactured at the arsenal, and a deep well had been drilled into the bedrock to dispose of contaminated water used in the process. The Army started pumping wastewater into the ground under high pressure just before the earthquakes began. When pumping stopped for a year, seismic activity also stopped; when pumping resumed, so did the swarms of small earthquakes. There was clearly a correlation between human activity and earthquakes.

The U.S. Geological Survey set up a unique experiment to find out how human beings had caused the earthquakes, and turned an abandoned oil field in Rangely, Colorado into an underground seismological laboratory. Rangely is in the thinly populated northwestern corner of Colorado, a safe place to experiment. Geologists pumped water into the abandoned oil wells to see if they could reproduce the Denver experience. Indeed, as soon as pressure reached a specific level, earthquake activity near Rangely increased. When water was pumped out of the ground, the pressure decreased, and so did the number of earthquakes. For the first time in history, human beings had turned earthquakes on and off at the flick of a switch.

Exactly what did the switch do? Water increases fault activity in two ways. As it passes into the ground, it enters small cracks and previously existing faults, reducing friction. This lubrication may allow faults to move under the conditions of stress already present, even though the strain level is below that required for faulting in dry, unlubricated rock. Further, the injection of water increases the pore pressures within the fractured rocks, adding to the stress to which they are subjected. We cannot cause earthquakes in unfaulted rocks, but we can make it easier for them to happen in rocks that already are fractured and in a state of accumulating strain.

Think of how this technique could be applied to the locked segments of major faults. After a severe earthquake had relieved most of the elastic strain in the rocks, periodic addition of water could cause small-scale displacements (and small-scale earthquakes) along an active fault zone. This would unlock the fault segment and release strain energy in many small increments rather than in a single, high magnitude event.

If this method is so successful, why are we not pumping water into the locked segment of the San Andreas fault that includes San Francisco? The answer is that the locked segment has been accumulating elastic-strain energy for more than 80 years; releasing the energy now would cause exactly what we are trying to prevent—a severe earthquake. Furthermore, we have no guarantee that, even after most of the energy has been released by the next naturally occurring earthquake, the injection of water will cause only small displacements. Many more years of experimentation in sparsely populated areas are needed before we can think of applying our earthquake-prevention methods to urban areas.

SEISMOLOGY

Despite our technological advances, we have yet to drill one half of one percent of the way from Earth's surface to its center. As a result, our concepts of the structure of Earth's interior come from remote sensing methods, of which seismology is one of the most powerful. As seismic waves move through the Earth toward the surface, they carry information about the rocks they pass through.

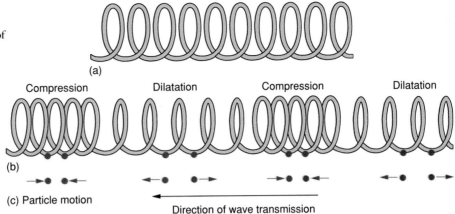

FIGURE 19–13

P waves (longitudinal waves). A Slinky toy illustrates transmission of a P wave. (a) Undeformed Slinky. (b) Zones of compression and dilatation pass through the spring. (c) Vibrational motion of an individual particle in the Slinky.

(a)

Compression   Dilatation   Compression   Dilatation

(b)

(c) Particle motion

Direction of wave transmission

Types of Seismic Waves

We saw in Chapter 15 that waves move across the surface of the ocean because of the interaction of water molecules, initially activated by the wind. Seismic waves move through the solid Earth in a similar manner, as particles of rock interact. Every particle of rock through which these waves pass moves in a specific path that causes collisions with other particles. Each particle transfers some of the wave energy to others, and each particle returns to its original position once the wave has passed. There are four different kinds of seismic waves. Two of these (P waves and S waves) transmit energy through the Earth and are called **body waves;** the other two transmit energy along the surface of the Earth and are called **surface waves.**

P Waves (Longitudinal Waves). The first waves to arrive at a seismograph after an earthquake are called **P waves** or primary waves. Their passage through rocks can be demonstrated with the familiar toy, Slinky (Figure 19–13). Rest a Slinky on a table and stretch it slightly to place its coils under tension. Then strike one end sharply. Pulses of energy can be seen passing through the coils as some are squeezed together (compression) and others are moved farther apart (dilatation). Energy passes from one end to the other as the areas of compression and dilatation move along the spring. It is important to realize that the ends of the spring do not go anywhere; only the *energy* travels any distance. Individual particles in the Slinky move in straight-line paths parallel to the di-

rection in which the wave is being propagated—half of the time in the same direction as the wave, half of the time in the opposite direction—Figure 19–13(b). A wave of this type is called a **longitudinal wave** because the back-and-forth motion of particles is *along* the direction of wave transport.

S Waves (Transverse Waves). The second waves usually to arrive at seismographs are called **S waves** because they are *s*econdary and because they involve a *s*hearing motion of rock particles. When S waves pass through a rock, the motion of particles is very different from that during passage of a P wave. To demonstrate this, tie one end of a rope to a doorknob and hold the other end so that there is some slack. A sharp snap of the wrist sends energy through the rope in a wave resembling a writhing snake—Figure 19–14(a). The motion of particles in the rope is a straight line—again, a back-and-forth motion—but this time *perpendicular to the direction in which the wave moves,* not parallel to that direction as in P waves—Figure 19–14(b). Waves of this type are called **transverse waves** because the particle vibration is at an angle to the direction of wave transport.

Surface Waves. Earthquake energy is transmitted along the surface by two types of surface waves, each named for a pioneer seismologist. **Love waves** are a form of transverse wave in which particle motion always lies *in* the surface along which the wave is propagated—Figure 19–15(a). **Rayleigh waves** differ from all other seismic waves in that particle

FIGURE 19–14

S waves (transverse waves). (a) Movement of a rope during passage of a transverse wave. (b) Particle motion in the rope is a back-and-forth vibration that is perpendicular to the direction of wave propagation. Solid arrows show particle motion at a single instant of time; dotted arrows show motion and waveform at a slightly later moment.

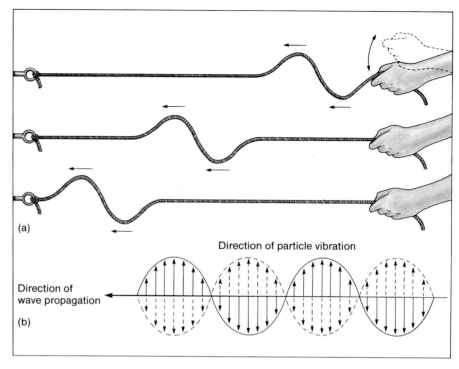

motion is not in a straight line, but in an elliptical orbit similar to the movement of water molecules in a wave—Figure 19–15(b).

Properties and Behavior of Seismic Waves

During an earthquake, P and S waves travel from the focus to the epicenter. Some of their energy is re-flected back into the Earth, but some is converted into Love and Rayleigh waves. Imagine what happens when all four waves pass through a city at the same time (Figure 19–16). P waves cause the ground to move up and down, S waves and Love waves cause it to shake from side to side, and Rayleigh waves make it move in small circular paths. It is no wonder that spectators at Candlestick Park during

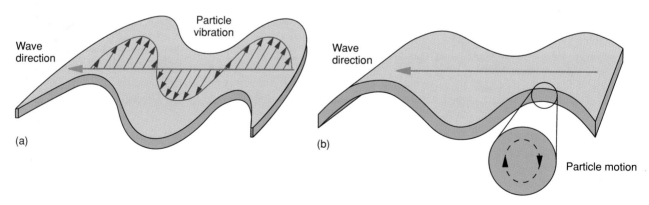

FIGURE 19–15

Surface waves. (a) Love waves are a special form of transverse wave in which particle motion lies in the surface along which the energy is transmitted. (b) Rayleigh waves are transmitted by a retrograde orbit of rock particles.

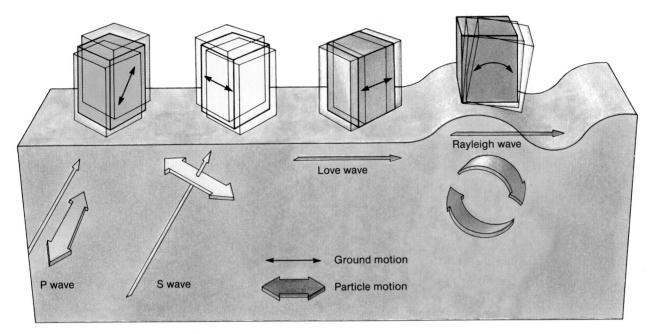

FIGURE 19–16
Ground motion when seismic waves strike. With all four types of waves passing beneath a building at the same time, the results can be disastrous.

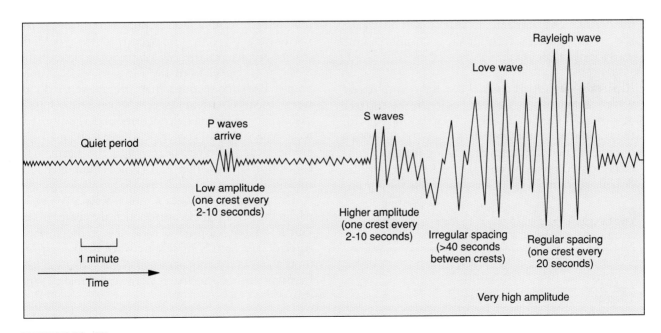

FIGURE 19–17
Simplified seismogram showing the distinction among the four types of seismic waves. Each wave type has a distinctive amplitude and periodicity (the spacing between adjacent crests).

the 1989 Loma Prieta earthquake described the parking lot as moving like an ocean wave, and others felt as if they were on a roller coaster. Few buildings can survive such ground movement.

Laboratory experiments show that each of the four waves behaves in a unique way when passing through rock. Thus, if we can identify the waves, we can interpret the nature of the rocks through which they have traveled. The first step is to identify the seismic waves from their traces on a seismogram. Fortunately, each type produces a characteristic record that makes identification possible (Figure 19–17). The four waves differ in their velocity, amplitude, and wavelength. P and S waves are faster than the surface waves, but the surface waves have greater amplitudes than the body waves and therefore cause more damage.

Velocity. The velocity of a seismic wave depends on the type of wave and the density and rigidity of the rocks through which it passes. Table 19–4 shows that P waves are faster than S waves in most representative rock types, regardless of their composition or texture. In general, P and S waves travel rapidly through low-density rocks and more slowly through denser rocks, but a rock's rigidity may counteract this effect. The more rigid a rock is, the more rapidly these waves can penetrate it. This is why ultramafic rock, which is denser than granite, transmits seismic waves faster than granite.

Reflection and Refraction. Like light and sound waves, seismic waves are reflected from the surfaces of objects—Figure 19–18(a). For seismic waves, these surfaces are the boundaries between materials of different densities, such as bedding planes or contacts between plutons and the rocks they intrude. Seismic reflection profiles are responsible for our un-

derstanding of the materials of the ocean floor shown in Figure 16–5.

Some seismic wave energy passes *through* these boundaries, and these waves are refracted or bent—Figure 19–18(b). The amount of refraction depends on the difference in density between the rock types; the greater the difference, the more the waves will bend. As a result of refraction, seismic waves passing through layered rocks do not travel in simple straight-line paths.

Propagation of Seismic Waves in Liquids. S waves can travel only through relatively rigid substances because only rigid materials can support transverse particle motion. As a result, S waves cannot be transmitted through a liquid such as water or magma. P waves, however, *can* pass through a liquid because the incompressibility of liquids permits the longitudinal particle movement required for this type of wave. This difference between P and S waves turns out to be very important in understanding the interior of the Earth.

Seismic Waves and the Earth's Interior

The energy released by an earthquake is recorded by hundreds of seismograph stations worldwide. If the Earth were perfectly homogeneous, we would be able to predict exactly how long it would take P and S waves to travel from the focus to each seismograph. Early predictions that attempted to do so were woefully inaccurate, even when seismologists calculated the changes in density and rigidity that were expected from known geothermal and geobaric gradients as one goes deeper into the Earth. The conclusion was obvious: Earth is not a homogeneous sphere.

The evidence indicates a global-scale layering in which the densest rocks are found at the center of the Earth and the least-dense rocks are at the surface. After detailed study of seismic wave behavior from thousands of earthquakes, geologists divided the Earth into three principal regions: crust, mantle, and core.

Identifying the Core and Mantle. To see how seismologists deduced the existence of a boundary between the mantle and the core, let us examine the

TABLE 19–4
Seismic-wave velocities in selected igneous rocks.

Rock type	P wave (km/s)	S wave (km/s)
Granite	5.94	3.53
Quartz diorite	6.46	3.69
Basalt	6.77	3.77
Ultramafic rock	7.54	4.28

FIGURE 19–18

FIGURE 19–18
Reflection and refraction of seismic waves. (a) Waves are reflected when they reach a contact between rocks of very different densities. (b) Some seismic-wave energy penetrates such contacts, but is bent (refracted) as it passes into a rock of different density.

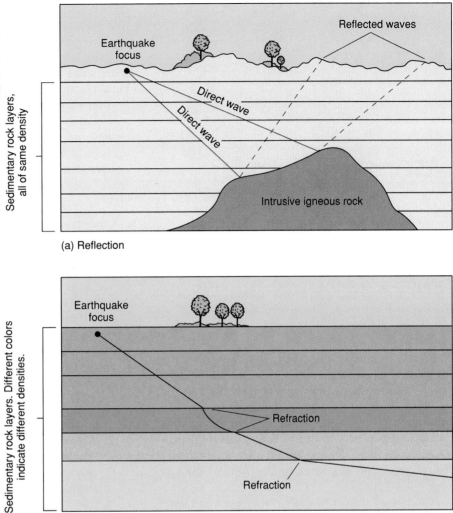

Sedimentary rock layers, all of same density

Earthquake focus

Reflected waves

Direct wave

Direct wave

Intrusive igneous rock

(a) Reflection

Sedimentary rock layers. Different colors indicate different densities.

Earthquake focus

Refraction

Refraction

(b) Refraction

arrival of P and S waves from a single earthquake at seismographs around the world (Figure 19–19). For the sake of simplicity, we will assume that the focus lies just below the surface at the North Pole (although the analysis would be the same regardless of focus location).

At angular distances of 0 to 103° from the epicenter, P waves predictably arrive at seismographs ahead of S waves because they travel faster than the S waves. Between 103° and 143°, however, neither P waves nor S waves arrive directly from the focus, as if this part of the surface were shielded from the waves by something within the Earth. Beyond 143°,

the P waves appear again, but S waves do not. The region between 103° and 143° is called the **P-wave shadow zone;** the region between 103° and 180° is the **S-wave shadow zone**.

To explain this "shadow" phenomenon, seismologists proposed that the Earth has a layered internal structure, as shown in Figure 19–20. From 0° to 103°, P and S waves follow smooth, curved paths caused by continuous refraction through rocks of *gradually* changing density. At a depth of about 2900 km, a *sharp* boundary exists between an outer region of relatively low-density rocks (the **mantle**) and an inner region of far denser material (the **core**).

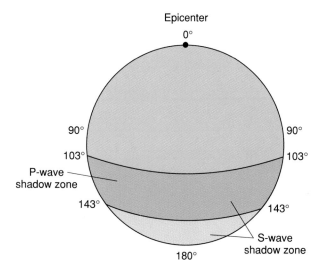

FIGURE 19–19
The seismic-wave shadow zones. No direct S waves arrive in the S-wave shadow zone between 103° and 180° from the epicenter, and no direct P waves arrive in the P-wave shadow zone between 103° and 143° from the epicenter.

P waves entering the core are sharply refracted because of the great density contrast at this boundary. The paths that they follow are controlled by the density differences in such a way that no P wave can emerge from the core on a path that will bring it to the surface between 103° and 143° from the earthquake epicenter. Refraction at the mantle-core boundary is thus the cause of the P-wave shadow zone.

The S-wave shadow zone is much larger, too large to be explained simply by greater refraction at the core-mantle boundary. The reason it is so large is that the S waves cannot penetrate a liquid, as P waves can. Imagine what would happen if the entire core were liquid: S waves could not be refracted through it the way P waves are, but would instead be reflected back toward the surface, as shown by the dashed lines in Figure 19–20(b). This explanation worked so well that seismologists initially believed the Earth to have an entirely liquid core. However, detailed analysis of P waves that pass through the core indicates that only the outer part of the core is liquid; the inner part apparently is solid.

Composition of the Core and Mantle. It is important to understand what seismology can *and cannot* tell us about the Earth's interior. It can reveal the size and density of the core and mantle, but can say nothing about what they are made of. For example, the core must have a density of 10 to 11 g/cm^3, which is far higher than that of any silicate mineral. We have never seen rocks from the core and thus can only speculate on what they might be, but our speculations must involve material of appropriate density. We hypothesize that both the liquid and solid parts of the core are a mixture of metallic iron and nickel. Two forms of evidence lead us to this conclusion:

1. Meteorites—meteorites are thought to be fragments of an Earthlike planet or group of planetoidal bodies, and many are similar in composition to typical Earth rocks. Some, however, are composed of a nickel-iron alloy that has the correct properties of rigidity and

FIGURE 19–20
Origin of the shadow zones. (a) P-wave shadow zone. Refraction prevents direct P-wave propagation. (b) S-wave shadow zone. S waves cannot penetrate the liquid outer core and are reflected back toward the surface.

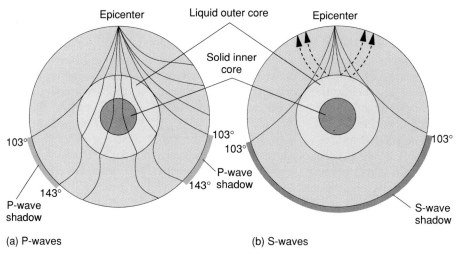

density required of the core by the seismic data.

2. Earth's magnetism—Earth acts like a giant magnet, a fact used by everyone who has used a compass for navigation. The source of the magnetism is most likely an electrically conductive material located in the core. The nickel-iron alloy found in meteorites is precisely the type of material needed.

The mantle is quite different. It is about 2900 km thick and ranges in density from 5.5 g/cm^3 at its boundary with the core to 3.3 g/cm^3 at its contact with the crust. These values are within the realm of what could be expected from rocks made of silicate minerals that are mixed with denser sulfides. We do have some information, both direct and indirect, as to what these minerals or rocks might be. Xenoliths of ultramafic rocks in some basalts are thought to come from the upper mantle. Melting experiments also suggest that the upper mantle—the source of basalt—is essentially a suite of ultramafic rocks made largely of olivine, pyroxene, and garnet. These minerals may be converted to even denser compounds in the lower mantle, where they may be mixed with sulfides and small amounts of metallic iron and nickel. Again, some meteorites have the mineralogic compositions predicted by our theoretical studies, and are thought to be derived from the mantle of a terrestrial planet.

The Boundary Between the Crust and the Mantle.
The crust is Earth's outermost layer, the part we live on and know most about. The fact that it is different from the mantle was proved by the Yugoslav geologist Andrija Mohorovičić in 1910. He discovered that *two* sets of P waves and S waves reach seismographs located within 800 km of the epicenters of shallow-focus earthquakes, but that only a single set of P and S waves reaches seismographs located farther away (Figure 19–21).

To explain this phenomenon, Mohorovičić proposed that a boundary exists between rocks of different densities, and that this boundary could cause refraction of the two body waves. Within 800 km of the epicenter, one set of P and S waves passes directly through the low-density crust, following a smooth, curved path. A second set arrives later because it follows a longer path; its P and S waves cross the crust-mantle boundary twice (entering the mantle and returning to the crust) and are refracted each time. The boundary must be a sharp one, but the density contrast is smaller than the contrast at the mantle-core boundary. Furthermore, because S waves are able to pass through this postulated boundary, there is no liquid involved. In honor of its discoverer, the boundary between the crust and mantle is called the **Mohorovičić discontinuity,** but is more commonly referred to by the shortened form, **Moho.**

Composition and Thickness of the Crust.
The seismologic evidence that defines the crust-mantle boundary also explains several of the differences between oceans and continents that we have seen in earlier chapters: they have different crusts (Figure 19–22).

The thickness of crust beneath a continent is greater, and varies more widely than that beneath an ocean. The average depth to the Moho beneath a

FIGURE 19–21
The crust-mantle boundary, as proposed by Mohorovičić from analysis of shallow-focus earthquakes.

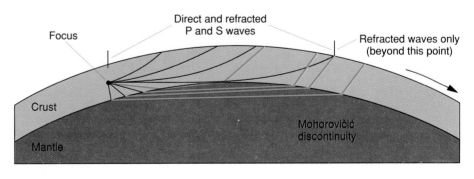

Focus

Direct and refracted P and S waves

Refracted waves only (beyond this point)

Crust

Mantle

Mohorovičić discontinuity

—— Direct P and S waves
—— Refracted P and S waves

FIGURE 19–22
The crust beneath continents and ocean basins. Continental crust is thicker and two-layered (sialic plus simatic), whereas oceanic crust has a single, thinner simatic layer.

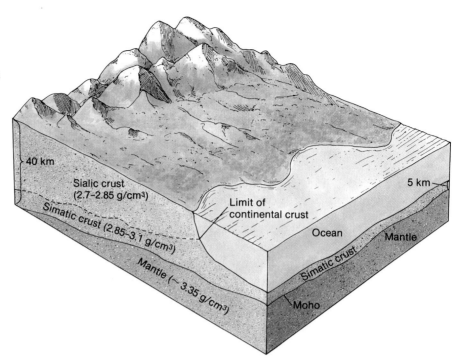

FIGURE 19–23
Distribution of earthquake epicenters throughout the world. After Barazangi and Dorman, 1969, "World Seismicity Maps Compiled from ESSA, Coast and Geodetic Survey Epicenter Data, 1961–1967," *Seismological Society of America Bulletin* 59(1): 369–80; used with permission.

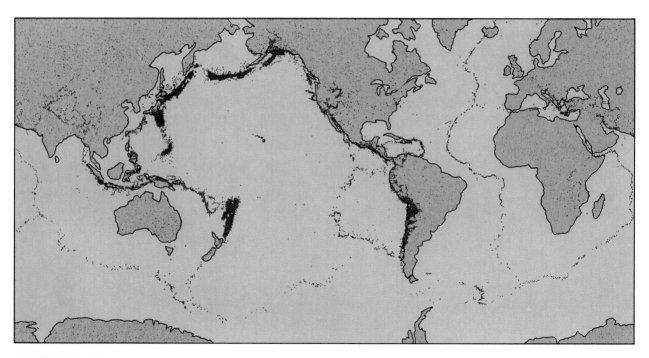

continent is 30 to 40 km. The crust is even thicker (up to 65 km) beneath major mountain ranges, is of average thickness beneath Precambrian shields and plains, and thins beneath the continental shelves. In contrast, ocean crust generally is only 5 km thick.

Continental crust consists of two parts—an upper layer with a density of 2.7 to 2.85 g/cm³, and a lower layer with a higher density of 2.85 to 3.1 g/cm³. The upper layer is made of igneous, metamorphic, and sedimentary rocks with an average composition similar to that of granite. It is this material that most commonly melts to produce granitic magma. This upper crustal layer is rich in silica (SiO_2), and alumina (Al_2O_3), and thus is referred to as the **sialic layer** of the crust, or simply as **sial** (*si + al*). The lower layer is similar to basalt in composition, and is rich in silica, iron, and magnesium. It is called the **simatic layer** of the crust, or **sima** (*si + ma*).

Ocean crust not only is thinner, but also is simpler, and consists only of the simatic layer. This explains the absence of granite from the oceans: there is no sialic layer that can melt to form granitic magma. The sialic layer of the continental crust extends beneath the continental shelves, but does not continue under the continental rise or slope. The true edge of a continent thus is the edge of its continental shelf, rather than the shoreline.

Seismology and Plate Tectonics

Earthquakes are a sign of an active, dynamic planet, and most are caused by modern plate tectonics processes. Seismologists have made several major contributions to plate tectonics theory, including the location of plate boundaries and the discovery of subduction zones and the asthenosphere.

Locating Plate Boundaries. When the locations of earthquake epicenters are plotted on a world map, a clear pattern emerges (Figure 19–23). Most earthquakes occur in narrow belts separating broad regions that are relatively aseismic (devoid of earthquakes). The belts of intense seismicity coincide with the regions of greatest volcanic activity, including the Pacific Ring of Fire (see Chapter 5). Plate tectonics theory holds that most seismicity should occur at plate boundaries where lithospheric plates collide, grind past one another, or are rifted apart. The seis-

mic zones shown in Figure 19–23 are thus essentially identical to the plate boundaries in Figure 1–6.

Locating Earthquakes. We take advantage of the different velocities of seismic waves to locate earthquake epicenters. The concept is simple. Imagine two cars (seismic waves) setting out on a journey, one traveling at 100 km/hr, the other at 80 km/hr. The faster car travels the first 100 km in 1 hour, but the slower car lags 15 minutes behind. After 200 km, the fast car is 30 minutes ahead; after 300 km, 45 minutes ahead. As the distance traveled increases, the slower car (or seismic wave) lags farther behind the fast car in the time of its arrival at the destination (the seismograph). In the case of earthquakes, we

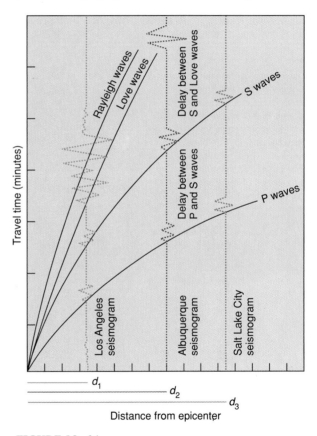

FIGURE 19–24

Travel-time curves. The seismograms show the same earthquake, recorded at three different seismograph stations. The curves show how the arrival time for each type of seismic wave changes with distance from the earthquake focus. By matching the arrivals of the waves from each seismogram on the travel-time curves, the distance of the stations from the epicenter can be calculated.

know the velocities of seismic waves and can measure from seismograms the delay between the arrival times of the different waves. The closer a seismograph is to an epicenter, the shorter will be the delay time between waves.

Figure 19–24 shows four **travel-time curves,** which are graphs showing how the delay between P-wave arrival and arrival of the later S-waves and surface waves changes with distance from an epicenter. To determine how far from the instrument the epicenter of this earthquake was located, we match the standard curves in Figure 19–24 against the observed delays among arrivals of several waves on the seismogram.

However, this process provides only the distance and cannot locate the epicenter exactly because it cannot indicate the *direction* from the instrument.

The epicenter could be anywhere on a circle drawn around the seismograph, a circle whose radius equals the calculated distance. To successfully locate the epicenter, data from a minimum of three stations must be used, as shown in Figure 19–25.

Finding Subduction Zones. The clue to finding subduction zones lies in deep-focus earthquakes. Not only are they rare, as shown in Table 19–1; they are restricted in geographic location (Figure 19–26). A comparison of Figure 19–26 with Figure 16–11 shows that most deep-focus earthquakes occur at convergent plate boundaries—the island arc–trench systems or continents adjacent to trenches, such as the west coast of South America.

Another pattern emerges when earthquake foci from a single island arc are plotted on a cross section

FIGURE 19–25
Locating an earthquake epicenter requires data from three seismographs. A circle is drawn around each seismograph station, with its radius equal to the distance from the seismograph station to the epicenter, corresponding to Figure 19–24 (*d*1, *d*2, *d*3). The intersection of the three circles marks the epicenter.

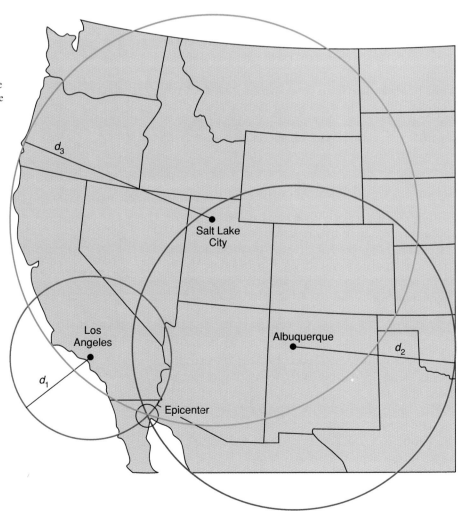

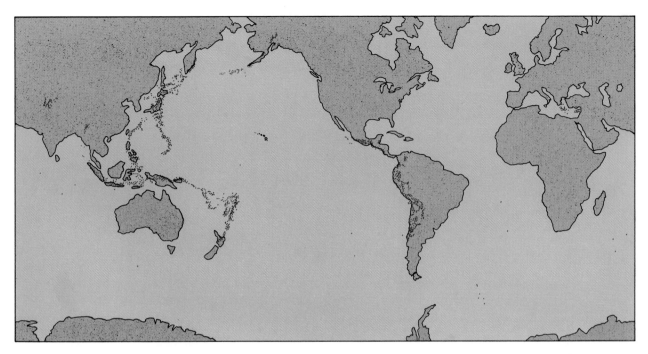

FIGURE 19–26
Distribution of deep-focus earthquakes (greater than 100 km). After Barazangi and Dorman, 1969, "World Seismicity Maps Compiled from ESSA, Coast and Geodetic Survey Epicenter Data, 1961–1967," *Seismological Society of America Bulletin* 59(1): 369; used with permission.

of the arc-trench system (Figure 19–27). Note that the foci increase in depth as one moves from the trench across the island arc; they cluster in a band about 20 km thick called a **Benioff zone**. Geologists interpret Benioff zones as regions where subduction is taking place. The progressively greater focal depths mark the downward progress of the subducted slab as it sticks and slips as it is thrust beneath the overlying plate.

Discovery of the Asthenosphere. One of the early arguments against the plate tectonics theory was that it is impossible for huge lithospheric plates to move through a rigid, solid Earth. Seismologists resolved this problem by discovering the region in the upper mantle that we have called the *asthenosphere.*

Seismic velocities generally increase with depth throughout most of the Earth, but P-wave and S-wave velocities appear to decrease at depths between 100 and 350 km (Figure 19–28). This region is therefore called the **low-velocity zone**. There is no evidence for a sudden change in composition or density at these depths, and our best explanation is that the region is one of unusually low rigidity. (Remem-

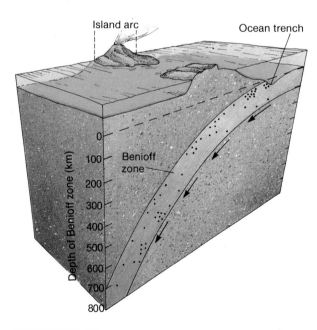

FIGURE 19–27
Benioff zones. The concentration of deep-focus earthquakes in regions called Benioff zones is thought to indicate movement of the subducted plate as it sticks and then slips against the upper plate.

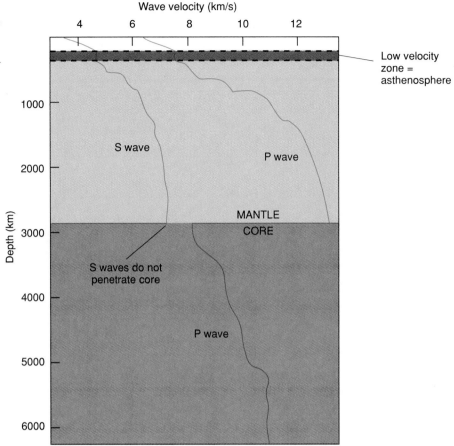

ber that seismic velocity decreases as rigidity lessens.) In fact, the rigidity is so low that the rocks must be extremely ductile; this is thought to be the zone on which lithospheric plates glide.

Rocks of the low-velocity zone are called the asthenosphere to distinguish them from the more rigid materials of the lithospheric plates above. The low rigidity probably is due to the widespread presence of small amounts of basaltic magma mixed with the solid rock. Thus, the asthenosphere not only is the "lubricating" layer that permits plates to move, but also is the source of all basalt.

SUMMARY

1. Earthquakes are episodes of ground motion caused by transmission of energy from underground faulting to the surface.

2. The strength of an earthquake can be measured by its intensity (the amount of damage that it causes) or by its magnitude (the amount of energy that it releases).

3. Damage during an earthquake is caused by a shaking of the ground that causes building failure; by vibration that causes liquefaction of sand deposits; by tsunamis (seismic sea waves) and landslides; by disruption of traffic and communications; by breakage of gas lines, which leads to fires; and by breakage of water lines, which hampers fire fighting.

4. Released energy is transmitted through the Earth by four kinds of seismic waves, each characterized by a unique type of particle vibration. P waves are caused by longitudinal vibrations. S waves and Love waves are caused by transverse vibrations. Rayleigh waves are the result of elliptical paths of particle motion.

5. Earth motion is recorded by instruments called seismographs. Detailed analysis of seismograph records (seismograms) helps us understand the internal structure of the planet.

6. The core is composed of an outer liquid part and an inner solid region. It is much denser than the mantle and is separated by a seismic discontinuity at which P waves are strongly refracted (causing the P-wave shadow zone) and S waves are reflected (causing the S-wave shadow zone).

7. The core is probably made of an iron-nickel alloy. The mantle is probably made of ultramafic rocks mixed with the iron-nickel alloy, particularly in the mantle's lower regions.

8. The crust is separated from the denser mantle by the Mohorovičić discontinuity or Moho, a discontinuity involving a smaller density contrast than the contrast between the mantle and core.

9. The crust beneath the oceans is thinner and simpler than that beneath the continents. It consists of a single simatic layer of essentially basaltic composition, whereas continental crust is composed of a granitic (sialic) layer as well as a simatic one.

10. Earthquake epicenters are concentrated in linear belts that mark regions of greatest crustal instability and delineate geographic boundaries between lithospheric plates.

11. Deep-focus earthquakes are associated with the ocean trenches and occur in thin zones which identify areas that are undergoing active subduction.

12. A shell of low rigidity lies in the upper mantle beneath the lithosphere, and is identified by the unique low velocity exhibited by seismic waves as they pass through it. This low-velocity zone is interpreted as the asthenosphere, the region on which lithospheric plates slide.

REVIEW QUESTIONS

1. Describe the different types of particle motion associated with the propagation of the different seismic waves.

2. Discuss the different things that can happen when a P wave reaches a boundary between rock types of markedly different densities.

3. What are the differences between the discontinuities that define the crust-mantle and mantle-core boundaries?

4. Why is the S-wave shadow zone larger than the P-wave shadow zone?

5. When P and S waves from a deep-focus earthquake reach a seismograph, one wave causes mostly up-and-down motion of the ground, whereas the other causes side-to-side motion. Which is which? Explain the differences by relating ground motion to the particle vibration associated with the two types of wave.

6. In a long block of attached apartment houses, earthquake damage caused by ground motion was greatest in the two end buildings, and least in the middle of the block. Suggest an explanation.

7. An earthquake felt by people living on the fifteenth floor of an apartment house might not be felt by a family living in a split-level ranch house on the next block. Why?

8. Show how the sporadic stick-slip nature of earthquakes can be used to help predict the location of earthquake epicenters.

9. P waves traveling at 6.1 km/sec arrive at a seismograph 3 minutes and 43 seconds before S waves (traveling at 2.84 km/sec). How far away was the earthquake that generated these waves?

10. Discuss the ways in which seismology has been instrumental in constructing the plate tectonics model of Earth behavior.

FURTHER READINGS

Bolt, B. A. 1978. *Earthquakes: A primer.* San Francisco: W. H. Freeman. (An excellent paperback combining theoretical principles with examples of major earthquakes.)

Kendrick, T. D. 1957. *The Lisbon earthquake.* Philadelphia: J. B. Lippincott. (A fascinating account of the 1755 earthquake, analyzing its effects on the religion, philosophy, and economy of eighteenth-century Europe.)

Oakeshott, G. B. 1976. *Volcanoes and earthquakes: Geological violence.* New York: McGraw-Hill Book Co. (A brief but very well-written explanation of the causes and effects of earthquakes.)

Press, F. 1975. Earthquake prediction. *Scientific American* 232(5):14–23. (A clear discussion for the layman of methods used to predict earthquakes.)

Thomas, G., and Witts, M. M. 1971. *The San Francisco earthquake.* Briarcliff Manor, NY: Stein & Day. (A nongeological account of what happened in 1906 when the "big one" struck.)

U.S. Geological Survey. 1966. *The Alaskan earthquake, March 27, 1964—field investigations and reconstruction effort.* Professional Paper 541. (See next reference.)

U.S. Geological Survey. 1970. *The Alaskan earthquake, March 27, 1964—lessons and conclusions.* Professional Paper 546. (This and the preceding reference form a two-part summary of research and analysis performed over several years on the effects of the "Good Friday" earthquake.)

20

Paleomagnetism and Plate Tectonics

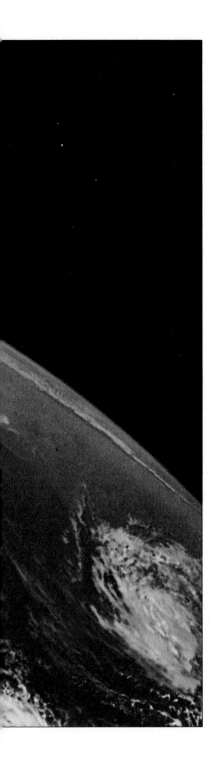

Landsat image of the Red Sea. Seafloor spreading beneath the Red Sea is splitting the Arabian plate from the rest of Africa. Photo courtesy of National Aeronautics and Space Administration.

In the first 19 chapters of this book we portrayed the Earth as a dynamic, ever-changing planet governed by plate tectonics processes. While you were reading these chapters, those processes made the Atlantic Ocean grow wider by about 2 cm, thrust part of the Pacific Ocean beneath South America, and moved Honolulu, Hawaii about 2.5 cm closer to Anchorage, Alaska. We have discussed several lines of evidence for the plate tectonics model, but have not yet presented the most convincing arguments—those based on study of the Earth's magnetic field.

In this chapter we will examine plate tectonics from several viewpoints. We will first show how magnetic evidence convinced geologists that plate tectonics is the best explanation of how our planet works. We will then review the basic concepts of the model, answering questions that we raised in earlier chapters. Then, proceeding as scientists should, we will use our newly acquired knowledge to raise new questions that lead us more deeply into the subject of plate tectonics and its causes.

EARTH MAGNETISM

To understand how magnetic evidence can be used to evaluate plate tectonics, we must first understand magnetism and the magnetic field that surrounds our planet. Magnetism is a force of attraction that affects only certain substances. Metals such as iron and nickel and minerals such as magnetite are affected strongly and may themselves act as magnets. However, other minerals, such as hematite, are attracted by a magnet but do not retain any magnetism once the magnet is removed (**paramagnetic**). Most substances are not affected at all by magnets.

The Earth behaves like a giant bar magnet, and is surrounded by an area called a **magnetic field** within which appropriate materials will be attracted, like a compass needle. The shape and extent of a magnetic field is defined by a series of lines of magnetic force. These lines of force can be revealed by shaking fine iron filings onto a sheet of paper placed over a bar magnet—Figure 20–1(a). The filings lie along the lines of force, demonstrating that the field emanates

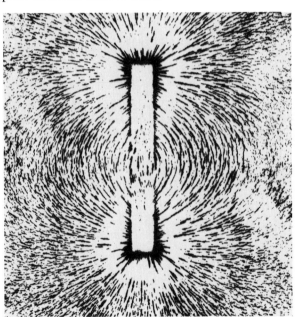

(a)

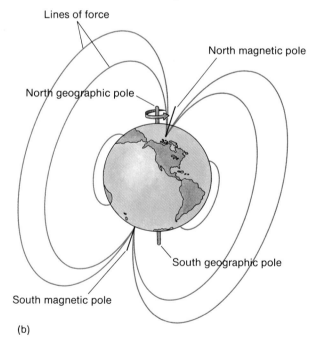

(b)

FIGURE 20–1

Magnetism and the Earth. (a) Lines of magnetic force around a bar magnet are shown by the orientation of iron filings sprinkled over the magnet. The lines of force emanate from the poles (ends) of the magnet, defining a dipolar (two-pole) magnetic field. (b) Earth has magnetic poles that are displaced slightly from its rotational poles. Photo courtesy of Education Development Center.

from opposite ends (poles) of the magnet. Bar magnets are therefore said to generate a **dipolar** (two-pole) magnetic field.

Earth also has two **magnetic poles** from which its lines of force emanate. These magnetic poles are located close to, but do not coincide with, the *geographic poles*—the ends of the axis about which the Earth rotates. The *magnetic* north pole is situated in northern Canada, approximately 11.5° from the *geographic* North Pole—Figure 20–1(b).

Measuring the Magnetic Field

When we study the Earth's magnetic field, we must measure both its strength and direction. The field has both vertical and horizontal components, called inclination and declination respectively.

The angle between the magnetic lines of force and the surface of the Earth is called **magnetic inclination.** We measure inclination with a magnetized needle that swings in a vertical plane (Figure 20–2). The needle swings until it comes to rest parallel to the local lines of force. Notice that the lines of force

in Figure 20–2 are perpendicular to the surface at the magnetic poles. Thus, the inclination at the magnetic poles is 90°. At the **magnetic equator,** lines of force are parallel to the surface and the inclination is 0°. Between the equator and the poles inclination ranges between 0° and 90°. We use inclination to tell us the local *magnetic latitude*—the location of a point between the magnetic pole and magnetic equator.

The compass needles with which most of us are familiar swing in a horizontal plane and point toward the north and south magnetic poles, thus indicating the local *horizontal* direction of the lines of force. In most places, this magnetic direction is either east or west of geographic north, or, as it is called, **true north**. The difference between true north and magnetic north is called **magnetic declination**. This difference is measured in degrees east or west of true north.

Declination varies depending on where it is measured (Figure 20–3). If you were navigating across the United States with a compass, you would have to adjust it constantly to account for declination, because declination changes dramatically: it is about 12° west of true north in New York City, 2° east in Detroit, and 16° east in Los Angeles.

Changes in the Magnetic Field

Earth's magnetic field is constantly changing, so its intensity, declination, and inclination vary over time. Intensity fluctuates on a daily basis, but these are very small, temporary changes. Intense solar activity often causes "magnetic storms" that bring about larger changes in the field, but these last only a few days. There are, however, long-term changes called **secular variations** produced by processes within the Earth. Since the strength of the magnetic field was first measured in 1830, it has decreased by about 6%. If this rate of decrease continues, the field will disappear in 2000 years. This rate of change in a global-scale phenomenon is remarkably rapid when compared with most geologic processes such as plate motions.

Origin of Earth's Magnetic Field

We do not completely understand why the Earth has a magnetic field. Our planet behaves like a dipolar magnet, but its field cannot have the same origin as

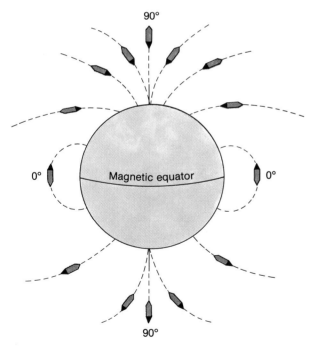

FIGURE 20–2
Magnetic inclination. Inclination is measured with a magnetic needle that is free to swing vertically. Inclination varies systematically from the magnetic equator to the magnetic poles.

FIGURE 20−3
Magnetic declination—the angular difference between magnetic north and true north—in North America.

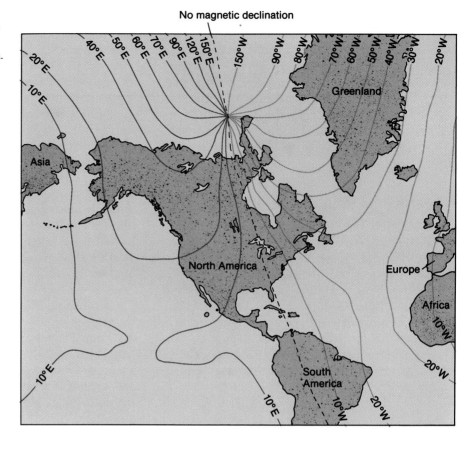

No magnetic declination

that of a bar magnet. A bar magnet is made by placing strongly magnetic material into a magnetic field. Magnetic materials contain atoms grouped into **magnetic domains** (Figure 20−4). Each atom within a domain exerts a force on adjacent atoms, compelling them to align magnetically. When such substances are placed in a magnetic field, the domains become aligned by the lines of force, and a strong magnet is created.

Earth's magnetic field cannot be caused by the alignment of domains in magnetic minerals because our planet's internal temperatures are too high. A magnet loses its magnetism when heated to a temperature called its **Curie point**. The exact temperature depends on what the magnet is made of and reflects the amount of heat energy needed to shift atoms in the magnetic domains to random positions. The Curie points for strongly magnetic earth materials are relatively low: 760°C for iron, 580°C for magnetite. At an average geothermal gradient of 30°C/km, these temperatures would be exceeded

within 25 km of the surface. Below 25 km it would be far too hot for these materials to be magnetic, and most of the iron in the Earth is in the core, thousands of kilometers below the surface. Magnetic minerals in the upper 25 km are not the source of the magnetic field because the crustal processes in which they take part are slow and do not match the rapid fluctuations that we observe in the field. There must be another explanation.

Most geologists believe that the Earth has a magnetic field because it behaves like a dynamo, a device used to generate electric current. A dynamo contains material that is an excellent conductor of electricity. This conductor is rotated through a magnetic field and, as it passes through or "cuts" the lines of force, an electric current is produced—Figure 20−5(a).

All the ingredients needed to make a dynamo are present in the Earth—Figure 20−5(b). The Sun generates a weak magnetic field and provides lines of force that surround the Earth. (We saw in Chapter 19 that the outer core is probably a molten iron-

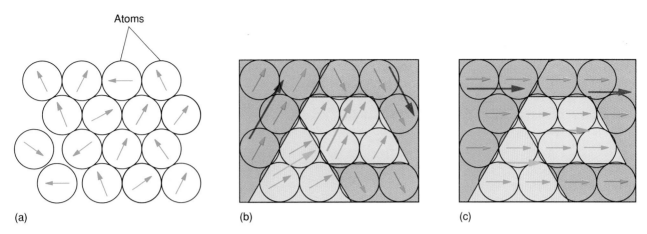

Atoms

FIGURE 20-4

Magnetic domains. (a) In nonmagnetic substances, the magnetism of each atom is randomly directed. (b) In substances susceptible to magnetism, atoms exert forces on their neighbors and small domains form in which the magnetism of several atoms is aligned. (c) A permanent magnet forms when the domains themselves are aligned by a magnetic field.

nickel alloy—an excellent electrical conductor.) Movement of the liquid iron-nickel alloy through the lines of force creates a weak electric current within the Earth. Every electric current flowing in a conductor generates a magnetic field around the conductor. (For example, a weak field surrounds each electric wire in your home, which is why you should keep magnetic computer diskettes away from electric wires and motors.) The Earth-dynamo's electric current is thus thought to be responsible for its magnetic field. This model appeals to geologists because the fluid nature of the outer core can explain the rapid field fluctuations that we observe.

FIGURE 20-5

The Earth as a dynamo. (a) A dynamo generates electricity when a rotating electrical conductor passes through the lines of force of a magnetic field, generating electric current in the conductor. (b) The Earth dynamo. Our planet's liquid outer core passes through lines of magnetic force from the Sun. This produces an electric current in the Earth, which in turn creates the magnetic field around the planet.

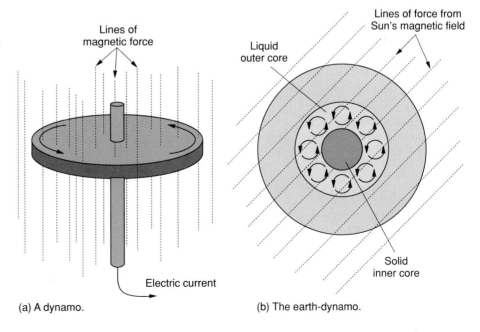

Lines of magnetic force

Electric current

(a) A dynamo.

Lines of force from Sun's magnetic field

Liquid outer core

Solid inner core

(b) The earth-dynamo.

PALEOMAGNETISM: KEY TO PLATE TECTONICS

Magnetic minerals record the Earth's ancient magnetic field, providing us with freezeframe pictures of the geography of the ancient Earth. The study of ancient magnetism is called **paleomagnetism** and is one of the newest types of evidence for the plate tectonics model.

Magnetic Memory

How do rocks record ancient magnetism? Imagine a magma chamber in which magnetite crystallizes as a minor mineral in an igneous rock (Figure 20–6). As the solidified rock cools below 580°C, the Curie point of magnetite, the domains in all the magnetite crystals are aligned by the Earth's magnetic lines of force. Each crystal becomes a permanent magnet that records the magnetic field direction that existed at the time that it cooled below its Curie point.

The record of the initial magnetism is called **remanent magnetism,** and the process just described is called ***thermo*remanent magnetization** because it is associated with heat. The initial direction of magnetic lines of force will remain in the mineral's "memory" unless the rock is reheated above the Curie point at some later time.

There are other ways in which remanent magnetism is imparted to a rock:

☐ **Chemical remanent magnetization** occurs when magnetic minerals form at temperatures below their Curie points, locking in the magnetic field at the instant of crystallization. This happens when hematite precipitates during sedimentation or magnetite forms during metamorphism.

☐ **Depositional remanent magnetization** occurs as clasts settle out during sedimentation. Magnetic minerals eroded from existing rocks tumble about during transport and lose their magnetic orientation. If the grains settle through quiet water, they can be aligned physically by the Earth's magnetic field, like tiny magnetic needles. They record the field at the time of deposition, not at the time that they first crystallized.

Magnetic Reversals

One of the first discoveries using paleomagnetism was so startling that it was not accepted for several years. Studies of thermoremanent magnetism in thick sequences of lava flows showed that the direction of the magnetic field in some flows was exactly opposite that in others, and opposite that of the magnetic field today. What is now the north magnetic pole appeared in those layers to once have been the south magnetic pole!

Geologists now accept the fact that the Earth's magnetic field is capable of reversing its polarity—that the north and south magnetic poles can, in effect, change places. Rocks with remanent magnetic polarity similar to that of the present field are said to have *normal* polarity; the others are said to be *reversed*. We do not yet know how or why such reversals happen, and they are one of the great mysteries in modern geology.

FIGURE 20–6
Thermoremanent magnetism. (a) Magnetite crystallizes in a magma, but the temperature is far above magnetite's Curie point. (b) The rock is solid, but still is hotter than magnetite's Curie point. (c) The rock has cooled below magnetite's Curie point. Magnetic domains in the magnetite are aligned in the magnetic field, recording its directional properties.

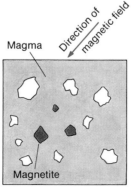

(a) 800° C

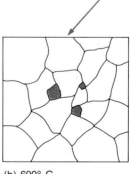

(b) 600° C

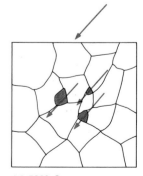

(c) 500° C

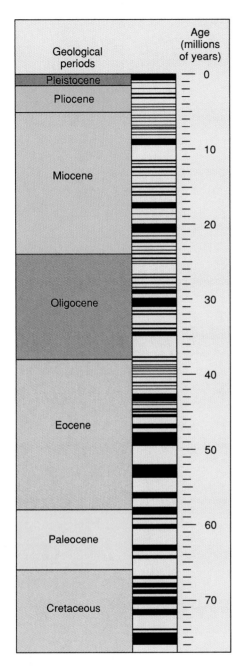

FIGURE 20–7
Magnetic time chart. Reversals of the magnetic field are shown
for the past 76 million years, based on data from the oceans.
Solid bars = times of normal polarity; white = times of
reversed polarity. After Heirtzler and others, 1968, "Marine
Magnetic Anomalies, Geomagnetic Field Reversals, and
Motions of the Ocean Floor and Continents," *Journal of
Geophysical Research* 73: 2119, redrawn with permission.

Radiometric dating of rocks using remanent magnetization shows us that reversals take place at irregular intervals ranging from 25,000 years to a few
million years (Figure 20–7). It may be that the intensity of the magnetic field decreases gradually until
it reaches zero at the moment of reversal. Perhaps the
gradually decreasing strength of today's magnetic
field will result in a polarity reversal in the future.

Magnetic Stripes and Seafloor Spreading

In the 1950s, ocean researchers discovered elongate
areas tens of kilometers wide and hundreds of kilometers long in which the magnetic field was either
stronger or weaker than the normal field (Figure
20–8). Such areas are called **magnetic anomalies;**
positive anomalies occur where the field is stronger
than usual and negative anomalies where it is weaker.
The elongate anomalies were quickly nicknamed
magnetic stripes. Stripes are of different widths, but
careful inspection shows that they are arranged sym-
metrically on opposite sides of ocean ridges.

It was not until the mid-1960s that the stripes
were explained by L. Morley, a Canadian geophysi-
cist, and the British team of F. Vine and D. Mat-
thews (Figure 20–9). They postulated that the measured magnetic intensity in a given area is the sum of
Earth's present field and the remanent magnetism
present in the rocks. If the rocks attained their remanent magnetism during a period of normal polarity
(such as today's field), the remanent magnetism reinforces the present field and produces a positive
anomaly—Figure 20–9(a) and left-hand arrows in
(b). If the remanent magnetism was acquired during
a period of reverse polarity, the ancient magnetism
would be the opposite of the present field, yielding a
lower-than-normal intensity, and thus a negative
anomaly—Figure 20–9(a) and right-hand arrows
in (b).

This not only explained the magnetic stripes but
also provided proof for the then-controversial hypothesis of seafloor spreading (Figure 20–10). As
ocean basins are rifted at an oceanic ridge, mid-ocean
ridge basalt (MORB) rises from the mantle and
erupts at the ridge crest. When the MORB crystallizes and cools below the Curie points of its magnetic
minerals, it is magnetized by the Earth's magnetic
field. As rifting continues, the cooled MORB is split
and the two halves are carried away from the ridge

FIGURE 20−8

Magnetic anomaly "stripes" in the
Pacific Ocean southwest of
Vancouver Island, British
Columbia. Colored stripes represent
positive anomalies; white stripes are
negative anomalies. Note
particularly the symmetrical
arrangement of anomalies about the
Juan de Fuca Ridge and Gorda
Ridge segments. After Figure 4
from Vine, "Magnetic Anomalies
Associated with Mid-Ocean
Ridges," in Phinney, R., ed., *The
History of the Earth's Crust,* © 1968,
with permission of Princeton
University Press.

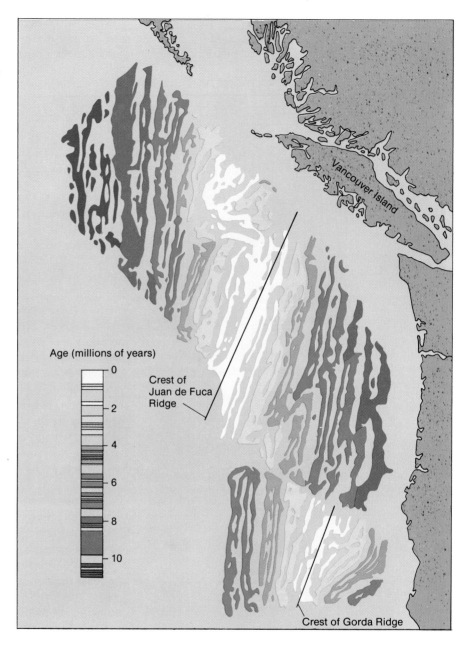

crest in opposite directions, to be replaced by new
upwellings of magma. If the magnetic field reverses
during this process, only the magma that is cooling
at the ridge crest will adopt the reversely magnetized
field direction; the earlier lavas will preserve their
normal remanent magnetism.

Magnetic anomaly stripes now have been found
at every ocean ridge, and the pattern of reversals is
the same at every ridge. The *width* of each anomaly is
different in every ocean, however, probably because
of different spreading rates: the faster the spreading,
the wider the anomaly. Thus, the Reykjanes Ridge
(the portion of the Mid-Atlantic Ridge that passes
through Iceland) seems to be spreading at a relatively
rapid rate (4.5 cm/year), whereas the East Pacific
Rise is spreading at less than half that rate.

FIGURE 20–9
The Vine-Matthews-Morley explanation for the magnetic stripes. (a) A simplified set of magnetic stripes. (b) The total measured field intensity is the sum of the present field intensity and the remanent magnetic intensity. This is illustrated for two conditions: the set of arrows on the left shows present and remanent fields of the *same* polarity, which makes the total measured intensity *greater* than that of the present field. The set of arrows on the right shows present and remanent fields of the *opposite* polarity, which makes the total intensity *less* than that of the present field.

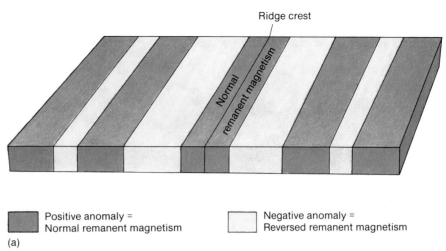

Ridge crest

Normal remanent magnetism

Positive anomaly = Normal remanent magnetism

Negative anomaly = Reversed remanent magnetism

(a)

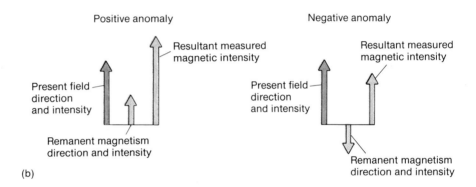

Positive anomaly

Resultant measured magnetic intensity

Present field direction and intensity

Remanent magnetism direction and intensity

Negative anomaly

Resultant measured magnetic intensity

Present field direction and intensity

Remanent magnetism direction and intensity

(b)

Polar Wandering and Continental Drift

Paleomagnetic data also show that continents have changed their positions on the surface of the Earth. The paleoinclination of a rock tells us how close it was to the magnetic poles at the time it acquired its remanent magnetization. The lower the inclination, the closer the rock was to the magnetic equator, and the steeper the inclination, the closer it was to one of the magnetic poles. Every continent shows some amount of shift over time. India shows one of the greatest changes, apparently moving from a south magnetic latitude to the magnetic equator, and then to north magnetic latitudes.

We usually show the sense of motion by plotting the ancient magnetic pole positions relative to the current positions of the continents (Figure 20–11). The change in pole position with time then defines a **polar wandering curve;** remember, however, that *it*

is not the pole that has moved, but rather the continents. When polar wandering curves from two continents are compared, as in Figure 20–11, they show that the two landmasses did not always follow the same paths. This indicates that the present relationship between the continents was not always the same—the plates carrying the continents may have diverged, converged, or rotated relative to one another.

THE PLATE TECTONICS MODEL: A REVIEW

This paleomagnetic evidence is what finally convinced most geologists that the plate tectonics model is valid. Before looking at possible causes of plate tectonics, we will briefly review the basic principles of the model described in the previous chapters:

FIGURE 20–10

Seafloor spreading and the origin of magnetic stripes. (a) Remanent magnetization is recorded. (b) Spreading continues and polarity reverses. Original basalt is split in two by spreading; new basalt is magnetized in the reverse direction at the ridge crest. (c) Continued spreading leads to magnetic-anomaly stripes.

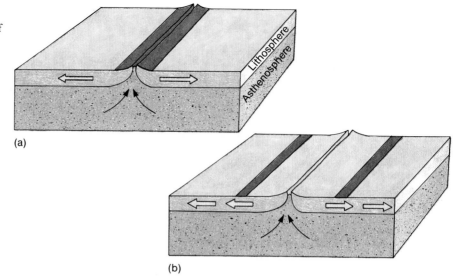

(a)

(b)

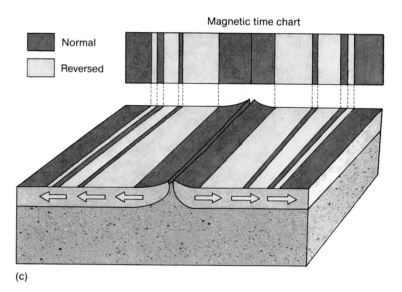

Magnetic time chart

■ Normal

□ Reversed

(c)

1. The Earth's surface is underlain by a small number of rigid rock masses called lithospheric plates (Figure 20–12) which are about 150 km thick.

2. Continents and oceans are merely passengers on these plates. An oceanic plate, such as the Pacific plate, has a one-layer simatic crust, whereas a continental plate, such as the Arabian plate, has a two-layered crust (sialic plus simatic). Some plates, such as the North American and African, contain both an ocean segment and a continent.

3. The lithospheric plates slide on the asthenosphere, a region of extremely low rigidity in the upper mantle.

4. Former positions of continents and oceans can be learned from paleomagnetic data and from fossils and rocks that record ancient climates. For example, fossils of tropical plants and animals have been found in frigid Antarctica, and tillites produced by widespread glaciation are preserved in tropical regions of South America, Africa, and India. The ancient climates are so drastically different from those of the cli-

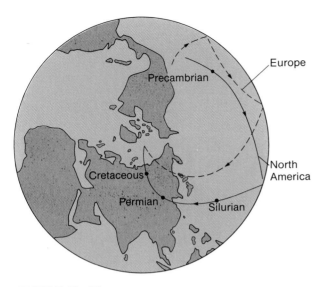

FIGURE 20–11
Polar wandering curves for North America and Europe. The solid curve shows the locus of pole positions based on North American data; the dashed curve is for Europe. The differences in the curves show that the two continents have changed position relative to one another through time.

matic zones in which the rocks are now found that they require major changes in the locations of these continents through geologic time.

5. There are three kinds of plate boundaries (Figure 20–13). *Divergent* boundaries are sites of tension where plates separate and move apart from one another. These are the oceanic ridges, places where new simatic crust is generated by the eruption of MORB. *Convergent* boundaries are sites of compression where plates collide and are subducted. These are the island arc–trench systems of the oceans, and Andean-type continental margins. Simatic crust is subducted back into the mantle at convergent boundaries, completing the tectonic cycle that began at the ocean ridges. *Neutral* boundaries involve neither generation of new simatic crust nor subduction of the old. These are transform faults, produced by shear stress as two plates grind past one another. Transform faults include fracture zones that cut the ocean ridges and major strike-slip faults on the continents such as the San Andreas fault.

6. Most mountains form at divergent boundaries (ocean ridges) and convergent boundaries (island arcs and fold-thrust mountains). Some can form far from plate margins, due to mantle hotspot activity (Hawaiian-type islands and seamounts). Old fold mountains now located at plate interiors form by collision of two continents formerly on opposite sides of an ocean basin when that ocean closed by subduction of its simatic crust.

7. New subduction zones or ocean ridges may form at any time. Continents may grow by plate collisions or be broken into smaller plates by rifting. As a result, no geographic features are permanent. The sizes, shapes, and positions of the oceans and continents today will not be duplicated in the future, and were not the same in the past.

TESTING THE HYPOTHESIS

The plate tectonics hypothesis is a unifying concept in geology that explains many things about the Earth. We have discussed several of these aspects in previous chapters: the worldwide distribution of volcanoes, mountains, and earthquakes; the causes of ice ages; the origin of the major topographic features of the oceans; and the apparent jigsaw-puzzle fit of the coastlines of Africa and South America.

There are many other questions about the Earth that we could not answer earlier. We will do so now, using the plate tectonics model and the information you have gained from the preceding chapters.

Continents Versus Oceans: Chemical Evolution of the Earth

Why Are the Ocean Floors Topographically Lower than the Continents? Continents are underlain by a two-layer crust composed of sialic and simatic rocks, whereas the crust beneath oceans contains a simatic layer only. The average density of sial plus sima is lower than that of sima alone, so that the density of continental crust is less than that of oceanic crust. Rocks of the continents therefore are buoyed isostatically above the denser rocks of the ocean basins.

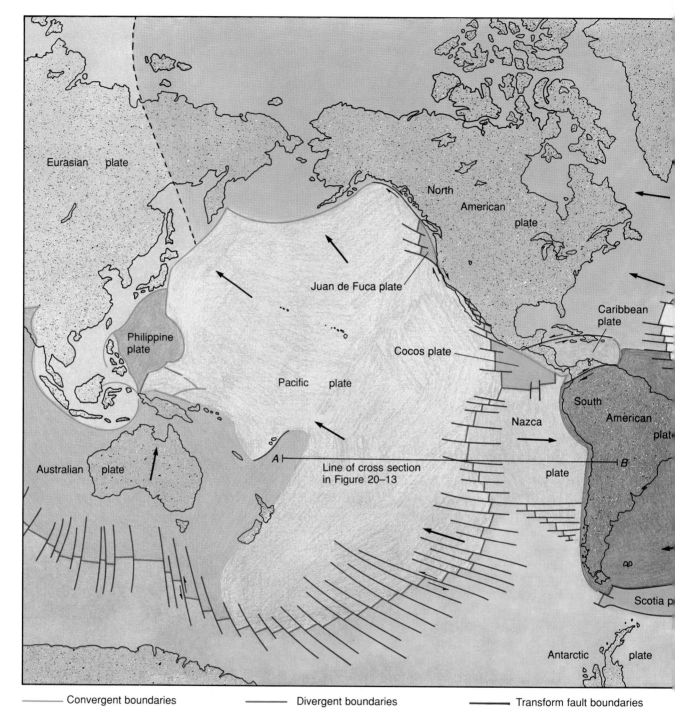

Convergent boundaries   Divergent boundaries   Transform fault boundaries

FIGURE 20–12
The Earth's major lithospheric plates. Arrows show approximate plate motions.

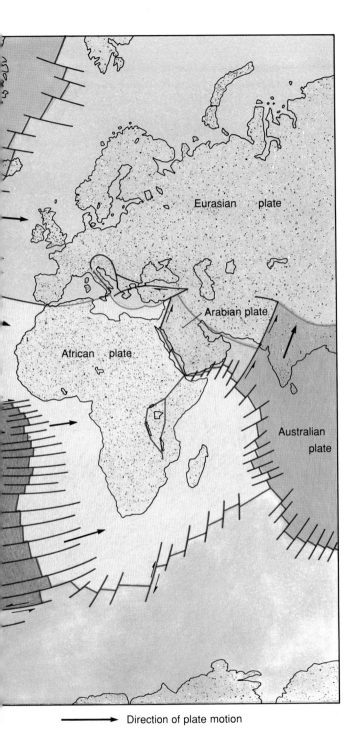

Direction of plate motion

Why Are the Ocean Floors So Young? Although the Earth is about 4.5 billion years old, the oldest rocks in the oceans are of Jurassic age, only 180 million years old. Is it possible that oceans are relatively new features? Were there were no oceans prior to the Jurassic?

The existence of Paleozoic and Precambrian rocks on the continents that are similar to sediments and lavas found today in the deep oceans argues strongly against this possibility. Fossils of marine organisms and the presence of ophiolites in several mountain ranges indicate that there were indeed ancient oceans. The reason all ocean rocks are so young is that any ocean floor existing before the Jurassic Period has been returned to the mantle by subduction. Some of this occurred at subduction zones that are still active today, but most took place in trenches that have long since ceased to exist. For example, an ocean once separated Asia from Europe, but a few obducted ophiolite fragments in the Ural Mountains are all that remain of that ocean and the subduction zone that destroyed it.

If the Oceans Are Young, Why Are So Many Continental Rocks So Much Older? Why have the continents not been recycled the way the oceans have? We saw in Chapter 19 that continental crust cannot be subducted into the mantle because the forces that drive one plate beneath another are unable to overcome the isostatic buoyancy of low-density sial. Once sialic rock has formed on a continent, it is too buoyant to return to the mantle; it may be buried beneath younger strata and hidden from view, but it will always be part of continental crust. The continents are the repository of all sialic material, and are thus the sites of the oldest rocks on the Earth.

Why Is Sial Found Only on the Continents? This question takes us back to the early days of Earth history. Uniformitarianism cannot help us here, because things were happening that have not happened since. We must consider the origin and evolution of the Earth, and how the first continents might have formed.

We learned in Chapter 19 that the Earth has evolved over 4.5 billion years into a planet that is layered according to density. In Chapter 1 we men-

FIGURE 20–13
Types of plate boundaries.

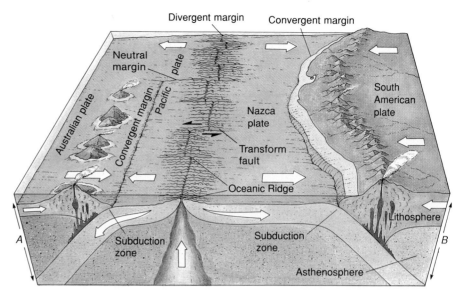

tioned that most geologists believe that the Earth was entirely molten early in its history, and in Chapter 4 showed that this molten Earth may have behaved like a giant magma chamber. Differentiation, primarily by gravity separation, caused elements such as nickel and iron to settle inward to form the core, whereas the low-density elements floated upward to form the atmosphere, hydrosphere, and a sialic crust (Figure 20–14). At a very early stage in this process, there probably were no oceans; the entire Earth was covered by a thin sialic shell.

Migration of the elements was more rapid at that time than it has been since because liquids are much more mobile than solid rocks. Once the Earth became mostly solid, major chemical differentiation became far more difficult. Most of the sial now found on the continents thus was probably part of continental crust by the early Precambrian. It has indeed been recycled, but this has been accomplished entirely within the crust by the rock cycle of erosion, sedimentation, metamorphism, and remelting (Chapter 3).

FIGURE 20–14
Chemical differentiation of the Earth. (a) Early in Earth history, concentration of dense iron and nickel formed the core, and segregation of light elements formed sialic crust. (b) Today, differentiation creates new simatic crust, facilitated by melting of the upper mantle.

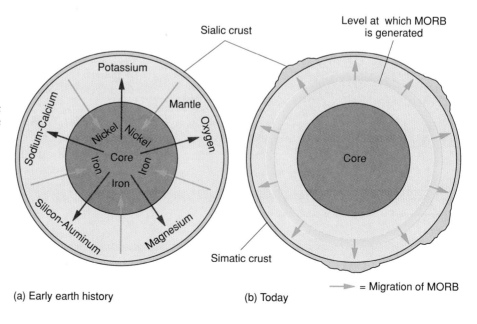

(a) Early earth history

(b) Today

Is New Sial Forming Today? Although Earth's chemical differentiation has slowed drastically since the early Precambrian, it has not stopped. Some new sial is being produced today, but only in minute amounts. Almost all of the sial that could be extracted from the deep Earth had become part of the crust during the Precambrian. Today, the lowest-density material that can be separated from the mantle and moved to the crust is MORB, not granite—Figure 20–14(b). The generation of MORB from mantle material thus is continuing the differentiation process. The change from sialic to simatic differentiation shows how much the chemical zonation of our planet has changed. With time, differentiation will proceed deeper and deeper into the mantle, progressively refining the separation of elements.

How Do Continents Grow? Figure 20–15 shows that North America appears to have a "nucleus" of old rocks surrounded by successively younger belts of strata. At one time it was thought that continents grew outward from old nuclei like the Canadian Shield by addition of mountain systems. We now know that neither North America nor any other continent is structured so simply or concentrically.

The discussion of fold mountains in Chapter 18 shows one way in which a continent may grow. Sub-

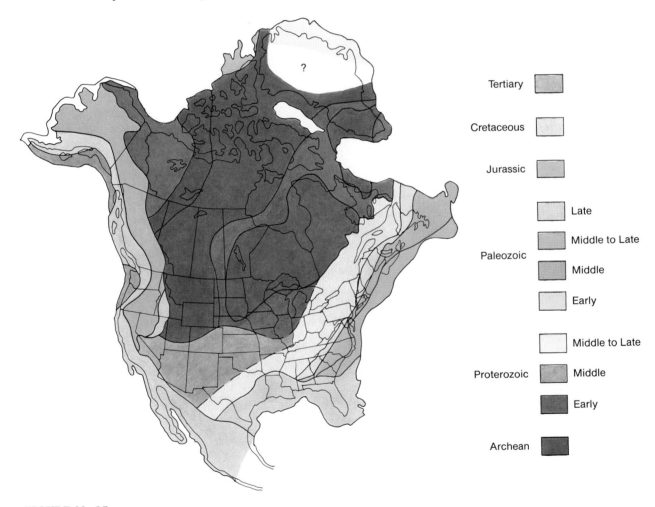

FIGURE 20–15
Ages of accretion of material to North America. In general, old rocks at the center of the continent are surrounded by progressively younger materials. After slide by Department of Earth Science, Memorial University of Newfoundland.

duction of an ocean that separates two continental plates eventually leads to collision between the continents, which produces a fold mountain system at their suture, and results in formation of a supercontinent. This process need not be concentric about a Precambrian shield; consider Focus 18–1 where all of the accretion of material to the South American continent occurred on its west coast.

Geologic studies of the northwestern United States, western Canada, and Alaska reveal another way in which continents grow. Western North America consists of many blocks of rock separated by large strike-slip faults (Figure 20–16). This region was problematic for many years because adjacent blocks contain rocks of such different type, age, and deformation history that they could not have evolved side-by-side as they are now. Paleomagnetic studies resolved this problem by showing that these blocks only recently have been sutured together. They evolved thousands of kilometers apart but, with time, some moved northward and others southward until they finally collided. Such blocks are called **exotic terranes** because they originally formed in totally different regions. Western North America is now viewed by most geologists as a collage composed of many exotic terranes.

Note that subduction is not required in this scheme. Two continents once separated by an ocean basin may be sutured by transform fault motion without trenches, Andean-type volcanoes, or fold mountains. Figure 20–17 shows how this process might work in the future to join Madagascar to the continent of Africa.

How Can Continents Be Broken Apart? The nearly perfect fit of Africa with South America and the continuation of Appalachian Mountain structures into Ireland, Scotland, and Scandinavia show that it is possible for continents to break apart as well as to get larger. We think that this process is happening now in Africa where the Dead Sea, Red Sea, and Gulf of Aden contain incipient ocean ridges whose spreading is beginning to separate Africa from the Arabian plate (see photograph that opens this chapter). How does a spreading center form beneath a continent?

Consider a continent flanked by oceans, as shown in Figure 20–18. Heat conducted from the interior of the Earth eventually reaches the surface and es-

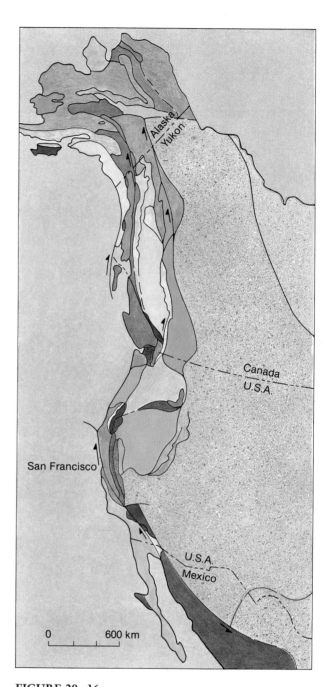

FIGURE 20–16
Exotic terranes of northwestern North America. Each of these terranes is thought to have been accreted by strike-slip motion along boundary faults. Redrawn by permission from *Nature*, Vol. 288, pp. 329–33. Copyright © 1980 Macmillan Magazines Limited.

FIGURE 20–17
A possible exotic terrane future for Madagascar. (a) Present geography, with line showing location of future transform fault. (b) Madagascar, transported along this fault for millions of years, is sutured to Somalia.

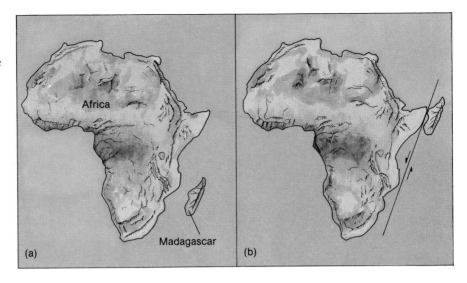

capes to the atmosphere. We know that the heat loss is slow because rocks are poor conductors of heat. The more rock that the heat must pass through, the slower the process will be. Heat rising beneath the continent is trapped beneath a thicker crust than that emerging beneath an ocean, and cannot reach the surface as quickly. Thus, heat builds beneath the continent. Heat energy may at first move by conduction,

but eventually convection also may play a role. A convection cell may form by the lateral transfer of heat between the hot basement beneath a continent and the cooler one below an ocean. If two convection cells have diverging motion beneath the continent, as shown in Figure 20–18, enough force may build to split the continent and begin the process of seafloor spreading.

FIGURE 20–18
Possible explanation for origin of a spreading center. (a) Heat rising to the surface is trapped beneath a continent, developing unequal heat distribution. (b) Lateral heat flow below the continent begins development of a convection cell.

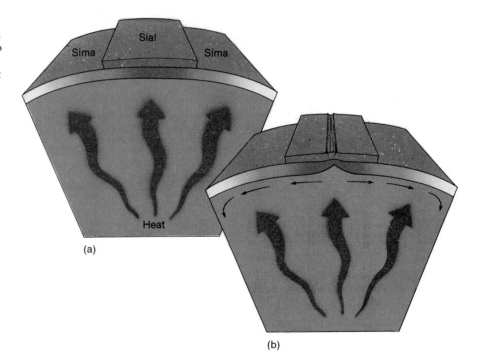

The thin, world-encircling sialic shell that may have existed early in the history of the Earth was probably broken into separate fragments—the first continents—by local heat concentrations such as these. As the shell broke, magma from below filled the cracks, producing the first ocean basins. Plate tectonics may well date back to the very first stages of the planet's development.

Tectonic Activity

Why Is North America's West Coast More Active Tectonically than the East Coast? Plate boundaries are the sites of Earth's most active tectonism, but coastlines are boundaries between ocean water and land, not necessarily between plates. Where plate boundaries and coastlines coincide, the land adjacent to the ocean will be tectonically active and experience seismicity and/or volcanicity, depending on the type of boundary. The west coast of North America is tectonically active because it lies close to several plate boundaries. The San Andreas transform fault causes earthquakes in western California. The East Pacific Rise causes earthquakes in the Gulf of Southern California, and subduction of the northern Pacific plate beneath the Aleutian arc-trench system causes volcanic activity and earthquakes in Alaska. Such coasts are called **active continental margins**.

When coastlines are situated far from plate boundaries, there is little cause for tectonism along the shore, and the result is a **passive continental margin**. The east coast of North America is a passive margin located well within the North American plate, thousands of kilometers from the plate boundary—the Mid-Atlantic Ridge.

If Tectonism Occurs at Plate Boundaries, Why Do Mid-Plate Earthquakes Occur, Such as New Madrid? A large plate like North America is not a simple, homogeneous mass, but is made of many fragments sutured together over hundreds of millions of years. Some of these sutures are faults, and just as a newly healed broken bone may break again at the same place, these suture/faults are zones of potential weakness. As the plate moves, stresses in the rigid lithosphere may reactivate some of these faults, causing earthquakes.

How Do Dome Mountains Form? As mentioned in Chapter 18, the plate tectonics model does not satisfactorily explain the origin of dome mountains. The vertical forces responsible for such mountains are not readily reconciled with the large-scale lateral motions typical of moving plates. Plate tectonics does not give an answer to this question, and we shall have to deal with it later.

Petrologic Problems

Why Are Andesites Found Only in Subduction Zones? Of the three major igneous rock types discussed in Chapter 4, andesite has the most restricted geographic distribution, occurring only in island arcs and Andean-type continental margins. Something about subduction must be responsible for making it. Remember that andesite is an intermediate volcanic rock composed largely of Ca-Na plagioclase feldspar and amphibole, with some pyroxene. The fact that andesites contain amphibole as their dominant ferromagnesian mineral instead of the pyroxene found in basalts is very important. Amphiboles contain the hydroxyl ion complex (OH^-) in their structure, whereas pyroxenes contain none. Hydroxyl ions are derived from water, indicating a more hydrous source rock for andesite than for basalt.

Andesitic magma probably forms by partial melting of subducted oceanic lithosphere composed of basalt and water-rich sediments and sedimentary rocks. The water promotes melting at relatively low temperatures (see Chapter 4) as the plate is subducted, and some of the water stays in the rock as hydroxyl ions in the amphiboles. The unique combination of melted sediment and partially melted basalt produces the composition of andesite.

Why Do Island Arcs Have Different Types of Basalt than Ocean Ridges? The source of MORB is at a single depth in the asthenosphere below ocean ridges. Relatively homogeneous rock melts or partially melts within a narrow range of temperature and pressure to produce the characteristic MORB magma (Figure 20–19).

Island arc and Andean-type basalts are much more varied than MORB because they form at several depths from a more varied source. Figure 20–19

FIGURE 20–19

Depths at which basalt magma forms beneath ocean ridges and island arcs. MORB forms at a single, shallow depth beneath ocean ridges, but island arc basalts melt at several depths.

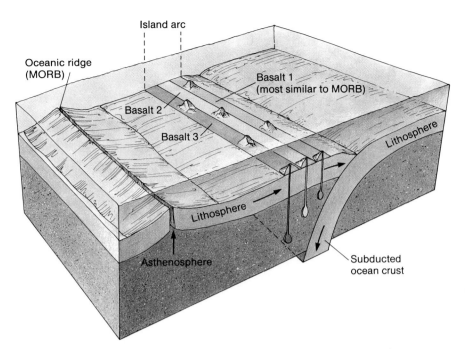

shows how the depth of melting along the subducted plate controls the type of basalt that forms. Rocks in the subducted plate change their mineralogy as they adjust by metamorphism to conditions deeper in the mantle. What melts at one depth thus is not the same as what melts at another, and different types of basaltic magma will melt at different depths. Melting at levels comparable to those at which magma forms beneath an ocean ridge produces rocks similar to MORB. Subduction basalts found farthest from the trench—those whose magma formed at the greatest depths—are very different from those of the ocean ridges.

Why Are There Different Regional Metamorphic Facies Series? Facies series that form during regional metamorphism were attributed to different geothermal gradients in Chapter 7. Figure 7–20 showed that low-temperature/high-pressure metamorphism occurs close to the trench during subduction; high-temperature/low-pressure metamorphism is found far from the trench. Both upper and lower plates in this subduction zone are made of oceanic crust. They contain the same amounts of radioactive elements and generate the same amount of heat. How does subduction cause the different facies series?

The answer lies in the way in which the subducted plate affects the distribution of heat in the mantle (Figure 20–20). A normal geothermal gradient is found at both sides of the diagram, where subduction has had no effect. Close to the trench, however, relatively "cold" crust has been subducted, lowering mantle temperatures. At point 1, for example, a temperature of 1600°C is reached at a depth of 200 km; close to the trench at point 2, the subducted slab has cooled the mantle so much that the same temperature is not reached until depths greater than 400 km. Directed pressures, however, are higher along line 2 than line 1 because of the plate collision. Low-temperature/high-pressure metamorphism therefore occurs near the trench along line 2. Along line 1, the geothermal gradient is higher and the directed pressure is lower than along line 2, so rocks along line 1 experience high-temperature/low-pressure metamorphism.

Continental Drift

How Much Has Geography Been Changed by Plate Motion? Enormously! We cannot trace the positions of plates throughout all of geologic time, but we can draw reliable paleogeographic maps from the begin-

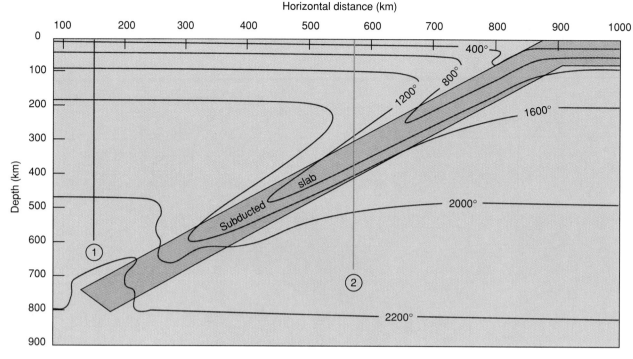

Horizontal distance (km)

FIGURE 20–20
Distribution of heat (in °C) around a subducted oceanic slab. The lines show how geothermal gradients are changed when "cold" ocean crust is subducted into a hotter mantle.

ning of the Mesozoic Era to the present (Figure 20–21). Rocks on every continent record major orogenies that occurred toward the end of the Paleozoic Era. These marked the closing of ancient oceans and the assembly of all continents into a single supercontinental plate called **Pangaea**. This continent was surrounded by a single ocean, known as Panthalassa.

This simple geography did not last long. Pangaea began to break up during the early Mesozoic, first splitting into two major plates, **Gondwanaland** and **Laurasia.** Gondwanaland contained what are now the southern continents of Africa, South America, Antarctica, and Australia, and what is now India; Laurasia encompassed Europe, Asia, North America, and Greenland. A new ocean, called the Tethys Sea, opened between Laurasia and Gondwanaland.

Further rifting and spreading split North America from Laurasia and South America from Gondwanaland, creating the Atlantic Ocean. The Tethys Sea eventually closed as Africa moved northward and collided with Eurasia, and as India completed its journey to be sutured to Asia. Changes continue to-

day, as rifting produces the Gulf of Aden and the Dead Sea—perhaps we are seeing the birth of a new, great ocean—and as ocean crust disappears beneath the world's island arcs.

CAUSES OF PLATE TECTONICS

Why Do Plates Move? Now that it seems proven beyond reasonable doubt that plates actually move, we must turn to the next questions—how they move and why. Discovery of the asthenosphere gave us part of the answer to "how": the asthenosphere is a zone of "lubrication" for plates to slide on. But what is the source of the incredible amount of kinetic energy needed to move a plate thousands of kilometers wide and over a hundred kilometers thick?

Two hypotheses have been proposed to explain the causes of plate movement. Both agree that the Earth's internal heat is a major factor in causing movement, but one holds that the plates are *pushed* by convection and the other that they are *pulled* by gravity:

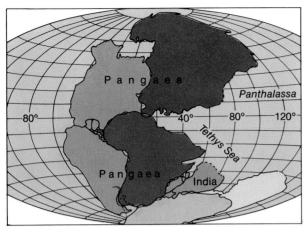

(a) 200 million years ago

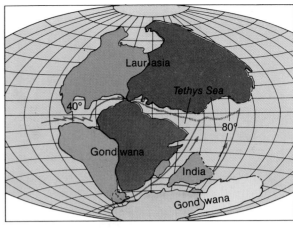

(b) 180 million years ago

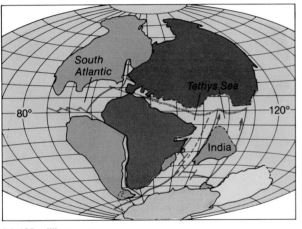

(c) 135 million years ago

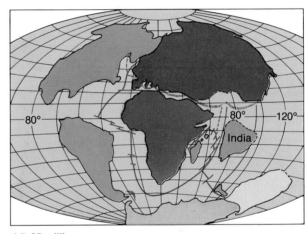

(d) 65 million years ago

FIGURE 20–21
The breakup of Pangaea. After "The Breakup of Pangaea," by R.S. Dietz and J. C. Holden.
Copyright © 1970 by *Scientific American, Inc.* All rights reserved.

Push: Convection (see Chapter 2 for a review) has long been regarded as the most practical mechanism for transferring heat energy from the deep Earth to the surface. We suggested earlier that mantle convection might be the cause of seafloor spreading. Could it be the cause of all plate motion? Convection in the upper mantle could cause material to flow fairly easily in the ductile asthenosphere, but friction would create strong stresses at the base of the rigid lithosphere (Figure 20–22). Seafloor spreading would occur where the tops of adjacent convection cells *diverged* (left side of Figure 20–22), and subduction where

two adjacent convection cells *converged* (right side of Figure 20–22).

Pull: Some geologists think that convective forces might be strong enough to cause the initial rifting at a ridge, but not to cause seafloor spreading or subduction. In this model, basalts formed at ocean ridges are later wedged apart by the eruption of newer lavas. The wedging creates a *weak* lateral force that could not be strong enough to drive one plate into the mantle beneath another thousands of kilometers away. Instead, these scientists see gravity as the principal cause of subduction (Figure 20–23).

CAUSES OF PLATE TECTONICS | 571

In 1756, a German theologian named Theodor Liliethal suggested that the apparent jigsaw-puzzle fit of the world's continents was not a coincidence, and that all had at one time been attached. Over the next 200 years, geologists laboriously pieced together the many fragments of information that led to the plate tectonics model, and finally concluded that he was right. Today, it is possible to measure plate motions more accurately than Liliethal could have dreamed.

Case I—The Himalayas

In 1913, British and Russian surveying teams completed detailed studies of the "Roof of the World," the Himalaya–Karakoram–Pamir–Hindu Kush mountain systems. Theirs was the first such expedition in history. Surveyors carried their heavy instruments to peaks never before seen, much less climbed by westerners. They produced the most accurate locations of the world's highest mountains, as well as the most precise elevation information ever collected in the region.

In 1980, a party equipped with modern instruments resurveyed a part of the region covered in 1913. Their goal: to determine whether the Indian plate is still moving northward into Asia by carefully measuring changes in the positions of some of the mountains. They occupied "only" 16 stations, but each was on a major peak, the lowest at nearly 3660 m (12,000 feet). The result: India is still on the move, grinding its way into the Asian continent at a little less than 1 cm per year.

Case II—The Great Alaska-Pacific Experiment (GAPE)

In 1987, preliminary reports were made of a different kind of measurement which also demonstrates modern plate motion. In 1984, radiotelescopes in western North America, the central Pacific, and Japan began to focus on a set of quasars so distant from the Earth as to be essentially stationary. After three years of measurement, it appeared that the distance between Hawaii and Fairbanks, Alaska had shortened at a rate of 52.3 ± 5.5 mm per year. Hawaii is apparently moving toward Japan at a much faster rate: 83 ± 8 mm per year.

As basalt cools, it contracts and thereby becomes denser; the denser rock must sink to achieve isostatic equilibrium. Movement of the basalts is due to two forces: wedging caused by convection at the ridge crest and sinking caused by gravity. The ridges are by far the highest features of the oceans, so the plate would be pulled "downhill" by gravity toward the trenches. This mechanism is similar to the process by which gravity causes soil creep on a hillside, except that it is on a much, much grander scale.

FIGURE 20–22
Convection cells push the lithospheric plates (vertical scale greatly exaggerated).

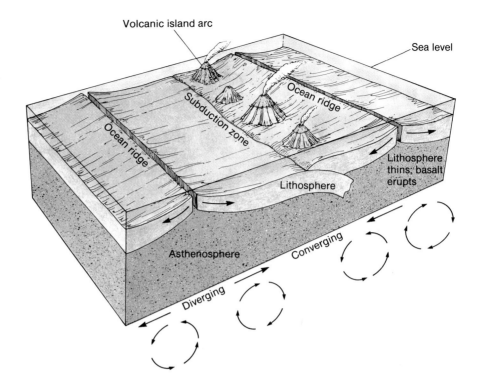

FIGURE 20–23
Gravity pulls plates. Movement of plates toward subduction zones may be caused by a weak push from the spreading center and a strong pull by gravity (vertical scale exaggerated).

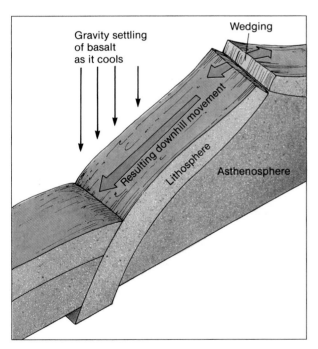

Which hypothesis is right? We do not know yet. Some computer models suggest that gravity pulling on a slab of ocean crust is enough to cause subduction, but others argue against a simple "slab-pull" mechanism. It is up to you—your generation of geologists-in-training—to resolve this problem.

QUESTIONING THE PLATE TECTONICS MODEL

It is easy to forget that plate tectonics is just a hypothesis and not a "fact." It explains so much about the modern Earth and about the way in which the Earth has evolved that it is accepted today by most geologists. Nevertheless, it is just another hypothesis that must be constantly checked, reevaluated, and revised to fit new observations.

After all, popularity among scientists, simplicity, and comprehensiveness are not valid reasons for accepting any hypothesis. At various times in our history, people believed that the Earth was the center of the universe; that it was created in 4004 B.C.; that glacial deposits were the results of Noah's flood; and that basalts and granites precipitated like halite from ocean water. These simple, once popular, ideas have long since been disproved.

Will plate tectonics suffer the same fate and be discarded one day? Most geologists would say no, that the model may not be perfect but that it answers too many questions to be seriously wrong. However, let us look at some of the imperfections, again phrased as a series of questions, and keep an open mind.

Convection in the Mantle

Some of the questions that have arisen about plate tectonics concern convection. Some doubts exist as to whether convection is possible; others deal with the manner in which deep mantle convection can affect rocks of the lithosphere.

Can Convection Cells Exist in the Mantle? Convection cells form in beakers of boiling water, and there is strong evidence that they develop in some magma chambers, but these media are both liquid. Some scientists do not think that convection can operate in the solid mantle, regardless of how ductile the rocks are. The asthenosphere is composed of extremely ductile materials and can move rapidly, but the postulated convection cells must operate *below* the asthenosphere in rocks whose seismic-wave velocities indicate significant rigidity. Furthermore, even if convection is possible in the mantle, no one has yet demonstrated that the cells can maintain the flow rates needed to drive plate motion at the velocities indicated by paleomagnetic studies.

Until the rate of convection—and even the very existence of convection—in the mantle can be proved conclusively, the causal mechanism of plate tectonics remains a serious problem.

How Large Are Convection Cells? Even if we assume that mantle convection is possible, there are still questions about its application to the lithosphere. Some geologists believe that convection cells are as large as the plates above them. If that is correct, the diameters of the cells beneath the South American, Nazca, and Pacific plates would have to be very different from each other (Figure 20–24). The three cells for these plates would then have to originate at very different depths and would indicate a far more complex temperature distribution within the mantle than physicists studying planetary objects can accept.

Perhaps there are many small convection cells instead of a few large ones—Figure 20–25(a). Unfortunately, this does not work either. It is difficult to see how a plate as large as the North American plate could be moved as a single mass without breaking up, or how cells driven by the smaller temperature differences implied by their smaller size could do the job required. Could there be just a few small cells adjacent to the subduction zone—Figure 20–25(b)? Again, it is difficult to see how these could do the work.

How Does Convection Cause Lithospheric Plates to Move? If we grant for the moment that mantle convection exists, there is still a critical problem for plate

FIGURE 20–24
Convection cells and plate size. If cells are as large as the plates above them, each must form at a different depth, leading to unlikely temperature distribution in the mantle.

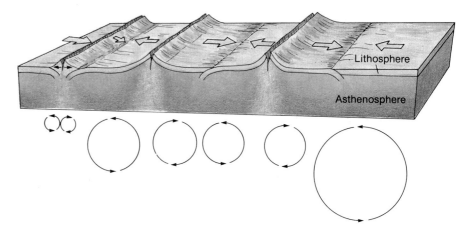

FIGURE 20–25
Other ways for convection cells to move plates. (a) Perhaps many small cells lie beneath large plates, and act like ball bearings; (b) the cells occur only at convergent or divergent plate margins. No cells exist at midplate locations.

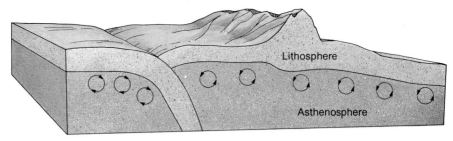

(a) Many small convection cells beneath plates

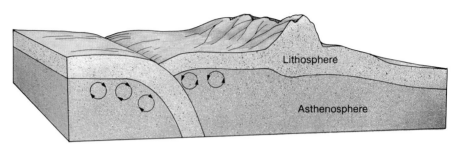

(b) A few small convection cells near the subduction zone

tectonicists. We have viewed the asthenosphere as a ductile zone separating the rigid lithospheric plates from the mantle below. How then do the convection cells transfer their divergent (or convergent) motion *through the asthenosphere* to affect the lithosphere? Again, we do not fully understand the answer to this question. This is another issue for the next generation of geologists to work on.

Seafloor Spreading

How Does Seafloor Spreading Start? Why Does It Stop? Earlier in this chapter, we showed how heat trapped beneath a continental plate may initiate the rifting process. Does that mean that a new ridge can never form in an already existing ocean basin? We do not know. The manner in which spreading can be stopped is perhaps easier to understand. If plates driven apart by a spreading center meet plates driven in the opposite direction by more powerful spreading centers, compression might bring about the extinction of the weaker ridge (Figure 20–26). This seems to be the best explanation because, if convection is indeed the principal cause of spreading, it is hard to understand how a convection cell can be turned on and off to match the stop-and-go

spreading-center histories indicated in the evolution of our older mountain systems.

Why Are Some Continents Completely Surrounded by Spreading Centers? This is one of the most puzzling problems facing plate tectonicists. Africa and Antarctica are completely or almost completely surrounded by spreading centers (Figure 20–12), and therefore should be under intense compression. Why has not subduction of the Atlantic or Indian oceans begun beneath Africa? Why is it that, instead of compression, *rifting* seems to be the dominant tectonic activity affecting Africa—in the East African Rift Zone and in the incipient ocean basin mentioned earlier that includes the Dead Sea, Red Sea, and Gulf of Aden? So far there is no satisfactory explanation.

EVALUATING THE PLATE TECTONICS MODEL

There are three ways to interpret the unanswered questions. One is to say that because they are not answered, we must reject the entire plate tectonics model. A second interpretation is that these questions address small details in the model which even-

FIGURE 20–26
One way to "turn off" a spreading center. The thickness of the arrows indicates rates of motion. Compression between the two converging plates eventually prevents further spreading at the center ridge (vertical scale exaggerated).

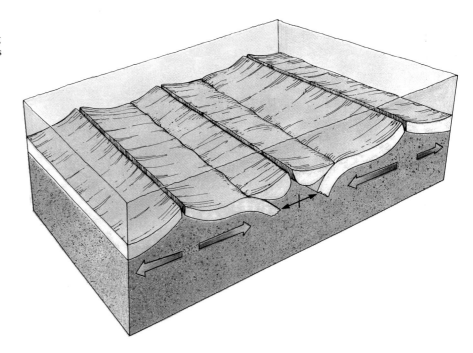

FIGURE 20–27
Density changes in the lower mantle may drive vertical tectonics. (a) Heavy ions (iron, nickel) migrate to the core. (b) Low-density material near the base of the mantle might rise to achieve isostatic equilibrium, warping the mantle and crust above it.

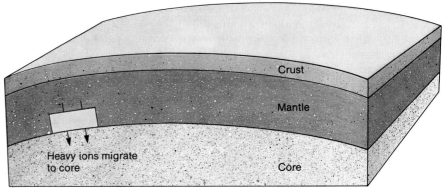

(a) Density of local mass in mantle decreases

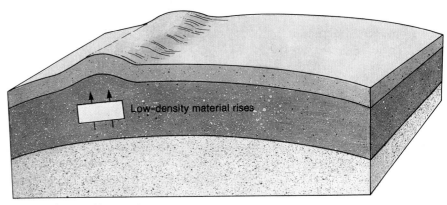

(b) Mass rises, arching crust

tually will be ironed out, but that are not so important that they challenge the validity of the entire theory. Both of these interpretations seem somewhat prejudiced.

A third view is more objective. The unanswered questions show that there is still some doubt about the model, and that it should not be accepted dogmatically. How, then, should we treat it? At the very least, the plate tectonics hypothesis gives geologists and geophysicists an organized framework within which the Earth can be studied. It points to the types of questions about Earth composition and processes that must be answered if we are to understand the evolution of our planet. At best, most plate tectonics concepts may be correct, suggesting that we already have figured out much about the workings of the Earth.

Correct or incorrect, the development, refinement, and constant testing of the plate tectonics model has brought challenge and excitement to geology over the past 35 years. It is a peaceful revolution in the Earth sciences similar to that brought about in biology during the nineteenth century by development of the theory of evolution.

ALTERNATIVE HYPOTHESES

Most earlier explanations for the behavior of the Earth have been discarded in favor of plate tectonics. However, some alternative models have been proposed. A few refute the entire concept of plate tectonics; others deal only with alternative energy sources for seafloor spreading. These ideas vary considerably in their plausibility. Some follow traditional uniformitarian reasoning; others call on nonrepetitive catastrophic events to explain the course of Earth history. We shall look briefly at three of these hypotheses to show the range of ideas.

Vertical Tectonics

Extensive lateral motion of plates is one of the basic concepts in the plate tectonics model. It is also a problem because, as we saw earlier, the mechanics and energy for plate motion are not completely understood. A small number of geologists believe that mountain building need not necessarily involve lateral movements. Instead, they propose that all major tectonic features of the Earth can be explained by large-scale *vertical* movement of rock material. Their ideas are called hypotheses of **vertical tectonics**.

According to these hypotheses, vertical movements are brought about by changes in density and volume in the mantle. Some are associated with the ongoing differentiation of the Earth. Elements are exchanged continually between mantle and core, effectively lowering the density of the lower mantle (Figure 20–27). It is unlikely that this process operates at the same rate everywhere in the mantle, so in some areas rocks may become less dense than those around them. They would then rise isostatically, and in so doing bring about mountain-building events in the overlying crust. In a series of classic experiments, Hans Ramberg showed that such "gravity tectonics" can produce many of the features of mountain ranges found on the continents, including folds and thrust faults normally attributed to lateral plate movements near subduction zones.

Metamorphism in the mantle also may bring about vertical tectonism. Most metamorphic reactions involve either an increase or decrease in volume. An increase in the volume of a large segment of the mantle could cause expansion, uplift, and rifting in the crust, and thus could produce ocean ridges, rift valleys, fault-block mountains, and upwarped regions such as the Colorado Plateau. A volume decrease could cause subsidence; continued subsidence might cause material to flow into the depression, resulting in compressional forces. Fold mountains, island arcs, and large regional thrust belts could form without any of the lateral motion required for plate convergence.

Vertical tectonics models satisfactorily explain simultaneous compression and tension in the Earth. They also explain nearly all of the major tectonic features and types of mountains found on the continents, but they are far less successful at explaining features of the oceans such as the magnetic anomaly stripes and Hawaiian-type islands. They do not begin to explain how tropical plant fossils can be found in Antarctica. Both the plate tectonics and vertical tectonics hypotheses leave some questions unanswered. Most geologists favor plate tectonics because they feel that lateral plate movements are now firmly demonstrated (see Focus 20–1), and that the problems of vertical tectonics are more serious than those associated with subduction and seafloor spreading.

An Alternative Energy Source

Our plate tectonics reconstruction has called upon the Earth's internal heat for the energy needed to bring about plate motions and orogeny. Some geologists believe that external energy sources also may be involved, in the form of impacts of large meteorites. Meteorite impacts certainly have been important in the evolution of other bodies in the solar system, and our study of the Moon, Mars, and Mercury has revealed many large impact craters. Could they cause plate tectonics events?

Most meteors burn up in the Earth's protective atmosphere, but a large one occasionally strikes the surface, producing a crater. These few large meteorites may have played important roles in Earth processes. For example, a large meteorite crashed into the Earth about 750,000 years ago, at a time coinciding with a magnetic pole reversal. In some way that we do not yet understand, the energy of that impact may have caused the reversal. Perhaps the sporadic impact of such meteorites explains the irregular timing of magnetic reversals throughout time.

An exceptionally large meteorite may even initiate rifting. The conversion of its considerable kinetic energy to heat would create local thermal instability and lead to magmatism, and the crust beneath the impact site might be weakened sufficiently to be rifted during the return to a normal geothermal regime. Meteorite impacts are not necessarily the cause of seafloor spreading but might be catalysts that help initiate some plate tectonics processes.

Colliding Planets: A Catastrophic Model

Plate tectonics, vertical tectonics, and even meteorite-assisted tectonism are all based on uniformitarian principles. Their supporters feel that processes acting today or in the very recent geologic past can explain all those ancient events for which evidence is preserved in rocks. In these models, Earth processes repeat themselves slowly, gradually, but inevitably.

Other hypotheses call upon extraordinarily large, rapid, worldwide events to bring about changes. Such events are termed **catastrophes,** and their proponents are called **catastrophists**. For several decades before his death in 1979, Immanuel Velikovsky was the leading supporter of the catastrophist viewpoint. Velikovsky suggested that the Earth's normal (uniformitarian) operation was interrupted occasionally by collisions or near-collisions with other objects in the solar system, possibly including comets, very large asteroids, or even the planet Venus. More recent ideas now postulate collisions or near-misses with black holes.

One effect of such a catastrophe could be to force the Earth to tumble in its orbit so that the surface would shift position and new rotational poles would develop. This could explain the coal in Antarctica and glacial deposits near the Equator. The vast rearrangement of rotational energy would lead to structural changes, worldwide orogenies, and probably extensive plutonism and volcanism.

Exciting as such catastrophic ideas are, few scientists accept them. Most physicists and astronomers believe that planetary collisions and near misses might have happened early in the history of the solar system, but are extremely unlikely during the later stages, as would be required by Velikovsky's ideas to explain recent geologic events. Furthermore, most geologists feel that evidence clearly recorded in rocks refutes Velikovsky's ideas entirely. Paleomagnetic data, for example, show that the continents have moved into their positions slowly, not catastrophically. Of the three alternative models discussed in this chapter, this one has received the least support from the scientific community.

SUMMARY

1. Earth's magnetic field is defined by magnetic lines of force, which emanate from its north and south magnetic poles.
2. Rapid changes take place in the intensity, direction, and polarity of the magnetic field. The rapidity suggests that the field is caused by the Earth functioning like a dynamo, with its liquid outer core rotating through the Sun's magnetic field, generating an internal electric current and consequent external magnetic field.
3. Remanent magnetism in minerals is a record of the magnetic field at the time the minerals cooled below their Curie points, were deposited as clasts, or crystallized during metamorphism and subsequent alteration.
4. The Earth's magnetic field has reversed its polarity frequently during the geologic past, at intervals of 25,000 to a few million years.
5. Linear magnetic anomalies in the ocean basins result from crystallization of MORB at ocean ridge crests during periods of normal and reverse polarity. The symmetry of the stripes about the ridges supports the concept of seafloor spreading, and the width of the bands indicates the spreading rates.
6. The plate tectonics model satisfactorily explains many aspects of Earth processes and behavior, including most kinds of mountain building, the origin and disappearance of oceans, and the distribution of what appear to be ancient climatic anomalies.
7. Precambrian and Paleozoic ophiolites suggest that plate tectonics processes have operated throughout most of Earth history. Oceanic crust—formed at ocean ridges and returned to the mantle in subduction zones—has been completely recycled since Jurassic times.
8. Earth is continuing to experience a chemical differentiation that began with rapid separation of sialic crust early in its history. Differentiation today involves the separation of MORB from the upper mantle.
9. Continents grow by collision with other continents or island arcs, and by accretion of exotic terranes along transform faults. Continents may break apart by rifting or by loss of terranes by transform fault displacement.
10. Andesites, found only in island arcs, result from partial melting of subducted oceanic crust and sediment. Island arc basalts are generated at several different depths along the subducted plate and are more varied than ocean ridge and hotspot basalts, which form at much more restricted depths.
11. Different regional metamorphic facies series develop at convergent margins because of the abnormal distribution of heat created in the mantle by subduction of a relatively cold slab into the hotter upper mantle.
12. About 250 million years ago, all landmasses on the Earth were collected in a single huge continent called Pangaea. Pangaea split, first into two large plates (Gondwanaland and Laurasia), and then into the continents of today.
13. Plate motion may result from convective push, gravitational slab pull, or some combination of the two.
14. Alternative models for Earth behavior include: (a) vertical tectonics, in which density variations in the mantle due to differentiation or metamorphism cause large-scale vertical adjustments in the crust, and (b) meteorite impacts as catalysts for inception of rifting.
15. Catastrophic Earth models that cite collisions or close encounters with other objects in the solar system as the causes of tectonism are not supported by the geologic record.

REVIEW QUESTIONS

1. Why is it unlikely that the Earth's magnetic field is caused by a permanent magnet located in the iron-nickel core?
2. Explain how the magnetic anomaly stripes in the oceans can be used (a) to support the concept of seafloor spreading, and (b) to indicate spreading rates.
3. Describe as many methods as you can, other than direct measurement, that geologists use to determine the direction and rate of plate motion.
4. Why is it that oceanic crust can be recycled through the mantle, but continental crust cannot?
5. Name as many places as possible where you could go today to observe (a) collision between two oceanic plates; (b) ocean-continent collision; (c) continent-continent collision; (d) rifting on a continent.
6. Explain how the exotic terrane concept provides a very different way for closing an ocean between two landmasses than does the subduction method.
7. Discuss the possibility that continents can shift position on the surface of the Earth, using evidence from as many areas of geologic research as you can.
8. What are the difficulties in attributing plate motion to the action of mantle convection cells?
9. What aspects of the present Earth does the plate tectonics model explain better than any other model? What aspects does it fail to explain?
10. What aspects of the present Earth does the vertical tectonics model explain better than other models? What aspects does it fail to explain?

FURTHER READINGS

Beloussov, V. V., "Why do I not accept plate tectonics"; Sengor, A. M. C., and Burke, K., "Some comments on: Why do I not accept plate tectonics." 1979. *Eos* 60:207–10. (An exchange of arguments between one of the leading vertical tectonicists and two leading plate tectonicists. Many of the points are at an advanced level, but a beginner can follow most of the argument.)

Sullivan, Walter. 1974. *Continents in motion*. New York: McGraw-Hill Book Co. (The history and evidence for continental drift and plate tectonics, written for the layman.)

Takleuchi, W. S., Uyeda, S., and Kanamori, H. 1967. *Debate about the Earth*. San Francisco: Freeman and Cooper. (Clear explanations of the geophysical evidence that led to the plate tectonics model.)

Uyeda, S. 1971. *A new view of the Earth*. San Francisco: W. H. Freeman. (An explanation of the Earth's features in terms of plate tectonics.)

Velikovsky, I. 1955. *Earth in upheaval*. New York: Dell. (A catastrophist's view of the Earth. A series of major catastrophes is called upon to explain ice ages, mountain building, and so forth.)

Wyllie, P. 1976. *The way the Earth works*. New York: John Wiley & Sons. (A well-thought-out discussion of tectonics processes and results.)

21
Resources

Kenecott copper mine, Bingham Canyon, Utah. Photo ©
Norman R. Thompson, 1990.

We always have depended on Earth materials for survival. Our ancestors used the basic resources: water for drinking and growing crops, land to hunt and raise food upon, and stone for building and making weapons. They learned to find the necessary water or obsidian—and were the first geologists. Our modern resource needs are very different: the variety of Earth materials we use in our everyday lives would bewilder the founders of our country. We have had to become better geologists, because each time chemists and physicists develop a new product, it is our job to find the raw materials needed for manufacturing it. In earlier chapters we showed how geologic processes produce useful minerals and rocks. Here, we will describe the different types of resources and consider the economic and environmental problems involved in their extraction and use.

POPULATION

Most of this chapter will deal with finding *supplies* of natural resources, but we also must examine the likely *demand* for these resources. Figure 21–1 shows how population has grown in the past 10,000 years. In geologic time this is a blink of the eye, but the enormous increase is straining Earth's ability to provide food, clothing, and shelter. We already use more than 500 kg of steel, 25 kg of aluminum, and 200 kg of salt *per person* each year. There are five billion people on Earth, and this number will double by the year 2100. Where will the additional resources come from?

The problem actually is worse than these numbers suggest because technologically advanced societies use more material and energy resources than primitive ones. Thus, as we strive for higher standards of living, our per capita resource need increases. The world of the future will contain many more people, each of whom will require more resources than a person today.

THE NATURE AND UTILIZATION OF RESOURCES

A **resource** is a material that is both *useful* and *available in sufficient amounts*. The usefulness of a substance often is temporary. For example, chert and obsidian once were valuable resources because of their use in tools and weapons; today they are curi-

FIGURE 21–1
Graph showing population growth with time. The arrow marks a sharp population drop caused by the Black Death, which struck Europe in 1348. Data from the Population Reference Bureau.

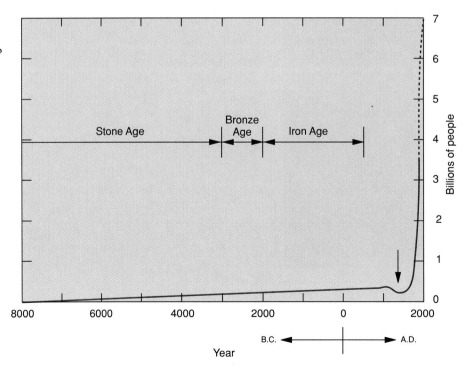

osities. Bauxite once was just a curiosity; it now is the principal source of aluminum.

Renewable resources are those that form faster than they are consumed. The rate of renewal depends on the particular resource. Water lost from the atmosphere by rainfall, for example, is restored in about a week, but a wooded area harvested for lumber requires 30 years of growth before it can be cut again. It takes several hundred years to form fertile soil and millions of years to make a bed of coal. Most resources are **nonrenewable** because, once consumed, they cannot be replaced within human lifetimes.

We can help some resources last longer and thus perhaps make them renewable. One way to do this is to *conserve*—to use less than we are now using; another is to *recycle*—to reprocess materials like copper wire, aluminum cans, and steel beams and use them again. However, recycling can work only for materials that are not actually consumed. There is no way to recycle heating oil, natural gas, or coal because when we use them there is nothing left to recycle.

Availability of Resources

Economic factors also must be considered in evaluating resources: if a resource cannot be used profitably, it will not be used at all. For example, a huge copper deposit was discovered in Maine several years ago, but the cost of mining is so great that it has not been developed. The portion of a resource that can be recovered profitably is called its **reserve**. To estimate how long it will take to exhaust a resource, we divide its proven reserves by the rate at which we are using it. This often proves to be a very inaccurate estimate because of rapid changes in both geologic and economic factors. Geologists may discover new reserves, adding to the life of a resource. Increased population, on the other hand, increases usage rates. New materials may replace old ones (such as the use of plastic instead of metal in cars), thus changing the **depletion** rates of both.

Geographic Distribution of Resources

Resources are not distributed equally over the Earth. A wide variety of geologic processes is employed to produce resources, so the more varied the geology of a country, the greater its resource potential will be.

Large countries such as the United States and Canada generally contain more types of geologic terrain than small ones and therefore have greater resource potentials. There are exceptions to this rule, of course. Small countries may not have a variety of resources but can be major suppliers of specific materials. For example, Guyana and Surinam, on the northeast coast of South America, are small nations but have some of the richest reserves of bauxite in the world.

NATURAL RESOURCES

We will now briefly study the resources required to maintain our civilization. Land and water are still as important as they were to our ancestors, so we shall look at them first, followed by materials and energy resources. In each case we will examine the environmental problems associated with extraction and use of the resource.

Land and Water

We use land for many things, including living space, agriculture, and mining. Population growth increases each of these demands and causes competition among conflicting land uses: Should farmland be converted to suburban housing? If so, where will we grow food? We cannot mine iron ore, build houses, and generate electricity on the same piece of land. The demand for land has caused us to extend development into areas that contain geologic hazards: we now farm on steep hills and live along unstable shorelines and in semiarid regions. The resulting problems have been detailed in Chapters 10–15.

Soil. Fertile soil is a vital resource because we grow most of our food in it. Loss of fertile soil is a serious problem because human activities have artificially increased rates of erosion. Recall from Chapter 9 that soil is generally thin (less than 2 m), so that erosion of just a few meters of the surface can change a prime growing area into a barren one. Soil is held in place by the roots of the plants that it nourishes; anything that removes the plant cover and exposes the soil will accelerate erosion. Table 21–1 shows how human use of land can increase soil erosion drastically. As homes replace fields in flat areas, farming must move

TABLE 21–1

Representative rates of erosion from various land uses.

Land Use	Metric Tons of Soil Loss per Km² per Year	Compared to Forest
Forest	8.5	(0)
Grassland	85	10 ×
Abandoned surface mines	850	100 ×
Cropland	1700	200 ×
Harvested forest	4250	500 ×
Active surface mines	17,000	2000 ×
Construction	17,000	2000 ×

to hilly or mountainous regions, where soil loss is accelerated because of mass movement and stream erosion.

Only a fraction of the land surface is farmable because both fertile soil and water must be present, and the two are not always found together. Irrigation brings water to arid parts of California and semiarid areas of the midwestern United States, and these are highly productive agricultural centers. However, not all effects of irrigation are beneficial. Much of the water piped into fields in arid regions evaporates back into the air, depositing its dissolved ions in the upper layer of the soil. A buildup of ions over a long period may be harmful to the crops that the water is supposed to nourish. These problems are beginning to arise in California, and more are to be expected.

Water. We consider water to be a renewable resource because its worldwide passage through the hydrologic cycle happens far faster than the rate at which it is consumed. Locally, however, it may be nonrenewable. For example, excessive pumping of groundwater in an arid region may remove water faster than it can be replenished by local infiltration.

The use of water may be **instream** (within a stream channel) or **offstream** (taken out of a stream or aquifer). Instream uses of water, such as transportation, generation of hydroelectric power, recreation, and wildlife management, usually do not consume water because it is not removed and remains available. However, offstream uses of water—irrigation, domestic and industrial purposes, and energy production—return some water to its source, but consume much of it. For example, water removed from a stream for food processing ends up in a can; some of the water used to cool a nuclear reactor is lost to the atmosphere as steam. The greatest offstream consumption occurs by evaporation during irrigation.

Both instream and offstream use may cause environmental problems. We outlined earlier the effects of dam construction on the Colorado River (Chapter 11) and the effects of overpumping of groundwater (Chapter 12). We now will examine one type of problem, thermal pollution, which can occur if offstream use of water significantly changes its temperature before it is returned to the source.

Large amounts of water are used in industrial processes. The water is used variously as an ingredient, a cleaner, or a coolant. Serious problems may arise if the unconsumed water is not treated before it is returned to the stream from which it came. The necessary treatment depends upon how the offstream water has been affected. Toxic chemicals may have to be removed, organic content may need to be reduced, and the temperature may have to be adjusted to natural levels. The harmful effects caused by hot water when it is returned to the hydrologic cycle are called **thermal pollution**. Heated water loses some of its ability to hold oxygen in solution, so the amount of dissolved oxygen decreases. The heat also increases the metabolic rates of animals that live in the water. They require more oxygen at a time when less is available, and many die.

Organic contaminants in water are decomposed by oxidation, further reducing the amount of dissolved oxygen. The amount of oxygen required for the life-needs of organisms and for oxidation of organic particles is called the **biological oxygen de-**

mand. As long as the dissolved oxygen and biological demand levels are balanced, there is no problem. However, consider a stream into which an industrial facility or sewage disposal plant discharges nontoxic but organic-rich wastewater. The organic waste increases the biologic oxygen demand and reduces the oxygen available to animals already living in the water. Many cannot survive.

Metallic Resources

A **mineral deposit** is a mass of naturally occurring mineral material having an economic value. Those minerals from which we extract metals are called **ore minerals.** The most important factor in determining the value of a metallic deposit is how much the metal has been concentrated relative to its typical abundance: the greater the enrichment, the more valuable the deposit. We listed the most important ore minerals in Chapter 3 and discussed their concentration by igneous, sedimentary, and metamorphic processes in Chapters 4, 5, 6, and 7. Here we will focus on examples of a few different types of deposits.

Hydrothermal Deposits. Ore minerals may be concentrated at any stage of igneous activity, such as during the gravity settling that has produced chromium and platinum deposits in South Africa (Chapter 4) or the eruption of sulfur from volcanic vents in Italy (Chapter 5). In the *last* stages of magmatic cooling, the magma is enriched in water that contains many dissolved ions. Hot water and steam penetrate fractures in the host rock and precipitate ore minerals as they cool. This kind of ore deposit is called a **hydrothermal deposit.** The large copper deposits at Butte, Montana and lead and zinc deposits near Joplin, Missouri formed this way. The potential for forming this kind of hydrothermal deposit is greatest in large granitic batholiths because, as we saw in Chapter 4, granite magma contains more water than any other kind. Vast deposits of hydrothermal gold and silver, for example, have been mined from the Sierra Nevada Batholith in California.

The "black smokers" described in Chapter 16 are a different type of hydrothermal deposit. The water there is not derived from the oceanic basalt magmas, but is instead cold seawater which is heated by cooling basalt and dissolves some of the metallic ions present. These ions are precipitated as rich metal-sulfide deposits when the water chills again.

Sedimentary Deposits. Some ore deposits are formed at the surface of the Earth by surficial processes. These include materials deposited in water, others separated from nonuseful minerals during transportation, and yet others that form as residual deposits during weathering.

Depositional Ores. One such deposit is the Precambrian banded iron ores that are found in the Lake Superior region. These ores, and similar ones of identical age in other parts of the world, supply most of the world's iron (Figure 21–2). Banded iron ores consist of alternating layers rich in iron oxides and silica. Their origin is controversial, but all geologists agree that they could not form under the conditions that now exist at the Earth's surface. Today, iron is quickly deposited as the insoluble ferric ion (Fe^{3+}) in our oxygen-rich atmosphere, but conditions during the Precambrian clearly were quite different. The atmosphere at that time contained less oxygen and larger amounts of gases such as methane and carbon dioxide. That kind of atmosphere would have made surface waters more acidic, favoring the easier transport of iron in the form of the more soluble ferrous ion (Fe^{2+}) which was eventually deposited in the iron-rich bands. When plants evolved that use photosynthesis, the oxygen content of the atmosphere increased and banded iron ores could no longer be formed.

Placer Deposits. The enrichment of mineral deposits by the action of a moving fluid (water or air) is called a **placer deposit.** Placer minerals tend to be harder, more resistant to chemical weathering, and, above all, denser than the common siliciclastic minerals (quartz, feldspars) that make up most surficial deposits. When the speed of a water or wind current drops, the denser grains are deposited, whereas the less dense but far more abundant siliciclastic minerals continue to move. This situation occurs commonly in these situations: on the inside of meander bends; where tributaries enter a larger river; on the upstream sides of outcrops or boulders extending upward from a stream bed; at the base of the windward

side of a sand dune; in potholes drilled into a stream bed (Chapter 11); in the surf zone; and along a beach.

Placers commonly may be observed along a beach after a storm (Figure 21-3). The high turbulence of the storm waves washes away the less-dense minerals and concentrates the denser ones, such as magnetite and garnet, to form black or purple sands. Important deposits of titanium, iron, tin, and even diamonds (Focus 21-1) are found in placer deposits.

FIGURE 21-3
Beach placer deposit composed of dense garnet grains. Photo by Nicholas K. Coch.

A single stream-placer deposit was responsible for starting the greatest migration in American history. Hydrothermal gold deposits exposed by weathering in the Sierra Nevada mountains of California were eroded by streams, and some of the gold was transported westward into lower areas. Gold nuggets and dust accumulated in the stream beds wherever water flow was impeded. A worker at Sutter's Mill in California found some of this placer gold in 1848, and this led to the migration of thousands to California in 1849 (the '49ers).

Nonmetallic Resources

Many of our most important resources are neither precious gems nor ores of metals. They are materials such as sand, gravel, and crushed stone used in construction; abrasives; minerals used to make industrial chemicals; and phosphate rock for fertilizer. We use more of these materials by weight than any other Earth resource, but they are less expensive than the metals and their total dollar value is relatively low. These minerals and rocks are produced by nearly every kind of rock-forming process.

Sand and gravel. Of all nonmetallic resources, 87% (by weight) consist of sand and gravel. These materials are used in construction to build roadbeds, provide suitably porous foundations for drainage systems, and for making concrete. Most sand and gravel comes from floodplains, river channels, and stratified glacial deposits such as eskers, kames, and outwash plains, but some is produced by dredging in coastal areas where other supplies are scarce. It is expensive to transport sand and gravel because of their weight, but these resources are widely distributed and usually are present near urban areas. As cities and suburbs expand, however, they encroach upon the old gravel pits. Land values rise and more profit can be made by building houses on land than by mining sand or gravel from it.

Cement and Clays. Modern buildings require cement, which is produced in areas having abundant deposits of both calcium carbonate minerals (generally limestone) and clay minerals (generally shale). Cement is "fired" at high temperatures during man-ufacture, causing the clay minerals in the shale to react with the calcium in the limestone to form "clinker," which is ground and mixed with small amounts of another nonmetallic resource, gypsum, to make Portland cement.

Clay minerals are used for many purposes. They are the principal ingredient in the manufacture of ceramics, such as brick, tile, pottery, and fine porcelains. There are many different clay minerals, and precise mixtures are required for different purposes. Flakes of the clay mineral kaolinite are used in the manufacture of high-grade paper. The flakes align themselves on the pulp fibers and make the paper smooth and white. We saw in Chapter 9 that kaolinite is a common product of chemical weathering, and its chief source in the United States is in the deeply weathered soils of Georgia. Clay minerals also are used, in different mixtures, to prepare special drilling muds used in petroleum exploration. These muds are designed to have specific properties of lubrication and electric conductivity to aid in well-drilling and in detecting oil and gas deposits.

Phosphate Rock and Halite. Phosphorus is a vital ingredient in fertilizers and many animal-feed supplements, and is so important that many scientists fear that its availability will be the ultimate control of how much food we can produce. Phosphate rock (rich in the mineral apatite) is obtained from three kinds of deposits: marine sedimentary rocks, bird droppings on tropical islands, and apatite-rich carbonatite plutons. Most is recovered from sedimentary rocks. Parts of Florida and coastal North Carolina are underlain by rich phosphate deposits, but, as is the case with the sand and gravel discussed above, the expanding population is competing for space with the mines. As Florida has quickly become the third most populous state, it is easy to see which economic pressure is greater.

The United States is the world's leading producer of halite, more commonly known as salt. Although we eat salt, its most important commercial uses are in the chemical industry to make chlorine gas and as a highway deicer in the northern United States. Most halite deposits formed as evaporite beds (see Chapter 6), and some of these beds are extremely thick. Underground mining of salt is a major industry in New York, Ohio, and Michigan.

DIAMONDS FROM THE SURF OF NAMIBIA

In Chapter 14 we described the Namib Desert, one of the driest and least populated on Earth (see Figure 14–2). One of the world's largest deposits of diamonds is being mined from placer deposits along the coastal border of the desert. The diamonds originally were formed in the mantle, in an ultramafic rock called **kimberlite,** and reached the surface in narrow vents far to the east in South Africa. The diamonds weathered out of the kimberlite and were transported westward toward the coast by the Orange River and its tributaries. The transported material was

FIGURE 1
Diamond mining along the Namibian coast. Diamonds found as clasts in the sediments are mined from depths below sea level. The bulge in the coastline is caused by dumping of sand that has been processed during mining. Photo © Anthony Bannister/Natural History Photographic Agency.

deposited along the ocean and later reworked into terraces (see Chapter 15) by sea level changes during the Pleistocene Epoch. The diamonds were concentrated in rock crevices and in gravel deposits at the base of the sandy beach terraces.

The process used to extract these diamonds is remarkable (Figure 1). Deep excavations, as much as 20 meters *below* sea level, are required to reach the base of the terrace deposits where the diamonds are concentrated. Earth-moving machines remove huge volumes of sand in the stripping operation. Massive sand walls and powerful pumps keep the ocean from flooding the excavation. Once the bottom of the terrace deposit is reached, the diamonds are literally swept out of the rock crevices by hand; some also are separated mechanically from the gravel deposits.

The Namibian diamond deposits are exceptional both in quality and quantity. Most of the diamonds obtained by mining kimberlite are of industrial grade and contain impurities and fractures that make them useless as gems. Most of the Namibian placer diamonds, however, are of gem quality (Figure 2). It is believed that the impure diamonds did not survive the rigors of weathering, transportation, recycling, and surf action as well as the flawless gem-quality stones. The quantities recovered are staggering: in 1977 alone, *2,001,217 carats* of diamonds were mined from the beaches of Namibia.

FIGURE 2
A small sample of gem-quality diamonds mined from the DeBeers Oranjemond Mine on the Namibian coast. This is the richest diamond mine in the world. Photo © Peter Johnson/Natural History Photographic Agency.

ENERGY RESOURCES

Our earliest ancestors needed little in the way of energy resources because their work was done by muscle power. Today, we are voracious consumers of energy for heating and cooling, lighting, machinery, manufacturing, and transportation. We tap the Earth in several ways for energy, utilizing the kinetic energy of moving air, streams, and ocean water; burning the remains of ancient animals and plants; directly using heat energy from within the Earth; and releasing the nuclear energy stored in atoms.

Fossil Fuels

Energy sources derived from the remains of plant or animal life are called **fossil fuels,** and may be solids (coal), liquids (petroleum), or gases (natural gas). Although there is some debate about a possible inorganic origin for petroleum and natural gas, most geologists agree that most of these resources are derived from organic processes.

Coal. Coal is a sedimentary rock derived from plant material that was buried and transformed into a black, carbon-rich fuel. The oldest coal dates only from the Devonian Period, because that was when land plants first evolved. The greatest accumulations of coal date from the Mississippian and Pennsylvanian Periods because conditions then were perfect for the formation of vast coal deposits in North America and Europe. Indeed, these two periods are combined on European time charts into the **Carboniferous Period**. During those times tectonic and

FIGURE 21–4
Dragline used in strip mining of coal at the Chinook Mine in Indiana. Photo courtesy of National Coal Association.

FIGURE 21–5
Stratigraphic oil traps. U = unconformity, F = facies. In both types of oil trap, oil is confined in reservoir rocks by an impermeable rock type.

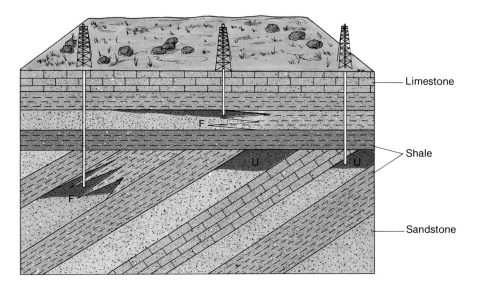

Limestone

Shale

Sandstone

climatic factors combined to produce extensive swampy areas bordered by lowlands, in a warm climate. The formation of coal requires an accumulation of plant material in an environment that is protected from oxidation that otherwise would destroy the organic tissue; a swamp is ideal. Adjacent areas should have little relief to minimize the input of siliciclastic sediment and thus make the coal more pure.

Plant debris collecting under the stagnant waters of a swamp undergoes several chemical changes as it is buried. Volatile gases such as carbon dioxide, methane, and water are driven off, and the percentage of carbon increases. The material undergoes a steady transformation that is marked by distinct stages, each of which is an energy resource: peat, lignite, bituminous coal, and anthracite coal.

Coal is mined from beds which are separated by other types of sedimentary rocks. Some coal is deep-mined in shafts and tunnels that extend for several miles in complex underground networks. Where the coal is close to the surface, strip mining is used instead (Figure 21–4). Enormous shovels called drag lines strip away the overlying rock to reveal the coal bed, and the coal can be broken up and loaded into trucks by the same drag lines.

Petroleum. Petroleum is a liquid believed to have formed from the burial and subsequent alteration of marine microfossils that originally lived in surface waters, but accumulated in the fine-grained sedi-ments on the seafloor after their death. Burial provided the heat and pressure that have formed the **oil** and/or associated **natural gas.** The oil and gas form in organic-rich **source rocks,** generally dark shales but sometimes other types, such as evaporites. Increased burial pressures and tectonic forces drive oil and gas out of their source rocks, and they migrate into porous and permeable **reservoir rocks,** such as sandstone or reef limestone, from which we recover them.

Oil and gas will migrate until their flow is impeded and they collect in what is called an **oil trap.** We already have seen how structural traps are formed by salt domes and anticlines (see Figure 17–30). **Stratigraphic traps** occur where permeable beds containing petroleum interfinger with adjacent impermeable rocks (Figure 21–5).

In older oil fields, after the more easily recovered oil has been removed, **secondary recovery** methods sometimes are used to free more oil; one such method involves pumping water into strata to force the oil out.

Natural Gas. Natural gas is found with petroleum in many oil fields. It also occurs alone in **gas fields.** The gas, mostly composed of methane (CH_4), is extracted from the ground by the same methods used for petroleum, and impurities such as sulfur are removed before the gas is transported by pipeline to users. In some underdeveloped countries where at one time there was no local use for natural gas, the gas was burned off. Today, it is liquefied under high pressure and transported in specially designed tankers to places where it can be used.

Oil Shales and Tar Sands. In some instances, oil has not migrated from its source rock and has formed **oil shales**. These shales develop from organic material accumulated in lakes or marine basins that have restricted circulation, preventing organic debris from being oxidized. The most important deposit of oil shales in the United States is in the Green River Basin of Wyoming, where a group of large lakes covered the area during the Eocene Epoch. It is more difficult to recover oil from impermeable oil shale than from porous and permeable reservoir rock. The shale first must be broken and crushed, and then heated until its oil is converted to a gas. The gas is condensed, and the liquid petroleum is treated the same as conventional oil.

Each petroleum deposit is a unique combination of several different hydrocarbons, and in **oil and tar sands** the mixture results in a heavy, viscous material that can be dislodged only with great difficulty. We recover this petroleum by crushing the sands and piping high-pressure steam through them. Large oil sand deposits occur in Alberta, Canada.

The high cost of processing oil shales and sands makes their use unprofitable at this time, considering the price of more conventional supplies of petroleum. When conventional supplies are depleted (and the price goes up), or when producers raise the prices sufficiently, production from oil shales and sands will become a money-making proposition and these materials will become a viable energy resource.

FIGURE 21–6
A nuclear reactor at St. Lucie, Florida. Photo courtesy of United States Council for Energy Awareness.

Alternative Energy Resources

There are serious environmental problems associated with the burning of fossil fuels (see below), as well as the long-term problem of eventual depletion. At present, none of the alternate energy sources provide the amount of energy that oil and natural gas supply, but their variety shows how important energy is to our civilization. Some of the alternatives can be only of local use, and serve as supplements to fossil fuels, but others may become the energy resources of the future.

Nuclear Fuels. Uranium minerals are found in a wide variety of geologic environments, including pegmatites, sandstones, lignite, and placer deposits. Most of the uranium ore now mined comes from sedimentary deposits formed when uranium minerals that were soluble under oxidizing conditions precipitated when placed in contact with organic matter present in the rock. The uranium ore is recovered, processed, and the uranium formed into fuel discs and rods for nuclear reactors (Figure 21–6). Uranium is thus a **mineral fuel**.

A reactor harnesses the heat produced from the fission of uranium atoms to drive dynamos that generate electricity. Water used to keep the reactor cool is converted to steam, and it is the steam that actually drives the turbines. Nuclear reactors solve some of the environmental problems of energy generated by burning fossil fuels, but create several new ones, as residents near Three Mile Island, Pennsylvania, and Chernobyl in the Soviet Union know only too well. The problems of radioactive waste disposal and public fear of nuclear power probably will limit its expanded use for years to come.

Solar and Wind Energy. Solar energy and wind energy are there for the taking—nothing need be mined—and are therefore attractive potential energy resources. Actually, wind has long been used to do work: sailing ships driven by wind power circled the Earth centuries ago, and sails on windmills turned millstones and sawblades. Today's windmills are very different from their predecessors (Figure 21–7). They are high-technology products designed to operate in very gentle winds and are attached to electrical generators rather than directly to machines. To be useful, wind must have a sufficient velocity (at least 7.2 km/hr), and must blow regularly. Only about 10% of the United States meets this requirement, but these areas include the energy-poor industrial Northeast and Hawaii; wind may become an important supplement to fossil and nuclear fuels in these areas.

FIGURE 21–7
Modern windmills at an experimental electricity generating station on Hawaii. Photo by Nicholas K. Coch.

We have recently learned how to collect and utilize energy from the Sun with photoelectric cells that convert sunlight to electricity. We also use solar energy on a small scale for water and space heating. The amount of sunlight in an area determines how suitable solar energy can be as a supplement to normal energy sources. The smaller the amount of sunlight, the larger the array of collecting cells must be, and the more complex the energy storage system. The cost is high, but we are experimenting with large-scale solar energy collection systems (Figure 21–8).

Hydroelectricity. The kinetic energy of water flowing in stream channels was one of the earliest energy resources. Waterwheels powered many mills and factories, and early in the twentieth century many small dams were built across streams in the United States to produce electricity locally. These were eventually replaced by larger, more cost-efficient central generating plants that burned fossil fuels, but we are turning again to **hydroelectricity** (electricity generated by running water) to satisfy our insatiable need for energy. In Canada, 75% of electrical generation is done by hydropower, compared to only about 15% in the United States.

The requirements for hydroelectric generation are simple: a stream, a dam, turbines to drive generators, and transmission lines to deliver the electricity to the consumers. Hydropower is clean and nonpolluting, but we already have seen some of the environmental problems that arise when any kind of dam is built. One response is to build the dams far away from population centers, but there has been increasing opposition to high-voltage transmission lines passing through heavily populated areas.

The rise and fall of the tides also can be used to generate power. At tidal power stations, water is impounded at high tide behind a containment dam and released through turbines at low tide to generate electricity. This method is feasible only where tidal range is high and where a generating station can be built across the estuary. The only major functioning tidal power plant today is on the Rance River estuary in France. The Bay of Fundy in Maritime Canada (between Nova Scotia and New Brunswick) has one of the greatest tidal ranges in the world (see Figure 2–11), and has long been considered as a possible source of energy. However, recent computer modeling of the effects of a containment dam suggests that there might be serious problems for communities along the Atlantic coasts of Canada and the United States, and the Fundy Tidal Power Project remains on the drawing board.

Geothermal Energy. Wherever hot rock occurs close to the Earth's surface, there are potential geothermal

FIGURE 21–8
A array of solar collecting panels at Natural Bridges National Monument in southeastern Utah. At the time that it was built, this was the largest solar-power installation in the world. Photo © Norman R. Thompson, 1990.

resources. In some cases, groundwater is heated naturally below ground and the escaping steam is used to generate electricity. In others, water is pumped into the ground to be vaporized, and the steam is returned to the surface to operate turbines. Geothermal power has been used in Italy since the early part of this century. The largest use of geothermal power is the Geysers power plant north of San Francisco, where geothermal power has been generated since 1960. Geothermal power can be a good supplementary heat source in tectonically active areas and volcanic terrains, such as those of Italy, California, New Zealand, and Iceland. In Iceland, geothermally heated water is used extensively for heating in the capital city of Reykjavik.

A different, experimental approach called **dry rock geothermal energy** may utilize Earth's geothermal gradient to produce usable energy anywhere it is needed. Even if there is no magma body close to the surface, we know that temperatures increase with depth (see Chapter 2). If we drill a hole 10 km deep and pump water into it, the heat of the rocks at the bottom (about 250°C) will convert the water into steam. The steam then could be pumped from the well and used to drive electric generators.

ENVIRONMENTAL PROBLEMS ASSOCIATED WITH RESOURCE USE

Along with the many beneficial uses of resources come problems associated with their extraction, processing, or utilization. These problems affect the quality of the atmosphere, hydrosphere, lithosphere, and biosphere. The wide range of environmental problems is causing a rapidly increasing concern as our population and need for resources grows. We will look now at just a few of these concerns.

Collapse

Underground mining can lead to collapse of the surface if insufficient support is left in the mine for the rocks above. This has become a serious problem in areas underlain by coal mines—areas such as the Appalachian coal fields of Pennsylvania, West Virginia, Kentucky, and neighboring states. Mining operations in the past used beams made of timber to support the mine ceilings so that as much of the coal as

FIGURE 21–9
Subsidence damage at a shopping center in western Pennsylvania. Mining of a 2-m-thick coal bed at a depth of about 90 m led to over 1 m of surface subsidence. Photo courtesy of Richard E. Gray/GAI Consultants, Inc., Pittsburgh, PA.

possible could be extracted. With the depletion of the mines and the passage of time, the timbers have rotted and there is insufficient support in some cases for the overlying rocks. This has resulted in collapse and significant damage to structures on the surface (Figure 21–9).

Even with modern mining methods—long steel bolts to hold the mine roof in place, and leaving pillars of coal to support the mine roof—subsidence remains a problem because of groundwater fluctuations. Extensive lowering of the water table weakens the support of overlying rocks and subsidence can occur.

Subsidence

In Chapter 12 we showed how the removal of groundwater and oil has caused extensive subsidence in the area south of Houston, Texas (Figure

FIGURE 21–10
Sea walls such as this one along the New Jersey shore are now needed to protect coastal cities that have subsided as the result of removal of underground water or petroleum. Photo by Nicholas K. Coch.

12–21), resulting in foundation and road cracking, and has caused an increase in coastal flooding because the area has such a low elevation. A similar situation has now occurred in Long Beach, California. Oil drilling since the early part of this century has caused enough subsidence that sea walls such as that shown in Figure 21–10 are now required to keep the ocean out.

Acid Mine Water

Coal mining can increase the acidity of the local water system. The iron mineral pyrite (FeS_2) is found in many coal deposits and in the dark shales that commonly overlie them. Blasting or digging removes the coal and leaves behind quantities of shale as a waste product both in surface (strip) mining and in underground operations. With time, groundwater reacts with the rock debris and the pyrite is oxidized to produce iron sulfate and sulfuric acid. This acid-rich, frequently rust-colored wastewater is referred to as **acid mine water**.

The acid water moves into the water table or sits in pockets in the debris until a heavy period of rain flushes it into the nearest body of water. The acid added to normal streams has been sufficient to cause a serious loss of aquatic life—Figure 21–11(a). The U.S. Bureau of Mines is experimenting with several novel methods for controlling such problems—Figure 21–11(b).

Soil Loss and Sediment Pollution from Strip Mining

Strip mining in some cases has resulted in destruction of soil and extensive erosion of the land surface. In strip mining, all of the soil, sediment, and rock overlying the resource must be removed to expose the resource. This material often is simply piled beside the mining operation. Upon completion of mining, the surface has become an irregular landscape on which nothing can grow because the small volume of soil that originally covered the bedrock has now been

FIGURE 21-11
Acid mine water. (a) Minerals present in the rocks at many mines combine with groundwater to produce dilute sulfuric acid, which dissolves metals from the surrounding rock. (b) Experiments by U.S. Bureau of Mines scientists show that artificial wetlands like these are effective in decontaminating the acid mine water. Photos courtesy of U.S. Bureau of Mines.

(a)

(b)

mixed with huge volumes of sterile rock. The irregular topography inhibits natural drainage and permits acid water conditions to develop if pyrite is present in the waste rock. Proper land restoration can alleviate this problem by careful removal and storage of topsoil when mining begins, refilling the excavation with waste rock, replacing the topsoil cover, and planting with grass and trees to retard erosion (Figure 21-12). Unfortunately, mining reclamation laws did not exist for many years, and aban-

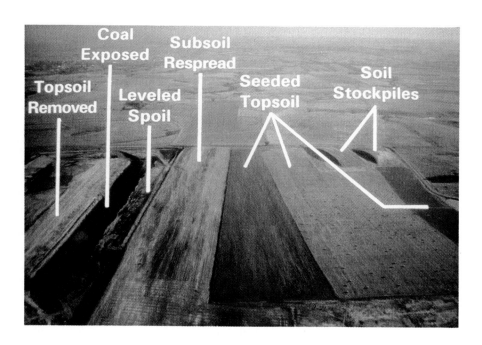

FIGURE 21–12
Steps taken in reclaiming land following strip mining.

doned strip-mined lands are a major environmental problem in some areas of the United States.

Problems from Burning Fossil Fuels

Increased combustion of fossil fuels is causing a serious decline in air quality. The air in many cities not only is less healthy to breathe, but also is having a corrosive effect on structures and is poisoning many water bodies. Two large-scale results of our dependence on coal, oil, and natural gas for energy are acid rain and the greenhouse effect.

Acid Rain. Burning of fossil fuels releases gases such as sulfur dioxide and carbon dioxide into the atmosphere. Sulfur dioxide combines with water vapor in the air to produce sulfuric acid, and results in **acid rain** when the vapor condenses and falls as rain. Acid rain dissolves soluble building stones such as marble, gypsum, and limestone (Figure 21–13), and accelerates deterioration of fabrics and painted surfaces. Acid rain falling into the watersheds of lakes acidifies them, resulting eventually in the elimination of animal life and much of the aquatic vegetation. This problem is most severe in lakes located in rocks such as granite, which releases acid-producing ions when it weathers. Other rocks, such as limestone, release ions that counteract the effects of acid rain; chemical

reactions occur which neutralize the acid. Lakes underlain by such rocks suffer far less than nearby lakes underlain by granite.

The Greenhouse Effect. The increased input of carbon dioxide into the atmosphere from combustion of fossil fuels creates an insulating blanket which prevents Earth's heat from being radiated into space, thus warming the atmosphere. The carbon dioxide acts much like the glass walls of greenhouses, and this process is therefore called the **greenhouse effect**. The greenhouse effect is increasing the temperature of the atmosphere and will cause a wide range of problems if the warming trend continues. Concerned scientists have predicted a considerable rise in sea level as melting of the polar icecaps is accelerated, as well as significant changes in climate in different parts of the world. Some of our major agricultural regions may become more arid and less productive as the climate warms. In addition, many of our coastal cities will be subjected to coastal erosion and flooding if sea level rises (see Chapter 15).

Problems from Waste Disposal

The disposal of waste materials from population centers is a rapidly increasing concern. This problem

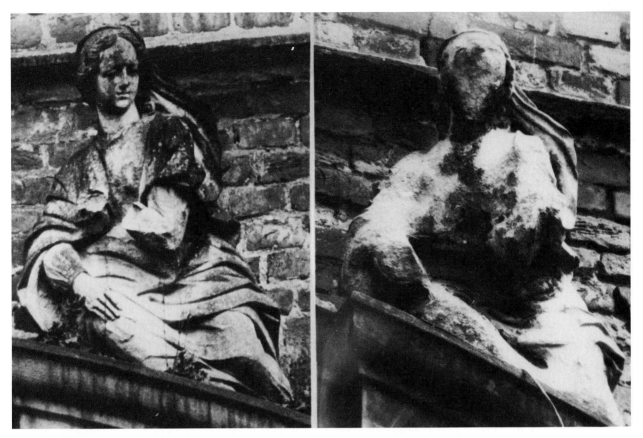

FIGURE 21–13
Accelerated decay of a statue at Herten Castle, Westphalia, Germany, caused by acid rain. The statue was carved in 1702 from a calcite-cemented sandstone. Left, photograph taken in 1908 shows that only limited damage had occurred in 206 years. Right, photograph taken in 1969 shows how solution by acid rain has nearly destroyed the statue in only 61 years. Photo courtesy of Mr. Schmidt-Thomsen, Provincial Office of Monuments, Westfalen-Lippe, Germany.

results from the short life of most manufactured goods, from wrapping materials made of substances that are not biodegradable, and from the lack of space in which to store the waste safely. Waste disposal problems fall into two categories—those involving solid wastes and those involving liquids.

Solid waste consists of rubbish, garbage, construction debris, manufactured goods, and any other solid material. Much of this is disposed of in landfills—piles of debris which are covered by a layer of sand each day to reduce odor and control pests (Figure 21–14). Some of these landfills have reached heights of several hundred feet and occupy many areas on the outskirts of major cities.

Many municipalities have run out of space for solid waste, and must ship it to other sites many miles away. The organic material within landfills reacts with infiltrating rain to produce gases like methane and a foul-smelling liquid called **leachate** which can migrate from the landfill into adjacent areas unless preventive measures are taken. Leachate kills vegetation and is a serious threat to both surface waters and groundwater supplies. New sanitary landfills are being designed with impermeable liners to inhibit the migration of leachate and with piping systems to collect methane gas. (Some cities, such as New York, are actually marketing the gas produced in their landfills.)

FIGURE 21–14
A sanitary landfill operation in California. Photo courtesy of Joseph J. Niland.

These problems would be greatly reduced by decreasing the amount of solid waste dumped into the landfills. To do so requires separation of those components that can be recycled from those that cannot. Many communities separate and recycle aluminum, ferrous metals, glass, and paper, thereby greatly reducing the volume of solid waste that must be stored. Some communities also incinerate that part of the remaining material that can be burned, but this leads to problems of atmosphere pollution.

The second type of waste is **liquid waste** produced as a byproduct of manufacturing goods and refining mineral resources. Some liquid waste is treated chemically and disposed of, and some is burned, but unfortunately, some is released illegally into the environment. Only a few years ago, liquid waste was simply poured on the ground or buried in drums in dumps. As the drums corroded, the liquid waste was released into the environment where it caused a variety of problems. Tragic consequences result when homes are built upon or in the vicinity of abandoned liquid waste dumps, as we found out in the case of Love Canal near Buffalo, New York.

Part of the problem is that we only now are becoming aware of the long-term consequences of such toxic wastes. Recognition of the problems associated with materials such as carcinogens is difficult because some health hazards only become noticeable after many years. Remedies are expensive, requiring chemical treatment of new liquid wastes to render

them harmless, and cleanup of old toxic waste dumps.

RESOURCES FOR THE FUTURE

The growth of population and our ever-increasing need for resources pose a problem for the future. We explained earlier why we cannot predict accurately the depletion dates for our resources, but it is clear that we have a serious problem.

We hope to discover new resources in areas that are poorly known geologically, such as the Amazon River Basin of South America, the Arctic regions of Canada and the Soviet Union, and Antarctica. The oceans have enormous potential for resources. Their vast fields of manganese nodules (Chapter 16) may be mined, and new techniques may make it feasible to profitably extract several elements from seawater. New methods of satellite monitoring are being developed that may enable us to assess the resource potential in remote regions and thus minimize exploration costs. We already are using gravity and magnetic prospecting methods and seismic profiling to delineate potential oil and gas deposits.

Eventually, most of the shallow-crustal, high-grade deposits will be located and depleted. We then will have to turn to low-grade deposits, and others that are buried much deeper. Extraction costs will rise, and everything made from these resources will become much more expensive. Each new method for extracting resources, each new area in which mining is begun, and each new processing plant will create more of the environmental problems that were discussed earlier. We must balance our hunger for more resources with an appropriate commitment to preserving the quality of our air, water, and land.

In the long run, we must utilize both conservation and recycling to meet the challenge of the future. Recycling not only saves the costs of new extraction, but also the energy required to refine the metals. For some resources, such as aluminum, the saving in energy that recycling provides already is considerable. It is not inconceivable that today's landfills may become the mining sites of the future.

Only when we understand the geologic basis and limitations of resources can we plan for a future in which growing populations can be assured of adequate resources with minimal environmental damage. This must be our goal.

SUMMARY

1. Resources are those geologic materials that are both useful and available.
2. Reserves are the portion of our resources that can be extracted profitably.
3. Renewable resources are those that are formed faster than they are consumed. Nonrenewable resources are those that cannot be replaced geologically as fast as we are using them.
4. The resource potential of a nation is a function of its size, the complexity of its geologic terranes, and its tectonic history.
5. Basic resources for human survival include land, water, and air. Arable land—that on which crops can be grown—requires fertile soil and a sufficient water supply.
6. Water is used as both an instream and offstream resource. Most offstream uses consume a portion of the water, and the greatest consumption is for irrigation purposes. Thermal pollution during offstream use may decrease the dissolved oxygen content of water returned to streams, and chemical effluent may increase nutrient levels, increasing the biologic oxygen demand.
7. A mineral deposit is a naturally occurring suite of minerals of some economic value. Ore minerals are those that have been concentrated by geologic processes to create abundances that make extraction and refining profitable.
8. Mineral deposits result from igneous, sedimentary, and metamorphic processes. Hydrothermal deposits precipitate from hot fluids associated with magmatic processes. Magmatic differentiation by gravity settling and filter pressing also concentrates individual minerals to economically important abundances. Sedimentary deposits include banded iron ores of Precambrian age, and those produced by processes occurring now, such as placer deposits in stream beds and bauxite deposits formed by deep tropical weathering.
9. The fossil fuels are formed by burial and transformation of organic material during burial. Coal is produced from plants that originally lived in swamps; petroleum and natural gas are produced from marine microfossils. Oil and natural gas form in organic-rich beds and commonly migrate into the porous and permeable reservoir rocks from which we recover them.
10. Alternatives to fossil fuels as energy resources include nuclear, solar, hydroelectric, tidal, wind, and geothermal energy. Many are of only local benefit.
11. Extraction and refining of geologic resources generally creates environmental problems by interfering with natural cycles. These problems include collapse, subsidence, acid mine water contamination, acid rain, and the greenhouse effect.
12. Additional problems occur because of the need to dispose of solid and liquid wastes without polluting groundwater or air.
13. Future resource needs will be met by exploring less geologically known areas, looking deeper within the Earth, making use of lower-grade ores, conserving reserves we already have identified, recycling, and developing new extraction methods.

REVIEW QUESTIONS

1. What are the basic resources that have been required since humans have evolved? How has the nature of these resources changed with time?
2. What are the problems in estimating when a resource may be depleted?
3. What is the distinction between a renewable resource and a nonrenewable resource? Give examples of each.
4. What factors determine whether a country will have extensive and varied resources or only limited geologic resources?
5. How is it possible for a country to be large but unable to develop an agricultural industry?
6. What steps can be taken to limit sediment pollution?

7. How do instream and offstream uses of water differ? Give examples of each.
8. What is meant by the term "water consumption?" In which types of water use is consumption greatest?
9. Describe how thermal pollution occurs. During which time of the year would you expect thermal pollution to be most severe? Explain.
10. What conflict might you expect between the use of dams on the Colorado River for both flood control and the generation of hydroelectric power?
11. Define and give examples of:
    a. Mineral deposits
    b. Ore minerals.
12. Describe the factors that must be taken into account in deciding whether or not to develop a particular mineral deposit.
13. Describe as many ways as you can by which metallic mineral deposits form.
14. What factors determine how and where a placer deposit will form?
15. What conditions are required for the formation of a coal deposit? Why are there no Cambrian coal deposits?
16. What is the difference between petroleum source rocks and reservoir rocks? What rocks would make good source rock? Reservoir rock?
17. What is the difference between a structural trap and a stratigraphic trap for petroleum? Describe different types of each kind of trap.
18. Why do alternative resources have only limited use in solving the energy crisis? Explain using several examples.
19. What is the greenhouse effect? How can we minimize its causes and effects?
20. What are the problems associated with solid waste disposal? How can we help to resolve them?

22
Planetary Geology

The full Moon, showing light-colored, cratered highlands and darker, less-cratered maria. Note the ejecta rays extending from the large crater to the south. Photo courtesy of National Aeronautics and Space Administration.

The course of history changed forever on July 20, 1969, when *Apollo 11* Astronaut Neil Armstrong took his first giant step onto the surface of the Moon. The entire world was excited, but geologists were ecstatic because a whole new world was opened for us to explore. No longer would we have to speculate about the Moon's composition (no, it is not made of green cheese) or whether there is water in its "seas." Since *Apollo 11,* several astronauts have visited the Moon, including geologist Harrison Schmitt; unmanned probes have landed on Venus and Mars; and the *Voyager* spacecraft have visited the outer planets. In Chapter 1 we introduced the entire **solar system** family. In this chapter we will look first at the methods used to study other **planets** and then describe our neighbors in space.

INVESTIGATING THE SOLAR SYSTEM

We know a lot about other planets even though we cannot visit them. A planet's size, density, moment of inertia, and magnetic field provide important information about its composition and tectonic activity. Its topography is a record of tectonism and surface processes such as weathering and erosion. Meteorites found on Earth and lunar samples returned by astronauts tell us what rocks exist on other planets. We will briefly look at how we use these features to analyze the planets, focusing mostly on the terrestrial planets.

Planetary Size

A planet's size (Table 22–1) tells us about its evolution, because size controls the amount of internal heat available for magmatic and tectonic activity. For example, large terrestrial planets are hotter than small ones. A planet's heat comes from gravitational effects during its formation and later from radioactive decay; large terrestrial planets have more mass (and thus more gravitational attraction) and more radioactive material than small ones.

Large planets also tend to cool more slowly because they have less surface area *in proportion to their size* than small ones. As a result, they dissipate their heat into space at a slower rate. Large planets are

more active tectonically than small ones because they get hotter and stay hot longer. The two largest terrestrial bodies, Earth and Venus, are geologically active today and presumably have been active throughout their histories. The smallest bodies apparently never were able to generate the heat needed to drive internal activity.

Density of a Planet

We calculate a planet's density from its size and the orbits of its satellites, and use the density to estimate its composition. We know that Earth's average density of 5.5 gm/cm^3 results from a combination of three different kinds of material: silicate rocks (density of about 3.0 gm/cm^3), iron-nickel alloy (8 gm/cm^3), and ice or water (1 gm/cm^3). Earth's density results from a mix of two-thirds silicate rock (crust and mantle) and one-third iron-nickel alloy (core). Water and ice account for less than 1% of the planet and do not affect the calculations significantly.

In Table 22–1, the "Bulk Density" column lists densities of the planets and large satellites. Those with densities between 3 and 8 gm/cm^3 are a mixture of rock and metal. Mercury, Venus, and Mars have densities similar to Earth's and probably contain slightly different mixtures of metal and rock. A density between 1 and 3 gm/cm^3 suggests a combination of silicate rock and either ice or water. The Jovian planets are mostly frozen gases (no rock at all), and satellites such as Callisto are probably composed of a rock core, a mantle of water, and an icy crust!

A Planet's Moment of Inertia

The tendency of a spinning object to keep on spinning is called its **moment of inertia**. The way the object spins is determined by where its mass is concentrated, a principle revealed when watching ice skaters. When they pull their arms close to their sides, they spin rapidly because their mass is then concentrated near the centers of their bodies. They slow the spin by extending their arms outward, thereby distributing mass farther from the body's center. Planets do not have arms, but some do have mass concentrated in their cores. We use a number called the *moment of inertia factor* to describe the distribution of mass on a planet. A planet whose mass is evenly distributed—one without a core—has a mo-

TABLE 22–1
Characteristics of the terrestrial planets and large satellites.

Planet	Large Satellite	Radius, km	Mass, 10^{22} Kg	Bulk Density, g/cm^3	Moment of Inertia Factor	Global Magnetic Field	Geologic Activity	Atmospheric Pressure at Surface, Bars	Atmosphere Constituents Chief	Atmosphere Constituents Minor
Earth		6378	598	5.5	0.333	Dipole	Active	1.0	N_2, O_2	CO_2, H_2O, Ar
Venus		6051	487	5.3	?	Nil	Active?	93	CO_2	N_2, Ar, H_2O, SO_2, O_2
Mars		3397	64	3.9	0.365	Nil	Past activity	0.005 to 0.01	CO_2	N_2, Ar, O_2, H_2O
	Ganymede	2638	15	1.9	?	?	Past activity	Vacuum	—	—
	Titan	2560	14	1.9	?	?	?	1.5	N_2	CH_4, C_2, H_2, HCN, C_2H_4
	Callisto	2410	11	1.8	?	?	Never active	Vacuum	—	—
Mercury		2440	33	5.4	?	Dipole	Never active	Vacuum	—	—
	Io	1816	9	3.5	?	?	Active	Vacuum	—	—
	Moon	1738	7	3.3	0.392	Nil	Ancient activity	Vacuum	—	—
	Europa	1563	5	3.0	?	?	Surface activity	Vacuum	—	—

The column headed Geologic Activity is a general assessment of internal geologic activity interpreted from surface features.

ment of inertia factor of 0.400; a planet with most of its mass in a core has a value less than 0.400.

The moment of inertia factor thus lets us infer the presence and size of a planet's core, and hints at whether chemical differentiation has taken place as it has on Earth. Unfortunately, we know the moment of inertia factor for only a few bodies. The value for Earth is 0.333, confirming the presence of a substantial core. Venus is nearly identical. The value for Mars, 0.365, suggests that Mars has undergone some differentiation and has a modest-sized core smaller in proportion to the planet's size than Earth's. The Moon (0.392) has at most an extremely small core because its moment of inertia factor is very close to that of a homogeneous sphere.

A Planet's Magnetic Field

Of all the terrestrial planets, only Earth and Mercury have magnetic fields. We saw in Chapter 20 that Earth's magnetic field is probably due to complex motions in the liquid outer part of its iron-nickel core. The presence or absence of a magnetic field thus tells us something about the nature of a planet's core.

Mercury's strong magnetic field suggests that it contains electrically conductive molten material. We do not have moment-of-inertia-factor data for Mercury, but its magnetism implies that it has a core similar to Earth's. On the other hand, the Moon does not have a magnetic field because, as shown by its moment of inertia factor, it lacks a substantial core. Venus's moment of inertia factor requires a core similar to Earth's, but it lacks a magnetic field, probably because it rotates only once every 243 *days* (compared to once in 24 *hours* for Earth). This slow rotation means that Venus cuts the Sun's magnetic lines of force much more slowly than Earth does—too slowly to generate much internal current—and consequently Venus has no magnetic field. Although Mars rotates fast enough, and has a core, it lacks a magnetic field, possibly because its core is entirely solid.

Topography

When we find topographic features similar to those on Earth, we infer that similar processes have occurred. We group all landforms into two categories: **constructive features**—such as volcanoes, mountain ranges, and ocean basins—which are produced by tectonic forces, and **destructive features,** which result from erosion. The landforms can tell us whether plate tectonics is operating (or has operated) on other planets. Are there large areas of high and low elevation that could represent continents and oceans? Are there elongate mountain chains in the low areas that might be spreading centers, or arcuate chains of volcanoes that could be island arcs? We will answer these questions later for each of the terrestrial planets.

Cratering

Interplanetary rock fragments that collide with planets and satellites are called **meteorites**. Some meteorites are microscopic but others are massive bodies 100 km across. Meteorites strike with great force, explosively carving bowl-like pits called **impact craters**. Material excavated by such an impact is thrown outward from the crater and settles on the surface as an **ejecta** deposit (see the photograph which begins this chapter). These deposits extend far from the crater in a series of light-colored **rays** that make the crater look something like a sunburst.

Planets like Earth and Venus are protected from most meteorite impacts by thick atmospheres. Meteorites smaller than about 10 meters in diameter burn up due to friction, but some big ones do get through (Figure 22–1). Airless and waterless planets have no such protection, and are as pockmarked as the surface of the Moon.

FIGURE 22–1

Meteor Crater near Flagstaff, Arizona. Note the blocks in the foreground, ejected from the crater by the impact. The crater is approximately 1220 m in diameter and 170 m deep from the crest of the rim to the crater floor. Photo by John Shelton.

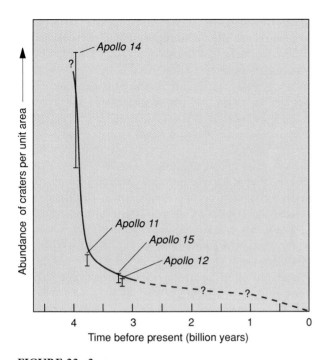

FIGURE 22–2

Crater density on the surface of the Moon as a function of age. There are no data to indicate crater density on surfaces older than 4.0 billion years. Differences in cratering at the different *Apollo* landing sites are indicated.

Crater Density—Age of a Planet's Surface. We use cratering to work out the sequence of events in the formation of a planet's surface. Young craters have distinct, often knife-sharp rims; as they age, their outlines become less distinct because of mass movements and partial burial by younger ejecta deposits. Overlapping crater walls (cross-cutting relationships) and superposition of an ejecta deposit on earlier ones enable us to determine the sequence of events.

The number of craters per unit area, called the **crater density,** is used to determine the relative ages of surfaces. Old surfaces have been exposed to more meteorite impacts than young ones, and so have a higher crater density. Radiometric dating of lunar rocks has pinpointed the ages of cratering events on the Moon (Figure 22–2). If the rate at which meteorites strike the Moon were constant through time, the curve in Figure 22–2 would be a straight line because a 3-billion-year-old surface would have three times the crater density of a 1-billion-year-old surface. The actual curve, however, shows a sharp de-

crease in crater density for surfaces younger than 4.0 billion years, indicating that many more meteorites struck the Moon early in its history than in recent times.

The 4-billion-year-old cratering probably marked the last stage in the formation of the planets and satellites—a sweeping of loose debris onto the larger bodies by gravitational attraction. If this hypothesis is correct, we are given a powerful tool for comparing events on different planets because the number of meteorites being swept up at any time should have been the same everywhere in the solar system. Each planet has a curve similar to that of Figure 22–2, and positions on this curve can help develop a relative time chart for the solar system in much the same way that index fossils are used on Earth.

A TOUR OF THE TERRESTRIAL PLANETS AND LARGE SATELLITES

Earth's Moon

The Moon is unique among satellites because it is so large compared to its planet: more than 25% the size of Earth. Its low density (3.3 gm/cm^3) indicates that it is composed almost entirely of silicate rock; if a core is present at all, it accounts for at most 5% of the Moon's total mass, compared with 33% for Earth's core.

Lunar Topography. The photograph that opens this chapter reveals two major types of terrain on the Moon. The dark areas are topographic lows called **maria** (singular, **mare,** the Latin word for *sea*) because early astronomers thought they were oceans. We know now that there is no water on the Moon and that the maria are enormous outpourings of basaltic lava. Light-colored areas in the photograph are mountainous areas called **highlands**. These contain basalts and gabbros, and igneous rocks called anorthosites that are nearly 90% plagioclase feldspar. As with ocean/continent ages on Earth, mare rocks are younger than those of the highlands. "Young" on the Moon, however, means something quite different from "young" on Earth. Highland rocks are 4.6 to 3.85 *billion* years old, older than nearly all dated rocks on Earth. The younger mare basalts are "only"

3.9 to 2.5 *billion* years old, 20 times older than our ocean basins.

Rilles. Steep-walled, linear depressions called **rilles** cut across the moonscape (Figure 22–3). Some, found in both maria and highlands, are nearly straight and are thought to be rift valleys or graben. Others, called sinuous rilles, look like meandering stream valleys and are restricted to the maria. Theories for their formation involving stream erosion were disproved by the total absence of water on the Moon's surface. We now think that the sinuous rilles are volcanic features, most probably enormous collapsed lava tunnels.

Craters. The lunar surface is dominated by impact craters (Figure 22–4). Indeed, mountain ranges on the Moon are circular, and are merely the walls of very large craters. Even the flat, apparently smooth plains of the maria contain craters, but craters in the highlands are much more abundant because of the difference in age of the two terrains. Highland craters are so numerous that they commonly overlap, the rim of one cutting across the rim of another. This permits us to unravel the sequence of lunar events.

The Surfaces of Earth and Moon. Earth's topography results from interaction between internal orogenic forces and surficial processes. The Moon is so small that it does not have enough internal heat to cause tectonism; as a result, major lunar landforms are caused by an *external* agent—meteorite impact.

On Earth, the atmosphere causes chemical weathering, and water, wind, and ice create our landforms. The Moon has none of these agents, yet its rocks are weathered and its surface exhibits a varied topography. Lunar weathering is largely caused by impacts with micrometeorites and by the **solar wind,** a

FIGURE 22–3
Rilles on the lunar surface in the eastern portion of Mare Imbrium. The sinuous Hadley Rille in the center meanders across the mare surface, passing close to the *Apollo 15* landing site. Straight rilles appear in the upper and left portions of the photograph. The mountains are the Apennine Mountains of the lunar highlands. Photo courtesy of National Aeronautics and Space Administration.

FIGURE 22–4

Closeup of the crater Euler on the Moon, showing features typical of young craters larger than 20 km in diameter. It has a flat floor, central peak complex, terraces, a continuous ejecta blanket, and more subtle ejecta rays. The floor and ejecta blanket of Euler are cut by younger, simpler craters with diameters up to 5 km. Photo courtesy of National Aeronautics and Space Administration.

stream of submicroscopic particles expelled from the Sun. The result is a type of physical weathering that affects the outermost layer of lunar rocks and forms a thin regolith.

Another major contrast is in the ages of the surfaces of Earth and Moon. Earth's landscape features are forming or being modified as you read these words. Even the present configuration of continents and oceans dates back only a few million years. The Moon's surface is ancient; even the "young" mare basalt plains are billions of years old.

Moonrocks. The *Apollo* astronauts collected 382 kg of rock and regolith samples, a treasure of information about the initial differentiation and subsequent evolution of the Moon. All the rocks collected from the maria are basalts, but there are chemical differences that clearly set them apart from any terrestrial basalt. Lunar basalts, like all of the Moon samples, contain far lower abundances than Earth rocks of the volatile elements—those with low melting points. For example, water, an extremely volatile compound, is totally absent from moonrocks. This means there can be no amphiboles or micas, limonite, or clay minerals. The mare basalts contain *greater*

amounts of iron, titanium, and nonvolatile elements than Earth basalts. Just as basalt mineralogy and chemistry indicate the depth of melting on Earth (see Figure 20–19), the composition of the mare lavas suggests that their magmas formed by partial melting of the Moon's mantle at depths between 150 and 400 km. This melting occurred billions of years ago—there is no recycling of basalt as there is in the plate-tectonic Earth.

Most samples from the lunar highlands are impact breccias (Figure 22–5) that formed by the lithification of material deposited during meteorite impacts. These rocks have fine-grained matrices that contain fragments of the lunar surface disrupted by the impacts. The impact debris is a mixture of rock and mineral fragments mixed with a solidified melt produced by the impact. The breccias are generally basaltic, but differ from the mare basalts by having more feldspar and less ferromagnesian minerals. Radiometric ages of the breccias cluster tightly, indicating that most impact events in the highlands took place between 3.9 and 4.05 billion years ago.

On Earth, the grains in a clastic rock tell us something about the source area from which the clasts were derived. The fragments in the impact breccias should provide an indication of the composition of

FIGURE 22–5

Impact breccia from the lunar
highlands. Note the mixture of fine-
grained matrix with rock and
mineral fragments. The clasts may
be lighter or darker than the matrix.
Photo courtesy of National
Aeronautics and Space
Administration.

the Moon's surface. One particular group of fragments is found in impact breccias collected from all over the Moon; these are pieces of plutonic rocks such as gabbro and anorthosite. Several of these rock fragments have been dated, and are among the oldest materials known—between 4.3 and 4.6 billion years old. They must have crystallized during the first few hundred million years of Moon history. The wide distribution of these rocks in the highlands suggests that the Moon, like the Earth, has a dominantly feldspathic crust. Earth's sialic crust, however, contains large amounts of potassium and sodium, whereas the Moon's crust is more simatic, largely made of calcium feldspars.

Moonquakes. Seismograph stations were set up on the Moon by the *Apollo* astronauts in an attempt to determine whether there is any current tectonic activity. They have recorded far fewer tremors than take place on Earth, and these "moonquakes" are concentrated at a depth of about 1000 km, much deeper than on Earth.

Evolution of the Moon. These data yield a preliminary picture of the evolution of our satellite. A feldspathic crust formed by igneous processes in the first few hundred million years of the Moon's history. This magmatism probably was caused by heat generated during the gravitational accretion of materials that formed the Moon early in the history of the solar system. The end of this first stage of lunar evolution was marked by the 4-billion-year cratering event in the highlands.

Brecciation of the highlands 4 billion years ago was the last significant geologic activity in that portion of the Moon. Surface activity nearly ceased, but there is evidence of continuing geologic activity at depth. The mare basalts are lava flows that erupted from depths of a few hundred kilometers between 3.9 and 3.2 billion years ago. These were eruptions of the flood basalt type (see Chapter 5), originating from large fissures rather than from volcanic cones and calderas. The Moon's surface has been inactive since eruption of the mare basalts, except for the formation of a thin veneer of regolith produced by impact.

Moonquakes show that there is still some internal activity at depths of about 1000 km, ten times deeper than the depth to the asthenophere on Earth. Earth's lithosphere is thin and easily broken into plates by seafloor spreading; the far thicker outer layer of the Moon (its "lithosphere") cannot be breached by forces that are weaker (a small planet has less radiogenic heat) and deeper seated. As a result, the surface of the Moon has not seen geologic activity for the past 3 billion years, and will not see any in the future.

Mercury

Mercury is a terrestrial planet about the same size as the Moon, but its density is much higher—nearly the same as Earth's. This suggests an internal structure similar to Earth's: a silicate outer layer surrounding an iron-nickel core. Mercury has a dipolar magnetic field that is tilted about 10° from its rotational axis. By analogy with Earth, the presence of the magnetic field suggests that at least part of Mercury's core is molten.

Mercury's surface is decidedly Moonlike (Figure 22–6), dominated by overlapping, shoulder-to-shoulder craters similar to the lunar highlands. Small, darker regions may be features like the lunar maria. The largest Mercurial craters are basins ringed by concentric circular ridges, and these seem to be the same as what are called multiringed basins on the Moon. Puzzling sinuous cliffs are found in some areas. These do not resemble rifts on Earth and are interpreted as the traces of large thrust faults.

Venus

Venus often is called Earth's twin; with almost exactly the same size and density, it is more like Earth than any other planet. Its atmosphere is very different, however. Venusian air is nearly 90 times denser than Earth's and is composed almost entirely of carbon dioxide (96%) with minor amounts of water vapor, nitrogen, and argon. Three layers of dense clouds completely surround the planet at heights between 48 and 68 km. This cloud cover creates a "greenhouse effect" that traps incoming solar radiation and keeps the Venusian surface at a sizzling 470°C. Even though there is water vapor in the Venusian atmosphere, the surface temperature is far too high for liquid water to exist. Thus, we do not expect to find evidence of activity by streams, oceans, groundwater, or glaciers.

Venusian Topography and Tectonics. The clouds prevent us from getting a clear view of the surface of Venus; as a result we know less about it than we do about the surface of any other terrestrial planet. However, radar can penetrate clouds, and starting in the late 1970s, radar telescopes were trained on Venus from Earth and from spacecraft that have orbited our twin—the American (*Pioneer Venus*) and Soviet (*Venera*).

FIGURE 22–6
The cratered surface of Mercury. This image was obtained by the *Mariner 10* spacecraft as it flew by the innermost planet in 1972. N and S indicate the geographic poles of Mercury. Photo courtesy of National Aeronautics and Space Administration.

Figure 22–7 is a map showing the major surface features of Venus. Most of the surface consists of rolling hills with a relief of about 1 km, but there are two continent-sized highland areas—Aphrodite Terrae and Ishtar Terrae—that rise nearly 4 km above the rolling terrain. The ground is different from that of a continent on Earth, consisting mostly of bedrock or areas covered by large blocks of rock (Figure 22–8) with only about 25% covered by regolith. There are some impact craters, but far fewer than on the Moon or Mercury—as we would expect, considering Venus's atmosphere.

Detailed studies of these "continents" by radar telescopy suggest that Venus may have experienced plate-tectonics deformation. Aphrodite Terrae, for example, is broken by a series of parallel linear troughs that are interpreted as rift zones. As on Earth, there are some volcanoes along these rifts. In many places the troughs are cut by linear fractures that are thousands of kilometers long and spaced a few hundred kilometers apart. These are strikingly similar to the patterns of Earth's oceanic fracture zones, and some geologists believe that Aphrodite Terrae exhibits signs of "Venus-crust spreading" complete with transform faults.

Mountains on Ishtar Terrae occur in long, linear belts and give a banded appearance common to Earth's fold mountains. Horizontal compressional forces may have acted there, as we would expect during plate collision. Other features have been interpreted as large strike-slip and thrust faults, and even as individual anticlines and synclines!

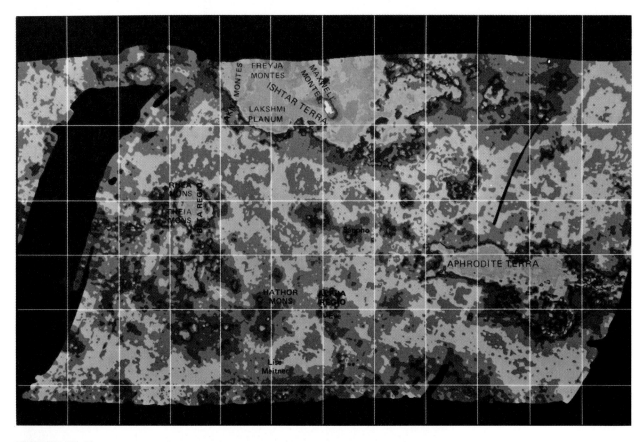

FIGURE 22–7
Map of the surface of Venus, showing its major physiographic features. The two large, light areas are the "continents" Ishtar Terrae and Aphrodite Terrae; they are darker regions of lower elevation. The small light area on the left is a volcanic cone. Photo courtesy of National Aeronautics and Space Administration.

FIGURE 22–8
Photograph of the surface of Venus taken by a Soviet robot lander. Note the large bedrock blocks and surrounding regolith. Photo courtesy of National Aeronautics and Space Administration.

There is plenty of evidence of volcanic activity on Venus, mostly in the form of small shield volcanoes. A few volcanoes, however, are enormous: Theia Mons stands 2.5 km above the surface and is over 300 km in diameter, with a summit caldera nearly 90 km across. Rhea Mons is just a bit smaller at 250 km in diameter. Both have lava flows up to 75 km wide radiating from their calderas, and extending for several hundred kilometers downslope. One of the Soviet robot landers radioed back radioactivity data suggesting that basalt makes up much of the Venusian shield volcanoes, just as it does on Earth. We have not identified large stratovolcanoes or any ex-

plosive pyroclastic volcanic activity, perhaps because this kind of eruption is prevented by the extremely high atmospheric pressure.

Mars

Mars is intermediate between Earth and the Moon in many aspects of interior and surface processes. Its density and moment of inertia factor suggest a planet with a rocky mantle and an iron-nickel core that is proportionally smaller than Earth's but larger than the Moon's. Some regions of the Martian surface are heavily cratered, suggesting a planet with little surface activity, but Mars has an atmosphere and glaciers and there is evidence of volcanism and tectonism in some areas.

The Martian atmosphere, like that of Venus, is composed mostly of carbon dioxide (95%), with minor amounts of oxygen, nitrogen, argon, and water vapor. Unlike Venusian air, it is only about 1% as dense as Earth's. This low-density atmosphere cannot cause a greenhouse effect, and the surface of Mars has frigid temperatures ranging from 0° to −130°C. Today, Mars is a cold, arid desert, but perhaps this was not always the case. Features such as dendritic valley patterns, outflow channels, and streamlined midchannel islands suggest that there were once rivers of liquid water (Figure 22–9). The

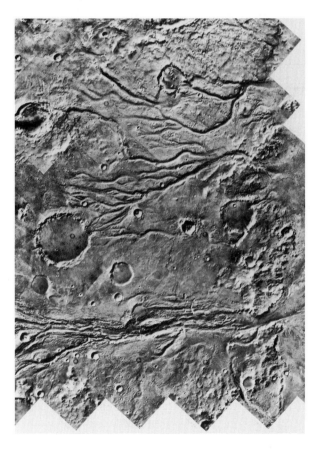

FIGURE 22–9
Possible stream channels on the surface of Mars. This photomosaic of the Maja Vallis region shows one of the best-developed dendritic channel systems on Mars, and suggests the presence of water on the Martian surface in the past. Photo courtesy of National Aeronautics and Space Administration.

channels are cut by meteorite impact craters, showing that it has been hundreds of millions of years since water last flowed through them.

The Martian surface consists largely of highlands and plains similar to the highlands and maria of the Moon. Like their lunar counterparts, the greater density of craters in the highlands indicates that they are older than the plains, and the low-lying areas are thought to be made of basaltic lavas. Cratering is more abundant even on the plains than on Earth, indicating that the Martian landscape is much older than ours, and suggesting less surface activity than on Earth.

Volcanic cones have been built on both highlands and plains. In the Tharsis region in particular (centered around 30°N Martian latitude, 90°W Martian longitude), many volcanoes have merged into an uplifted volcanic plateau. Impact craters are very rare in the Tharsis region, indicating that these volcanoes are among the youngest geologic features on the planet. Mars, like Venus, has some very large volcanoes; Olympus Mons, the largest, would extend from Boston to Washington, DC if transported to Earth. Some of the smaller volcanoes appear to be arranged in chains reminiscent of hotspot tracks like the Hawaiian islands, but most Martian tectonism seems to be dominated by vertical movements. Broad regions of what look like horst-and-graben structures are common, suggesting tensional forces. One graben, called Valles Marineris, is 4000 km long, up to 240 km wide and 6 km deep! The Grand Canyon is tiny compared to this "valley."

We have not detected activity in any Martian volcano, and current surface processes are limited to wind and glacial scouring. These processes are somewhat different from those on Earth. Vast regions of Mars are covered with dune fields, and every spring a dust storm forms in the northern hemisphere and spreads southward over almost the entire planet. The storm lasts for months. The north and south poles of Mars are covered by ice caps, but these contain both water ice and "dry ice"—frozen carbon dioxide. Each winter, a seasonal polar ice cap composed of CO_2 frost forms and expands from the poles to latitudes as low as 50°. There is no evidence of plastic ice flow in either ice cap, indicating that they are very different from those of Antarctica and Greenland.

The critical problem in Martian surface geology is what happened to the water that is widely inferred to have carved the channels. Sedimentary features similar to levees have been found in some areas of the highlands, and these have been suggested to ring large lakes. The levees are breached in a few places, perhaps as a lake burst through its confining walls. The fact that these features are limited to the highlands implies that they are very old. Some geologists think that the channels were not carved by streams, but rather by muddy debris flows—sort of a continental turbidity current. Others have argued that some of the water may be trapped today as permafrost beneath the Martian surface.

Data from the *Viking* lander suggests that the Martian surface is rock-strewn like that of Venus, but has more regolith. Basalt is abundant, but its iron has been oxidized to hematite, producing the planet's red color. To the dismay of science fiction fans, attempts to identify life on the surface of Mars have yielded negative results. As far as we know, there are no Martians.

The Asteroids

We have not yet visited the rocky fragments of the **asteroid belt,** yet we know more about them than we do of any extraterrestrial rock other than those collected on the Moon. Like the planets, asteroids orbit the Sun; most do so between the orbits of Mars and Jupiter. Collisions among asteroids sometimes knock them out of their orbits so that they intersect the orbit of the Earth. Those that land on Earth are called meteorites. Attempts to backtrack individual meteorites and meteorite showers to their source regions have invariably led to the asteroid belt, but there is evidence that some meteorites were derived from the *surfaces of other planets* (see Focus 22–1). More than 3000 meteorites now have been found on the Earth; some are like Earth or Moon rocks, but many are unlike anything we have seen before.

Types of Meteorites. The most common meteorites are the **chondrites,** stony objects that contain small rounded grains called **chondrules.** Chondrules are small spheres composed of olivine or pyroxene, with lesser amounts of plagioclase feldspar and iron-nickel metal. Some chondrites contain substantial amounts of carbon and graphite, and a complex mixture similar to organic compounds on Earth. These **carbon-aceous chondrites** also are unique in another way: they contain water which has reacted with the original minerals to form clays similar to those formed by weathering on Earth.

Chondrites have been radiometrically dated at about 4.6 billion years, indicating that they are samples of material that existed during the earliest days of the solar system. Furthermore, their chemical composition is similar to that of the Sun, except that they have much lower concentrations of the gaseous elements. This suggests that the chondrites may not have experienced any heating, partial melting, or significant chemical differentiation. It is possible that they are pieces of the primitive matter that originally accreted to form the planets.

Achondritic meteorites are stony objects that do not contain chondrules. They are similar to mafic and ultramafic igneous rocks such as gabbro and peridotite, and their chemical compositions are quite different from average solar abundances of the elements. They obviously have undergone processes of igneous differentiation that the more primitive chondrites were not subjected to. **Iron meteorites** consist of a coarse-grained iron-nickel metallic alloy with minor amounts of iron sulfide. **Stony-iron meteorites** consist of an intergrowth of iron-nickel metal with silicate minerals, or metal mixed with bits of achondritic meteorite. Achondrites, iron meteorites, and stony-irons are billions of years old, but are younger than the chondrites and thus date from a later stage in solar system evolution.

Taken together, the three nonchondritic meteorite types provide a sampling of what we expect from the differentiation of a large terrestrial planet, and the chondrites represent the basic undifferentiated ingredients. As a result, we have used the meteorites to model what Earth's interior might be like.

The Galilean Satellites of Jupiter

Jupiter has 16 satellites, several of which have densities suggesting similarity to the terrestrial planets. The four largest satellites — Io, Europa, Ganymede, and Callisto—were discovered by Galileo in 1610 and have been named **Galilean satellites** in his honor. Although they can be seen with a telescope, they are so far from Earth that we knew very little about them until the *Voyager* spacecraft passed

THE MARTIANS ARE COMING, THE MARTIANS ARE COMING!

Well, perhaps not live ones; and perhaps they have been here for a long time; and maybe they are not from Mars. . .but we certainly got your attention! No, this is not a *War of the Worlds* warning from Orson Welles, but rather one of the most exciting advances in planetary geology that has come *from studying rocks found on Earth*. Geologists are certain that the vast majority of meteorites come from the asteroid belt, but evidence is mounting that a few rare meteorites, only eight of the more than 3000 known, actually have come from Mars.

The eight meteorites in question are achondrites that belong to three classes: shergottites, nakhlites, and chassignites, called SNCs (snicks). They have puzzled geologists for years because they are so similar to terrestrial basalts, and so unlike anything we would expect from asteroid-sized bodies because small asteroids could not have produced as differentiated a material as basalt. The SNCs are unique among meteorites in yielding relatively young radiometric ages (1.3 billion years instead of 4.5 billion), and in having oxygen isotope compositions different from anything we have sampled from the Earth, the Moon, or other meteorites.

A solution to the puzzle began to appear when the *Viking* landers measured concentrations of gases in the Martian atmosphere that proved to be nearly identical to gases trapped in one of the shergottites. Finally we had something comparable to the SNCs, but how could a piece of Mars find its way to Earth? The best explanation seems to be that a large-scale meteorite impact with Mars provided enough energy to blast the SNCs off the planet's surface with the velocity needed to escape from Mars entirely. Estimates of the size of the impacting meteorite, the energy needed to launch the SNCs, and the time at which all this happened are still in progress, but the likelihood of a Martian origin has become stronger in the last few years. Studies of moonrocks have demonstrated that other achondrites probably came from the Moon rather than from the asteroid belt, so a Martian ancestry for the SNCs is plausible.

through the Jovian system in 1979. *Voyager* images of these satellites (Figure 22–10) reveal that each is a very different world with its own geologic features and history.

Ganymede and Callisto. The fourth and fifth satellites of Jupiter are about the same size as Mercury but have densities less than 2 gm/cm^3. This suggests that they are made of ice and rock, probably with a

FIGURE 22–10
Composite image of the four Galilean satellites of Jupiter, obtained by the *Voyager* spacecraft in 1979. The satellites Io (upper left), Europa (upper right), Ganymede (lower left), and Callisto (lower right) are shown in correct relative size, color, and brightness. Each satellite has different surface features and a different history. Photos courtesy of Jet Propulsion Laboratory.

A TOUR OF THE TERRESTRIAL PLANETS AND LARGE SATELLITES | 621 silicate core and a mantle and crust made of ice and/or water. Surface relief on both is small, on the order of 1 km, perhaps because an ice/water structure is not strong enough to support larger features. All of Callisto and much of Ganymede exhibit a dark surface, suggesting that a thin layer of rock dust or debris covers the bright, reflective ice.

Callisto's surface is covered uniformly by overlapping and shoulder-to-shoulder craters that are even more abundant than in the lunar highlands, suggesting great age. Callisto also has several huge circular features that consist of about 20 low, concentric ridges arranged in a bullseye pattern. These features have very few craters in their centers and seem to be the icy planet's version of multiringed basins found on Mercury and the Moon.

In contrast, Ganymede has two types of surface terrain: cratered and furrowed (Figure 22–11). Cratered regions are similar to those on Callisto, but the furrowed areas are lighter colored and consist of sets of furrows 5 to 10 km across and about 100 m deep. The furrowed terrains appear to separate heavily cratered regions. In places, the belt of furrows widens abruptly in a pattern reminiscent of midocean ridges and transform faults. The furrows are thought to be spreading centers in which water rises from the mantle of Ganymede into rift zones and then freezes to form the furrowed areas. Are plates still moving on Ganymede? The furrowed terrain is itself cratered with about the same density as the lunar maria, indicating an age of billions of years. This suggests that, if plate tectonics operated on Ganymede, the activity stopped long ago.

Europa. The third satellite of Jupiter is about the size of our Moon. Its density indicates a silicate rock interior, but it also has a smooth, highly reflective surface probably made of ice. Estimates based on size and density suggest an ice thickness of about 100 km. Europa has very few craters, fewer even than Earth, yet it does not have a protective atmosphere. This implies a very active surface on which the ice is renewed regularly and rapidly to eradicate impact features.

Io. The second satellite of Jupiter, Io is one of the most interesting bodies in the solar system. In size and density it is very much like our Moon, but in other ways it is very, very different. Io is brightly colored, in tones of red, orange, white, and brown, probably because of extensive deposits rich in sulfur. Io has no impact craters but instead displays a surface marked with hundreds of volcanic calderas and lava flows. We are no longer surprised to find extinct volcanoes on other planets, but those on Io are still active. Cameras on *Voyager* detected eight active volcanoes on Io, each ejecting gas and dust to heights as much as 250 km above the surface (Figure 22–12). This volcanism is unique, for it is not caused by radiogenic heat. Instead, enough *frictional* heat is generated from tidal effects caused by Jupiter's enormous mass to cause melting.

THE JOVIAN PLANETS

Jupiter, Saturn, Uranus, and Neptune are far larger than the terrestrial planets, but much lower in density. They have atmospheres very different from the

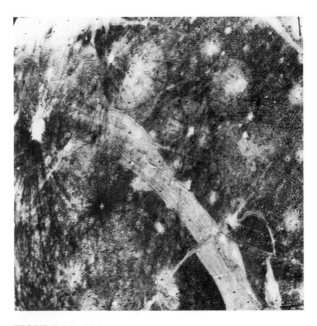

FIGURE 22–11
Closeup of the surface of Ganymede obtained by *Voyager I* in 1979. This view shows a region 1200 km across with two types of terrain—dark cratered areas and light furrowed areas. Note the transverse fault cutting the furrowed terrain in a manner similar to the way a transform fault cuts an oceanic ridge on Earth. Photo courtesy of National Aeronautics and Space Administration.

terrestrial planets, consisting mostly of hydrogen with 10% helium and smaller amounts of methane and ammonia. Their interiors also are different from the terrestrial planets, apparently containing "mantles" and "cores" of liquid hydrogen and helium. The surfaces of the Jovian planets are covered by layers of dense clouds, and the very rapid rotation rates of these giants (10 hours for Jupiter) cause beautiful, complex cloud patterns (Figure 22–13). Similar patterns now are known from Saturn, Uranus, and Neptune.

Another major difference between Jovian and terrestrial planets is rings: Jovian planets have them; terrestrial planets do not. Saturn's rings (Figure 22–14) have been known for hundreds of years because they are light colored and reflect light brilliantly, but it was not until the 1979 and 1989 *Voyager* trips that we discovered rings around the other Jovian planets. Saturn's rings probably are ice; those

of Jupiter probably are fine-grained rock dust. *Voyager II* revealed that Uranus and Neptune also have ring systems.

PLUTO

Pluto is the planet farthest from the Sun and was the last to be discovered. Its small size and great distance make it the most enigmatic planet; only in the last few years have we begun to strip away some of the mystery. Pluto's orbit is unique because it is more elliptical than that of the other planets, the orbital plane is different, and, at its closest approach to the Sun, Pluto is actually inside the orbit of Neptune. As a result, most geologists believe that Pluto is an orphan captured by the Sun's gravitational field, and that it did not evolve along with its sister planets.

FIGURE 22–12
Volcanic eruption on Io, one of the satellites of Jupiter. This photograph was obtained by the *Voyager* spacecraft as it passed through the Jovian system, and shows a plume extending 280 km above the surface of Io. Photo courtesy of National Aeronautics and Space Administration.

Pluto is the smallest planet, even smaller than the Galilean satellites of Jupiter. Its thin atmosphere is mostly methane and it has polar icecaps of frozen methane. Preliminary evidence suggests that the icecaps expand and contract throughout the Plutonian year (but keep in mind that this "year" corresponds to nearly 250 Earth years, and we have been observing Pluto only since its discovery in 1930).

We have discovered a satellite orbiting Pluto and named it Charon after the ferryman over the River Styx in Greek mythology. Infrared studies of Pluto and Charon in 1986 and 1987 have greatly refined our ideas about these bodies. For the first time, we can attempt density estimates, although the best so far indicates an average density for the Pluton-Charon system: 1.84 gm/cm^3. This suggests that, if the two bodies are of similar composition, each is made of nearly equal amounts of silicate rock and ice. Although ice on Pluto is mostly frozen methane, geologists were startled to discover that Charon's ice cover fits the properties of frozen water—a puzzle that waits to be solved.

FIGURE 22–13
Jupiter, showing latitudinal bands of clouds and swirling activity in the dense atmosphere. The large round object in the lower central part of the planet is the Great Red Spot, an atmospheric storm that has been observed from Earth for the past 300 years. Photo courtesy of Jet Propulsion Laboratory.

FIGURE 22–14
Mosaic image of Saturn obtained by the *Voyager* spacecraft in 1980. The soft, velvety appearance is due to a layer of haze around the planet, but latitudinal cloud cover is clearly visible. The width of the rings at the center of the disk is estimated to be about 10,000 km. Photo courtesy of Jet Propulsion Laboratory.

WHERE DO WE GO FROM HERE?

Twenty years ago all of this would have been science fiction. Today it is science. Geology has been exposed to two incredible revolutions in the second half of this century: the plate tectonics revolution that has caused us to completely rethink the way in which the Earth works, and the application of geologic principles to the rest of the solar system in an attempt to find out how the other planets work. It is an exciting time for geology; we hope you have enjoyed this introduction to it, and that some of you will make the next geological revolution happen.

SUMMARY

1. The objects that orbit the Sun in our solar system consist of terrestrial and Jovian planets, their satellites, and asteroids.
2. The internal structure, composition, and tectonic activity of a planet can be studied by examining its size (as an indicator of the amount of internal heat available for orogenesis), moment of inertia factor (whether it has a core or not), density (is it made of rock, metal, ice, or a combination?), and magnetic field (is there a liquid metallic core?).
3. The topography of a planet indicates whether it has active surface processes and tectonically produced

landforms, or whether it is generally inactive except for meteorite impacts.

4. Planets with dense atmospheres have few impact craters, but those with little or no atmospheric protection usually are heavily cratered. Most cratering probably occurred early in solar system history, as particles of varying sizes were swept up by already accreted planets. Curves showing crater density versus time therefore can indicate the age of a planet's surface.

5. The Moon is a heavily cratered body with two strikingly different kinds of surface features: *highlands* (uplifted, heavily cratered remains of the rims of large ancient craters) and *maria* — (large crater basins filled with extensive basalt flows). Both features exceed 2.5 billion years in age, indicating a lack of surface processes such as erosion and weathering.

6. Rocks collected on the Moon are basalts, gabbros, anorthosites, and complex impact breccias. There is no water on the Moon, so there can be no amphiboles or micas comparable to those on Earth, and no clay minerals or other hydroxides.

7. Mercury is a terrestrial planet the size of the Moon, and like the Moon it shows no signs of internal tectonism. Its surface is heavily cratered but shows no evidence of volcanic landforms.

8. Venus is in many ways Earth's twin, particularly in size. It has a dense, carbon dioxide atmosphere that shields the surface from meteorites and elevates the surface temperature to over 400°C.

9. Venus appears to have two "continents" that stand thousands of meters above lower-lying terrain. One of the continents is broken by a series of rift valleys, whereas the other exhibits fold-mountain topography. These may indicate crustal spreading and plate collision, respectively. Internal heat also is indicated by several large volcanoes.

10. In contrast to Venus, Mars has a very thin atmosphere and surface temperatures lower than those on Earth. Some areas of the Martian surface are heavily cratered, but others show signs of large-scale volcanism.

11. Surface processes on Mars are dominated by seasonal expansion and contraction of polar ice caps, and by intense windstorms that produce large-scale sand dune fields. Dendritic channels appear to be the result of running water, but there is no sign of water today and the channels may date from the earliest days of Martian history.

12. Most meteorites that land on Earth come from the asteroid belt and tell us something about the formation of terrestrial planets. Chondrites are probably assemblages of very primitive solar system material, per-haps even samples of the matter of which the planets were constructed. Achondrites are less primitive, and indicate planetary differentiation. Iron meteorites are metallic, and suggest enough differentiation to form a planetary core.

13. The Jovian planets are uniformly much larger than any of the terrestrial planets (all of the latter easily could be encompassed by Jupiter alone), and are made of frozen gases rather than rock and metal. They have numerous satellites, and are associated with ring systems of varying size and composition. The well-known rings of Saturn are probably made of ice, whereas those of Jupiter, Uranus, and Neptune seem to be fine-grained dust particles.

14. Pluto, the farthest planet from the Sun, remains something of a puzzle. It is probably made of a silicate rock core and a crust of frozen methane.

REVIEW QUESTIONS

1. What are the major differences between the inner (terrestrial) and outer (Jovian) planets? How is Pluto somewhat unique in this classification scheme?

2. How can the state of tectonic activity or inactivity of a planet be determined?

3. What can be told about a planet's structure and composition from its density? Its moment of inertia factor? Its magnetic field?

4. Critically discuss the evidence that plate tectonics processes may be active on other objects in the solar system.

5. Why is the volcanic activity on Io so different from that on Earth?

6. It is much easier for meteorites to strike the surface of the Moon or Mars than that of Earth or Venus, yet surface features of the Moon and Mars are older and better preserved than those on Earth and Venus. Explain how this is possible.

7. What role do meteorites play in our understanding of the history and structure of the Earth?

8. Contrast and compare the geologic evolution of the terrestrial planets, and explain the differences.

9. Imagine that you are a Martian geologist who views Earth through a telescope. Using a uniquely Martian version of uniformitarianism, what would you be able to deduce about the nature of the Earth?

10. Which processes of weathering and erosion would you expect to find on Venus? Which would not occur?

Appendixes

TABLE A–1
Periodic table of the elements.

Legend:

22	← Atomic number
Ti	← Chemical symbol
Titanium	← Element

= Most important elements in solid earth materials

1 H Hydrogen																	2 He Helium
3 Li Lithium	4 Be Beryllium											5 B Boron	6 C Carbon	7 N Nitrogen	8 O Oxygen	9 F Fluorine	10 Ne Neon
11 Na Sodium	12 Mg Magnesium											13 Al Aluminum	14 Si Silicon	15 P Phosphorus	16 S Sulfur	17 Cl Chlorine	18 Ar Argon
19 K Potassium	20 Ca Calcium	21 Sc Scandium	22 Ti Titanium	23 V Vanadium	24 Cr Chromium	25 Mn Manganese	26 Fe Iron	27 Co Cobalt	28 Ni Nickel	29 Cu Copper	30 Zn Zinc	31 Ga Gallium	32 Ge Germanium	33 As Arsenic	34 Se Selenium	35 Br Bromine	36 Kr Krypton
37 Rb Rubidium	38 Sr Strontium	39 Y Yttrium	40 Zr Zirconium	41 Nb Niobium	42 Mo Molybdenum	43* Tc Technetium	44 Ru Ruthenium	45 Rh Rhodium	46 Pd Palladium	47 Ag Silver	48 Cd Cadmium	49 In Indium	50 Sn Tin	51 Sb Antimony	52 Te Tellurium	53 I Iodine	54 Xe Xenon
55 Cs Cesium	56 Ba Barium	57 La Lanthanum †	72 Hf Hafnium	73 Ta Tantalum	74 W Tungsten	75 Re Rhenium	76 Os Osmium	77 Ir Iridium	78 Pt Platinum	79 Au Gold	80 Hg Mercury	81 Tl Thallium	82 Pb Lead	83 Bi Bismuth	84 Po Polonium	85* At Astatine	86 Rn Radon
87* Fr Francium	88 Ra Radium	89 Ac Actinium ‡	104* Rf Rutherfordium	105* Ha Hahnium	106*												

†Lanthanides

58 Ce Cerium	59 Pr Praseodymium	60 Nd Neodymium	61* Pm Promethium	62 Sm Samarium	63 Eu Europium	64 Gd Gadolinium	65 Tb Terbium	66 Dy Dysprosium	67 Ho Holmium	68 Er Erbium	69 Tm Thulium	70 Yb Ytterbium	71 Lu Lutetium

‡Actinides

90 Th Thorium	91 Pa Protactinium	92 U Uranium	93* Np Neptunium	94* Pu Plutonium	95* Am Americium	96* Cm Curium	97* Bk Berkelium	98* Cf Californium	99* Es Einsteinium	100* Fm Fermium	101* Md Mendelevium	102* No Nobelium	103* Lw Lawrencium

Elements with an asterisk (*) are artificial.

TABLE B–1
Conversion of metric and English units.

Metric–English			English–Metric		
To convert	to	Multiply by	To convert	to	Multiply by
Units of Length					
centimeters (cm)	inches (in)	0.3937	in	cm	2.54
meters (m)	feet (ft)	3.2808	ft	m	0.3048
m	yards (yd)	1.0936	yd	m	0.9144
kilometers (km)	miles (mi)	0.6214	mi	km	1.6093
Units of Area					
cm^2	in^2	0.1550	in^2	cm^2	6.452
m^2	ft^2	10.764	ft^2	m^2	0.0929
m^2	yd^2	1.196	yd^2	m^2	0.8361
km^2	mi^2	0.3861	mi^2	km^2	2.590
m^2	acres	2.471×10^{-4}	acres	yd^2	4.840
			acres	m^2	4047.
Units of Volume					
cm^3	in^3	0.0610	in^3	cm^3	16.3872
m^3	ft^3	35.314	ft^3	m^3	0.02832
m^3	yd^3	1.3079	yd^3	m^3	0.7646
liters (l)	U.S. quart	1.0567	U.S. quart	l	0.9463
l	U.S. gallon	0.2642	U.S. gallon	l	3.7853
Units of Mass					
grams (g)	Avoirdupois ounces	0.03527	Avoirdupois ounces	g	28.3495
g	Troy ounces	0.03215	Troy ounces	g	31.1042
kilograms (kg)	Avoirdupois pounds	2.2046	Avoirdupois pounds	kg	0.4536
Units of Density					
g/cm^3	lb/ft^3	62.4280	lb/ft^3	g/cm^3	0.01060
Units of Pressure					
kg/cm^2	lb/in^2	14.2233	lb/in^2	kg/cm^2	0.0703
bars	atmospheres	0.98692	atmospheres	bars	1.01325
kg/cm^2	atmospheres	0.95784			
kg/cm^2	bars	0.98067			
Units of Velocity					
km/h	mi/h	0.6214	mi/h	km/h	1.6093
km/h	cm/s	27.78	mi/h	in/s	17.60
Units of Temperature					
°C	°F	$(9/5)(°C) + 32$			
°F	°C	$(5/9)(°F - 32)$			

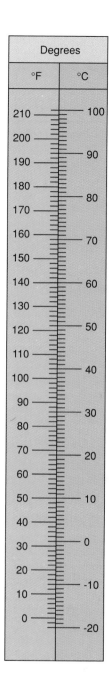

The identification of an unknown mineral follows a logical sequence of steps. First, the physical properties of the "unknown" are determined as completely as the specimen permits. Then, all minerals whose physical properties do *not* match those of the unknown are excluded. This is done by following the *decision tree* shown; at each branch, more minerals are eliminated as possibilities, and the field of suspects narrows. Finally, the unknown is compared closely with those few minerals which remain as valid candidates.

This appendix is divided into two parts. Table C–1 groups minerals into categories that match the sequence of branches on the decision tree: first by luster, and then by hardness, streak, and cleavage, so that these easily determined properties can be used for identification. Table C–2 is an alphabetical list of the more common rock-forming, accessory, and ore minerals which can be used for the detailed comparison once the possibilities have been narrowed. Table C–2 also indicates the type of process (igneous, metamorphic, sedimentary) by which the minerals form.

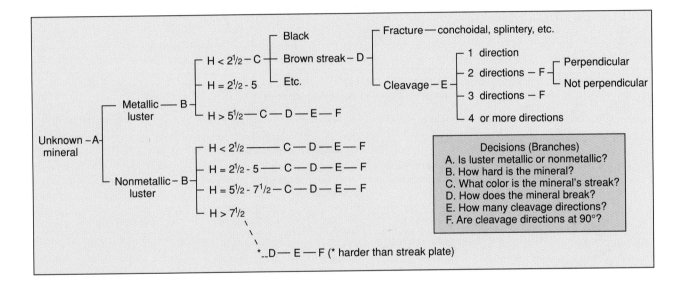

Decisions (Branches)
A. Is luster metallic or nonmetallic?
B. How hard is the mineral?
C. What color is the mineral's streak?
D. How does the mineral break?
E. How many cleavage directions?
F. Are cleavage directions at 90°?

TABLE C–1
Determinative table.

I—Minerals with a Metallic or Submetallic Luster

(a) Hardness less than 2½ (can be scratched by a fingernail)

Streak:	Black	Gray	Red	Red-Brown	Yellow-brown
	Graphite (1)	Galena (3⊥)	Cinnabar	Hematite	Limonite
	Molybdenite (1)				Goethite
	Stibnite (1)				

(b) Hardness between 2½ and 5½ (cannot be scratched by a fingernail; can be scratched by a knife)

Streak:	Black	Coppery red	Yellow
	Bornite	Copper	Sphalerite (6)
	Pyrrhotite		Gold
	Chalcopyrite		
	Chromite		

(c) Hardness greater than 5½ (cannot be scratched by a knife)

Streak black:	Pyrite	Arsenopyrite	Ilmenite
	Magnetite	Marcasite	

II—Minerals with a Nonmetallic Luster

(a) Hardness less than 2½ (can be scratched by a fingernail)

Streak:	Orange	Yellow	White or colorless	
	Realgar	Sulfur	Muscovite (1)	Bauxite
		Orpiment	Chlorite (1)	Sylvite (3⊥)
			Talc (1)	Serpentine
			Gypsum (1)	Kaolinite

(b) Hardness between 2½ and 5½ (cannot be scratched by a fingernail; can be scratched by a knife)

Streak:	Red, red-brown	Yellow-brown	White or colorless		Green	Blue
	Cinnabar	Limonite	Halite (3⊥)	Calcite (\3)	Malachite	Azurite
	Hematite	Siderite (\3)	Barite (1)	Dolomite (\3)		
		Sphalerite (6)	Anhydrite	Serpentine		
			Aragonite	Kyanite (2)		
			Fluorite (4)	Rhodochrosite (\3)		
			Sphene	Apatite (1)		

(c) Hardness between 5½ and 7½ (cannot be scratched by knife; leaves streak on streak plate)

Streak:	Brownish-black	Colorless			
	Rutile	Sillimanite	Kyanite	Andalusite	Quartz
		Cordierite	Garnet	Turquoise	Potassic feldspar (2⊥)
		Amphiboles (\2)	Pyroxenes (2⊥)	Olivine	Plagioclase feldspar (2⊥)
		Tourmaline	Cassiterite	Rutile	

(d) Hardness greater than 7½ (harder than streak plate)

Streak:	(None)		
	Topaz	Diamond (4)	Corundum
	Beryl	Zircon	Spinel

Symbols in parentheses indicate direction of cleavage and number of cleavage directions: ⊥ = cleavage at 90°   _ = cleavage not at 90°
E.g., 2⊥, \3.

TABLE C–2
Alphabetical listing of rock-forming and ore minerals.

Mineral	Composition	H	Specific Gravity	Color	Cleavage	Comments	Occurrence*	Uses
Actinolite	$Ca_2(Mg,Fe)_5(Si_8O_{22})(OH)_2$	5–6	3.0–3.2	Green	2/56,124°	An amphibole	M	Fibrous variety as asbestos; variety of jade
Aegirine	$NaFeSi_2O_6$	6	3.4–3.55	Green	2/87,93°	A pyroxene	I	
Alabaster	$CaSO_2 \cdot 2H_2O$	2	2.32	Varied	–	Massive gypsum	Evaporite	Sculpture
Albite	$NaAlSi_3O_8(An_0)$	6	2.62	Gray-white	2/93,87°	A plagioclase; striations	I, M, S	
Almandine	$Fe_3Al_2(SiO_4)_3$	7	4.25	Red	Conchoidal fracture	A garnet	M (i)	Abrasive, jewelry
Amethyst	SiO_2	7	2.65	Purple	Conchoidal fracture	Variety of quartz	I	Jewelry
Analcime	$Na(AlSi_2O_6) \cdot H_2O$	5	2.27	White		Crystals like garnet; a zeolite	I, M	
Andalusite	Al_2SiO_5	7.5	3.16–3.20	Gray, Pink		Elongate prisms	M	Spark plugs, ceramics
Andesine	$An_{30}–An_{50}$	6	2.69	Gray	2/87,93°	A plagioclase; striations	I, M	
Andradite	$Ca_3Fe_2(SiO_4)_3$	7	3.75	Brown, Yellow	Conchoidal fracture	A garnet	M	Abrasives, jewelry
Anorthite	$CaAl_2Si_2O_8(An_{100})$	6	2.76	Gray	2/93,87°	A plagioclase; striations	I	
Apatite	$Ca_5(F,Cl,OH)(PO_4)_3$	5	3.15–3.2	Green, Brown	3 (poor)	Hexagonal prisms	I, M, S	Source of phosphate fertilizer
Aquamarine	$Be_3Al_2Si_6O_{18}$	8	2.77	Green		Variety of beryl	Pegmatites	Jewelry
Aragonite	$CaCO_3$	3.5–4	2.95	White, Yellow		Polymorph of $CaCO_3$	M, S	
Arsenopyrite	$FeAsS$	5.5–6	6.05	Silver		Black streak	Hydrothermal	Ore of arsenic
Augite	$Ca(Mg,Fe,Al)(Al,Si)_2O_6$	5–6	3.2–3.4	Green	2/93,87°	A pyroxene	I (m)	
Azurite	$Cu_3(CO_3)_2(OH)_2$	3.5–4.0	3.77	Blue		Fine-grained crusts; reacts with HCl	Alteration of copper ores	Minor ore of copper
Barite	$BaSO_4$	3.0–3.5	4.5	Clear, White	1 good 2 poor	Unusually high specific gravity for nonmetallic mineral	S; hydrothermal	Source of barium; drilling mud additive
Bauxite	Mixture of clays	–	2.0–2.5	Gray, Brown		In small spheres	Weathering	Ore of aluminum
Beryl	$Be_3Al_2Si_6O_{18}$	8	2.77	Varied		Hexagonal crystals	Pegmatites	Source of beryllium; gemstone

*I, M, S: Common in igneous (I), metamorphic (M), or sedimentary (S) rocks; i, m, s: Rare in igneous (i), metamorphic (m), or sedimentary (s) rocks. An_x: Plagioclase feldspars often are described by the amount of their anorthite end member. For example, $An_{30} = 30\%$ anorthite and 70% albite.

TABLE C-2
continued

Mineral	Composition	H	Specific Gravity	Color	Cleavage	Comments	Occurrence*	Uses
Biotite	$K(Mg,Fe)_3(AlSi_3O_{10})(OH)_2$	2.5–3.0	2.8–3.2	Green, Brown	1	A mica	M,I	
Borax	$Na_2B_4O_7 \cdot 10H_2O$	2.0–2.5	1.7	White	1	Bitter taste	Evaporite	Antiseptic, cleanser
Bornite	Cu_5FeS_4	3	5.07	Bronze		Metallic luster; colorful tarnish	Hydrothermal	Copper ore
Bytownite	An_{70}—An_{90}	6	2.72–2.75	Gray	2/93,87°	A plagioclase; striations	I	
Calcite	$CaCO_3$	3	2.72	Varied	3/not 90°	Chief mineral in limestone, marble	S,M (i)	Cement, fertilizer
Carnotite	$K_2(UO_2)_2(VO_4)_2 \cdot nH_2O$	—	—	Yellow		Yellow powder	Alteration of uranium ores	Major source of uranium
Cassiterite	SnO_2	6–7	6.8–7.0	Dark brown		White streak	Pegmatite	Ore of tin
Chalcopyrite	$CuFeS_2$	3.5–4.0	4.2	Brassy yellow		Bronze tarnish; black streak	Hydrothermal	Ore of copper
Chlorite	$(Fe,Mg,Al)_6(Al,Si)_4O_{10}(OH)_8$	2–2.5	2.6–2.9	Green	1 (perfect)	Similar to micas	M	
Chromite	$FeCr_2O_4$	5.5	4.6	Black		Dark brown streak; submetallic luster	I (ultramafic)	Only ore of chromium
Chrysocolla	$CuSiO_2 \cdot 2H_2O$	2–4	2.2	Blue-green	Conchoidal fracture		Hydrothermal	Minor ore of copper
Cinnabar	HgS	2.5	8.10	Crimson		Scarlet streak	Hydrothermal	Ore of mercury
Copper	Cu	2.5–3.0	8.9	Copper-red		Dendritic masses; malleable, ductile	I	Wire, jewelry
Cordierite	$Mg_2Al_3(AlSi_5O_{18})$	7–7.5	2.65	Blue-gray	1 (poor)	Vitreous luster	M	Minor gemstone
Corundum	Al_2O_3	9	4.02	Varied	Parting common	Hexagonal crystals	M, I	Abrasive, gemstone
Cuprite	Cu_2O	3.5–4	6.0	Red	4	Red-brown streak; adamantine luster	Altered copper minerals	Minor ore of copper
Diamond	C	10	3.5	Varied		Adamantine luster	I, M	Abrasives, jewelry
Diopside	$CaMg(Si_2O_6)$	5–6	3.2–3.3	Green	2/87,93°	May show parting	M	Minor gemstone
Dolomite	$CaMg(CO_3)_2$	3.5–4	2.85	White, Pink	3/not 90°	Powder reacts in cold HCl	S	Special cements, ornamental stone
Emerald	$Be_3Al_2Si_6O_{18}$	8	2.7–2.8	Green		Variety of beryl	Pegmatites	Gemstone
Enstatite	$MgSiO_3$	5.5	3.2–3.5	Brown, Green	2/87,93°	A pyroxene	I, M	
Epidote	$Ca_2Al_3O(SiO_4)(Si_2O_7)(OH)$	6–7	3.4	Apple green		Often in fine granular coatings	M	
Fayalite	Fe_2SiO_4	6.5	4.14		Conchoidal fracture	An olivine	I	
Fluorite	CaF_2	4	3.18	Varied	4	Cubic crystals	S; hydrothermal	Flux in steelmaking
Forsterite	Mg_2SiO_4	6.5	3.2	Green	Conchoidal fracture	An olivine	I, M	

Mineral	Composition	H	G	Color	Cleavage	Remarks	I, M, S*	Uses
Galena	PbS	2.5	7.5	Gray	3/90°	Metallic luster; cubic crystals		Major ore of lead
Garnet	$X_3^{2+}Y_2^{3+}(SiO_4)_3$ ($X = Ca$, Fe^{2+}, Mg^{2+}, Mn^{2+}) ($Y = Al^{3+}$, Fe^{3+}, Cr^{3+})	7	3.5–4.3	Varied	Conchoidal fracture	Mineral family	M (i)	Abrasives, jewelry
Glaucophane	$Na_2(Mg,Fe)_3Al_2(Si_8O_{22})(OH)_2$	6.5	3.1	Blue-gray	2/56,124°	An amphibole	M	
Goethite	$HFeO_2$	5–5.5	4.37	Earthy brown		Earthy luster; yellow-brown streak	Weathering	Ore of iron
Gold	Au	2.5–3	15–19.3	Yellow		Malleable	I	Jewelry, dentistry
Graphite	C	1–2	2.3	Black	1 (perfect)	Greasy feel	M	Lubricant, pencils, electrodes
Grossularite	$Ca_3Al_2(SiO_4)_3$	7	3.53	Green, Yellow, White, Red	Conchoidal fracture	A garnet	M	Abrasives, jewelry
Gypsum	$CaSO_4 \cdot 2H_2O$	2	2.32	Colorless, White	1		Evaporite	Construction material (paster, plasterboard)
Halite	$NaCl$	2.5	2.16	Colorless, White	3/90°	Cubic crystals	Evaporite	Nutrient, chemical industry
Hematite	Fe_2O_3	5.5–6.5	5.26	Red-brown, Black		Red-brown streak	I, M, S.	Major ore of iron
Hornblende	$Ca_2Na(Mg,Fe)_4(Al,Fe)_3(Si_8O_{22})(OH)_2$	5–6	3.25	Green	2/56,124°	A common amphibole	I, M	
Hypersthene	$(Mg,Fe)SiO_3$	5.6	3.4–3.5	Bronze	2/87.93°	Bronze pyroxene	I, M	
Ice	H_2O	1.5	0.92	Colorless, White		Hexagonal crystals	Snowfall	Snowballs, cooling
Idocrase	$Ca_{10}(Mg,Fe)_2Al_4(Si_2O_7)_2(SiO_4)_5(OH)_4$	6.5	3.40	Brown		Striated four-sided crystals	M	Minor gemstone
Ilmenite	$FeTiO_5$	5.5–6	4.7	Black		Black streak; may be magnetic	M, I	Ore of titanium
Iron	Fe	4.5	7.3–7.9			Magnetic; metallic luster; malleable	Meteorites	Manufacture of steel
Jadeite	$NaAlSi_2O_6$	6.5–7	3.4	Green	2/87,93°	A pyroxene	M	Jade
Kaolinite	$Al_2Si_2O_5(OH)_4$	2–2.5	2.62	White, Gray	1 (perfect)	A claylike sheet silicate	Alteration of feldspars	Brick, tile, ceramics
Kyanite	Al_2SiO_5	5 and 7	3.61	Blue-gray	2/90°	Directional hardness is diagnostic	M	Refractory porcelains (spark plugs)
Labradorite	An_{50}–An_{70}	6	2.71	Dark gray	2/87,93°	Play of colors	I, M	A plagioclase feldspar
Leucite	$KAlSi_2O_6$	5.5–6	2.45	Grayish tan		Garnetlike crystals; a feldspathoid	I	

*I, M, S: Common in igneous (I), metamorphic (M), or sedimentary (S) rocks; i, m, s: Rare in igneous (i), metamorphic (m), or sedimentary (s) rocks. An_x: Plagioclase feldspars often are described by the amount of their anorthite end member. For example, An_{30} = 30% anorthite and 70% albite.

TABLE C–2
continued

Mineral	Composition	H	Specific Gravity	Color	Cleavage	Comments	Occurrence*	Uses
Limonite	FeO(OH)·nH$_2$O	—	—	Yellow-brown		A mineraloid; amorphous	Alteration of ferromagnesians	Pigment (yellow ochre)
Magnetite	Fe$_3$O$_4$	6	5.18	Black		Black streak; octahedral crystals	I, M	Major iron ore
Malachite	Cu$_2$(CO$_3$)(OH)$_2$	3.5–4	3.9–4	Green		Reacts with HCl	Altered ores of copper	Ore of copper
Marcasite	FeS$_2$	6–6.5	4.89	Silver, Gray		Black streak; metallic luster	S; hydrothermal	Source of sulfur for sulfuric acid
Microcline	KAlSi$_3$O$_8$	6	2.55	Gray, Flesh, Salmon, Green	2/90°	A potassic feldspar	I, M	Porcelain, ceramics
Molybdenite	MoS$_2$	1–1.5	4.67	Black	1 (excellent)	Greasy feel; black streak	I, M	Ore of molybdenum
Muscovite	KAl$_2$(AlSi$_3$O$_{10}$)(OH)$_2$	2–2.5	2.76–3.10	Colorless	1 (perfect)	A colorless mica	I, M	Electrical insulator
Nepheline	(Na,K)AlSiO$_4$	5.5–6	2.55	Colorless, White	3/not 90°	Often greasy luster	I	Glass, ceramics
Oligoclase	An$_{10}$–An$_{30}$	6.1	2.65	Gray	2/87,93°	A plagioclase	I, M	Ceramics
Olivine	(Mg,Fe)$_2$SiO$_4$[Fo$_0$–Fo$_{100}$]	6.5–7	3.27–4.37	Green	Conchoidal	A mineral family fracture	I, M	Minor gemstone
Opal	SiO$_2$·nH$_2$O	5–6	1.9–2.2	Varied	Conchoidal fracture	Milky or fiery luster	I, S	Gemstone
Orpiment	As$_2$S$_3$	1.5–2	3.49	Yellow	1	Resinous luster; yellow streak	Hydrothermal	Source of arsenic
Orthoclase Perthite	KAlSi$_3$O$_8$ Plagioclase plus potassic feldspar	6	2.57	Varied	2/90°	A potassic feldspar	I, M (s) I, M	Ceramics, porcelain
Platinum	Pt	4–4.5	14–19	Gray			I (ultramafic)	Chemical apparatus, jewelry
Prehnite	Ca$_2$Al$_2$(Si$_3$O$_{10}$)(OH)$_2$	6–6.5	2.8–2.9	Green			I, M	
Pyrite	FeS$_2$	5	4.3	Brassy yellow		Black streak; in cubes	I, M, S	Source of sulfur for sulfuric acid
Pyrope	Mg$_3$Al$_2$(SiO$_4$)$_3$	7	3.51	Deep red	Conchoidal fracture	Twelve-sided crystals; a garnet	M (i)	Abrasive, jewelry
Pyrrhotite	Fe$_{1-x}$S(x = 0–0.2)	4	4.62	Bronze		Black streak; magnetic	I (m)	Source of iron and associated nickel
Quartz	SiO$_2$	7	2.65	Varied	Conchoidal fracture	Hexagonal crystals	I, M, S	Glass, gemstones, optical lenses
Realgar	AsS	1.5–2	3.48	Red-orange		Orange streak	Hydrothermal	Ore of arsenic
Rhodochrosite	MnCO$_3$	3.5	3.45–3.6	Pink	3/not 90°	Soluble in hot HCl; a carbonate	Hydrothermal	Minor ore of manganese

Mineral	Composition	Hardness	Specific Gravity	Color	Cleavage/Parting	Other Properties	Occurrence	Uses
Ruby	Al_2O_3	9	4.02	Red	Parting common	Six-sided crystals (variety of corundum)	M	Gemstone
Rutile	TiO_2	6–6.5	4.18–4.25			Submetallic luster; four-sided prisms	I, M	Coatings for welding rods
Sapphire	Al_2O_3	9	4.02	Blue	Parting common	Variety of corundum	M	Gemstone
Serpentine	$(Mg,Fe)_3Si_2O_5(OH)_4$	2–5	2.2	Green, Yellow		Platy or fibrous	M	Asbestos
Siderite	$FeCO_3$	3.5–4	3.83–3.85	Brown, Yellow	3/not 90°	Soluble in hot HCl	S, I	Ore of iron
Sillimanite	Al_2SiO_5	6–7	3.23	White	1	Prismatic or fibrous	M	
Silver	Ag	2.5–3	10.5	White		Malleable, ductile	Hydrothermal	Jewelry, coinage
Sphalerite	ZnS	3.5	3.9–4.1	Varied	6 (rare)	Light-yellow streak	Hydrothermal	Ore of zinc
Spessartine	$Mn_3Al_2(SiO_4)_3$	7	4.18	Red-brown	Conchoidal fracture	A garnet	M	Abrasive, jewelry
Sphene	$CaTiSiO_5$	5–5.5	3.4–3.55	Brown, Green, Black, Yellow	2/not 90°		I, M	Ore of titanium
Staurolite	$Fe_2Al_9O_7(SiO_4)_4(OH)$	7–7.5	3.65–3.75	Red-brown		Commonly in cross-like crystals	M	Gemstone
Stibnite	Sb_2S_3	2	4.52–4.62	Gray-black	1	In slender prisms	Hot springs	Ore of antimony
Sulfur	S	1.5–2.5	2.07	Yellow	Conchoidal fracture		S; volcanoes	For producing sulfuric acid
Sylvite	KCl	2	1.99	Varied		Bitter salty taste	Evaporite	Source of potassium
Talc	$(Mg,Fe)_3Si_4O_{10}(OH)_2$	1	2.7–2.8	Green	1		M	Lubricant
Topaz	$Al_2(SiO_4)(F,OH)_2$	8	3.5	Varied	1	Elongate prisms	I	Gemstone
Tourmaline	$XY_3Al_6(BO_3)_3Si_6O_{18}(OH)_1$	7–7.5	3.0–3.25	Blue, Red, Green, Black, Yellow		Striated trigonal crystals; varicolored	I, M	Gemstone
Tremolite	$Ca_2Mg_5Si_8O_{22}(OH)_2$	5–6	3.0–3.33	White, Pale green	2/56,124°	Sometimes fibrous; an amphibole	M	Fibrous variety for asbestos
Turquoise	$CuAl_6(PO_4)_4(OH)_8 \cdot 4H_2O$	6	2.6–2.8	Blue-green		Often as crusts	Alteration of volcanic rock	Gemstone
Uraninite	UO_2	5.5	9.0–9.7	Black		Black streak	I	Uranium ore
Uvarovite	$Ca_3Cr_2(SiO_4)_3$	7	3.77	Bright green	Conchoidal fracture	A garnet	M	Gemstone
Vanadinite	$Pb_5Cl(VO_4)_3$	3	6.7–7.1	Red, Orange, Yellow, Brown		Rounded crystals; globular masses	Hydrothermal	Ore of lead; vanadium
Wolframite	$(Fe,Mn)WO_4$	5–5.5	7.0–7.5	Black	1		Pegmatites	Ore of tungsten
Wollastonite	$CaSiO_3$	5–5.5	2.8–2.9	White	2/90°		M	Ceramics, porcelain, tile
Wulfenite	$PbMoO_4$	3	6.8	Red-orange		White streak	Hydrothermal	Minor ore of molybdenum
Zircon	$ZrSiO_4$	7.5	4.68	Varied		Four-sided prisms	I, M	Gemstone; ore of zirconium

*I, M, S: Common in igneous (I), metamorphic (M), or sedimentary (S) rocks; i, m, s: Rare in igneous (i), metamorphic (m), or sedimentary (s) rocks. An_x: Plagioclase feldspars often are described by the amount of their anorthite end member. For example, An_{30} = 30% anorthite and 70% albite.

Two major soil classifications are in use in the United States. The older classification, based on soil orders and groups, has been in use for many years. The newer classification, *The Seventh Approximation,* has been adopted by the U.S. Department of Agriculture and many state soil surveys. Proponents of the new classification claim that it is superior to the old one because it is more precise, with 10 major orders and 47 suborders, permitting a detailed classification of soils. Opponents of the new classification claim that it is not needed and is much too complex for use by anyone except soil specialists. Needless to say, the topic of soil classification is controversial at the present time. In the interests of objectivity, we present both major classifications. We have cross-referenced the two and have indicated which soil types in each classification fall under the major soil types, the pedalfers and pedocals, described in Chapter 9.

TABLE D–1
Soil classification by orders and groups.

ARID, SEMIARID, AND SUBHUMID CLIMATES (PEDOCALS)				HUMID CLIMATES (PEDALFERS)	
Azonal Order[1]	Intrazonal Order[2]	Zonal Order[3]	(Transitional)	Zonal Order[3]	Intrazonal Order[2]
Lithosols Stony, thin mountain soils that either have had little time to develop or are on slopes steep enough so that thicker sections of soil cannot accumulate. **Regosols** Very poorly developed soils with no horizon development that have formed on recently deposited sediments.	**Saline Soils** Soils without horizons that contain an excess of soluble minerals. Formed during the desiccation of fine-grained sediments and saline waters in basins with internal drainage.	**Chernozem Soils** Dark surficial layer (A) of highly organic soil derived from decay of grass parts. Underlain by a B layer which is brown to yellowish brown. Reduced rainfall results in little leaching and formation of calcium carbonate nodules in the B horizon. Common in the eastern parts of the Dakotas. Nebraska, central Kansas, and western Oklahoma and Texas. **Chestnut Soils** Similar to chernozem soils but form in a drier climate and are less organic, lighter in color, and have more abundant calcium carbonate nodules in the B horizon. Common in eastern Montana, the western parts of the Dakotas and Nebraska, northeastern Colorado, and western Kansas. **Brown Soils** Drier versions of the chestnut soils. Have a light-brown color and a B zone with a distinctive columnar structure. Common on the plains abutting the Rocky Mountains. **Gray Desert Soils** Sandy, pale grayish to reddish-gray soils with little organic material. Form in cooler mid-latitude desert areas of Nevada, Arizona, Utah, and New Mexico. Grayish-red soils characterize subtropical deserts such as those in southern Arizona and New Mexico. **Red Desert Soils** Red-colored soils in tropical deserts.	**Prairie Soils** Soils transitional between the pedocals (chernozems) and pedalfers (podzols). Among the most fertile soils because they contain sufficient organic material, such as in the chernozems, and form in sufficient rainfall, such as in the podzols. Common in northern Illinois, Iowa, northwestern Missouri, eastern Kansas, Oklahoma, and north-central Texas.	**Podzols** Soils with organic-rich accumulations in the top of the A zone underlain by a strongly leached, ash-white horizon in the lower part of the A zone. The B zone is clay-rich. The type of soil that forms in humid subarctic climates. Common in northern Wisconsin, Minnesota, Michigan, and northern New England. **Gray-brown Podzols** Similar to the podzol but less intensely leached. The base of the A zone is grayish brown rather than ash-white as in podzols. Thick, dark-brown B zone rich in clay minerals. Widespread in the humid temperate northeastern United States. **Red-yellow Podzols** Similar to gray-brown podzols but has a B zone enriched in hydroxides of iron and aluminum. Common in the humid subtropical climates of the southeastern and Gulf Coastal states. **Latosols** Deep brownish-red surface deposits with only a thin cover of organic debris. Soil lacks horizons and becomes lighter in color with depth. Forms by intense chemical weathering in hot humid climates. Silica is leached out, and the soil becomes enriched in hydroxides of iron, aluminum, or manganese, depending on the composition of the parent material. **Tundra Soils** Soils which form in areas underlain by permafrost. Mixture of organic material and physically weathered rock debris. Chaotic structure resulting from perennial freezing and thawing. Common in arctic and subarctic climates.	**Bog Soils** Dark-brown water-saturated and partially decomposed peaty material. Can form in arctic, temperate, or tropic climates wherever plant material accumulates under standing water. **Meadow Soils** Dark, organic-rich upper layers beneath which is a bluish-gray clay horizon. **Planosols** Thick and very dark organic-rich soil formed on flat surfaces between stream valleys where soil erosion is limited.

Increasing aridity →

Increasing temperatures →

[1] Soil horizons are poorly developed or absent.
[2] Soil characteristics are determined by local conditions such as poor drainage.
[3] Well-developed soil horizons corresponding to the climatic and vegetative zones in which they are found.

SOURCE: Modified with permission from A. N. Strahler, *Physical Geography*, New York, Wiley, 1960.

TABLE D-2
Soil classification (Seventh Approximation).

Soil Order	Characteristics	Some Areas Where It Is Common	Approx. Equiv. in Great Soil Groups
Entisols	Soils without horizons; soil-forming processes have not had sufficient time to produce horizons.	Wide geographic range, from desert sand dunes to frozen ground of sub-arctic zones.	Azonal soils
Inceptisols	Weakly developed soil horizons; soil horizon A is developed.	Wide geographic range wherever soils have just begun to develop on newly deposited or exposed parent materials such as volcanic or glacial deposits.	Lithosols (mountain soils) Regosols (recently deposited sediments)
Spodosols	Humid forest soils with a gray leached A horizon and a B horizon enriched in iron or organic material leached from above. Commonly under coniferous forests.	New England, northern Minnesota, and Wisconsin.	Podzolic and brown podzolic soils (pedalfer)
Alfisols	Soils with clay enrichment in the B horizon. Lower organic content than mollisols. Medium to high base supply. Commonly under deciduous forests.	Western Ohio, Indiana, lower Wisconsin, northwestern New York, Central Colorado, western Montana.	Gray-brown podzolic soils (pedalfer)
Mollisols	Grassland soils with a thick, dark organic-rich surface layer. High base supply (calcium, sodium, and potassium).	Widespread in central and northern Texas, Oklahoma, Kansas, Nebraska, North and South Dakota, and Iowa.	Chestnut, chernozem, and prairie soils (pedocal)
Aridisols	Desert and semiarid soils; low organic content along with concentration of soluble salts within soil profile.	Widespread in desert and semiarid areas of Nevada, California, Arizona, New Mexico.	Desert soils (pedocal)
Ultisols	Deeply weathered red and orange clay-enriched soils on surfaces that have been exposed for a long time.	Humid temperate to tropical soils. Widespread in southeastern United States east of Mississippi Valley.	Red and yellow podzolic soils, certain lateritic soils (pedalfer)
Oxisols	Intensely weathered soils consisting largely of kaolin, hydrated iron and aluminum oxide. Bauxite forms in these soils.	Warm tropical areas with high rainfall.	Most lateritic soils (pedalfer)
Histosols	Organic soils and peat.	Mississippi delta, Louisiana; Everglades, Florida; local bogs in many areas.	Bog soils (pedalfer)
Vertisols	Swelling soils with high clay content which swell when wet and crack deeply when dry.	Southeast Texas; local areas	Swelling clays

SOURCE: *USDA* Soil Conservation Service, 1960.

Glossary

Aa. Highly viscous, blocky lava.

Abrasion. The grinding of mineral and rock particles against each other or against bedrock.

Absolute age. The age of Earth materials in years. Absolute age usually is determined by using radioactive isotopes.

Absolute time scale. A time scale subdivided on the basis of radiometric dates.

Abyssal hills. Extinct submarine volcanoes partially buried by sediments on the abyssal plains.

Abyssal plains. Flat depositional oceanic features extending from the continental rise toward the oceanic ridges.

Accessory minerals. Minerals present in small amounts in rocks.

Accretionary wedge. A mass of melange and other sedimentary rocks found in the area between an oceanic trench and the adjacent volcanic island arc.

Accumulation horizon. The B zone of a soil profile in which clays and oxides of iron and aluminum tend to be concentrated.

Achondritic meteorites. Stony meteorites that do not contain chondrules. They are similar in composition to gabbro and peridotite.

Acid mine water. Groundwater which has been acidified by the addition of sulfuric acid derived from the oxidation of pyrite associated with coal deposits.

Acid rain. Rain which has been acidified by addition of sulfur dioxide produced by industrial emissions.

Active continental margins. Coasts where shoreline and plate boundaries coincide so that the area experiences tectonism frequently.

Active volcano. A volcano that is erupting currently.

Aeration zone. That part of the regolith and underlying rock in which the pore spaces are not completely filled with water.

Aftershocks. Earthquakes, generally of gradually diminishing magnitude, that follow major earthquakes.

Agate. A banded variety of quartz.

Agents of metamorphism. Heat, lithostatic pressure, directed pressure, and chemically active fluids which, when applied to rocks, bring about metamorphic changes.

Agglutinates. Particles of the lunar soil fused together by the heat of micrometeorite impacts on the Moon's surface.

Aggradational stream. A stream that is actively filling its valley with alluvium because of a rise in base level.

A horizon. The soil zone from which material such as clay and oxides of iron and aluminum are leached.

Alluvial basin. A topographically low area underlain by sediments deposited by a stream. (See *alluvium.*)

Alluvial fan. A triangular-shaped stream deposit extending out from the bases of mountain ranges, most commonly in arid and semiarid regions.

Alluvium. The general name given to all sediment deposited by streams.

Alpha decay. A nuclear reaction in which a particle composed of two protons and two neutrons (an alpha particle) is expelled from the nucleus of an atom.

Alpha (α) particle. A particle composed of two protons and two neutrons that is ejected from a nucleus during a nuclear reaction.

Alpine glacier. See *mountain glacier.*

Amber. Hardened tree sap in which fossils are commonly preserved.

Amethyst. A purple variety of quartz.

Amino acid dating. Dates determined by changes in the ratio of two different forms of amino acid with time.

Amphibole family. A group of minerals containing double chains of silicon-oxygen tetrahedra.

Amphibolites. Unfoliated metamorphic rocks composed mostly of amphiboles.

Amplitude. One-half the distance between the crest and trough of a waveform.

Andesite. A fine-grained intermediate igneous rock with the same chemical composition as diorite, usually found in subduction zones.

Angle of repose. The maximum slope angle to which particles of a given size can be built up without slumping.

Angular unconformity. A contact between layers of sedimentary rock where beds above and below the unconformity are not parallel with each other.

Anhydrite. (a) An evaporite mineral with the composition $CaSO_4$; (b) A chemical sedimentary rock composed of crystals of anhydrite.

Anion. A negatively charged ion formed when an atom gains one or more electrons.

Anion complex. A group of anions and cations that have a negative charge, e.g., the carbonate $(CO_3)^{2-}$ and hydroxyl $(OH)^-$ complexes.

Annual ring. The portion of a tree that has grown in one year. Counting successive annual rings can give an age for the tree.

Anthracite coal. A hard, black coal with a semimetallic luster and a carbon content between 92 and 98%. Anthracite is the highest-ranking coal and burns with a short blue flame without smoke.

Anticline. A fold in which the limbs dip away from one another so that old rocks are arched upward in the axial region of the fold.

Aphanitic. An igneous texture so fine grained that individual mineral grains cannot be seen with the naked eye.

Aquiclude. A sediment or rock through which groundwater cannot pass. (Cf. *aquifer*.)

Aquifer. A sediment or rock through which groundwater can move. (Cf. *aquiclude*.)

Arenite. A sandstone with less than 15% matrix, in which the particles are held together by a crystalline cement.

Arête. A narrow, knife-sharp ridge formed as an erosional remnant between two cirques.

Arkose. See *arkose sandstone*.

Arkose sandstone. A sandstone that has greater than 25% feldspar.

Artesian spring. A spring under artesian pressure which flows on the land surface or from the bottom of a body of water.

Artesian well. A well in which the water rises to the surface under its own pressure, without being pumped.

Asbestos. The name given to fibrous varieties of serpentine or amphibole.

Ashfalls. Tephra deposited by gravity settling from the air after volcanic eruption.

Ashflows. Tephra erupted from a volcanic vent and deposited from nuée ardente clouds.

Assemblage. The combination of minerals that makes up a metamorphic rock.

Assimilation. Incorporation in a magma of ions from surrounding host rocks or xenoliths.

Asteroid. Body of rock a few kilometers to 1000 km in diameter which orbits the Sun between the orbits of Mars and Jupiter.

Asteroid belt. Clusters of asteroids between the orbits of Mars and Jupiter.

Asthenosphere. A region in the mantle 100 to 250 km beneath the surface; it is composed of materials of abnormally low rigidity and corresponds to the seismic low-velocity zone. In the plate tectonics model, this is the region of decoupling of lithospheric plates from the lower mantle.

Atmosphere. The gaseous envelope composed mostly of nitrogen and oxygen that surrounds the solid Earth.

Atoll. A circular coral reef with a lagoon in the center.

Atom. The smallest particle that possesses the chemical properties of an element; it is composed of a central nucleus and orbiting electrons.

Atomic energy. (See *nuclear energy*.)

Atomic mass. The number of protons and neutrons in the nucleus of an atom.

Atomic mass unit (amu). The unit by which the mass of an atom is measured; equivalent to the mass of a proton or a neutron.

Atomic number. The number of protons in the nucleus of an atom; also the number of electrons orbiting the nucleus of an atom.

Authigenic minerals (and sediments). Minerals (sediments) that form in place in the sedimentary basin of deposition.

Avalanche. Extremely rapid form of mass movement flow.

Axial plane. An imaginary plane that connects fold axes from all the different layers of rock in a fold.

Bajada. A sloping plain at the base of a mountain front which is formed by the merging of adjacent alluvial fans.

Bankfull discharge. Stream discharge sufficient to fill the stream channel completely.

Bar. A ridgelike accumulation of coarse sediment deposited by flowing water.

Barchan. A crescent-shaped sand dune, the tips of which point downwind.

Barrier island. An elongate body of coastal sand separated from the mainland by a body of water.

Barrier reef. A reef separated from the land by a body of water.

Barrier rollover. Erosion of the front side of a barrier island and deposition of washover fans on the backside, resulting in migration of the barrier island toward the mainland with time.

Basalt. A fine-grained mafic igneous rock with the same composition as gabbro.

Base flow. The portion of stream discharge that is supplied by groundwater.

Base level. The lowest level to which a stream can erode its channel.

Basin. (1) A geologic structure in which all beds dip inward toward a central point; (2) a lowland flanked by uplifted blocks during normal faulting. See also *horst and graben.*

Basin and range. See *horst and graben.*

Batholith. A large, irregularly shaped body of intrusive igneous rock with an outcrop area of at least 75 km^2.

Bauxite. A mixture of aluminum oxides and hydroxides formed during extensive chemical weathering of aluminum-rich rocks in hot, humid climates; the chief ore of aluminum.

Beach. A gently sloping accumulation of coarse sediment along a lake or ocean shoreline. The beach extends from the low-water line landward to where there is a change in morphology, such as a cliff, dunes, or permanent vegetation.

Bed. (a) A layer of sedimentary rock more than 1 cm thick. (b) The bottom of a stream channel.

Bedding plane. A discrete physical break in a sedimentary rock which separates adjacent beds and allows them to be distinguished from one another.

Bed load. All sedimentary particles which are carried at the bottom of a moving fluid such as water or air.

Benioff zone. The region of concentration of deep-focus earthquakes associated with ocean trenches. Focal depths increase systematically from the trench toward and beneath the adjacent island arc or continent.

Bentonites. Clays that have the ability to absorb great quantities of water and to swell up to eight times their original volume.

Bergschrund. A large crevasse that separates a cirque glacier from its headwall.

Beta decay. A form of nuclear reaction in which an electron is expelled from the nucleus of an atom.

Beta particle. A particle, essentially the same as an electron, ejected from the nucleus of an atom during a nuclear reaction.

B horizon. That portion of the soil profile in which dissolved ions are precipitated. *Synonym:* Zone of accumulation.

Biogenic rock. Lithified biogenic sediment.

Biogenic sediment. Sediment produced by organic processes or composed of organic remains.

Biological oxygen demand. The oxygen required for animal respiration and the oxidation of organic debris in water.

Bituminous coal. A dark brown-to-black coal with about 15 to 20% volatile matter. Bituminous coal ranks between lignite and anthracite in the coal-forming process and is the most abundant form of coal.

Black smokers. Chimneylike forms on oceanic ridges from which hot mineral-rich waters erupt.

Block fields. Areas underlain by frost-wedged bedrock masses up to 10 m on a side, produced by periglacial processes.

Body waves. Seismic waves that transmit energy through the Earth (P waves and S waves).

Bonding. The electrostatic attachment of one atom to another or to several others.

Bottomset beds. The fine-grained, organic-rich, and horizontally layered beds that extend out from the base of a delta.

Boulder. A sedimentary particle with a diameter greater than 256 mm.

Bowen's Reaction Series. A diagrammatic representation of the sequence in which igneous minerals crystallize from a magma.

Braided channel; braided stream. A stream whose channel splits into numerous intersecting channels separated by islands or sand bars.

Breaker zone. The portion of the nearshore zone where wave velocity and wavelength decrease and the waves get higher and become progressively asymmetrical.

Breccia. (a) Sedimentary rock composed of angular particles of pebble, cobble, or boulder size; (b) Angular chunks of rock in a fine-grained matrix produced by faulting (fault breccia).

Brittle behavior or deformation. Breaking of rock during deformation to produce faults, fractures, and joints.

Brittle zone. The upper zone of a glacier in which motion of ice produces breakage.

Brown clays. Pelagic deposits, reddish-brown in color, produced by the oxidation of iron in the sediment.

Burial metamorphism. Metamorphism caused predominantly by lithostatic pressure in a thick pile of sedimentary or volcanic rocks, or both.

Butte. A flat-topped erosional remnant in dry regions which is similar in origin to a mesa but has a less extensive summit area.

Calcareous ooze. Biogenic sediment composed of the skeletons of microscopic organisms made of calcium carbonate.

Calcareous rocks. Limestones, dolostones, and their metamorphosed equivalent, marble.

Calcite. A common rock-forming mineral in sedimentary and metamorphic rocks ($CaCO_3$).

Calcium carbonate compensation depth (CCCD). The depth in the oceans at which calcareous sediment dissolves as it falls because of the greater solubility of $CaCO_3$ in the colder bottom waters.

Calc-silicate. (a) A mineral containing calcium and silica, e.g., wollastonite, $CaSiO_3$. (b) A rock containing those minerals formed by metamorphism of quartzose limestones and dolostones.

Caldera. Huge, irregular volcanic craters formed by explosion, collapse, or a combination.

Caliche. A type of pedocal soil in which calcium carbonate accumulates in the upper part of the soil profile as a hard crust.

Capacity. The ability of a transporting agent to carry material, as measured by the amount of load carried at a given point per unit of time.

Carbonaceous chondrites. Chondritic meteorites that contain substantial amounts of carbon and graphite, and a complex mixture of organic molecules.

Carbonate. (1) An anion complex composed of carbon and oxygen ions: $(CO_3)^{2-}$; (2) minerals containing the carbonate anion complex.

Carbonation. A type of chemical weathering in which the parent material reacts with carbon dioxide dissolved in water (i.e., a weak carbonic acid solution).

Carboniferous Period. In European usage, this is comprised of the Mississippian and Pennsylvanian periods of the United States.

Carbonization. The process of fossilization in which organic material is transformed into carbon imprints, preserving the form of the original material.

Cast. A mineral deposit that fills a cavity (mold) within a sediment or rock and preserves the external features of a plant or animal fossil that formerly filled the cavity.

Catastrophism. A view of Earth history that states that major Earth changes, such as orogenies, are produced by sudden, abnormal processes rather than by slow evolution, as suggested by uniformitarianism.

Catastrophists. People who believe in catastrophisim as a principal mode of geologic change.

Cation. A positively charged ion formed when an atom loses one or more electrons.

Cave. A natural underground open space. The most common type of cave is formed by groundwater solution of soluble rocks, principally limestone.

Cavern. An underground space similar to a cave but larger in size.

Cementation. Precipitation of crystalline material in pore spaces during lithification.

Central rift valley. A steep-sided valley found at the crest of every ocean ridge and presumably formed by normal faulting (tension).

Chalk. A soft, white limestone composed mostly of skeletons of microscopic marine plants and animals.

Chemical energy. Energy released or absorbed during chemical reactions.

Chemical load. Sediment which is carried as dissolved ions in water.

Chemically active fluid. Water containing large amounts of dissolved ions; these ions interact with rocks to cause metasomatism and may create hydrothermal mineral deposits.

Chemical reaction. Interactions between atoms involving their outer electrons.

Chemical remanent magnetization. The record of the Earth's ancient magnetism imprinted on minerals that form below their Curie points during diagenesis or metamorphism.

Chemical rocks. Sedimentary rocks which are formed from the precipitation of crystals from solutions.

Chemical weathering. The decomposition of rock by chemical reactions in which some of the minerals are destroyed, new ones may form, and soluble ions are removed by water flowing through the weathered material.

Chemical weathering. The decomposition of rocks as they are exposed to the atmosphere, hydrosphere, and biosphere.

Chert. A chemical sedimentary rock composed of silica, SiO_2.

Chondritic meteorites. Stony meteorites that contain small round grains called chondrules.

Chondrules. Small rounded bodies composed of the minerals olivine or pyroxene, with minor amounts of plagioclase feldspar.

Cinder cone. A volcano made entirely of tephra.

Cirque. A bowl-shaped rock basin which was eroded by a mountain glacier above the snowline.

Cirque glacier. A small alpine glacier which occupies a cirque basin above the snowline.

Clastic. (a) A type of particle formed by weathering of preexisting rocks or by the breaking of shell material. (b) A type of sedimentary rock composed of mineral or rock particles, and matrix or cement or both.

Clay. Sedimentary particles with diameters less than 1/256 mm.

Clay minerals. Minerals with a layered structure that may be hydrous silicates or nonsilicates.

Cleavage. (a) The breakage of a mineral along smooth flat surfaces parallel to zones of weak bonding in the mineral structure. (b) Close-spaced fractures developed in folded rocks, generally parallel to the axial planes of the folds.

Cleavage plane. Flat, smooth surfaces along which mineral breakage occurs because of planes of bond weakness in a mineral.

Climatic desert. An area that is arid because it is located at a place where atmospheric circulation patterns result in descending dry air sweeping across the surface.

Coastal aquifer. Aquifer that lies beneath coastal regions and is infiltrated partially by saltwater.

Coastal desert. Area that is arid because it is located at a coastline bordered by cold oceanic currents that generate cool, moisture-deficient winds.

Cobble. A sedimentary particle with a diameter between 64 and 256 mm.

Cohesion. The ability of rock and mineral particles to attract and hold one another.

Col. A mountain pass formed when cirque glaciers erode through opposite sides of an arête.

Colluvium. Sediments deposited by mass movements.

Columnar joints. Systematic fractures produced in fine-grained igneous rocks during cooling that result in polygonal smooth-sided shapes.

Comet. Body of rocky dust and ice approximately 1 km in diameter that orbits the Sun at greater distances and with more eccentricity than a planet.

Compaction. Rearrangement of particles and decrease in volume caused by the weight of overlying material during lithification.

Competence. A measure of the largest-size particle that can be carried by a transporting agent.

Competent rock. Rock that resists changes in shape or volume during deformation until applied stresses become intense.

Composite cone. A steep-sided volcano composed of alternating layers of lava and tephra.

Compressional stress. A type of stress produced by deforming forces that squeeze rocks from opposite sides.

Concentration. The amount of dissolved ions per volume of water.

Conchoidal fracture. Breakage of a mineral along smooth, curved surfaces.

Concordant. A type of igneous intrusion whose contacts parallel major planar features in the rocks they intrude.

Conduction. Transfer of heat from one object to another object with which it is in contact.

Cone of depression. A conically shaped depression in the water table which occurs when groundwater is pumped from a well at a faster rate than it is replenished.

Confined aquifer. An aquifer, underlain and overlain by aquicludes, in which the groundwater is under pressure and therefore rises toward the surface without pumping.

Conglomerate. A sedimentary rock composed of rounded particles of pebble, cobble, or boulder size.

Constructive feature. A landform produced by deposition of sediment (e.g., an alluvial fan) or of lava or tephra (e.g., a volcano).

Consumption. Uses of water in which the water is evaporated, transpired from plants, incorporated into plants or animals, boiled into steam, or otherwise changed from a usable liquid form.

Contact aureole. The area around an intrusive body affected by the heat emanating from the cooling igneous rock.

Contact metamorphism. Metamorphism around a body of intrusive igneous rock caused by the heat given off during magmatic cooling.

Contact metasomatic ore. Mineral deposits formed adjacent to intrusive bodies by hydrothermal fluids.

Continent. A large body of land underlain by a two-layer crust (sial and sima).

Continental accretion. A hypothesis that states that continents appear to grow outward from an old, stable region.

Continental glacier. A mass of glacial ice that covers most of the topographic features of a region of continental or subcontinental proportions.

Continental rise. A gently sloping part of the seafloor at the base of the continental slope formed by deposition of continent-derived sediment.

Continental shelf. A very gently sloping surface that extends outward from the shoreline and is underlain by a two-layer crust (sial and sima).

Continental slope. A relatively steeply sloping part of the seafloor between the continental shelf and the continental rise or abyssal plain.

Continuous melting. A melting process common in solid solution series in which a mineral melts gradually over an interval of temperature.

Continuous reaction. A progressive series of reactions between early-formed crystals and residual magma.

Convection. Transfer of heat from one place to another by the movement of the heated material.

Convection cell. (a) The circular path followed by materials transferring heat. They rise when hotter than their surroundings and sink when cooler. (b) The possible source of movement of lithospheric plates.

Convergent margins or plate boundaries. Subduction zones where two plates collide with one another and lithosphere material is subducted back into the asthenosphere.

Coordination number. The number of ions that can surround ions of the opposite charge.

Coquina. A sedimentary rock composed largely of abraded and transported fossil debris.

Coral reef. Massive linear, wave-resistant structures made largely of carbonate-secreting organisms such as corals.

Cordilleran ice sheet. A Pleistocene continental glacier centered in the Rocky Mountains, which spread westward toward the Pacific and eastward beyond the Rocky Mountains.

Core. The innermost chemical region of the Earth; a combination of very high-density materials—probably a mixture of iron and nickel.

Coriolis effect. The deflection of moving water or wind toward the left in the Southern Hemisphere and toward the right in the Northern Hemisphere because of the Earth's rotation.

Correlation. The demonstration that rock units from different areas were formed at the same time.

Coupled ionic substitution. Simultaneous substitution of two ions brought about to maintain the electrical neutrality of a mineral structure.

Covalent bonding. The joining of atoms by combining their outermost electrons in a special molecular orbit.

Crater. (a) A bowl-shaped depression at the summit of a volcano. (b) A landscape feature formed by impact of a meteorite.

Crater density. The number of meteorite craters per unit area.

Creep. Very slow downslope movement of surface soil and rock material under the influence of gravity.

Crest. The highest point of a waveform.

Crevasse. A fracture formed in the brittle zone of a glacier during glacial movement.

Critical angle of repose. The maximum angle at which particles of a given size can be built before the slope fails.

Cross-bedding. A type of stratification in which the layers within the bed are not horizontal but are inclined to the horizontal upper and lower surfaces of the bed. Also called cross-stratification.

Cross-cutting relationships. A principle of relative-age dating which states that any geologic feature that cuts through another must be younger than the feature it truncates.

Crust. The outermost layer of the solid Earth. It is composed of low-density silicate and nonsilicate minerals.

Cryptocrystalline texture. Crystals too small to be seen without magnification.

Crystal. A mineral specimen occurring in a regular geometric shape that reflects its internal structure.

Crystalline structure. The ordered internal atomic structure characteristic of minerals.

Crystallization point. The temperature at which a mineral crystallizes from magma.

Crystal settling. Separation of early-formed minerals from magma by sinking to the bottom of the magma chamber.

Cubic packing. The type of particle packing which is the loosest and has the greatest pore space. (Cf. *rhombohedral packing*.)

Curie point. The temperature at which a substance loses its magnetism when heated.

Current ripple marks. Asymmetrical ripple marks formed by currents of air or water moving along the sediment surface.

Darcy's Law. A law that expresses the rate of water migration through sediments or rocks as: $V = Kh \div l$, where V is the velocity of water migration, K is a coefficient of permeability which varies for different materials, and $h \div l$ is the vertical drop in the water table (h) over a given horizontal distance (l) between any two points.

Daughter element. The element formed as the product of a nuclear reaction.

Debris avalanche. Very rapid to extremely rapid downslope movement of relatively dry material.

Debris flow. Rapid downslope movement of relatively dry material.

Decarbonation. A type of metamorphic reaction in which carbon dioxide is driven out of a rock.

Decay rate. The rate at which atoms of radioactive isotopes break down in nuclear reactions.

Declination. The angular difference at a given location between the geographic north and magnetic north poles.

Deep currents. Subsurface ocean-water circulation caused by seawater-density differences. Deep currents involve vertical movement as well as horizontal displacement.

Deflation. The erosion of sediment particles from the ground by wind.

Deflation basin. A depression formed as wind blows sand off a surface unprotected by vegetation.

Dehydration. The removal of water from a mineral or sediment.

Delta. A deposit formed when a stream flows into a large body of standing water.

Dendritic stream pattern. A stream network similar to the branching pattern of the veins in a leaf. Dendritic stream patterns are common in areas underlain by rocks that have the same general resistance to erosion.

Depletion. The decrease in availability of a resource.

Depositional remanent magnetization. A record of ancient Earth magnetism preserved by alignment of grains during deposition of sediment.

Desert. A land area that receives less than 25 cm of precipitation per year.

Desert pavement. A layer of coarse particles left at the surface in arid regions after wind, water, or both have removed all the finer grains.

Desert varnish. Dark-colored iron and manganese oxide coatings that form on rocks exposed to the air in dry climates.

Destructive feature. A landform carved by agents of erosion, e.g., a stream valley, cirque, or sinkhole.

Diatremes. Cylindrical plutons composed of ultramafic rock that form by explosive upward boring from great depths.

Differential weathering. Weathering that occurs at different rates as a result of variations in composition of the parent materials.

Dike. A flat, tabular mass of intrusive igneous rock that cuts across previously existing structures in its host rock.

Diorite. A coarse-grained intermediate igneous rock composed of a Ca–Na plagioclase feldspar and ferromagnesian minerals such as amphiboles and pyroxenes.

Dip. A measure, in degrees, of how much a planar geologic structure is displaced from the horizontal, and the general direction of the tilt. It is used with *strike* to describe the three-dimensional attitude of planar structural features.

Dipolar field. A magnetic field defined by lines of force that emanate from two ends (magnetic poles) of an object.

Dip-slip fault. A fault in which the displacement has been largely up or down the dip of the fault plane; i.e., there is little or no horizontal motion.

Directed pressures. Pressures that are greater in some directions than others when applied to rocks; they generally are due to tectonic activity.

Discharge. The volume of water passing a point in a stream or aquifer in a given amount of time.

Disconformity. An unconformity in which sedimentary rocks of significantly different age are in contact and the beds above and below the unconformity are parallel.

Discontinuous crystallization. See *discontinuous reaction.*

Discontinuous melting. A melting process in which a substance first melts partially to form liquid and a new mineral, followed by gradual melting of the residual mineral.

Discontinuous reaction. A crystallization process during which an early-formed mineral interacts with residual magma to form a new mineral.

Dissection. Erosion through a land surface by a glacier or stream.

Dissociation. The separation of a molecule into its component ions. A water molecule (H_2O), for example, dissociates into hydrogen (H^+) and (OH^-) ions.

Distributary drainage pattern. A network pattern in which the streams branch out in the downstream direction.

Distributary network. See *distributary drainage pattern.*

Divergent margins or plate boundaries. Spreading centers on land or in the oceans (ocean ridges) in which two lithospheric plates move apart from each other and new lithosphere is formed.

Dolomite. A mineral, $CaMg(CO_3)^2$.

Dolostone. A rock composed of the mineral dolomite.

Dome. A structural feature in which all layers dip outward from a central point.

Dome mountain. A mountain with the internal structure of a dome.

Double chain. A form of silicate mineral structure in which half of the silicon-oxygen tetrahedra share two oxygens and half share three with other tetrahedra.

Down-ice. The direction toward which a glacier is moving.

Drainage basin. The area drained by a stream.

Drainage divide. A topographically high area that separates the drainage basins of two streams.

Drift. A general term for glacial deposits.

Driving forces. Those factors which tend to aid in mass movements.

Drumlin. A streamlined hill composed largely of till deposited by continental glaciers.

Dry valley. A valley not currently occupied by a stream because of climatic change, stream capture, or the diversion of surface water underground in areas of soluble rocks.

Ductile deformation or behavior. A type of deformation in which the strain produced is not proportional to the amount of stress applied. Ductile deformation is permanent; when stresses are relaxed the rock retains its deformed appearance.

Ductility. The property of a substance in which it can be drawn into thin wires.

Dunes. Streamlined hills composed of sand-sized particles deposited by wind.

Dynamic equilibrium. A system in which changes in one part are balanced by changes in another.

Dynamic metamorphism. Metamorphism caused by directed pressures in fault zones.

Ebb currents. Tidal currents that move from estuaries and wetlands into the ocean as the tidal level is falling.

Ecliptic. The plane traced by the Earth's orbit.

Eddies. Swirling, whirlpool-like masses of water which extend down through the streamflow and move with the current.

Effluent streams. Streams that receive a portion of their discharge from groundwater flow. (Cf. *influent stream.*)

Ejecta. Debris surrounding meteorite impact craters that was produced by the meteorite impact.

Elastic deformation or behavior. A type of deformation in which stress and strain are proportional and in which the effects of deformation disappear once the deforming forces are relaxed.

Elastic rebound hypothesis. An explanation for earthquakes in which built-up strain is released as a rock breaks.

Electric dynamo. A device used to generate electricity; in a dynamo, a rotating electrical conductor passes through magnetic lines of force, which induce a current in the conductor.

Electric energy. Energy associated with the flow of electrons through a substance.

Electron. A negatively charged subatomic particle which orbits the nucleus of an atom.

Electron capture. A form of nuclear reaction in which an electron is taken into the nucleus of an atom from the inner electron shells.

Electron shell. A group of electrons at approximately the same distance from the nucleus of an atom and containing the same energy.

Electrostatic forces. Forces of attraction and repulsion between positively and negatively charged ions or subatomic particles.

Element. An element consists of all atoms that have the same atomic number.

Emerald. The deep-green gem-quality variety of beryl.

End member. A principal component of a solid-solution series, e.g., forsterite, Mg_2SiO_4, and fayalite, Fe_2SiO_4, are end members of the olivine solid-solution series—$(Fe,Mg)_2SiO_4$.

End moraine. A moraine deposited near the terminus of a glacier.

Energy. The capacity to do work or cause activity.

Entrenched meander. A sinuous stream channel bordered by steep valley walls and a valley width which is close to that of the stream channel.

Eon. The largest subdivision of geologic time, Eons are subdivided into eras, periods, and epochs.

Epeirogeny. An episode of gentle regional uplift without extensive folding or faulting.

Ephemeral stream. A stream that flows only for a short time after a rainfall before the water evaporates or infiltrates into the ground.

Epicenter. The point on the Earth's surface directly above the focus of an earthquake.

Epoch. A span of geologic time. (See *era.*)

Era. A large subdivision of geologic time. Geologic eras are distinguished from one another on the basis of marked evolutionary changes in the fossil record. Eras are subdivided into *periods,* which in turn are subdivided into *epochs.*

Erosion. The removal of sediment, rock, or both by gravity, ice, water, or wind.

Erosional mountains. Mountains formed by broad regional upwarping and subsequent deep erosion.

Erratics. Rock fragments carried by glaciers far from their source and deposited on bedrock of a completely different type.

Esker. Stratified drift deposited by meltwater streams in channels carved into the surface of a glacier or in tunnels beneath a glacier and left as a narrow sinuous ridge after the ice has melted.

Estuary. A branch of the sea with tidal flow and saline-to-brackish water.

Eustatic changes in sea level. Variations in the ocean surface elevation resulting from increasing or decreasing volumes of ocean water.

Evaporation. Conversion of water to water vapor.

Evaporites. Chemical sedimentary rocks deposited from concentrated solutions.

Exfoliation. The separation of curved sheets of rock from rock exposures at the Earth's surface, brought about by the release of pressure.

Exfoliation domes. Rounded landforms that result from exfoliation.

Exotic terrane. Rocks underlying a large geographic area that have been moved great distances from the place where they first formed, and are now in contact with rocks that evolved in a totally different place.

Experiment. Scientific procedure carried out to test a hypothesis.

External drainage. A drainage system in which streams draining high-precipitation areas traverse a desert.

Extinct volcano. A volcano which has not erupted in historic time.

Extrusive rock. See *volcanic rock.*

Facies. (a) In sedimentary rocks, a group of rocks with distinctive characteristics that were deposited in a specific sedimentary environment. (b) In metamorphism, all rocks subjected to a particular set of pressure-temperature conditions.

Facies series. The sequence with which metamorphic facies show increased metamorphic grade in a given region.

Fall. A mass-movement process which involves sediment and rock falling through the air and accumulating at the base of a slope.

Fault. A fracture formed when rocks break and materials on opposite sides of the fracture are displaced relative to one another.

Fault-block mountains. Mountains formed by tension-induced normal faulting.

Fault gouge. Broken and crushed rock material produced in fault zones by movement of the fault blocks.

Faunal succession. The change in fossils from the bottom of a sequence of rocks to the top that reflects the evolution of life during the time span represented by the rocks.

Feeder. See *feeder conduit*.

Feeder conduit. A cylindrical channel through which lava rises to the surface to form a volcano.

Feldspar. A large family of framework silicate minerals, including the plagioclase and potassic feldspar groups.

Feldspathic sandstone. A sandstone that has a feldspar content between 10 and 25%.

Felsic rocks. Igneous rocks composed largely of potassic and plagioclase feldspar, often with quartz.

Fetch. The distance over which a wind blows to generate waves. The greater the fetch, the higher the waves which may be generated.

Filter pressing. Separation of crystals from magma when a magma chamber is deformed and the liquid is forced through narrow fissures, leaving the solid mineral grains behind.

Firn. A granular type of snow formed by partial melting of the edges of snowflakes and refreezing of the meltwater.

Fission. A type of nuclear reaction in which the nucleus of an atom splits into two or more parts.

Fission track dating. Dates obtained by determining the track density on unit areas of crystals.

Fjords. Steep-sided U-shaped valleys that have been drowned by the postglacial rise in sea level.

Flank eruption. Eruption of lava from the side of a volcanic slope rather than at the summit.

Flood basalt. Large expanses of basalt erupted from a series of fissures.

Flood currents. Tidal currents moving from the ocean into estuaries and wetlands as tidal level rises.

Floodplain. The area adjacent to a stream channel that is covered periodically with water when the stream overflows its banks.

Flow. A mass-movement process in which sediment or rock exhibits a continuity of motion in a plastic or semifluid state.

Flow lines. The paths individual particles make in moving through a fluid such as air or water.

Fluorescence. The color of a mineral under ultraviolet light, often quite different from its appearance under normal light.

Focus. The actual site of fault displacement and source of seismic-wave energy.

Fold. A flexure in rock formed when rocks bend in response to external deforming forces.

Fold axis. A line that traces the maximum curvature of a folded layer and separates the two limbs of the fold.

Fold mountains. Mountains that form as a result of intense compression and hence are characterized by tightly folded rock.

Fold nose. The convergence of bands of rock that results when plunging folds are eroded.

Foliation. Parallel alignment of planar minerals in a rock.

Footwall block. The fault block that lies beneath an inclined fault plane.

Foreset bed. An inclined layer of sandy material deposited upon or along an advancing and relatively steep frontal slope, such as the outer margin of a delta or the lee (downwind) side of a dune.

Fossil. The preserved remains or traces of a plant or animal that lived in the geologic past.

Fossil fuel. Energy-producing materials formed by partial decomposition of ancient organic materials, e.g., coal, oil, natural gas.

Fracture. (a) Breakage in minerals that occurs in random directions relative to the mineral structure. (b) Irregular, randomly oriented breaks in rocks.

Framework silicate structure. A type of mineral structure in which each silicon-oxygen tetrahedron shares all four of its oxygens with other tetrahedra.

Fringing reef. A reef built up against a coastline.

Frost creep. The downslope movement of particles resulting from repeated freezing and thawing of slope materials.

Frost wedging. A process of physical weathering in which water in rock openings freezes and expands, fracturing the rock.

Fumarole. A volcanic vent that erupts gases.

Fusion. A nuclear reaction in which two nuclei join.

Gabbro. A coarse-grained mafic igneous rock composed of plagioclase feldspar and ferromagnesian minerals such as olivine and pyroxenes.

Galilean satellites. Four large, planet-sized satellites of Jupiter (Callisto, Ganymede, Europa, and Io).

Garnet. A family of minerals containing independent silicon-oxygen tetrahedra:
$(Ca, Fe^{2+}, Mn^{2+}, Mg^{2+})_3(Al, Fe^{3+}, Cr^{3+})_2(SiO_4)_3$.

Geobaric gradient. The rate at which lithostatic pressure increases with depth in the Earth.

Geochronology. The science of obtaining radiometric ages by the analysis of the radioactive elements in a mineral, rock, or organic material.

Geologic cycle. A series of geologic processes in which Earth materials are recycled.

Geologic structures. Features that reveal the deformation history of a rock, such as folds or faults.

Geologic time. A view of time relative to the formation of the Earth, 4.6 billion years ago, rather than historical time.

Geometric variables. The set of variables which describe the shape of a stream channel.

Geomorphology. The study of surface landforms and the processes that form them.

Geophysics. The application of physics to the study of the Earth, including seismologic, gravity, magnetic, and heat-flow studies.

Geothermal energy. Energy obtained by tapping the heat contained within the Earth.

Geothermal gradient. The rate at which temperature in the Earth increases with depth.

Geyser. Groundwater that emerges from the ground as a jet of steam and hot water.

Ghyben-Herzberg ratio. A ratio describing the static relation of fresh groundwater and saline groundwater in coastal areas. For each meter the water table is above sea level, the saltwater surface is displaced 40 m below sea level, i.e., in a ratio of 1:40.

Glacial economy. The balance between accumulation of snow and ice in the snowfield and loss of snow and ice along the length of the glacier. The glacial economy determines whether glaciers will advance, retreat, or stagnate.

Glacial striations. Elongate scratches carved in rock surfaces by rock particles carried at the base of a glacier.

Glacier. Mass of ice which forms on land by the compaction and recrystallization of snow and moves downslope or outward in all directions under the pressure of its own weight.

Glass. An igneous rock that cooled so quickly that no minerals had a chance to form.

Gneiss. Metamorphic rock that is both layered and foliated.

Gneissossity. A featured of layered metamorphic rocks in which alternating layers are foliated and nonfoliated and commonly color-banded as well.

Gondwanaland. A supercontinent composed of what are now the Southern Hemisphere continents plus India.

Gouge. Crushed and broken rock held together by a finer-grained matrix. Gouge results from the grinding of two fault blocks against one another during deformation.

Graben. See *horst and graben*.

Graded bedding. A variety of stratification in which the particle size decreases systematically upward in a bed.

Gradient. The difference in elevation of a stream bed or slope over a given horizontal distance.

Granite. A coarse-grained felsic igneous rock composed mainly of quartz, potassic feldspar, and plagioclase feldspar, with minor amounts of ferromagnesian minerals (biotite, hornblende) and muscovite.

Granoblastic texture. Metamorphic texture in which grains are oriented randomly; i.e., there is no foliation or lineation.

Granofels. A metamorphic rock with a granoblastic texture (nonfoliated and nonlineated).

Granule. A particle of sediment with a diameter between 2 and 4 mm.

Gravity. A force of attraction that exists between all substances because of their mass.

Gravity anomaly. An area of the Earth's surface where the measured force of gravitational attraction is significantly higher (positive anomaly) or lower (negative anomaly) than the average.

Greenhouse effect. The process by which increased carbon dioxide in the atmosphere prevents heat from escaping into space. The result is an increase in temperature at the surface of the Earth.

Greenschist. (a) Foliated, low-grade metamorphosed mafic rock. (b) Low-grade facies of regional metamorphism.

Greenstone. Metamorphosed mafic rock with no foliation.

Groin. Rock, concrete, or wood structure built into the surf to trap sand and build out beaches.

Groundmass. The fine-grained matrix of a porphyritic igneous rock.

Ground moraine. Till deposited as a sheet beneath a moving glacier.

Groundwater. Water that is accumulated in sedimentary pores and rock fractures beneath the ground surface.

Guyot. A flat-topped seamount formed by subsidence and wave erosion of a volcano.

Gypsum. (a) An evaporite mineral with the chemical composition $CaSO_4$. (b) A chemical sedimentary rock composed of the mineral gypsum.

Habit. The typical crystal shape of a mineral.

Half-life. In nuclear reactions, the amount of time necessary for 50% of the total number of parent atoms to decay into daughter atoms.

Halide. A compound containing one of the halogen element anions (chlorine, fluorine, etc.); e.g., halite—NaCl; fluorite—CaF_2.

Halite. (a) A mineral with the composition NaCl. (b) A chemical sedimentary rock composed of crystals of halite; commonly called rock salt.

Hanging valley. A preglacial tributary valley not eroded as deeply by valley glaciers as the main stream valley, and as a result left isolated well above the main stream valley when the ice melts.

Hanging wall block. The fault block that lies above an inclined fault plane.

Hardness. The resistance of a mineral to being scratched.

Hawaiian-type islands. Basaltic island chains associated with hotspots rather than ocean ridges or trenches.

Headwall. The steep, nearly vertical slope formed at the upslope side of a cirque.

Headward erosion. Erosion of previously undissected areas at the upstream end of a stream or the upvalley end of a glacier.

Heat energy. A form of kinetic energy in which the movement is on the atomic scale.

Herringbone cross-bedding. Successive cross-beds in which the laminae dip in directions which differ by as much as 180°. This type of bedding is diagnostic of the reversing currents (flood-ebb) in tidal environments.

High-grade metamorphism. Metamorphism in which rocks change drastically from their original states.

Highlands. Lunar topographic features that stand above the flatter, lower maria.

High-pressure/low-temperature metamorphism. Metamorphic facies series produced by more rapid increases in pressure than in temperature.

High-temperature/low-pressure metamorphism. Metamorphic facies series produced by more rapid increases in temperature than in pressure.

Hill. A topographic high standing less than 300 m above its surroundings.

Horn. A steep-sided faceted mountain peak formed by headward erosion of several cirque glaciers.

Hornblende schist. Foliated metamorphic rock composed mostly of the amphibole mineral hornblende.

Hornfels. Granoblastic rock produced by contact metamorphism.

Horst and graben. A type of structure developed in regions of tension in which uplifted blocks (horsts) are separated from downdropped blocks (graben) by normal faults.

Hotspots. Fixed sources of heat in the mantle below the asthenosphere that are the cause of volcanism in Hawaiian-type islands.

Hydration. The addition of water to a mineral or rock, producing new minerals.

Hydraulic variables. Variables such as velocity and discharge that describe streamflow.

Hydroelectricity. Electric energy produced by the passage of running water through a turbine.

Hydrologic cycle. The continued transformation of water into its different forms by evaporation, melting of ice, etc.

Hydrolysis. The reaction between minerals and the H^+ and $(OH)^-$ ions formed during dissociation of water.

Hydrosphere. The liquid outer covering of the Earth, encompassing the oceans, rivers, and lakes.

Hydrothermal deposit. See *hydrothermal solution*.

Hydrothermal solution. Hot water formed either by heating of groundwater by plutons or by expulsion from a cooling magma. Such solutions contain ions that may be deposited in rocks to form economically valuable mineral ores called *hydrothermal deposits*.

Hypersaline water. Water with more than 40 parts per thousand of dissolved solids.

Hypothesis. In the scientific method, a tentative explanation for an observed phenomenon.

Iceberg. A floating mass of ice broken off from the terminus of a glacier along a coast.

Ice cap. A glacier so thick that it covers most of the topographic features of an entire region.

Ice centers. Huge snowfields that spawned continental glaciers.

Ice-rafting. Transport of sediment into the deep ocean basin by debris-laden icebergs which then melt and drop the sediment to the seafloor.

Ice sheet. A very large and thick glacier covering an area of subcontinental or continental size.

Ice shelf. A mass of floating ice consisting of those parts of a glacier which advanced beyond the shoreline and float because of their low specific gravity.

Ice-wedge cast. A structure formed when an ice wedge melts during postglacial thawing and leaves an open space which subsequently fills with washed-in fine sediment.

Igneous rock. Rock formed by solidification of molten rock material.

Impact craters. Bowl-like pits excavated in the surface as a result of meteorite impacts.

Impact metamorphism. Metamorphism caused by meteorite impact.

Inclination. The angle between magnetic lines of force and the Earth's surface, ranging from 90° at the magnetic poles to 0° at the magnetic equator.

Inclined fold. A fold with an axial plane that is neither horizontal nor vertical.

Inclusions. A principle of relative-age dating which states that a rock containing fragments of another rock must be younger than the fragments that it contains.

Incompetent rock. Rock that changes shape easily during deformation, with only little stress applied.

Independent tetrahedron structure. Silicate mineral structure in which silicon-oxygen tetrahedra do not share any oxygen ions.

Index mineral. A mineral whose presence in a metamorphic rock can be used to estimate the relative intensity of metamorphism.

Inert elements. Elements with filled electron shells that do not enter into chemical reactions.

Infiltration. The migration of a fluid through openings in a sediment or rock.

Infiltration capacity. The rate at which water may pass through surface materials.

Influent stream. A stream that is above the local or regional water table and thus contributes water to the groundwater reservoir. Compare with *effluent stream*.

Inselberg. A mountain left as an erosional remnant in arid regions; it is surrounded by gently sloping pediments.

Instream use. Human use of water within a stream channel.

Intensity. A measure of the strength of an earthquake based on the damage to human-built structures; it is measured by the modified Mercalli intensity scale.

Interglacial. A relatively warm period of time separating episodes of continental glaciation.

Intergranular slip. A mechanism of deformation in which grains slide past one another to new positions, changing the shape of the deformed substance.

Intermediate rocks. Igneous rocks with compositions between those of felsic and mafic rocks.

Intermittent streams. Streams that are dry for part of the year but receive part of their flow from the water table when it is high enough.

Internal drainage. A type of drainage system in which streams flow into topographic lows bounded by mountains and consequently never connect with the ultimate base level—the oceans.

Intrusive rock. See *plutonic rock*.

Ion. A charged particle formed when an atom gains or loses an electron in a chemical reaction.

Ionic bonding. The joining of atoms that have become electrically charged by gaining or losing electrons.

Ionic substitution. The ability of an ion to take the place of another in a mineral structure because the two have similar sizes and charges.

Iron meteorite. A meteorite composed predominantly of an alloy of nickel and iron.

Island arc. An arc-shaped volcanic island chain found on the concave (landward) side of oceanic trenches.

Isoclinal fold. A fold whose limbs dip in the same direction at the same angle.

Isograd. The boundary between two metamorphic zones along which a new index mineral appears. Ideally, metamorphic conditions are equal everywhere along an isograd.

Isostasy. The buoyancy of low-density rock material in higher-density material.

Isostatic rebound. Uplift of an area due to the removal of mass from the surface, e.g., following melting of a continental glacier or removal of a mountain mass by erosion.

Isotope. Isotopes of an element are atoms that have the same atomic number but different atomic masses.

Isotopic time scale. A time scale subdivided on the basis of radiometric dating.

Jade. A semiprecious stone that is either the amphibole tremolite or the pyroxene jadeite.

Jetties. Rock or concrete structures built into the surf zone on either side of a tidal inlet to prevent the inlet from closing through deposition of sand.

Joint. Systematically aligned fractures along which there has been no displacement. Joints are produced by cooling in igneous rocks, unloading due to erosion, and deformation.

Jovian planets. The fifth through eighth planets—Jupiter, Saturn, Uranus, Neptune—large, low-density bodies composed of hydrogen, helium, ammonia, and methane.

Kame. A conical hill composed of stratified drift formed by meltwater deposition in the crevasses of a stagnating glacier.

Kame terrace. A flat-topped meltwater stream deposit formed between a stagnating valley glacier and the valley walls.

Karst topography. Topography characterized by sinkholes, underground drainage, caverns, and a few hills that exist because they are resistant to groundwater solution.

Karst towers. Residual masses of limestone that stand above the surrounding lowlands.

Kettle or kettle hole; kettle lake. A bowl-shaped depression in outwash plains formed by the melting of a block of stagnant ice that had been buried by outwash. It is called a kettle lake when filled with water.

Kilobar. A unit used to measure pressure in the Earth; equivalent to approximately 1000 times atmospheric pressure.

Kimberlite. An ultramafic igneous rock that is the most common source of diamond crystals.

Kinetic energy. The ability of a moving object to induce activity in other objects.

Laccolith. A flat-bottomed pluton having a dome-shaped top that arches the overlying rocks.

Lag time. The time difference between the center of mass of rainfall and center of mass of the stream discharge.

Laminae. Rock and sediment layers less than 1 cm thick.

Laminar flow. A type of fluid flow characterized by parallel, horizontal movement with little or no vertical movement of the flow lines.

Lateral continuity. A principle of stratigraphy which states that, as a sedimentary rock layer is deposited, it extends outward horizontally until it either thins and disappears or until it terminates against the boundaries of the basin in which it accumulates.

Lateral crevasse. Crevasse formed at the sides of a valley glacier.

Lateral erosion. Erosion of materials on the side of a stream or valley glacier.

Lateral moraine. Rock fragments carried at the sides of a valley glacier and deposited as a low ridge when the ice melts.

Laterite. A soil formed in tropical climates with high rainfall that is enriched in oxides of iron and/or aluminum.

Laurentide ice sheet. A Pleistocene continental glacier which covered most of Canada and extended deeply into the northern United States.

Lava. Molten rock material that has extruded onto the Earth's surface.

Lava dome. Bulbous bloblike shape characteristic of the eruption of extremely viscous lava.

Lava plateau. A broad upland region underlain by lava.

Lava tube. An elongate cavity in a lava flow.

Leachate. A liquid formed as rain soaks through, and reacts with, the garbage in a landfill.

Leached horizon. The A zone of a soil profile from which clays and iron oxides are removed by percolating groundwater.

Leaching. The removal of soluble material (generally from soil) by water percolating through the soil.

Left-lateral fault. A strike-slip fault in which displacement of one block has been to the left, as viewed from the other block.

Lignite. A brownish-black coal that is an intermediate stage between peat and subbituminous coal in the coal-forming process.

Limbs. The opposite sides of a fold.

Limestone. A sedimentary rock composed of calcite, $CaCO_3$.

Lineation. Parallel alignment of rod-shaped minerals in a rock.

Liquefaction. The loss of cohesion and structural strength in sands or clays brought about by water coating grain surfaces and reducing friction drastically.

Liquid waste. Harmful liquid byproducts of domestic, agricultural, and industrial activities.

Lithic wacke. A sandstone composed of rock fragments and more than 15% matrix.

Lithification. The process of transforming a sediment into a sedimentary rock.

Lithographic limestone. A very fine-grained rock derived from the lithification of muds produced by calcareous algae.

Lithosphere. (a) The uppermost 100 km of the Earth, composed of relatively rigid rock material incorporating both crust and part of the upper mantle. (b) The rock material of which moving plates are made.

Lithostatic pressure. Pressure that is equal in all directions on a rock buried beneath a cover of other rocks and sediments.

Load. The sedimentary materials carried by a transporting medium. The physical load is carried as particles; the chemical load is transported as ions in solution.

Loess. Silt derived from deflation of glacial outwash or desert sediments and deposited downwind of the source.

Longitudinal dune. A massive linear sand dune oriented parallel to the dominant wind direction.

Longitudinal profile. A cross section showing the gradient of a stream from its headwaters to a point downstream.

Longitudinal wave. A seismic wave (P wave) propagated by the straight-line, back-and-forth motion of particles parallel to the direction of wave propagation.

Longshore currents. Currents moving parallel to the coast in the nearshore zone.

Longshore drift. Particles of sediment that are moved along a shoreline by wave action and longshore currents.

Lopoliths. Funnel-shaped plutons generally composed of mafic rock.

Love wave. A seismic surface wave propagated by straight-line transverse particle motion that lies in the surface along which the wave is moving.

Low-grade metamorphic rocks. Rocks that have not changed much from their original state during metamorphism.

Low-velocity zone. A region in the upper mantle in which seismic-wave velocities decrease, probably because of decreased rigidity of the mantle material.

Luster. A physical property of minerals that describes the manner in which light passes through them.

Mafic gneiss. Foliated metamorphic rock composed largely of ferromagnesian minerals.

Mafic rocks. Igneous rocks rich in ferromagnesian silicate minerals such as olivines and pyroxenes but low in potassic feldspar and quartz.

Magma. Molten rock material.

Magma chamber. The region within the Earth where magma crystallizes.

Magmatic differentiation. Crystallization of magma during which early-formed minerals are separated from the remaining melt.

Magnetic anomalies. Areas of the Earth's surface where the magnetic field strength is abnormally high (positive anomaly) or low (negative anomaly).

Magnetic dating. Dates determined from changes over time in the intensity and polarity of the Earth's magnetic field.

Magnetic declination. The angular difference between the magnetic and geographic poles.

Magnetic domains. Regions within a substance characterized by parallel alignments of the magnetic fields of atoms.

Magnetic equator. An imaginary line connecting all those points on the Earth's surface at which the magnetic lines of force parallel the surface.

Magnetic field. The area around a magnetic substance affected by its force of magnetic attraction.

Magnetic inclination. Angular relationship between Earth's magnetic lines of force and its surface.

Magnetic poles. Points on the Earth's surface (or on any magnetic object) from which magnetic lines of force emanate.

Magnetic reversals. Periods during which the Earth's magnetic field reversed its polarity.

Magnitude. A measure of the amount of energy released during an earthquake, based on the amplitude of seismic waves; calculated according to the Richter magnitude scale.

Malleability. The ability of a substance to be pounded into flat sheets.

Manganese nodules. Authigenic concretions formed largely from manganese oxides precipitated from seawater.

Mantle. The intermediate region of Earth's interior lying between the crust and the core. It is composed largely of high-density silicate minerals.

Marble. Metamorphic rock formed by recrystallization of limestone or dolomite.

Mare. A large, dark-colored, relatively flat basin on the Moon filled with basalt (plural, *maria*).

Marginal sea. A shallow sea that lies between a continent and an island arc.

Maria. See *mare*.

Mass movement. Downslope movement of regolith and rock at the Earth's surface caused by gravity.

Mass wasting. See *mass movement*.

Matrix. (a) Fine-grained particles that fill spaces between coarser grains in a sedimentary rock. (b) Fine-grained crystals in a porphyritic igneous rock.

Mature soil. A soil which has developed sufficiently so that it exhibits several soil horizons.

Meander. A curve in the course of a stream.

Meandering stream. A stream with a channel composed of a series of sinuous curves.

Medial crevasse. Crevasse which forms on the central part of the surface of a valley glacier.

Medial moraine. A low ridge of rock debris formed when the lateral moraines of two merging valley glaciers join.

Melange. A sedimentary rock composed of debris eroded from island arcs. The debris is enclosed in a shale matrix. This mixed rock is commonly deformed as it is being deposited.

Melting. The change from the solid state to the liquid state.

Melting point. The temperature at which a substance passes from the solid to the liquid state.

Mesa. A flat-topped erosional remnant bordered on its sides by steep slopes or cliffs and capped with resistant and nearly horizontal rocks (commonly lava flows or sandstones).

Metallic bonding. Joining of atoms by the sharing of all outer shell electrons among all the nuclei present.

Metamorphic grade. A measure of the relative intensity of metamorphism (e.g., low grade).

Metamorphic facies. All metamorphic rocks produced at the same condition of metamorphism.

Metamorphic facies series. The sequence with which one metamorphic faces passes into another in an area.

Metamorphic minerals. Minerals produced during metamorphic reactions.

Metamorphic rocks. Rocks formed within the Earth when previously existing rocks are subjected to temperatures and pressures different from those under which they first formed.

Metamorphic zone. A mapped area in which metamorphic intensity was approximately the same, as shown by the presence of a particular metamorphic index mineral.

Metamorphism. Changes in the mineralogy, texture, and composition of a rock caused by physical or chemical conditions different from those under which the rock first formed.

Metasomatism. Metamorphism in which the composition of the affected rock changes significantly by addition or removal of ions.

Meteorite. A rock fragment from interplanetary space that has collided with the Earth or another planetary body.

Mica. A family of sheet silicate minerals characterized by excellent cleavage in one direction.

Microcline. A form of potassic feldspar.

Migmatite. A very high-grade metamorphic rock composed of solidified material that had melted from the metamorphic rock mixed with the initial rock's unmelted residue.

Milankovitch theory. A theory to explain cyclical fluctuations in Pleistocene climate by changes in the ellipticity of the Earth's orbit and the inclination of the Earth's rotational axes.

Mineral. A naturally occurring inorganic solid with an ordered internal arrangement of atoms or ions and a relatively fixed chemical composition.

Mineral deposit. A concentration of minerals, rocks, or organic materials such that mining and processing them is physically and economically possible.

Mineral fuels. Fuels produced by geologic processes and stored in rocks and minerals (e.g., uranium, thorium).

Mineraloid. A substance similar to a mineral but lacking either an ordered internal structure or a relatively fixed chemical composition (e.g., opal or limonite).

Modified Mercalli intensity scale. A scale used to measure the strength of an earthquake; based on the amount of damage that has occurred.

Moho. See *Mohorovičić discontinuity*.

Mohorovičić discontinuity. The boundary between the Earth's crust and mantle, identified by a sudden increase in seismic-wave velocity in the upper mantle.

Mohs hardness scale. A series of reference minerals used to estimate the hardness of minerals.

Mold. A three-dimensional impression of a fossil organism. Casts are formed when sedimentary particles or crystalline material fills a mold.

Molecule. The smallest amount of a compound that can exist.

Molecular shell. In covalent bonding, the combination of outer shell electrons that orbits the atoms that are bonded together.

Moment of inertia. The tendency of a spinning object to keep on spinning.

Moonstone. A translucent, semiprecious gemstone variety of either albite or orthoclase.

Moraine. A landform, usually composed largely of till, which is deposited at the margin of a glacier.

MORB. *Mid-ocean ridge basalt.* A distinctive type of basaltic lava formed at ocean ridges. MORB is different in composition from basalts extruded on continents, island arcs, and Hawaiian-type oceanic islands.

Mountain. A topographic high that stands more than 300 m above its surroundings.

Mountain glacier. Any glacier in a mountain range except an ice cap or ice sheet. It usually originates in a cirque and may flow down into a valley previously carved by a stream. Also referred to as an *alpine glacier*.

Mountain range. A group of mountain peaks that have been formed at approximately the same time by the same process.

Mountain root. The downward projection of relatively low-density rocks beneath major mountain ranges.

Mountain system. A group of mountain ranges.

Mud cracks. Polygonal patterns of downward-closing fractures formed by the shrinking which occurs in the drying out of a clay-rich sediment.

Mudflow. Downslope movement of fine-grained material that contains large (up to 30%) amounts of water.

Multiple working hypotheses. In the scientific method, the practice of proposing as many hypotheses as can fit the observations and data, and testing them simultaneously.

Mylonite. A rock with a mylonitic texture.

Mylonitic texture. A texture produced in dynamic metamorphism in which original rock grains have been stretched, flattened, crushed, and smeared out into a well-developed foliation.

Native state. An element whose atoms occur in combination with one another, rather than combined with atoms of other elements.

Natural bridge. An archlike rock formation. One type of natural bridge forms when a stream abandons a meander and cuts through the narrow meander neck; in a limestone area, a natural bridge represents the remnant of the roof of an underground cave or tunnel that has collapsed.

Natural levees. Low ridges of sediment that parallel stream channels and are deposited when streams overflow their banks.

Nearshore zone. The body of water between the beginning of the breaker zone and the beach; includes the breaker and surf zones.

Neutral plate boundaries. Transform faults along which two plates slide past one another, involving neither creation of new lithosphere nor destruction of old.

Neutron. An uncharged subatomic particle found in the nucleus of an atom. It may be thought of as being composed of a proton and an electron.

Nonconformity. An unconformity in which eroded intrusive igneous rocks or metamorphic rocks are overlain by sedimentary rocks.

Nonmagnetic materials. Substances that are not affected by a magnetic field.

Nonplunging fold. A fold with a horizontal axis.

Nonradiogenic. An element or isotope which has not been produced by radioactive decay of another element or isotope.

Nonrenewable resource. A resource that is being depleted more rapidly than it is being formed.

Normal fault. A dip-slip fault in which the footwall block has moved upward relative to the hanging wall block as a result of tensional stresses.

Nuclear energy. Energy released from a nucleus during a nuclear reaction.

Nuclear reaction. Alteration of atoms involving changes in their nuclei.

Nucleus. The central part of an atom in which the protons and neutrons are located.

Nuée ardente. A fiery cloud composed of tephra and gases that rises from a volcano and moves rapidly downslope like a superheated density current.

Oasis. An area of vegetation and standing water in a desert formed by artesian springs rising from any break in a confined aquifer below.

Obduction. The scraping of part of a subducted plate and incorporation of the scraped material in the upper plate by thrusting during subduction.

Obsidian. A glassy igneous rock with a composition similar to that of granite.

Ocean. A large body of water whose basin is underlain by a one-layer simatic crust.

Ocean fractures. Linear topographic lows that segment and offset the ocean ridges, and correspond to transform faults.

Ocean ridges. Large elongate mountain ranges found in every ocean; they are the sites of earthquake activity and basaltic volcanism.

Oceanic trenches. Deep, arc-shaped depressions in the seafloor in which the greatest water depths of the oceans are recorded. They are the sites of extensive earthquake activity and correspond to subduction zones.

Oceanography. The study of the oceans, including their physical, chemical, biologic, and geologic aspects.

Offstream use. Human use of water that has been taken from its natural reservoir and transported to its place of use (e.g., for cooling a power plant, irrigation, etc.).

O horizon. The uppermost horizon in soils of humid regions. The O horizon is rich in fresh and decomposed plant matter.

Oil. A liquid composed of compounds of carbon and hydrogen which is believed to form from diagenetic changes in marine plant and animal matter that accumulated in fine-grained sediments.

Oil and tar sands. Sands or sandstones with pore spaces filled with highly viscous asphalt.

Oil shales. Shales with an exceptionally high content of organic matter that yield oil when heated.

Oil trap. A geologic setting in which oil is concentrated because of structural (e.g., anticlinal crests, salt domes) or stratigraphic (e.g., facies interfingering) relationships involving permeable and impermeable rocks.

Olivine group. A solid-solution series of independent tetrahedral ferromagnesian silicates with the general composition $(Fe,Mg)_2SiO_4$.

Ooid. Spherical, concentrically banded grains believed to form from inorganic precipitation of calcium carbonate in an agitated marine environment.

Oolite. Sedimentary rock comprised of ooids.

Oolitic limestone. A rock containing ooids of calcium carbonate cemented together by calcite.

Ophiolite suite. A sequence of ultramafic-through-basaltic plutonic and volcanic rocks capped by deep-ocean sediments that is found in mountains produced by intense compression. It is thought to result from obduction of oceanic crust.

Ore minerals. Minerals that have been naturally concentrated to form metallic deposits.

Original horizontality. A principle of relative-age dating which states that sediments are deposited in horizontal or nearly horizontal layers.

Orogeny. An episode of mountain building characterized by extensive folding and faulting.

Orthoclase. A type of potassic feldspar.

Oscillation ripple marks. Symmetrical ripple marks formed by wave action.

Outwash. Sediments deposited by meltwater streams well beyond the terminus of a glacier.

Outwash plains. Flat and gently sloping plains underlain by meltwater stream deposits built beyond the margins of glaciers.

Overwash. Washing of sand from the beach across a barrier island by storm waves.

Oxbow lake. An arc-shaped lake formed when water fills cutoff meander loops.

Oxidation. A type of chemical weathering in which oxygen from the atmosphere interacts with minerals and rocks.

Packing. The arrangement of particles in sediment.

Pahoehoe. Highly fluid, ropy lava.

Paleo-. A prefix meaning *ancient,* as in *paleocurrent* (ancient current), *paleoclimate* (ancient climate), or *paleowind* (ancient wind).

Paleomagnetism. Earth's ancient magnetic field as recorded by magnetic minerals and rocks.

Pancaking. A form of earthquake-induced building collapse in which the upper floors of a building fall onto the lower ones.

Pangaea. A single "supercontinent," composed of all the separate continents as we now know them, which may have existed prior to the Jurassic opening of the modern Atlantic Ocean.

Parabolic dune. A crescent-shaped sand dune whose tips point upwind.

Paramagnetic materials. Substances that are affected by application of a magnetic field but do not act magnetically once the field is removed.

Parent element. A radioactive element before it decays to form a new element.

Parent material. A term used to describe the original material (rock, sediment, or soil) which is being weathered.

Partial melting. Melting of a portion of a source rock to produce magma.

Passive continental margin. Coastlines where the shoreline is located far from the nearest plate boundary so that earthquake and volcanic activity are absent.

Patterned ground. Surface features such as bands, stripes, polygonal networks of rock debris, and polygonally fractured soil formed by the freezing and thawing of the upper part of permafrost.

Peat. An unconsolidated deposit of semicarbonized plant remains with a high moisture content; an early stage in the development of coal.

Pebble. A particle of sediment with a diameter between 4 and 64 mm.

Pedalfer. A type of soil produced in humid regions characterized by the absence of calcium carbonate and the accumulation of clay minerals and iron oxides in the B horizon.

Pediment. A gently sloping surface carved into the bedrock at the base of a mountain range in arid or semiarid environments.

Pedocal. A type of soil common in arid regions where calcium carbonate is accumulated rather than leached from the soil as it is in humid regions.

Pegmatite. An extremely coarse-grained igneous rock.

Pelagic ooze. Very fine-grained, continent-derived oceanic sediment.

Pelagic sediments. Deposits which form from particles that have settled through the water.

Pelitic rock. Rock containing large amounts of alumina, Al_2O_3, e.g., shales, slate, and schist.

Peneplain. A large, featureless, flat land surface caused by extensive stream erosion.

Perched water table. A local zone of saturation which exists at a level higher than the regional water table because of the presence of an underlying layer of impermeable rock or sediment.

Perennial stream. A stream which flows year-round and receives most of its discharge from groundwater flow.

Peridot. The gem variety of olivine.

Periglacial climate. The cold climate that occurs at the periphery of continental glaciers.

Period. A span of geologic time. (See *era*.)

Permafrost. Perennially frozen ground.

Permeability. A measure of the ease with which a fluid can move through a rock or sediment.

Phaneritic. An igneous texture coarse enough for individual mineral grains to be seen with the naked eye.

Phenocryst. See *porphyritic texture*.

Phosphorite. A chemical sedimentary rock composed of crystals of complex phosphates of calcium.

Photic zone. The upper few tens of meters of the oceans to which sunlight can penetrate.

Phyllite. Medium-grained foliated metamorphic rock having aligned micas which produce a dull sheen.

Physical load. The sediment carried as discrete particles within a fluid.

Physical properties. Those attributes of minerals which are caused by mineral composition and structure (e.g., hardness, color, cleavage, etc.).

Physical weathering. The disintegration of a rock into progressively smaller pieces without altering the chemical composition of the minerals present.

Piedmont glacier. A large glacier formed when two or more trunk glaciers merge at the foot of a mountain range and flow across the adjacent plains.

Pillar. A column formed by the merging of a stalactite and stalagmite in a cave or cavern.

Pillows. Globular masses of volcanic rock formed when lava flows into a body of water.

Placer deposit. A surficial mineral deposit concentrated by water flow in streams by wave action in the near-shore zone, or by wind in dunes.

Plagioclase feldspar. A complete solid-solution series between the framework silicates albite, $NaAlSi_3O_8$, and anorthite, $CaAl_2Si_2O_8$.

Planet. A spheroidal body that revolves around the Sun in a fixed, nearly circular orbit.

Planetology. The study of the planets, their satellites, and other bodies in the solar system.

Plastic zone. The lower zone of movement in a glacier in which the pressure from the overlying snow and ice enables the glacier to move by plastic flow.

Plateau. An extensive flat area with little internal relief that stands prominently above its surroundings.

Plate. Rigid block composed of crust and upper mantle material which rests upon, and can move across, the asthenosphere.

Plate tectonics. A model for Earth behavior in which the outer 100 km of the planet (the lithosphere) consists of a small number of individual masses called plates which can move independently of one another. Separation and collision of plates results in the formation of ocean basins and mountains.

Playa lake. A temporary lake formed in the low areas of desert basins after a rainfall.

Plucking. A process of glacial erosion in which meltwater at the base of a glacier flows into rock crevices, refreezes, and loosens the rock so that it is incorporated in the advancing ice.

Plug. A small, irregularly shaped body of intrusive igneous rock.

Plunging fold. A fold whose axis is not horizontal.

Pluton. A body of intrusive igneous rock.

Plutonic rock. An igneous rock formed by the solidification of magma within the Earth.

Pluvial climate. Cool, wetter climate that affected presently arid areas during the Pleistocene Epoch.

Pluvial lakes. Lakes that formed in what are now arid areas during periods of pluvial climate.

Point bars. Arc-shaped sand bars deposited on the inner parts of meander loops.

Polar desert. A cold desert which forms in polar areas where there is low precipitation.

Polar compound. A compound, such as water, H_2O, that has positive and negative ends because of the arrangement of its constituent atoms/ions, even though the compound is electrostatically neutral.

Polar wandering curve. A diagram showing the apparent migration of the magnetic poles with time, relative to a landmass.

Polymorphs. Minerals that have the same chemical composition but different internal structures (e.g., diamond and graphite).

Pools. Relatively deep-water portions of streams with straight channels.

Pore. A space between grains in a rock or sediment.

Porosity. That portion of the volume of a sediment or rock which is made up of open spaces.

Porphyritic texture. An igneous texture in which there are a few coarse grains (phenocrysts) found in a generally finer-grained groundmass.

Porphyroblast. A large metamorphic mineral surrounded by finer grains.

Potassic feldspar. Any of the framework silicate minerals (orthoclase, microcline, or sanidine) in which potassium is the dominant large cation, $KAlSi_3O_8$.

Potential energy. Energy in storage prior to being used.

Potholes. Circular cavities in a stream bed carved by rock particles that have been swirled around in eddies.

Precambrian shield. An extensive terrain underlain by rocks of Precambrian age.

Precision Depth Recorder (PDR). An instrument that measures water depth by timing the passage of sound energy from a ship to the seafloor and back.

Preferred orientation. Alignment of particles in a specific direction.

Proglacial lakes. Lakes fed by glacial meltwater which accumulated behind dams of ice or glacial deposits.

Prograde metamorphism. Metamorphism that occurs as heat and pressure increase toward their maximum values for a given metamorphic event.

Progressive metamorphism. Gradual readjustment of a rock to continuously changing temperature, pressure, and chemical conditions.

Protolith. The initial rock that has been converted by metamorphism to a metamorphic rock.

Proton. A subatomic particle found in the nucleus with 1832 times the mass of an electron, and a positive charge.

Pumice. A vesicular (porous), glassy, light-colored igneous rock with a composition similar to that of granite.

P wave. A longitudinal seismic body wave which, because of its high velocity, is generally the first wave to reach a seismograph after an earthquake occurs.

P-wave shadow zone. That area on the Earth's surface which does not receive direct P waves after an earthquake.

Pyroclastic texture. Igneous texture characterized by broken particles ejected into the air from a volcano. (See *tephra*).

Pyroxene. A group of minerals containing single chains of silicon-oxygen tetrahedra.

Pyroxene granofels. High-grade unfoliated metamorphic rock composed largely of pyroxene and plagioclase feldspar. Derived from a mafic protolith.

Quartz. A framework silicate mineral composed entirely of silicon and oxygen—SiO_2.

Quartz arenite. A sandstone composed of quartz grains and crystalline cement with less than 15% matrix. (See also *arenite*.)

Quartzite. Metamorphosed quartz sandstone.

Quartzose sandstone. See *quartz arenite*.

Quartz wacke. A sandstone composed of quartz grains and more than 15% matrix. (See also *wacke*.)

Quick clay. Clay deposits that undergo rapid liquefaction.

Radial pattern. A stream network in which streams either drain away in all directions from a central high point or drain into a low point from all directions.

Radiant energy. Energy forms transmitted as electromagnetic waves, e.g., light, X-rays.

Radiation. Transfer of heat energy by conversion to a form of radiant energy (commonly infrared energy).

Radioactive isotopes. Those isotopes of elements that undergo nuclear reaction.

Radioactivity. The spontaneous nuclear breakdown of an atom.

Radiogenic. A term used to describe the product of a nuclear reaction.

Radiometric age. The age determined for a mineral, rock, or organic material based on an analysis of the radioactive elements present.

Ranges. Fault blocks uplifted above downdropped basins by normal faulting during application of tension to an area.

Ray. Ejecta deposit extending radially from meteorite impact craters.

Rayleigh wave. A seismic surface wave propagated by a retrograde circular orbital motion of rock particles.

Recessional moraine. An end moraine which is deposited by a glacier as it melts back from a terminal moraine.

Recharge. The natural or artificial addition of water to the groundwater system.

Recrystallization. Metamorphic change in which grain size increases or decreases without a change in mineralogy.

Rectangular pattern. A stream network in which tributaries flow into main streams at right angles. Streamflow generally is controlled by fractures in the bedrock.

Recumbent fold. A fold whose axial plane is horizontal.

Refraction. A change in the direction of a seismic wave as it passes from rocks of one density into those of another.

Regional metamorphism. Widespread metamorphism caused by directed and lithostatic pressures, heat, and chemically active fluids.

Regolith. The unconsolidated layer of soil, sediment, and rock debris that partially or completely covers bedrock.

Rejuvenated stream. A stream that begins downcutting actively, following prolonged lateral cutting, as a result of a lowering of its base level.

Relative age. The age of a rock or geologic event in comparison (i.e., older or younger) to that of others.

Relative geologic time scale. A time scale that places the stages in evolution of life and geologic events and products in a chronologic sequence relative to each other.

Relict feature. A feature in metamorphic rocks inherited from the original rocks.

Relief. The relative difference in elevation between the highest and lowest points in an area.

Remanent magnetism. Traces of ancient Earth magnetism preserved in a rock or mineral.

Renewable resource. A resource for which the processes of formation regenerate the material faster than it is consumed, and thus make it continually available.

Reorientation. Metamorphic change in which elongate and platy grains become aligned in response to directed pressures.

Replacement. A process of fossilization in which the hard parts of a plant or animal are replaced by minerals precipitated from solutions circulating through the enclosing sediment.

Reserve. The portion of a resource that can be extracted profitably and legally with current technology.

Reservoir rock. A porous rock in which fluids, such as oil, can accumulate and be recovered from.

Residual regolith. Regolith formed by the physical or chemical breakdown of the underlying materials.

Resisting forces. Those factors which act against the downslope movement of materials.

Resource. A geologic process or material that is useful to humans and is available for use.

Retrograde metamorphism. Metamorphism that occurs as temperatures and pressures decrease after the highest intensities attained during a metamorphic event.

Reverse fault. A dip-slip fault in which the footwall block has moved downward relative to the hanging wall block.

Rhombohedral packing. The type of packing that is tightest and has the least pore space. (Cf. *cubic packing*.)

Rhyolite. A fine-grained felsic igneous rock with the same composition as granite.

Rhythmic bedding. Stratification consisting of an alternation of two different types of sediment, implying a sequential alternation of two depositional conditions.

Richter magnitude scale. A scale for measuring the strength of an earthquake, based on the amount of ground motion and energy released.

Riffles. Shallow barlike areas located at regular distances along straight stream channels.

Right-lateral fault. A strike-slip fault in which displacement of one block has been to the right as viewed from the other block.

Rill. A shallow, temporary channel formed in an area during early stages of stream system development.

Rilles. Linear, steep-walled depressions on the lunar maria and in the lunar highlands; probably large lava tubes whose roofs have collapsed.

Ring silicate. Silicate mineral structure in which silicon-oxygen tetrahedra share two oxygens with adjacent tetrahedra, forming rings composed of 3, 4, or 6 tetrahedra.

Rip currents. Currents moving seaward through the surf zone.

Ripple marks. Regularly spaced small ridges of sand resembling ripples of water and formed on the bedding surface of a sediment.

Rock. An aggregate of minerals, or many grains of a single mineral.

Rock cycle. The processes by which one type of rock (igneous, sedimentary, metamorphic) may be converted into the other types.

Rockfall. The fall of rock particles from the face of a cliff.

Rock-forming minerals. Minerals that are the most abundant in common rocks.

Rock glacier. A lobe-shaped accumulation of boulder rubble formed by frost wedging in a periglacial climate.

Rockslide. Downslope movement of rock along planar surfaces.

Root wedging. A process of physical weathering in which the growth of plant roots into fractures in a rock dislodges pieces of the rock.

Roundness. The degree to which clastic sedimentary particles develop rounded surfaces.

Ruby. The red gem-quality variety of corundum.

Salinity. The concentration, in parts per thousand (ppt), of dissolved solids in seawater.

Saltation. The process by which particles are picked up and transported by wind or water, with intermittent contact with the bottom.

Salt dome. An upward projection of halite, formed by extreme ductile deformation, that arches and sometimes penetrates overlying sedimentary rocks.

Saltwater encroachment. The infiltration of seawater into coastal aquifers.

Sand. Sedimentary particles with diameters between 1/16 and 2 mm.

Sandstone. Sedimentary rock made up of sand-sized particles.

Sanidine. A type of potassic feldspar.

Sapphire. A gem variety of corundum.

Satellites. Small bodies which orbit planets.

Saturation zone. That portion of the regolith and underlying rock in which all pore spaces are filled with water.

Schist. Strongly foliated metamorphic rock composed mainly of platy minerals.

Schistosity. Foliated metamorphic texture in which the aligned platy minerals are micas.

Scientific method. An orderly, logical method of analysis used by scientists to determine the most reasonable explanation for natural phenomena. The scientific method involves observation, hypothesis, experiment, revision, and further experimentation.

Scoria. A vesicular (porous) dark-colored igneous rock with a composition similar to gabbro.

Sea arch. An archlike opening along a rocky coastline resulting from wave erosion cutting through opposite sides of a rocky headland.

Sea cave. A cavity formed along a rocky shoreline when mass movements remove a section of the cliff.

Seafloor spreading. In the plate tectonics model, the opening and enlargement of an ocean by continuous rifting at ocean ridges, accompanied by extrusion of basalt.

Seamounts. Submerged volcanoes that rise at least 1000 m above the seafloor.

Sea stack. A mass of rock isolated from others by wave erosion.

Seawall. A wall built against a coast to protect it from wave erosion.

Secondary recovery. The pumping of water under pressure into reservoir rocks in order to free more oil.

Secular variation. Long-term variation in the strength of the Earth's magnetic field.

Sediment. Unconsolidated debris formed at the Earth's surface. (See *clastic*, *siliciclastic*, *biogenic*, and *chemical* sediments.)

Sedimentary environment. An area with distinctive physical, chemical, and biologic conditions in which a unique type of sediment accumulates.

Sedimentary mineral. A mineral formed by sedimentary processes.

Sedimentary rocks. Rocks formed at the Earth's surface by deposition and lithification of particles eroded from previously existing rocks, biogenic materials, or dissolved ions.

Sedimentary structure. A feature in a sedimentary rock that formed either contemporaneously with deposition or subsequent to deposition.

Seed crystal. A small cluster of ions that eventually grows to form a mineral grain.

Seismic waves. Pulses of energy released during an earthquake.

Seismogram. The printed record of ground motion measured by a seismograph.

Seismograph. An instrument that detects and records ground vibrations produced by seismic waves.

Seismology. A branch of geology that studies earthquakes and the passage of earthquake energy through the Earth.

Semiarid area. An area that receives between 25 and 50 cm of precipitation per year.

Series decay. A sequence of nuclear reactions in which the parent element breaks down into a radioactive daughter which in turn undergoes nuclear reaction until a nonradioactive daughter is produced.

Shadow zones. Regions on the Earth's surface that do not receive direct P-wave or S-wave transmission after an earthquake.

Shale. Sedimentary rock composed of clay-sized particles.

Shards. Small fragments of volcanic glass hurled into the air during volcanic eruptions.

Shear. A type of stress produced by forces that do not oppose one another directly or are of unequal intensity in opposing directions.

Sheeted dikes. Closely spaced basaltic dikes found as part of the ophiolite suite.

Sheetflow. Movement of water as sheets on the ground surface rather than in channels.

Sheet silicate structure. Mineral structure in which all silicon-oxygen tetrahedra share three of their four oxygens with adjacent tetrahedra.

Shelf break. The seaward edge of the continental shelf.

Shelf valleys. Submarine channels cut into the continental shelves by stream erosion during a period when sea level was lower than it is today.

Shield volcano. A volcano with very gentle slopes composed mostly of fluid lava.

Sial. The uppermost layer of the continental crust, composed of rocks with the average composition of granite.

Silica. SiO_2 in mineral (as in quartz) or amorphous form.

Silicate minerals. Minerals that contain silicon and oxygen.

Siliceous ooze. Biogenic oceanic sediment composed of the skeletons of organisms made of silica.

Siliciclastic. Particles composed in part of silica, formed from the weathering of preexisting rocks.

Silicon-oxygen tetrahedron. Coordination structure in which a silicon ion is surrounded by four oxygen ions; the basic building block of the silicate minerals.

Sill. A tabular mass of intrusive igneous rock that parallels structures in its host rock.

Silt. Sedimentary particles with a diameter between 1/256 and 1/16 mm.

Siltstone. Sedimentary rock composed of silt-sized particles.

Sima. The lower layer of continental crust (and the only layer of oceanic crust), composed of rocks with the average composition of basalt.

Simple melting. A melting process in which a solid passes directly into the liquid state at a single temperature.

Single-chain structure. Elongate silicate structure in which each silicon-oxygen tetrahedron shares two of its oxygen ions with adjacent tetrahedra.

Sinkhole. A circular depression in the land surface caused by solution of the underlying rock and collapse of the land surface.

Slate. Fine-grained, foliated low-grade metamorphic rock produced from shale.

Slickensides. Polished and finely grooved surfaces caused by movement of pulverized material in fault planes. The grooves caused by relatively large grains indicate the sense of displacement (dip-slip, etc.).

Slide. A mass-movement process in which rock or sediment moves downslope along a planar surface.

Slip face. The steep, downwind side of a dune.

Slope failure. The downslope movement of materials underlying a slope.

Slump. A type of slide in which the sediments or rocks move downslope along a curved surface.

Snowline. In mountainous regions, the lowest altitude at which snow covers the ground throughout the year.

Soil. Surficial material, weathered sufficiently to support plant growth.

Soil horizons. Layers in the soil profile that are distinguished by color, organic content, mineralogy, or grain size as a result of weathering of the parent material.

Soil water zone. The thin belt of partially saturated pores within the root systems of plants.

Solar system. A family of objects held in orbit around a star by gravitational attraction.

Solar wind. Electrically charged particles that are generated by nuclear reactions in the Sun and stream outward throughout the solar system.

Solid-solution series. A group of minerals that represents a mixture by ionic substitution of two or more end members; e.g., the olivine family, $(Mg,Fe)_2SiO_4$, a solid solution series between the end members forsterite, Mg_2SiO_4, and fayalite, Fe_2SiO_4.

Solid waste. Solid byproducts of domestic, agricultural, or industrial activities.

Solifluction. Slow downslope movement of water-saturated surface sediment; most common in areas where the underlying sediments are frozen.

Sorting. The degree of uniformity of size among sedimentary particles. Particles all of the same size are well sorted; if the particles cover a range of sizes, the sediment is poorly sorted.

Source rock. The original rock from which sedimentary particles were derived.

Specific gravity. The ratio of the density of a material to the density of water.

Spheroidal weathering. The formation of rounded boulders as a result of chemical weathering during which the outermost layer of rock decomposes, crumbles, and falls off.

Spit. An elongate and commonly curved bar of coarse material built by longshore drift from headlands or sandy islands.

Spreading center. An ocean ridge segment which, according to the plate tectonics model, is the site of active rifting and emplacement of basalt from the mantle.

Spring. A natural discharge of groundwater onto the land surface, or as a discrete flow into a body of surface water.

Stalactite. Projection extending downward from the roof of a cave, formed by evaporation of groundwater and precipitation of dissolved ions.

Stalagmite. Projections extending upward from the floor of a cave, formed by evaporation of groundwater and precipitation of dissolved ions.

Steam tubes. Elongate cavities in volcanic rock which trace the path of gas bubbles that escaped from the lava.

Stock. A small, irregular intrusive igneous body.

Stony-iron meteorite. A meteorite composed of an intergrowth of iron-nickel alloy and silicate minerals.

Stoping. A process by which magma rises; upward projections of magma surround and detach blocks of host rock which then sink into the magma. The magma then rises to fill the space from which the blocks had come.

Storm surge. The elevation of the sea surface, resulting from the low pressure and high winds associated with a storm.

Straight channel. A segment of a stream that has no curves.

Strain. The amount that a rock changes in shape or volume in response to deforming forces.

Stratification. A sequence of layers of sediment marked by differences in particle size, color, or other characteristics.

Stratified drift. The stratified deposits produced by glacial meltwater streams.

Stratigraphic trap. Oil or gas reservoir caused by the interfingering of porous and permeable rocks with impermeable rocks.

Stratovolcano. See *composite cone*.

Streak. The color of a mineral's powder.

Stream. Water moving on the Earth's surface, confined in a channel.

Stream capture. The intersection and incorporation of one stream's drainage by another stream which has eroded headward through the drainage divide that originally separated the two streams.

Stream depth. The vertical distance between the water surface and the bed, measured in the center of the channel.

Stream network. Large numbers of interconnected streams.

Stream terraces. Flat erosional remnants of valley fill sediment left behind when a stream begins downcutting.

Stream width. The distance from bank to bank, measured at the surface of a stream and at right angles to its flow.

Stress. The interaction between deforming forces and the internal cohesive forces in a rock; a measure of the intensity of deformation.

Striations. Elongate scratches and grooves carved into rock by debris carried in a glacier.

Strike. The compass orientation of a horizontal line drawn on any planar geologic structural feature (e.g., N25°E). Strike is used with *dip* to describe the three-dimensional attitude of planar geologic structures such as bedding planes and faults.

Strike-slip fault. A fault in which displacement has been essentially horizontal, parallel to the strike of the fault plane.

Structural trap. Concentration of oil or gas caused by geologic structures, e.g., anticlinal crests or salt domes.

Subatomic particle. Particles such as protons, neutrons, and electrons that make up atoms.

Subduction. A process in which one lithospheric plate is thrust downward into the asthenosphere beneath a second plate with which it has collided.

Subduction zone. The region in which one lithospheric plate is thrust into the asthenosphere beneath another plate.

Submarine canyons. Deep V-shaped valleys carved into the continental slope by underwater erosion.

Subsidence. Sinking of the ground surface because of removal of large amounts of water or petroleum from the pores of underlying sediment or rocks.

Superposition. A principle of relative-age dating that states that unless there has been tectonic overturning, the oldest rock in a sequence of sedimentary rocks will be at the bottom, and the youngest at the top.

Surface creep. The lateral movement of sand particles along a surface, resulting from the impact of wind-transported particles. Surface creep is the main mechanism which moves sand grains up the windward face of a dune.

Surface currents. Circulation of ocean water caused by the prevailing wind system in a region.

Surface tension. A force acting parallel to the water surface which enables fluids to remain as thin films within the pores of rocks and sediments.

Surface waves. Seismic waves that transmit energy along the Earth's surface. (See *Love waves* and *Rayleigh waves*.)

Surf zone. The portion of the nearshore zone where over-steepened waves topple over and crash onto the beach.

Surging glacier. A glacier which moves with a high velocity (several kilometers per year rather than several meters) and whose rapid movement is short lived (from a few months to a few years).

Suspended load. Those sedimentary particles which are carried within the body of a transporting agent (wind, water, or ice).

S wave. A transverse seismic body wave which is propagated more slowly than P waves and cannot pass through a liquid.

S-wave shadow zone. That area on the Earth's surface that does not receive direct S-wave energy following an earthquake.

Swelling clay. Mixtures of clay minerals that can expand to several times their normal volume by absorbing water.

Swell. Regular, long-period and long-wavelength wave which has moved away from the area in which it was generated.

Syncline. A fold in which the limbs dip toward one another so that the young rocks are found in the axial region of the fold.

Talus. The accumulation of debris by rockfall at the base of a cliff.

Tar sands. See *oil and tar sands*.

Tectonic cycle. A geologic cycle in which materials are transferred from the mantle to the crust and then back again as a result of seafloor spreading and subduction.

Tectonics. A branch of geology that studies the development and structure of the outer part of the Earth.

Temperature. The average kinetic energy of atoms in a substance.

Tension. A type of stress produced when deforming forces pull a rock apart.

Tephra. All solid particles formed from material erupted into the atmosphere during volcanic eruptions (synonym: *pyroclastic debris*).

Terminal crevasse. Crevasse formed at the down-ice end of a glacier.

Terminal moraine. A deposit formed at the maximum extent of a valley or continental glacier.

Terminus. The furthest extent of a glacier.

Terrestrial planets. The four inner planets (Mercury, Venus, Earth, Mars) that are relatively small but denser than the outer, larger Jovian planets. They are composed of silicate minerals and iron-nickel alloy.

Terrigenous sediment. Sedimentary particles derived from landmasses.

Texture. A property of a rock encompassing its grain size, grain shape, and the manner in which the individual grains (or crystals) are related to each other.

Theory. In the scientific method, a theory is a hypothesis that has been proved, or has withstood all experimental testing.

Thermal metamorphism. See *contact metamorphism*.

Thermal pollution. Environmental problems resulting from the return of heated water to a body of surface water or groundwater.

Thermal springs. Springs that bring heated water to the surface.

Thermoremanent magnetization. Traces of ancient Earth magnetism recorded by the minerals of a cooling igneous rock as they pass below their Curie points.

Thin section. An extremely thin slice of a rock mounted on a glass slide so that it can be studied with a microscope.

Threshold velocity. The lowest velocity at which sand grains begin to move in water or air.

Thrust fault. A gently dipping form of reverse fault with extensive horizontal displacement perpendicular to the strike of the fault plane.

Tidal delta. A body of sand transported through a breach in a barrier island and deposited in the bay behind it.

Tidal flat. Lower portion of wetlands consisting of nearly horizontal and generally unvegetated expanses of fine-grained sediments which are covered completely by water at high tide and emerge at low tide.

Tidal inlet. Opening in the shoreline through which estuarine and oceanic waters move back and forth during ebb and flood tides.

Tidal marsh. Higher portions of wetlands covered by salt-tolerant vegetation and partially or completely covered during high tide.

Tidal wetland. Portion of the coastal zone which is covered partially or completely by water during part of the tidal cycle.

Tides. The regular daily rise and fall of ocean level caused by interaction of the gravitational fields of the Earth, Moon, and Sun.

Till. Unstratified drift deposited by actively moving glacial ice.

Tillite. An unstratified and poorly sorted rock of glacial origin.

Tombolo. Bars of sediment which connect offshore islands with the mainland or with each other.

Top-and-bottom structures. Sedimentary structures that can be used to determine the original tops and bottoms of beds.

Topographic desert. An area that is arid because it is separated by mountain ranges from sources of moisture.

Topset beds. The thin and horizontally layered sandy sediments covering the top surface of a delta.

Trace fossils. Sedimentary structures resulting from the life activities of organisms. Some common trace fossils are tracks or footprints on the surface of a bed and burrows within the bed.

Track (within crystals). Linear area formed by the passage of high-energy particle resulting from the decay of radioactive isotopes such as uranium-238.

Track density (within crystals). The number of tracks per unit area.

Traction. Transportation of particles by wind or water where the particles move in constant contact with the bottom of the stream or the ground surface.

Transform fault. A type of strike-slip fault that appears to offset ocean ridge segments, island arcs, or other transform faults. Actual displacement is far more complex than in other strike-slip faults.

Transform fault margin. A neutral plate boundary along which ocean crust is neither created nor destroyed.

Transported regolith. Sediment deposited on bedrock by water, wind, ice, or mass wasting.

Transverse dune. A sand dune that forms at right angles to the dominant wind direction.

Transverse wave. A seismic wave (S wave) propagated by the straight-line, back-and-forth motion of particles in directions at right angles to the direction of wave propagation.

Travel-time curves. Charts plotting the relative travel times of the different seismic waves against distance from earthquake epicenters. They are used to calculate the location of epicenters.

Travertine. A banded form of calcium carbonate characteristic of cavern floors.

Tree-ring dating. Dates determined from counting the number of annual rings in trees.

Trellis pattern. A stream network composed of long major streams and short tributaries that join the main stream at high angles; these patterns develop best in areas of folded rocks.

Trough. The lowest point of a waveform.

True north. The direction toward the north *rotational* pole.

Trunk glacier. A type of mountain glacier formed by the merging of two smaller valley glaciers.

Tsunami. A seismic sea wave (often incorrectly called a tidal wave) generated when vertical displacement of the seafloor occurs.

Tuff. A volcanic rock formed from aggregates of shards.

Turbidite. A graded bed deposited by a turbidity current.

Turbidity current. Submarine downslope movement of a dense water-sediment mixture.

Turbulent flow. A type of fluid flow in which there is both horizontal and vertical movement.

Twin-tetrahedron structure. Mineral structure in which pairs of silicon-oxygen tetrahedra share an oxygen.

Ultramafic rock. Rock composed almost entirely of ferromagnesian minerals with little or no feldspar.

Ultramylonite. A glassy metamorphic rock formed by extreme grinding and pulverizing of grains during dynamic metamorphism.

Unconfined aquifer. An aquifer in which the water is at atmospheric pressure.

Unconformity. A substantial gap in the local rock record during which rocks either were not deposited or were removed later by erosion.

Uniformitarianism. A principle and a conceptual view of Earth history that holds that processes active on and in the Earth today were active in the geologic past and thus can be used to interpret the features found in rocks.

Unstable isotope. An isotope that breaks down by nuclear reaction to form another element.

Unstable slope. A slope along which there is a potential for mass movement.

Unstratified drift. A glacial deposit which lacks stratification and which generally is composed of poorly sorted and angular particles.

Up-ice. The direction toward the source of a glacier.

Upright fold. A fold with a vertical axial plane.

U-shaped valley. A valley with a U-shaped cross section resulting from erosion by a valley glacier.

Valley glacier. A glacier that forms in mountainous areas and extends like a tongue down mountain valleys.

Van der Waals' bonding. A weak electrostatic attraction between atoms caused by positioning of their orbiting electrons.

Varve. A type of rhythmic stratification, common in the deposits of glacial meltwater lakes, which consists of a yearly couplet of a coarser-grained (summer) bed and a finer-grained (winter) bed.

Vent. The opening at the surface of a volcanic feeder conduit.

Ventifacts. Faceted rock fragments shaped and smoothed by sedimentary particles carried by the wind.

Vertical erosion. Downcutting of the channel of a stream or of the ground at the base of a glacier.

Vertical exaggeration. A deliberate increase in the vertical scale of a topographic section, relative to the horizontal scale, in order to show subdued features more clearly. The vertical exaggeration (V. E.) can be calculated by dividing the horizontal scale by the vertical scale (both in the same units) and is expressed as 2×, 4×, etc.

Vertical tectonics. A theory which states that mountain-building events can be explained by dominantly vertical movements of deep mantle materials rather than the lateral movements involved in plate tectonics.

Vesicular. Porous igneous rock texture caused by escaping gas.

Viscosity. A measurement of the sluggishness with which a fluid moves. The opposite of fluidity.

Volcanic neck. Erosional remnant of a volcano, consisting of the resistant feeder conduit.

Volcanic rocks. Igneous rocks formed by solidification of magma at the Earth's surface (synonym: *extrusive rocks*).

Volcano. A mountain formed by the accumulation of erupted lava and/or tephra.

Wacke. Sedimentary rock composed of sand-sized particles in a finer-grained matrix.

Washover fans. Lobe-shaped deposits of sand eroded from the ocean side of a shoreline and deposited inland.

Water cycle. A geologic cycle describing the journey of water molecules from the oceans to the atmosphere and back to the oceans via streams and groundwater.

Water mass. Large volume of ocean water, having distinctive characteristics, which moves through the ocean basins.

Water table. The surface that separates the groundwater aeration zone from the saturation zone. Below the water table, all pores are filled with water.

Wave base. The downward limit of orbital water movement in a wave. Wave base occurs at a depth equal to about one-half the wavelength of the wave.

Wave-built terrace. A gently sloping depositional surface extending seaward from a coastal cliff.

Wave crest. The highest part of a wave.

Wave-cut bench. A gently sloping erosional platform extending toward the sea from the base of a coastal cliff.

Wave-cut notch. An indentation cut at the base of a coastal cliff by wave erosion and abrasion from wave-transported debris.

Wave fronts. The crest lines of moving waves.

Wave height. The vertical distance between the wave crest and trough.

Wavelength. The horizontal distance between a similar point on two waves, such as between two wave crests.

Wave normal. A line at right angles to the wave crest which indicates the direction of wave movement.

Wave period. The time required for two wave crests to pass a given point.

Wave refraction. The bending of wave fronts as they enter shallow water and pass over submerged topography. Wave refraction concentrates wave energy on headlands, tending to smooth out the shoreline.

Wave trough. The lowest part of a wave.

Weathering. Physical and chemical changes that occur in sediments and rocks when they are exposed to the atmosphere and biosphere.

Welded tuff. Volcanic rock formed from fine-grained tephra particles that have been fused together by their own heat.

Wentworth grade scale. A series of terms descriptive of clastic sediment particles, e.g., pebble, sand, and silt.

Whole-rock date. Radiometric age determined by the parent:daughter ratio in an entire rock rather than in a single mineral.

Wind shadows. Zones of quiet air formed in front of and behind an object that obstructs the airflow.

Xenolith. An inclusion of host rock in an igneous rock.

Index

667

Contact aureole, 165
  of Hartland pluton, 176–77
  index minerals in, 176–77
Contact metamorphism, 164–65
  facies of, 177–78
Continental accretion, 565–66
Continental drift, 559–61, 569–71
Continental glaciers, 337–38
Continental rise, 436
Continental shelf, 433
Continental slope, 434–36
  submarine canyons on, 435
Continents
  breakup of, 566–68
  crust of, 543–45, 563–65
  drift of, 559–61, 569–71
Continuous crystallization, 87–89
  in Bowen's Reaction Series, 90–91
Convection, 34–35
Convection cells, size of, in mantle, 574–75
Convection currents and plate motion, 570–73
Convergent plate boundary, 561
Coordination number, 49
Coquina, 146
Coral reef, 413–14
Cordilleran ice sheet, 357
Core, 30
  composition of, 542–43
  discovery of, 540–42
Coriolis effect, 371
Correlation, 198
Coupled ionic substitution, 52
Covalent bonding, 27–28
Crater density, 611
  time scale, 611
Crater Lake, Oregon, 119–21
Craters
  impact, 610–11
  lunar, 612–13
  volcanic, 110, 112–13
Creep, 249
Crevasses, 341–42
Cross-bedding
  in Coconino sandstone, 195
  cross cutting relationships, 188
  use in determining bed orientation, 154–55, 157
  formation of, 154–55
  principle of, 188
Crust
  composition of, 31, 543–45
  thickness of, 30, 543–45
Crystal growth, role of, in physical weathering, 218
Crystal settling, 91–92
Crystallization of magma, 87–90

Crystalline structure, 47–50
Crystals, 47
  growing at home, 58–59
Cubic packing, 244
Curie point, 554
Currents
  oceanic
    deep, 447
    hot and cold, 447–48
    saline, 448–49
    turbidity, 449
  surface, 446
    generated by major wind systems, 446–47
    longshore, 402
    rip, 402
    tidal, 403
Cuvier, Georges, 190
Cycles
  atmosphere, 17
  rock, 17
  tectonic, 16
  water, 17

Darcy's Law, 305–6
Darwin, Charles
  theory of evolution, 190
  theory of formation of atolls, 415
Daughter element in nuclear reactions, 199
Death Valley
  evaporites in, 137
  precipitation in, 374
Debris avalanche, 259
Debris flows, 258
Declination, magnetic, 553–54
Decomposition, 216
Deep oceanic currents, 447–48
Deep-Sea Drilling Project (DSDP), 432–33
Deflation, 386
Deformation, 460–69
Delta, 290
Dendritic stream network, 293
Denver, Colorado, 1962 earthquake, 538
Deposition
  coastal, 407–9
  deposits of mass movements, 259
  by glaciers, 349–56
  by groundwater, 316–20
  of sediment. See Sediment, transportation and deposition of
  by streams, 283–91
  by turbidity currents, 449
  by winds, 381–86
Depositional ore deposits, 587–88
Depositional remanent magnetism, 556

Desert aquifers, 315
Desert pavement, 386
Desert varnish, 376
Deserts. See also specific deserts, for example: Sahara Desert
  characteristics of, 370
  erosional cycle in, 381
  soils in, 376
  stream activity in, 378–81
  temperatures in, 373–74
  types of, 370–73
  wind activity in, 381
Devil's Postpile National Monument, 93
Devil's Tower, Wyoming, 124
Diamond, 49
  cutting of, 67–69
  structure of, 50–51
Diatreme, 87
Differential weathering, 225–26
Dike, 92
Diorite, 81, 84
Dip, 470–471
Dip-slip faults, 475, 479, 480–81
  normal, 475, 479, 480
  reverse, 475, 481
  thrust, 481
Directed pressure, 163–64
Discharge, 266–69
Disconformity, 191
  in Grand Canyon, 194
Discontinuous reactions, 88–90
  in Bowen's Reaction Series, 90–91
Disintegration, 216
Dissolved load, 272
Distributary stream network, 293
  on a delta, 289
  on alluvial fans, 381
Distributary streams, 293
Divergent plate boundary, 561
Dolomite, 66, 70
Dolostone, 148–49
Dome mountains, 511
Domes, 472
Double-chain silicate, 64
Drainage basin, 292
Drainage in deserts, 376–78
Drainage divide, 292
Drift, glacial, 349
Drumlins, 351–52
Dry valleys, 318
Ductility, 56
Dunes
  in the geologic record, 392
  migration of, 389
  origin of, 388–89
  sediment characteristics of, 391–92
  types of, 389–91
Dynamic equilibrium in streams, 268